Wilhelmus H. A. Schilders
E. Jan W. ter Maten
Stephan H. M. J. Houben
*Editors*

# Scientific Computing in Electrical Engineering

Proceedings of the SCEE-2002 Conference held in Eindhoven

Springer

*Editors*

Wilhelmus H. A. Schilders
Philips Research Laboratories
IC Design/Digital Design and Test
Building WAY, Room 4.77
Prof. Holstlaan 4
5656 AA Eindhoven
The Netherlands
Email: wil.schilders@philips.com

Stephan H. M. J. Houben
Magma Design Automation
TU/e Dommelgebouw
Den Dolech 2
5612 AZ Eindhoven
The Netherlands
Email: stephanh@magma-da.com

E. Jan W. ter Maten
Philips Research Laboratories
IC Design/Electronic Design and Tools
Building WAY, Room 3.073
Prof. Holstlaan 4
5656 AA Eindhoven
The Netherlands
Email: jan.ter.maten@philips.com

Library of Congress Control Number: 2004109917

Mathematics Subject Classification (2000):
65-06, 65Lxx, 65Mxx, 65Nxx, 65L06, 65L12, 65L15, 65L60, 65L80, 65M06, 65M60, 78-06

ISBN 978-3-642-62478-0 ISBN 978-3-642-55872-6
DOI 10.1007/978-3-642-55872-6



Springer is a part of Springer Science+Business Media

springeronline.com


Softcover reprint of the hardcover 1st edition 2004



Typeset by the authors using a Springer TEX macro-package
Cover design: *design & production* GmbH, Heidelberg
Printed on acid-free paper 46/3142LK - 5 4 3 2 1 0

# Preface

The fourth international conference on *Scientific Computing in Electrical Engineering (SCEE)* was held at the Eindhoven University of Technology, from 23rd to 28th June, 2002. It was sponsored by Philips Research Laboratories Eindhoven, the Eindhoven University of Technology, Computer Simulation Technology (CST) from Darmstadt, ABB Corporate Research, Thales Netherlands, the European Consortium for Mathematics in Industry (ECMI), the University of Rostock (organiser of SCEE-2000), the European network for Mathematics, Computing and Simulation for Industry (MACSI-net), the Royal Netherlands Academy of Arts and Sciences (KNAW), and the Scientific Computing Group of the Eindhoven University of Technology.
The Program Committee consisted of:

- Dr. Alain Bossavit, Electricité de France, Clamart, France.
- Dr. Uwe Feldmann, Infineon Technologies A.G., Munich, Germany.
- Prof.Dr. Leszek Demkowicz, University of Texas at Austin, USA.
- Dr. Michael Günther, Universität Karlsruhe, Germany.
- Prof.Dr. Ulrich Langer, Johannes Kepler Universität, Linz, Austria.
- Dr. Jan ter Maten, Philips Research Laboratories Eindhoven, The Netherlands.
- Prof.Dr. Ursula van Rienen, Universität Rostock, Germany.
- Prof.Dr. Jaijeet Roychowdhury, University of Minnesota, USA.
- Prof.Dr. Wil Schilders, Technische Universiteit Eindhoven and Philips Research Laboratories Eindhoven, The Netherlands.
- Prof.Dr. Thomas Weiland, Technische Universität Darmstadt, Germany.

As on all previous occasions there was widespread support both from the industrial sector and academia. This is considered to be an essential feature of this series of conferences, because it guarantees the relevance of the work to practical situations, while at the same time ensuring that long term basic research is strongly represented. For this reason, the interaction between electrical or electronic engineers and the mathematics community is one of the main aims of the SCEE conferences. Computational electromagnetics, computational electrodynamics, circuit simulation and coupled problems are areas that are covered.
At SCEE-2002 Invited Speakers were:

- Prof. Antonio Di Carlo (Universita degli Studi "Roma Tre", Italy): "G. Lamé versus J.C. Maxwell: how to reconcile them?"
- Prof.Dr. Susan C. Hagness (University of Wisconsin-Madison, USA): "Frontiers in FDTD theory and applications".
- Dr. Patrick Joly (INRIA, Le Chesnay, France): "Variational methods for FDTD computations in time dependent electromagnetism".

- Dr. Tom A.M. Kevenaar (Philips Research Laboratories, Eindhoven, The Netherlands): "Methods and approaches for RF circuit simulation and electromagnetic modelling".
- Dr.-Ing. Rolf Schuhmann (TU Darmstadt, Germany): "Recent advances in Finite Integration Technique for high frequency applications".
- Dr.-Ing. habil. Peter Schwarz (Fraunhofer-Institut für Integrierte Schaltungen, Dresden, Germany): "Continuous simulation of coupled systems".
- Prof.Dr. Igor Tsukerman (University of Acron, Ohio, USA): "Generalized finite element method in electromagnetic analysis: benefits and hurdles".

In total there were 29 contributed oral presentations and 38 poster presentations.

It has always been the policy of these conferences to encourage participants from all countries, with an emphasis on Europe. On this occasion this has been remarkably successful, there were 111 participants from 13 countries. Thus, the series of SCEE conferences has grown into truly international events. This will continue to be the policy of the series.

A new feature of the SCEE-2002 conference was the *Industry Day*, organized on Tuesday, June 25th. This was a very successful event, with 8 renowned speakers from industry and academia, discussing the needs of the electronics industry, now and in future:

- Prof.Dr. Robert M.M. Mattheij, Technische Universiteit Eindhoven, The Netherlands.
- Ir. Gerard F.M. Beenker, Philips Research Laboratories Eindhoven, The Netherlands.
- Dr. Frank Demming-Janssen, Computer Simulation Technologies (CST), Darmstadt, Germany.
- Dr. Wilhelm Dürr, Siemens Medical Solutions, Erlangen, Germany.
- Dr. Koen van Eijk, Magma Design Automation Inc, Eindhoven, The Netherlands.
- Dr. Marc Swinnen, Sequence Design, Paris, France.
- Prof.Dr. A. Peter M. Zwamborn, TNO Physics and Electronic Laboratory, The Hague, The Netherlands.
- Dr. Isaac Shpantzer, CeLight, Silver Spring, MD., USA.

The Industry Day meant a break away from technical detail, with talks presenting a bird's eyes view of scientific computing in electrical engineering. Besides the registered conference participants, the *Industry Day* attracted quite a number of additional participants from industry.
Another new feature of the conference was the short oral introduction of posters. In two one-hour sessions, contributors of posters were given a maximum of 2 minutes to advertise their work. The general feeling was that this was a very successful initiative, to be continued in future SCEE conferences.

The papers appearing here fall into two categories. The first are the invited papers given by the keynote speakers. Secondly, the Program Committee carefully refereed the contributed papers. The latter section contains both the accepted papers presented orally at the conference, as well as the papers corresponding to the posters.

It is a pleasure to thank all of the people and instances, both named here and others, whose enthusiasm and hard work ensured the success of this conference SCEE-2002.

Eindhoven, November 2003

Wilhelmus H.A. Schilders
E. Jan W. ter Maten
Stephan H.M.J. Houben
Local Organising Committee SCEE-2002

# Table of Contents

# G. Lamé vs. J.C. Maxwell: How to Reconcile Them? *

Antonio DiCarlo

Università degli Studi "Roma Tre",
Mathematical Structures of Materials Physics at DiS,
Via Vito Volterra, 62 I-00146 Roma, Italy

**Abstract.** Nowadays, after more than a century of inconsiderate divergence between electromagnetic and mechanical field theories, we find it hard to bring them together. This can be best exemplified by the problematic status of the electrodynamics of deformable media. The blame can be laid mainly on the limitations of the underlying theoretical frameworks and on the practitioners' education, too narrow to bridge the gap between them. I would like to concentrate here on the first problem—even though I am convinced that the second one carries more weight.

## 1 Lamé's Treatise on Three-Dimensional Elastic Solids

I happen to have a copy of the second edition of Gabriel Lamé's "Leçons sur la Théorie Mathématique de l'Elasticité des Corps Solides" [1] on my bookshelf. It was published in 1866. Its first edition had been published in 1852. Compare with the dates of the electromagnetic trilogy by Maxwell: "On Faraday's lines of force" appeared in 1856, "On the physical lines of force" in 1861–2, and "A dynamical theory of the electromagnetic field" in 1865.

These lecture notes on the mathematical theory of elastic solids differ strongly from any present-day book with a similar title. I do not mean by this simply that its notions and notations are somewhat outdated, or its mathematics obsolete, which would be quite trivial. The main difference is that Lamé was much bolder and more oriented towards fundamental physics than any of his modern followers in elasticity or solid mechanics. Secondly, and as a consequence, his main aim was to study the vibrations and waves of a peculiarly thin and outlandish solid medium: the all-pervading elastic ether.

### 1.1 Mathematical Physics as a New Science

In the terse preface to the first edition, Lamé held the opinion that—at the moment of his writing—the only well-founded chapters of proper Mathematical Physics were the electrostatics of conducting bodies, the theory of heat conduction and the mathematical theory of three-dimensional elasticity. Insisting that these creations—contrary to Rational Mechanics of old—belonged exclusively to his own century, Lamé was unfair to the "Geometers"

* Invited paper at SCEE-2002

of the Baroque period (Euler and the Bernoullis, to name the most prominent ones), at least as much as a quantum physicist of the twentieth century could be to him and his contemporaries in elasticity.

Elasticity—the most difficult and the least developed of the three chapters of Mathematical Physics—was also the most *useful*, he emphasized. It is thought-provoking to read that Lamé already viewed that his own time was willing to assess the importance of a mathematical theory through the immediate benefits it could provide industrial practice. We could with all the more reason appropriate his judgment (think of our "Industry Day"), even though we would not put elasticity at the top of our list of most useful theories. However, Lamé did not focus on the engineering side of elasticity—the most prominent nowadays, mainstream physics having repudiated it long ago.

## 1.2 Surrogate Sciences vs. Rational Physics

In Lamé's eyes, novel Mathematical Physics—while far deeper and wider in scope than older Rational Mechanics—shared its standards of rigour and precision. This quality differentiated Rational Physics from the host of empirical treatments based on doubtful principles and *ad hoc* hypotheses, whose only merit was practicality, and whose function was essentially provisional:

> Malgré leur utilité actuelle, qui est incontestable, toutes ces théories empiriques et partielles ne sont que des sciences d'attente. Leur règne est essentiellement passager, intérimaire. Il durera jusqu'à ce que la Physique rationnelle puisse envahir leur domaine.

I cannot be as confident as Lamé in the final victory of the one Rational Physics over the myriad of special, empirical theories devised to cope with practical problems for which a truly scientific treatment is not yet available. This is for two reasons: firstly, I know that make-do sciences actively reproduce, after swallowing morsels of true science; secondly, I doubt whether there is only one Rational Physics. However, I find his lucid depiction of the subtle rivalry between technical and scientific thinking in the age of technology a great contribution from him, and I stand on his side in this confrontation.

## 1.3 Elasticity and Molecular Mechanics

At variance with modern textbooks in continuum mechanics, elasticity is defined by Lamé in terms of molecular interactions, even though in a rather vague way. In the very first paragraph of Chap. 1, Sect. 1, Lamé calls elastic the restoring forces which tend to bring molecules back to their equilibrium positions. Notice that Lamé did not use his naïve molecular picture as a convenient pedagogic cartoon. Strange as it may appear to us, he and other founding fathers of *continuum* physics strongly believed in the *necessity* of an underlying *discrete* structure of matter. Lamé criticized Navier's method for establishing the general equations of three-dimensional elasticity, on the grounds that it presupposed matter to be continuous, which he considered flatly absurd and inadmissible.

The subtle role played in the nineteenth century by the molecular interpretation of elasticity is best expounded in the masterly historical account of structural mechanics by Benvenuto [2, Sect. 14.2]:

> Elasticity represented the most promising line of inquiry, not only because of its extraordinary practical usefulness and the accuracy of the theoretical synthesis that it permitted, but also because of the implications of its general principles and equations. The molecular interpretation of elastic behavior that Navier, Cauchy and Poisson promoted led many scientists to attempt finally to unify and explain all forces operating in Nature in the light of a universal law of attraction and interatomic repulsion, like that foreseen by Boscovich. [...] From this perspective, the laws of elasticity are by no means restricted to a specific class of bodies, but express an inherent property of matter itself.

Lamé shared that same perspective. Indeed, his definition of elasticity concludes along very similar lines (third paragraph of Chap. 1, Sect. 1):

> L'élasticité est donc une des propriétés générales de la matière. Elle est, en effet, l'origine réelle ou l'intermédiaire indispensable des phénomènes physiques les plus importants de l'univers. C'est par elle que la lumière se répand, que la chaleur rayonne, que le son se forme, se propage et se perçoit, que notre corps agit et se déplace, que nos machines se meuvent, travaillent et se conservent, que nos constructions, nos instruments échappent à mille causes de destruction.

## 1.4 Lamé's View of Ether

The main focus of Lamé's treatise on elasticity is undoubtedly the study of small vibrations and linear waves. Out of one hundred and thirty-four sections, only the eight sections of Chaps. 12 and 16—making up twenty-eight pages in all—are devoted to equilibrium problems of three-dimensional elasticity. The notion of wave speed, with emphasis on the classification into longitudinal and transversal waves, is introduced in Chap. 11. From Chap. 17 on—one hundred and eleven pages in all—Lamé strives to explain light waves through elasticity theory, starting from Fresnel's birefringence.

This endeavour leads him to the following conclusion: the phenomena of light propagation in space, diffraction, and birefringence prove the ubiquitous existence of ether beyond all conceivable doubt. While crediting the mathematical theory of elasticity for this important and rigorous result, Lamé was confident that accounting properly for the interaction between ethereal and ponderable matter would have disclosed the secrets of a host of mysterious and incomprehensible beings, ranging from caloric, electricity, magnetism, universal attraction, cohesion, to chemical affinities.

## 2 Maxwell's Electromagnetism vs. Lamé's Elasticity

Lamé was wrong with his elastic ether. As is clear to us now, Maxwell was right with his electromagnetic field theory, shaped on Faraday's unorthodox ideas. This makes me appreciate Maxwell's penchant for understatement and the "absurd and infuriating modesty" which Dyson reproached him with in a witty and enlightening short essay [3]. But was Maxwell against ether?

### 2.1 Maxwell on Ether

By no means. Consider the passing mention of *his own* theory that Maxwell uttered in the presidential address he gave at the annual meeting of the British Association for the Advancement of Science in 1870:

> Another theory of electricity which I prefer denies action at a distance and attributes electric action to tensions and pressures in an all-pervading medium, these stresses being the same in kind with those familiar to engineers, and the medium being identical with that in which light is supposed to be propagated.

However, despite the kinship of "stresses" vaguely advocated by Maxwell, his electromagnetic ether deeply differs from Lamé's.

### 2.2 The Key Difference between Maxwell's and Lamé's Theories

As we have just seen, this is not in being for or against the existence of ether (*cf*. the entry "Ether" in the ninth edition of the Encyclopædia Britannica, written by Maxwell). Also their different invariance properties are not as discriminating as is commonly adduced. After all, "classical" (*i.e.*, non-relativistic and non-quantistic) field theories of mechanics behave well under the action of *slow*[1] Lorentz changes in observer—which is all to be expected.

The distinguished feature that makes the real difference between Lamé's ether and Maxwell–Faraday's ether is *topological* in nature. In electromagnetic field theory there are plenty of *real*-valued physical quantities associated with geometric objects (*cells*) of various dimensions, embedded in four-dimensional *space*-time (*cf.* the discussion by Tonti in [8, Sect. 5.1]). On the contrary, in continuum mechanics there is only one *bona fide* real-valued quantity, namely *work*, associated with the cells of highest dimension in the *body*-time manifold and their boundaries (beware: body-time, not

[1] After my talk, Alain Bossavit assured me that this notion could be made precise using well-established mathematical tools [4,5].

space-time!). Mechanical work is backed up by a team of *vector*-valued (and *covector*-valued) physical quantities, associated with cells of different dimensions. In electromagnetics, on the contrary, there are no *bona fide* vector quantities (distrust appearances!).

**Space-time cells.** I espouse the approach pionereed by Tonti [6–8] and expounded by Mattiussi [9,10], among others. In the following, I consider space-time as the product of a 3-dimensional space manifold times a 1-dimensional time line, thus adopting the viewpoint of an observer. Space-time vectors decompose accordingly into space and time components. A *space vector* is, by definition, a space-time vector with null time component (and viceversa). The dichotomy (space, time) is invariant under *slow* Lorentz changes in observer. It should *not* be confused with the Minkowskian trichotomy (spacelike, lightlike, timelike), which is invariant under a *general* Lorentz transformation [11]. All space vectors are spacelike, and time vectors timelike—but the converse does not hold true.

Consider the hierarchy of parallelepipedal cells in space-time, ranging from 0-dimensional (an event, *i.e.* a place at an instant) to (3+1)-dimensional (a chunk of space times a time lapse). A nondegenerate *k-cell* (*i.e.*, a $k$-dimensional cell) has $k$ independent edges, which are space-time vectors. More precisely, a $k$-cell (with $k>0$) is the equivalence class of all $k$-parallelepipeds sitting on the same 0-cell, lying in the same $k$-plane and having the same $k$-volume (as a basic reference for this gadgetry, see [12, Chap. 4]). When talking about cell edges, I really mean the edges of *at least one* of its representative parallelepipeds.

Let me call *plumb* all nondegenerate cells whose edges are either space or time vectors; *slant* the other ones. Most cells are slant, but all of them can be obtained as linear combinations of plumb cells. Among plumb cells, I single out *time-dipped cells*, which have one time edge (they cannot have more); the remaining ones, having no time edges, I call simply *space cells*. At the two extremes of the hierarchy, all nondegenerate cells are plumb: at the bottom ($k=0$) all of them are space cells; at the top ($k=3+1$) all nondegenerate cells are time-dipped. In between, most cells are slant; however, they decompose (in a unique way) into space and time-dipped components.

The distribution (in space-time) of a real-valued quantity associated with $k$-dimensional cells is properly gauged by a (real-valued) *k-form*, which is—by construction—the integrand that makes sense to integrate on a $k$-cell, yielding the amount of the gauged quantity contained in that cell. The *co*-space and *co*-time components[2] of a $k$-form are singled out by integrating it on space and time-dipped $k$-cells, respectively.

**The basic structure of electromagnetic field theory.** Maxwell's play—when staged in this transcription—has two leading characters: the *electromag-*

[2] To be read space-conjugate and time-conjugate components, respectively.

*netic 2-form* and the *charge-current 3-form*. Maxwell's equations establish the existence of their companion potentials: the *electromagnetic potential* (a 1-form) and the *charge-current potential* (a 2-form). To sum up, we have a nice symmetric panoply, centered on dimension 2: one 1-form, two 2-forms, and one 3-form. All of them are real-valued, none of them sit at the extremes (0 and 4).

Down-to-earth electric and magnetic quantities are elicited by evaluating the co-space and co-time components of the above $k$-forms: the scalar-valued *charge density* $\rho$ and the vector-valued *current density* $\mathbf{J}$ represent respectively the co-space and co-time components of the charge-current 3-form; the vector-valued *electric flux density* $\mathbf{D}$ and *magnetic field intensity* $\mathbf{H}$ represent respectively the co-space and co-time components of the charge-current potential 2-form; the vector-valued *electric field intensity* $\mathbf{E}$ and *magnetic flux density* $\mathbf{B}$ represent respectively the co-time and co-space components of the electromagnetic 2-form; the *scalar potential* $V$ and the *vector potential* $\mathbf{A}$ represent respectively the co-time and co-space components of the electromagnetic potential 1-form.

It should be stressed that all of the above scalar fields and (space) vector fields are but proxies of the thing-in-itself. Much structural information is obliterated when those pallid substitutes are taken at face value: in fact, while $V$ is a true scalar field, $\rho$ is a *pseudo*scalar field (a space density); only $\mathbf{J}$ and $\mathbf{D}$ should be regarded as true vector fields, while $\mathbf{A}$ and $\mathbf{E}$ are *co*vector fields, $\mathbf{B}$ is a *pseudo*vector field, and $\mathbf{H}$ a *pseudoco*vector field.

When referred to space vectors, the dichotomy (true, pseudo-) is synonymic with (polar, axial). The independent dichotomy (covector, vector) is related to the alternative between inner and outer orientation (in this order!). These distinctions were well known to Maxwell [13]. However, they never entered the physics vulgate—or evaporated early on. The electric and magnetic fields $\mathbf{E}$, $\mathbf{B}$, *et cetera* can obviously be extended to any space dimension—as bare vector fields. This is a futile exercise, however, since the delicate underlying structure can *not* be exported, as I now formally state.

*Proposition.* Take for granted that time is one-dimensional. Then the only forms having the same number of strict co-*time* and strict co-*space* components are those sitting exactly halfway from the extremes: if space has even dimension, there are none; if space has dimension $2k-1$, $k$-forms will do.

*Proof.* Trivial: a $k$-form in $n$ dimensions has $n!/(n-k)!\,k!$ strict components.

*Proposition.* Assume that space has dimension $2k-1$. Then $k$-forms have $2k-1$ strict co-time (and co-space) components if and only if $k = 1, 2$.

*Proof.* Check that only $k=1$ and $k=2$ solve $(2k-1)!/(k-1)!\,k! = 2k-1$.

*Proposition.* A nontrivial electromagnetic theory requires $k > 1$.

*Proof.* If space is one-dimensional, electricity and magnetism do not couple, electromagnetic waves do not exist, the Poynting vector field $\mathbf{S} = \mathbf{E} \times \mathbf{H}$ (another true vector field, like $\mathbf{J}$ and $\mathbf{D}$) is null.

In conclusion, Maxwell's play can only be staged on a $(3+1)$-dimensional space-time.

**The basic structure of mechanical field theory.** Let me now transcribe Lamé's play into the same unifying language. The first major change in the stage setting is that mechanical quantities are associated with cells embedded in an $(n+1)$-dimensional *body-time* manifold, defined as the product of an $n$-dimensional *body* manifold $(0<n\leq 3)$ times a 1-dimensional time line. It should be noted that introducing a body manifold does not privilege any observer. Singling out a time line from the space-time continuum does call for an observer, however.[3] The cases $n=1,2$ describe corporeal curves and surfaces respectively. Following Lamé, I will concentrate here on the mechanics of space-filling bodies ($n=3$). Even in this case it is of the essence to keep body quite distinct from space.

Body-time cells are isomorphic to space-time cells, so only terminology needs adapting. In particular, space cells translate into *body cells.* Not to confuse the issue, I distinguish carefully between elements of the space manifold—which I call *places*—and elements of the body manifold—which I call *points.* A 0-cell is now a *point* at an instant, a body 1-cell a *line element,* a body $(n-1)$-cell a *facet,* and a body $n$-cell a *bulk element.* At the top of the cell hierarchy sit body-time lumps (bulk elements times time lapses). Distributions of physical quantities associated with $k$-dimensional body-time cells are gauged by $k$-forms, endowed with *co*-body and *co*-time components.

The important difference with respect to electromagnetic field theory is that most mechanical quantities—and hence the corresponding forms—are *not* real-valued. The notion of vector-valued forms surfaces in a passing remark—entitled "A glimpse of other physical theories"—in [9]. Vector- and covector-valued forms are explicitly introduced in [14,15], where their use is rightly advocated for the evaluation of electromagnetic forces. The way these papers treat mechanics is, however, far from satisfactory.

The fundamental mechanical descriptor is the *placement* 0*-form* $p$, a *place*-valued field attaching to each point at each instant—*i.e.*, to each 0-cell—a place in space. At variance with electromagnetic forms, the very definition of placement calls for an observer: a different observer sees the *corresponding* placement $\widetilde{p}$, as decreed by the action of the group of changes in observer on space-time. The restriction of a placement to all *simultaneous* 0-cells is required to be an *embedding.* The (exterior) differential of the placement

[3] A *proper*-time line may be attached to each body point independently of any observer. But an observer is required to trivialize the proper-time bundle, *i.e.* to identify lines attached to different points with each other (see [11, Sect. 1.4]).

0-form is the (*space*) *vector*-valued *displacement* 1-*form*, whose co-time and co-body components are represented respectively by the *velocity* $\dot{p}$ and the *body gradient* of placement $\nabla p$. To focus on the essentials, I will stick here to an *affine* space-time, whose tangent bundle has a canonical global connection, whereby all tangent spaces are trivially identified with each other. Handling vector-valued forms requires an easy extension of the rules valid for real-valued forms (see [17, Def. 6.3.11]).

In dynamics, a key role is played by *test velocities*, vector-valued 0-forms (*zero* should be emphasized[4]) sharing the physical dimensions of $\dot{p}$. No differential compatibility is required between a test velocity $\mathbf{v}$ and the placement $p$, the equality $\mathbf{v} = \dot{p}$ selecting the one velocity *realized* along $p$.

*Work* is the chief mechanical quantity which can be properly integrated over body-time lumps. It is a *real*-valued quantity, stemming from a duality between vector- and covector-valued forms.[5] The basic features of each individual dynamical theory are encoded in the structure of this distinguished duality. The standard model of continuum mechanics—encompassing, in particular, Lamé's theory of elasticity— is founded on the following assumption.

The (so-called "virtual") work done on a test velocity $\mathbf{v}$ over an $(n+1)$-dimensional body-time cell is the sum of two contributions: an integral over the cell itself, and another over its boundary (which consists of $n+1$ pairs of $n$-cells: $2n$ facets $\times$ the time edge of the lump, plus its bulk element $\times$ the two ends of its time edge). The real-valued $(n+1)$-form to be integrated over the lump is the sum of two exterior products: the vector-valued 0-form $\mathbf{v}$ times the *impulse-supply form*—a covector-valued $(n+1)$-form—*minus* the differential of $\mathbf{v}$ (a vector-valued 1-form) times the *impulse-flux form*—a covector-valued $n$-form. The real-valued $n$-form to be integrated over the cell boundary is the exterior product of $\mathbf{v}$ times the *boundary-impulse form*—a covector-valued $n$-form living on the cell boundary.

An overriding *balance principle* (see [18–22]) commands that the total ("virtual") work done on any test velocity over any body-time lump should be zero, implying that: (i) the impulse-supply form and the exterior derivative of the impulse-flux form should add up to the null covector-valued $(n+1)$-form; (ii) the boundary-impulse form on any $n$-cell should match with the impulse-flux form (a body-time version of the celebrated Cauchy stress theorem).

In comparison with Maxwell–Faraday's gossamer edifice, continuum mechanics has much more robustness than fineness in it: granted that time is one-dimensional, its basic structure can easily accommodate for any space

[4] My interpretation is that a test velocity attaches to each 0-cell $\{\mathsf{b}\}\times\{\mathsf{i}\}$ ($\mathsf{b}$ a point, $\mathsf{i}$ an instant) the difference $p_\varepsilon(\mathsf{b},\mathsf{i}) - p(\mathsf{b},\mathsf{i})$ between the place assigned to it by a *juxtaposed* placement $p_\varepsilon$, for vanishingly small $\varepsilon$, and that assigned by $p$, it being intended that $\lim_{\varepsilon\to 0} p_\varepsilon = p$ (*cf.* [18]). In other words, test displacements develop in an extra subsidiary time dimension, parameterized by the *pseudo-time* $\varepsilon$.

[5] Playing with the (real-valued) time variable, other real-valued work-related ancillary quantities may be introduced, as the time rate of work, namely the *power* (or *working*), or its time integral, namely the *action*.

dimension. Also body dimension may be chosen freely—provided it does not exceed space dimension.

## 3 Electromagnetic-Mechanical Coupling

On the surface, coupled electromagnetic-mechanical problems are mostly taken as issues in computer-aided design—of a quite difficult nature, admittedly.[6] This is no surprise, thanks to the lasting division between Maxwell's and Lamé's disciples, further aggravated in their pet computer codes, laboriously developed by niche experts.

However, the heart of the matter lies much deeper. First of all, lumping electromagnetic and mechanical effects into *distinct* components of the same device is not always possible. Therefore, loose coupling between pre-existent specialized simulators may not suffice. Second, and more important: the electromagnetic and the mechanical response of a medium cannot in general be characterized independently of each other. Therefore, a unified understanding of both disciplines cannot be dispensed with—at least in principle.

### 3.1 A Sound Basis for the Electrodynamics of Deformable Media

"[T]he issue of force densities in material media is the most controversial, the least investigated, and the least understood topic of classical electromagnetism." This is how it is put by the editor of the Academic Press Electromagnetism series, Isaak Mayergoyz, in his foreword to the recent book [25] by the late Scipione Bobbio.

In this book Bobbio details several specific materials. In all cases, the key point is that sound constitutive assumptions for the free-energy density—ascribed to matter—and for the electromagnetic energy—ascribed to ether[7]—can *not* be laid down independently of each other. This point is touched upon also in [16, Sect. 1], where Henrotte and Hameyer write: "In order to define specifically *electromagnetic forces*, we have to use *electromagnetic* energy functionals instead of *total* energy functionals." Then they admit that defining "such restricted functionals" is not at all obvious, *e.g.*, for a magnetostrictive material.

In the case of an electrically linear fluid dielectric, Bobbio shows that Helmholtz's and Kelvin's formulas for the *electric* force density—which notoriously disagree—correspond to a different splitting of the same total energy density into ether- and matter-related terms—hence, to different prescriptions for the *mechanical* force density, *i.e.*, for pressure. A long-lasting—and quite idle—controversy between supporters of either formula is dispelled this way. See [25, Sect. 4.7] for a review of the experimental side of the dispute.

[6] Such difficulties were vividly reported at the workshop. See in particular [23,24].

[7] This is my own wording, inspired also by a recent mind-teasing contribution by Ericksen [26]. Bobbio uses the neuter term "field".

**The basic structure underlying Maxwell's stress tensor.** The Maxwell stress tensor represents the fundamental coupling between electromagnetism and mechanics. I need not belabour this point, which is so neatly expounded by Henrotte and Hameyer [16]. I agree with them and dissent from [22, Sect. 4.2].

Let us consider a placement $p$, the place-valued field defined in Sect. 2.2. It is now convenient to associate with $p$ the mapping $(\mathsf{b},\mathsf{i}) \mapsto (p(\mathsf{b},\mathsf{i}),\mathsf{i})$, which embeds body-time into space-time ($\mathsf{b}$ is a point, $\mathsf{i}$ an instant). By a slight abuse of language, I will denote also this trivial extension by $p$. Let $\mathsf{c}$ be a body-time $k$-cell, and $\varphi$ a spacetime-conjugate $k$-form (Fig. 1). Of course, $\varphi$ cannot be integrated on $\mathsf{c}$, but it makes sense to integrate it on its *push-forward* by $p$, namely $p_\star\mathsf{c}$. By definition, the same result is obtained integrating on $\mathsf{c}$ the *pull-back* of $\varphi$ by $p$, namely $p^\star\varphi$:

$$\langle p^\star\varphi, \mathsf{c}\rangle = \langle \varphi, p_\star\mathsf{c}\rangle ,$$

where integration is aptly denoted as a duality between $k$-forms and $k$-cells. Given $\mathsf{c}$, the value of the integral depends on both $\varphi$ and $p$, which may be varied independently of each other. In particular, a test dispacement (*cf.* footnote 4 on page 8) may be added to the placement $p$ and the spacetime-conjugate form $\varphi$ changed, *so as to keep its body-time pull-back $p^\star\varphi$ fixed.* This kind of operation is central to a proper definition—and an efficient computation—of Maxwell's stress tensor and related force densities, as clearly pointed out by Henrotte and Hameyer [16] (*cf.* my Fig. 1 with their Fig. 1).

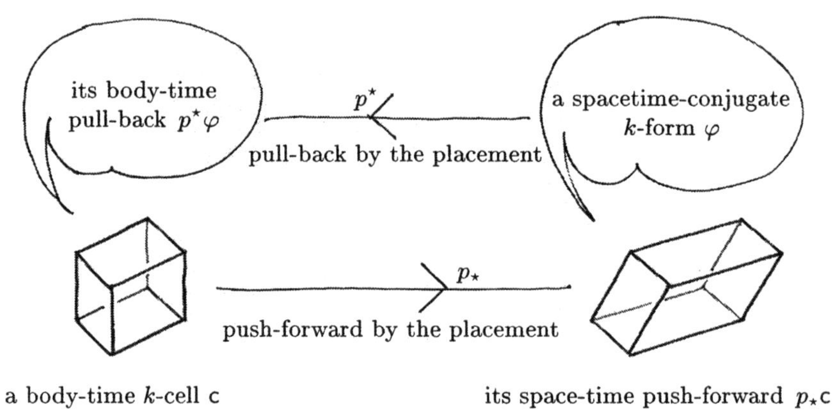


**Fig. 1.** Placement-related cells and forms

**What is to be done?** In the closing paragraph of [3], Dyson praises Maxwell as the forerunner of mainstream twentieth-century physics:

> The ultimate importance of the Maxwell theory is far greater than its immediate achievement in explaining and unifying the phenomena of electricity and magnetism. Its ultimate importance is to be the prototype of all the great triumphs of twentieth-century physics.

My personal view is that Maxwell's theory, while paving the way to quantum mechanics, interacted early on with the true Newtonian chapters of classical mechanics, *i.e.*, the classical theories of inertia and gravitation. This conceptual integration produced the special and general theories of relativity, well within the first two decades of the twentieth century. On the contrary, the integration of Maxwell's electromagnetic theory with the main body of classical mechanics, *i.e.*, the mechanics of deformable continuous media,[8] still remains to be done, nearly a century later.

Should we succeed in this endeavour, we could ascribe our accomplishment exclusively to the "Geometers" of our own century—as justly (or as unjustly) as Lamé did with his *Physique mathématique, proprement dite.* While skeptical with regard to our ability to counter parochialism in science and education, I am quite confident that the present strong trend towards miniaturization—down to the nano-scale—could make a unified electro-mechanics of deformable media one of the most *useful* physical theories in the near future.

**Acknowledgements**

My work was partly supported by the Italian GNFM-INdAM (Gruppo Nazionale di Fisica Matematica). I feel indebted to many people, but I am especially grateful to the workshop organizers, who had the heart to invite an outsider and the patience to bear with his slowness and longwindedness; to my old professor, who donated the collection of books where I picked out Lamé's *Leçons*; and to my new copy-editor, who redressed my English when faltering or simply wrong. Without them this paper would have never been written.

## References

1. Lamé, G.: Leçons sur la théorie mathématique de l'élasticité des corps solides, deuxième édition, Gauthier-Villars, Paris, 1866
2. Benvenuto, E.: An introduction to the history of structural mechanics, Springer, New York, 1991
3. Dyson, F.J.: Why is Maxwell's theory so hard to understand? Reprint distributed (with no further reference) at the Fourth International Congress on Industrial and Applied Mathematics (ICIAM 99), Edinburgh, 1999

[8] As starkly put by Truesdell and Toupin in the exordium of [27]: "Only pædagogical custom has hindered general realization that *as a physical theory, continuum mechanics is better than mass-point mechanics.* [...] As Hamel stated, *classical mechanics is the mechanics of extended bodies.*"

4. Saletan, E.J.: Contraction of Lie groups. J. Mathematical Physics **2** (1961) 1–21
5. Le Bellac, M., Levy-Leblond, J.-M.: Galilean electromagnetism. Il Nuovo Cimento **14 B** (1973) 217–32
6. Tonti, E.: A mathematical model for physical theories. Rend. Accademia dei Lincei **52** (1972) 175–181 (1st part) & 350–356 (2nd part)
7. Tonti, E.: On the geometrical structure of electromagnetism. In: Gravitation, Electromagnetism and Geometrical Structures (G. Ferrarese, ed.), Pitagora, Bologna (1996) 281–308
8. Tonti, E.: Finite formulation of the electromagnetic field. In: PIER 32 – Geometric Methods for Computational Electromagnetics (F.L. Teixeira, ed.), EMW Publishing, Cambridge, MA (2001) 1–44
9. Mattiussi, C.: Edge elements and cochain-based field function approximation. In: Proceedings of the 4th International Workshop on Electric and Magnetic Fields (Marseille, May 12–15, 1998), A.I.M., Liège (1998) 301–306
10. Mattiussi, C.: The finite volume, finite element, and finite difference methods as numerical methods for physical field problems. Advances in Imaging and Electron Physics **113** (2000) 1–146
11. Naber, G.L.: The geometry of Minkowsky spacetime: an introduction to the mathematics of the special theory of relativity, Springer, New York, 1992
12. Crampin M., Pirani F.A.E.: Applicable differential geometry, Cambridge University Press, Cambridge, 1986
13. Maxwell, J.C.: Remarks on the mathematical classification of physical quantities. Proceedings of the London Mathematical Society **3** (1871) 224–232
14. Henrotte, F., Legros, W.: Elasticity and electromagnetic force calculation with differential forms. In: Proc. 4th International Workshop on Electric and Magnetic Fields (Marseille, May 12–15, 1998), A.I.M., Liège (1998) 295–300
15. Henrotte, F., Hameyer, K.: A mathematical framework for the finite element modelling of electromechanical problems. In: Scientific Computing in Electrical Engineering (U. van Rienen et al., eds.), Springer, Heidelberg (2001) 359–366
16. Henrotte, F., Hameyer, K.: Electromagnetic force densities in a continuous medium. In these Proceedings
17. Abraham, R., Marsden J.E., Ratiu, T.: Manifolds, tensor analysis, and applications, Addison-Wesley Publishing Company, Reading, MA, 1983
18. Nardinocchi, P., Teresi, L., Tiero, A.: A direct theory of affine rods. European Journal of Mechanics A/Solids **21** (2002) 653–667
19. DiCarlo, A.: A non-standard format for continuum mechanics. In: Contemporary Research in the Mechanics and Mathematics of Materials (R.C. Batra, M.F. Beatty, eds.), CIMNE, Barcelona (1996) 263–268
20. Germain, P.: Sur l'application de la méthode des puissances virtuelles en mécanique des milieux continus. Comptes Rendus de l'Académie des Sciences de Paris **274** Série A (1972) 1051–1055
21. Germain, P.: The method of virtual power in continuum mechanics. Part 2: Microstructure. SIAM Journal on Applied Mathematics **25** (1973) 556–575
22. Maugin, G.A.: The method of virtual power in continuum mechanics: application to coupled fields. Acta Mechanica **35** (1980) 1–70
23. Schwarz, P.: Continuous simulation of coupled systems. In these Proceedings
24. Rothemund, K., van Rienen, U.: Coupled simulation of electromagnetic fields and mechanical deformation. In these Proceedings
25. Bobbio, S.: Electrodynamics of materials: forces, stresses, and energies in solids and fluids, Academic Press, San Diego, 2000

26. Ericksen, J.L.: Electromagnetic effects in thermoelastic materials. Mathematics and Mechanics of Solids **7** (2002) 165–189
27. Truesdell, C., Toupin, R.A.: The classical field theories. In: Handbuch der Physik, Vol. 3/1 (S. Flügge, ed.), Springer, Berlin (1960) 226–793

# Variational FDTD-like Methods for Maxwell's Equations *

Patrick Joly

INRIA Rocquencourt BP105 Le Chesnay France

**Abstract.** In this article, we describe two types of conservative variational techniques that aim at improving the use of FDTD methods for the treatment of complex geometries with time dependent Maxwell's equations.

## 1 Introduction

Although very old, the finite differences time domain methods (FDTD in the electromagnetic literature) remain very popular and are widely used for time dependent numerical simulations of electromagnetic wave propagation.These methods allow us to get discrete equations whose unknowns are generally field values at points of a regular mesh with spatial step $h$ and time step $\Delta t$. For Maxwell's equations, the Yee scheme [12], [13] is a prototype of such a scheme. In 1D, it concerns the following system:

$$\frac{\partial u}{\partial t} + c\,\frac{\partial v}{\partial x} = 0, \qquad \frac{\partial v}{\partial t} + c\,\frac{\partial u}{\partial x} = 0, \qquad x \in \mathbb{R}, \quad t > 0, \tag{1}$$

Without any mesh refinement, the equations of the scheme are

$$\frac{u_j^{n+1} - u_j^n}{\Delta t} + c\,\frac{v_{j+\frac{1}{2}}^{n+\frac{1}{2}} - v_{j-\frac{1}{2}}^{n+\frac{1}{2}}}{h} = 0, \quad \frac{v_{j+\frac{1}{2}}^{n+\frac{1}{2}} - v_{j+\frac{1}{2}}^{n-\frac{1}{2}}}{\Delta t} + c\,\frac{u_{j+1}^n - u_j^n}{h} = 0, \tag{2}$$

where the discrete unknowns are evaluated on a staggered uniform grid. There are several reasons that explain the success of Yee type schemes, among which are their easy implementation, their efficiency linked to the use of a uniform regular grid together with an explicit time discretization, and the fact that a lot of properties of continuous Maxwell's equations (energy conservation, free divergence property,...) are respected at the discrete level. The stability and accuracy properties of such a scheme are well known. As a consequence of its explicit nature, the scheme is stable under the C.F.L. condition ($c$ denotes the propagation velocity and $d$ the space dimension)

$$\frac{c\Delta t}{h} \leq \frac{\sqrt{d}}{d}. \tag{3}$$

This implies that the time step cannot be too large but this is not restrictive in practice since a sufficient accuracy requires a small time step. On the other

* Invited paper at SCEE-2002

hand, it must not be too small either because, as is well known, the numerical dispersion, roughly speaking the error committed on the propagation velocity of waves, increases when the ratio $c\Delta t/h$ decreases.

The counterpart of the nice properties of FDTD schemes is a lack of "geometrical flexibility" which makes the use of such finite difference schemes not obvious in the case of computational domains of complicated shape (consider here the diffraction of electromagnetic waves by an obstacle as a target problem). It may also be difficult (at least with a theoretical guaranty of stability) to treat boundary conditions and variable coefficients or to be able to do mesh refinement. To overcome such difficulties there exist at least two attractive solutions:

(i) The variational methods, in particular the finite element methods.
(ii) The finite integration technique [4] and the finite volume methods [11].

These are "natural" extensions in the sense that, for instance, the Yee scheme can be interpreted as a particular mixed finite element method or a particular finite volume method on a uniform grid. My objective in this paper will be to review briefly two recent works that aim as making the treatment of complex geometries with FDTD schemes possible while preserving the nice properties of these methods:

- The data of the problem remain (mostly) structured,
- The time discretization remains (essentially) explicit,
- The stability condition is not affected by the geometry of the domain.

## 2 Fictitious domain methods

Let us consider the model problem of the scattering of an incident electromagnetic wave by a perfectly conducting obstacle. The idea of the method consists in artificially extending the solution inside the obstacle - which makes the use of a 3D uniform regular grid for the electromagnetic field possible - and to introduce at the same time a conforming surface mesh for the boundary of the obstacle to handle the boundary condition. On this mesh, one computes an auxiliary unknown that can be interpreted as a Lagrange multiplier associated to the boundary condition and coïncides, in this case, to the surface electric current. The challenge is then to let these two "independent" meshes communicate in a clever way. This can be done through the use of a mixed variational formulation in which the boundary condition is taken into account in a weak sense. The stability of the method in ensured through a discrete energy conservation and the stability condition is the one of the pure FDTD scheme. The only additional computational cost (with respect to the standard FDTD scheme) is reduced to the boundary mesh: a sparse positive definite linear system has to be solved at each time step.

## 2.1 The fictitious domain formulation

Our model problem is the scattering of an electromagnetic wave (in a homogeneous medium with $\varepsilon = \mu = 1$) by an obstacle $\mathcal{O}$, with exterior $D$. We impose the perfectly conducting boundary condition on $\gamma = \partial D = \partial \mathcal{O}$:

$$\begin{cases} \dfrac{\partial \boldsymbol{e}}{\partial t} - \operatorname{rot} \boldsymbol{h} = 0, & x \in D, \\ \dfrac{\partial \boldsymbol{h}}{\partial t} + \operatorname{rot} \boldsymbol{e} = 0, & x \in D, \\ n \times (\boldsymbol{e} \times n) = 0, & \text{on } \gamma = \partial D. \end{cases} \tag{4}$$

We assume that the incident wave is generated by initial conditions (omitted here) at time $t = 0$. In order to have a finite computational domain, the classical technique consists in bounding the domain $D$ and in imposing absorbing conditions on the exterior boundary ( [12], [6]). For the sake of simplicity, a perfectly conducting condition is assumed on the exterior boundary as well, and, for our purpose, we choose the geometry of the external boundary (which does not interest us here) to be rectangular or parallelepipedic. We denote by $\Omega$ this bounded domain and the box $\Omega \bigcup \mathcal{O}$ by $C$ .

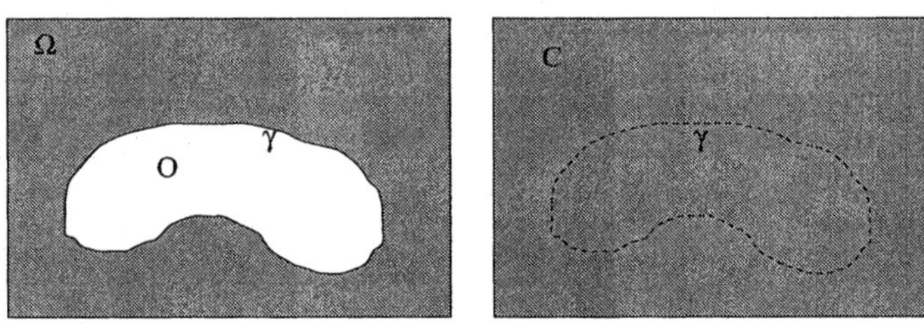


**Fig. 1.** The geometry of the problem.

The main idea of the FDM is to extend the electromagnetic field $(e, h)$ from $\Omega$ to the enlarged domain $C$ by 0 inside $\mathcal{O}$. Introducing as an additional boundary unknown the electric current:

$$\jmath = h \times n \,|_{\gamma} , \tag{5}$$

it is easy to see that the extended unknowns $(e, h)$ satisfy in the sense of distributions:

$$\begin{cases} \dfrac{\partial \boldsymbol{e}}{\partial t} - \operatorname{rot} \boldsymbol{h} = \jmath \, \delta_\gamma, & \text{in } C, \\ \dfrac{\partial \boldsymbol{h}}{\partial t} + \operatorname{rot} \boldsymbol{e} = 0, & \text{in } C, \\ n \times (\boldsymbol{e} \times n) = 0, & \text{on } \partial C, \end{cases} \tag{6}$$

where $\delta_\gamma$ is the Dirac measure supported by $\gamma$. To get the variational fictitious domains formulation, we multiply the first equation of (6) by a test field $\boldsymbol{e}_t$ (which only depends on the space variable $x$ - the subscript $t$ here holds for "test functions" and has no relationship with the time variable) and integrate over $\Omega$. Similarly, we multiply the second equation of (6) by a test field $\boldsymbol{h}_t$, integrate over $\Omega$ apply an integration by parts. One completes the formulation by writing the boundary condition in the weak sense. The equation is multiplied by a surface test field $\jmath_t$ and one integrates over $\gamma$. The resulting formulation is the following:

$$\text{Find } (\boldsymbol{e}, \boldsymbol{h}, \jmath) : \mathbb{R}^+ \longrightarrow U \times V \times L \text{ such that}$$

$$\begin{cases} \dfrac{d}{dt}(\boldsymbol{e}, \boldsymbol{e}_t) - b(\boldsymbol{e}_t, \boldsymbol{h}) + c(\jmath, \boldsymbol{e}_t) = 0, & \forall\, \boldsymbol{e}_t \in U, \\ \dfrac{d}{dt}(\boldsymbol{h}, \boldsymbol{h}_t) + b(\boldsymbol{e}, \boldsymbol{h}_t) = 0, & \forall\, \boldsymbol{h}_t \in V, \\ \quad c(\jmath_t, \boldsymbol{e}) = 0, & \forall\, \jmath_t \in L. \end{cases} \tag{7}$$

The various bilinear forms appearing in (7) are defined by:

$$(\boldsymbol{u}, \boldsymbol{u}_t) = \int_\Omega \boldsymbol{u} \cdot \boldsymbol{u}_t \, dx, \; b(\boldsymbol{e}, \boldsymbol{h}) = \int_\Omega \boldsymbol{h} \cdot \text{rot}\, \boldsymbol{e} \, dx, \; c(\jmath, \boldsymbol{e}) = \int_\gamma \jmath \cdot \pi_\tau \boldsymbol{e} \, d\gamma, \tag{8}$$

where $\pi_\tau$ denotes the trace map $\pi_\tau \boldsymbol{u} = n \times (\boldsymbol{e} \times n)_{|\Sigma}$. The appropriate functional spaces are:

$$\begin{cases} U = H_0(\text{rot}, \Omega) = \{\boldsymbol{u} \in L^2(\Omega)^3 \;\|\; \text{rot}\, \boldsymbol{u} \in L^2(\Omega)^3, n \times (\boldsymbol{u} \times n) = 0 \text{ on } \partial\Omega\} \\ V = [L^2(\Omega)]^3, \quad L = \mathbf{H}_{\|}^{-1/2}(\text{div}_\gamma, \gamma), \quad \text{(see [3] for a precise definition).} \end{cases}$$

One of the main differences between this approach and a standard conforming finite element approach lies in the fact that the boundary condition is taken into account in a weak sense instead of being imposed in the functional space. In the formulation, there is no mention to the geometry of the problem, namely the boundary $\gamma$, in the functional spaces for the volume unkonwns $\boldsymbol{e}$ and $\boldsymbol{h}$. The geometry only appears in $c(\cdot, \cdot)$ and $L$. We can also interpret the unknown $\jmath$ as a source term distributed on $\gamma$. If this source term were known, we simply would have to solve the wave equation is a square with a right hand side: the FDTD approach makes sense in such a geometric situation. In fact, $\jmath$ is unknown and becomes a contrôl variable in order to make $\boldsymbol{e}$ satisfy the boundary condition on $\gamma$. The above approach is referred to as the fictitious domain method (FDM).

There is an analogy between the FDM and the integral equations for scattering problems [1]. Indeed, in this kind of method $\jmath$ typically is the quantity that is chosen as the unknown. Nevertheless let us point out a very important difference between our approach and these methods. Integral equations are known to lead, after discretization, to the solution of full linear systems in $\jmath$; as will be shown later, this will not be the case for the FDM.

## 2.2 Finite element approximation and time discretization

*Space discretization.* One obtains the semi-discrete (in space) scheme by replacing in (7) the functional spaces $U$, $V$ and $L$ by appropriate finite dimensional spaces $U_h$, $V_h$ and $L_H$ defined as follows. We consider a uniform mesh $\mathcal{T}_h$ made of equal cubes $K \in \mathcal{T}_h$ of side $h$. The appropriate approximation spaces $U_h$ and $V_h$ are:

$$\begin{cases} U_h = \{ \boldsymbol{u}_h \in U \mid\mid \forall\, K \in \mathcal{T}_h, \boldsymbol{u}_h|_K \in \mathcal{Q}_{0,1,1} \times \mathcal{Q}_{1,0,1} \times \mathcal{Q}_{1,1,0} \} \\ V_h = \{ \boldsymbol{v}_h \in H(\mathrm{div}, \Omega) \mid\mid \forall\, K \in \mathcal{T}_h, \boldsymbol{v}_h|_K \in \mathcal{Q}_{1,0,0} \times \mathcal{Q}_{0,1,0} \times \mathcal{Q}_{0,0,1} \} \end{cases} \tag{9}$$

where $H(\mathrm{div}, \Omega) = \{\boldsymbol{u} \in L^2(\Omega)^3 \mid\mid \mathrm{div}\, \boldsymbol{u} \in L^2(\Omega)^3\}$ and where $\mathcal{Q}_{p_1,p_2,p_3}$ is the set of polynomials of three variables whose degree with respect to the $i^{th}$ variable is less than or equal to$p_i$. In addition, one uses adapted quadrature formulas for approximating the various integrals appearing in (8).

One can notice that, contrary to what one could expect, we do not use a space of completely discontinuous elements to approximate $V = L^2(\Omega)^3$; we use vector fields whose normal component is continuous across each face of the mesh. The spaces defined by (9) are known as edge elements for the electric field and face elements for the magnetic field (see [10]). In particular a set of degrees of freedom is given by (see figure 2):

- For the space $U_h$ (degrees of freedom for the electric field) : the (constant) tangential component of the vector field along each edge of the mesh. In the sequel, we shall denote by $\mathbf{e}$ the vector of degrees of freedom of $\boldsymbol{e} \in U_h$
- For the space $V_h$ (degrees of freedom for the magnetic field) : the (constant) tangential component of the vector field on each face of the mesh. We denote by $\mathbf{h}$ the vector of degrees of freedom of $\boldsymbol{h} \in V_h$.

For the space $L_H$, we consider an approximation $\gamma_H$ of the surface $\gamma$ made of a piecewise linear surface with triangular facets of diameter less than or equal to $H$ (see figure 2 for a sphere) and consider the corresponding Raviart-Thomas tangential face element space described in [1] for instance. Let us simply recall that this space is made of tangential surface vector fields which are piecewise linear, have a constant normal component on each edge of the mesh. These fluxes constitute of the degrees of freedom for this surface unknown (see figure 2). We shall denote $\mathbf{j}$ the corresponding vector of degrees

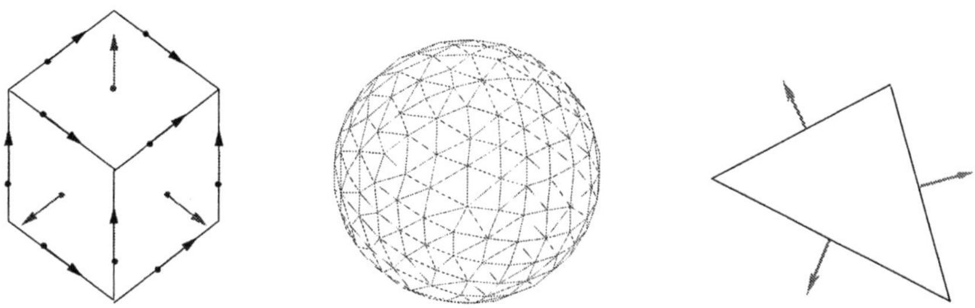

**Fig. 2.** The degrees of freedom for edge and face elements (left) and for boundary elements (right)

of freedom. The space semi-discretization results into an algebraic-differential system of the following type:

$$\begin{cases} \mathcal{M}^E \dfrac{d\mathbf{e}}{dt} - \mathcal{B}\,\mathbf{h} - \mathcal{C}^*\,\mathbf{j} = 0, \\ \mathcal{M}^H \dfrac{d\mathbf{h}}{dt} + \mathcal{B}^*\,\mathbf{e} = 0, \\ \mathcal{C}\,\mathbf{e} = 0. \end{cases} \tag{10}$$

- The two mass matrices $\mathcal{M}^E$ and $\mathcal{M}^H$ are diagonal.
- The two matrices $\mathcal{B}$ and $\mathcal{B}^*$ represent discrete rotational operators.
- The matrix $\mathcal{C}$ represents a discrete tangential trace operator.

For the time discretization, we introduce a constant time step $\Delta t > 0$ and apply a standard staggered grid centered scheme:

$$\begin{cases} \mathcal{M}^E \dfrac{\mathbf{e}^{n+1} - \mathbf{e}^n}{\Delta t} - \mathcal{B}\,\mathbf{h}^{n+\frac{1}{2}} - \mathcal{C}^*\,\mathbf{j}^{n+\frac{1}{2}} = 0, \\ \mathcal{M}^H \dfrac{\mathbf{h}^{n+\frac{1}{2}} - \mathbf{h}^{n-\frac{1}{2}}}{\Delta t} + \mathcal{B}^*\,\mathbf{e}^n = 0, \\ \mathcal{C}\,\mathbf{e}^n = 0 \end{cases} \tag{11}$$

For the practical computations, we remark that the first equation of (11) permits to compute explicitly $\mathbf{h}^{n+\frac{1}{2}}$ from $\mathbf{e}^n$ and $\mathbf{h}^{n-\frac{1}{2}}$. Furthermore, if $\mathbf{j}^{n+\frac{1}{2}}$ is known, the first equation of (11) allows to compute $\mathbf{e}^{n+1}$. It remains to determine the equation to compute $\mathbf{j}^{n+\frac{1}{2}}$. This is where we use the third equation of (11) that we combine with the first one to obtain:

$$\mathcal{C}\,(\mathcal{M}^E)^{-1}\,\mathcal{C}^*\,\mathbf{j}^{n+\frac{1}{2}} = -\,\mathcal{C}\,(\mathcal{M}^E)^{-1}\,\mathcal{B}\,\mathbf{h}^{n+\frac{1}{2}} \tag{12}$$

For each time step, the algorithm consists in:

- compute two steps of explicit scheme, i.e. the first two equations of (11), which are nothing but the standard Yee scheme with surface source term.

– solve the linear system (12), reduced to the surface unknown $\mathbf{j}^{n+\frac{1}{2}}$.

We see that, in comparison with a standard FDTD procedure inside $C$, the only additional cost is the inversion of the matrix $\mathcal{Q} = \mathcal{C}\ (\mathcal{M}^E)^{-1}\ \mathcal{C}^*$. This cost is marginal, due to the properties of $\mathcal{Q}$ (see section 2.3). In fact, from the computational point of view, the difficult step in the implementation lies in the construction of the matrix $\mathcal{B}$ (and then of $\mathcal{Q}$): in 3D, it implies the determination of the intersections between the cubic mesh and the surfacic mesh. See [7], for instance, for more details.

### 2.3 Theoretical issues

*Existence of the discrete solution.* One easily observes observe that

– $\mathcal{Q}$ is symmetric and positive (by construction !),
– $\mathcal{Q}$ is a "small" matrix (since "reduced to the surface"),
– $\mathcal{Q}$ is a sparse matrix with narrow bandwidth.

The (crucial) remaining question is the definiteness of $\mathcal{Q}$ which corresponds to the fact that the kernel of the matrix is equal to 0, or equivalently that $\mathcal{B}$ is surjective from $U_h$ onto $L_H$. This suggests that the space $L_H$ must not be too large, or in other words that one must not impose too many "boundary" constraints to the discrete solution. In fact, as for any mixed method, there is a compatibility condition with the two spaces $U_h$ and $L_H$ that can be reduced to a compatibility relation between the two meshes of $C$ and $\gamma$ : the volume mesh cannot be too large with respect to the boundary mesh or, roughly speaking, the ratio $H/h$ must be large enough. In this sense, the two meshes cannot be completely independent [5], [7].

*Stability analysis.* The numerical scheme is stable under the same CFL condition than without the obstacle. This is a consequence of the conservation of the following discrete electromagnetic energy (the proof is straightforward):

$$E^n = \frac{1}{2}\left(\mathcal{M}^E\ \mathbf{e}^n \cdot \mathbf{e}^n + \mathcal{M}^H\ \mathbf{h}^{n+\frac{1}{2}} \cdot \mathbf{h}^{n-\frac{1}{2}}\right). \tag{13}$$

Finally, one can prove that this quantity is a "true" energy, namely a positive quantity that constitutes a norm of the discrete solution, if (3) holds.

*Convergence analysis.* For simplicity, let us restrict ourselves to the case of the semi-discrete problem (the treatment of the time discretization does not pose any specific difficulty). The convergence of the method requires not only the invertibility of the matrix $\mathcal{Q}$ but also a uniform bound (uniform with respect to the approximation parameters $h$ and $H$) of an appropriate norm of its inverse $\mathcal{Q}^{-1}$. The general theory ( [2]) says that this is equivalent to verifying the so called uniform inf-sup condition whose verification leads to a compatibility condition between the sizes of the two meshes namely (see [8] or [9] for instance) $H \geq Ch$ for some $C > 0$. It is difficult to get an explicit value for $C$ but in practice it suffices to take $C$ slightly greater than 1.

*Accuracy of the fictitious domain method.* The counterpart to the good properties of the fictitious domain method, in terms of simplicity and robustness, is its limited accuracy. The limitation is due to the fact that the regularity of the exact solution in $C$ - or inside an element $K$ of the volume mesh - is limited, independently of the smoothness of the data of the problem, since the tangential component of $h$ presents a jump across $\gamma$. As a consequence, the method is only of order 1, i.e. the error is bounded by a constant times $h+H$. However, numerical experiments (see [5] for instance) demonstrate that the fictitious domain method provides a better accuracy than the FDTD method combined a staircase approximation of the boundary of the obstacle. In this last case, the approximation of the geometry induces artificial singularities in the surface of the scatterer: these singularities provoke spurious diffractions that are not present with the fictitious domain method.

# 3 Space-time mesh refinement and domain decomposition

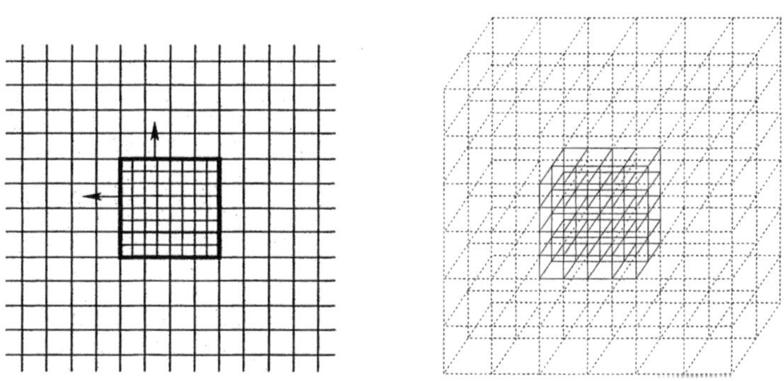

**Fig. 3.** 2 D slice of $\Omega_f$ and $\Omega_c$ (left) - 3D view of the refinement (right)

An alternative approach (that can be combined with the fictitious domain method) consists in refining the mesh at the neighborhood of the obstacle. I will present below some recent research about conservative space-time mesh refinement methods. When one works with regular grids, the transition between a coarse and a fine grid is necessarily "non conforming". Moreover, for efficiency and accuracy considerations, one would like to use a local time step in order to keep the ratio time step/space step constant. Interpolation methods that can be traditionally found in the literature can lead to non standard instability phenomena. Here, we shall propose two alternative methods based on the reformulation of the problem as an artificial domain decomposition problem. The key issue of these methods is that their stability is guaranteed from the theoretical point of view through the conservation of an appropriate discrete energy. The first method involves the introduction of a

Lagrange multiplier on the interface coarse grid / fine grid (as the so-called mortar element method), the second does not. As for the fictitious domain method, both methods lead to the inversion of a small sparse positive definite linear system on the interface. We shall also show how spurious numerical phenomena due to a change of grid can be analyzed and controlled.

### 3.1 The domain decomposition approch

As a model problem, we consider Maxwell's equations in the whole space $\mathbb{R}^3$ (with $\varepsilon = \mu = 1$):

$$\left\{\begin{aligned} &\frac{\partial \boldsymbol{e}}{\partial t} - \mathrm{rot}\, \boldsymbol{h} = 0, \quad x \in \mathbb{R}^3 \\ &\frac{\partial \boldsymbol{h}}{\partial t} + \mathrm{rot}\, \boldsymbol{e} = 0, \quad x \in \mathbb{R}^3 \end{aligned}\right. \tag{14}$$

Our goal is to solve numerically this problem by domain decomposition using locally a two times finer grid than in the rest of the domain. To be more precise, we consider the fine grid domain $\Omega_f$ with boundary $\Sigma$ (with outward and unit normal vector $n$), the coarse grid domain $\Omega_c$ denoting the exterior of $\Omega_f$ . The domain $\Omega_f$ is the one that we shall discretize with a fine grid of stepsize $h$ and $\Omega_c$ is the one that we shall discretize with a coarse grid of step $2h$ (see figure 3). In what follows, $(\boldsymbol{e}_c, \boldsymbol{h}_c)$ (resp. $(\boldsymbol{e}_f, \boldsymbol{h}_f)$) will denote the restriction to $\Omega_c$ (resp. $\Omega_f$) of $(\boldsymbol{e}, \boldsymbol{h})$.

Saying that $(\boldsymbol{e}, \boldsymbol{h})$ is solution of (14) is equivalent to saying that $(\boldsymbol{e}_c,\ \boldsymbol{h}_c)$ and $(\boldsymbol{e}_f\ \boldsymbol{h}_f)$ are solutions of the same equations (14) respectively in $\Omega_c$ and $\Omega_f$ and are coupled by the following transmission conditions (the continuity of the tangential traces of the two fields across $\Sigma$):

$$\left\{\begin{aligned} &n \times (\boldsymbol{e}_c \times n) = n \times (\boldsymbol{e}_f \times n) \quad \text{on } \Sigma, \\ &\boldsymbol{h}_c \times n = \boldsymbol{h}_f \times n \quad \text{on } \Sigma. \end{aligned}\right. \tag{15}$$

### 3.2 A variational formulation with interface unknown

This formulation is similar to the fictitious domain formulation we presented in section 2. We introduce as an additional unknown, namely the common tangential trace of the magnetic fields $\boldsymbol{h}_c$ et $\boldsymbol{h}_f$ on the interface:

$$\jmath = (\boldsymbol{h}_f \times n)|_{\Sigma} = (\boldsymbol{h}_c \times n)|_{\Sigma}. \tag{16}$$

Note that $\jmath$ is nothing else but the surface electric current on $\Sigma$. We can then reformulate our problem as follows. Assuming for a moment that $\jmath$ is known,

$(\boldsymbol{e}_c, \boldsymbol{h}_c)$ and $(\boldsymbol{e}_f, \boldsymbol{h}_f)$ are the respective solutions of the following decoupled problems, in which $\jmath$ appears as a (boundary) source term:

$$\begin{cases} \dfrac{\partial \boldsymbol{e}_f}{\partial_t} - \operatorname{rot} \boldsymbol{h}_f = 0, \quad x \in \Omega_f \qquad (i) \\ \dfrac{\partial \boldsymbol{h}_f}{\partial_t} + \operatorname{rot} \boldsymbol{e}_f = 0, \quad x \in \Omega_f \qquad (ii) \\ \boldsymbol{h}_f \wedge n = \jmath, \quad x \in \Sigma = \partial \Omega_f \end{cases} \tag{17}$$

$$\begin{cases} \dfrac{\partial \boldsymbol{e}_c}{\partial_t} - \operatorname{rot} \boldsymbol{h}_c = 0, \quad x \in \Omega_c \qquad (i) \\ \dfrac{\partial \boldsymbol{h}_c}{\partial_t} + \operatorname{rot} \boldsymbol{e}_c = 0, \quad x \in \Omega_c \qquad (ii) \\ \boldsymbol{h}_c \wedge n = \jmath, \quad x \in \Sigma = \partial \Omega_c \end{cases} \tag{18}$$

For a given $\jmath$, by construction, the continuity of the tangential magnetic field is ensured. The idea of the method is to consider $\jmath$ as a contrôl variable in order to ensure the continuity of the tangential electric field

$$n \wedge (\boldsymbol{e}_f \wedge n)\,|_\Sigma = n \wedge (\boldsymbol{e}_c \wedge n)\,|_\Sigma. \tag{19}$$

To derive our variational formulation, we use in each domain the same mixed formulation as in section 2.1: once again, only the two equations involving the rotational of the magnetic field (cf. (17)-(ii) and (18)-(ii)) are integrated by parts: this makes appear $\jmath$ in the boundary term and leads to the following abstract formulation:

Find $(\boldsymbol{e}_c, \boldsymbol{h}_c, \boldsymbol{e}_f, \boldsymbol{h}_f, j) : \mathbb{R}^+ \longrightarrow U_c \times V_c \times U_f \times V_f \times L$ such that

$$\begin{cases} \dfrac{d}{dt}(\boldsymbol{e}_c,\, \boldsymbol{e}_{c,t})_c - b_c(\boldsymbol{e}_{c,t},\, \boldsymbol{h}_c) + c_c(\jmath,\, \boldsymbol{e}_{c,t}) = 0, & \forall\, \boldsymbol{e}_{c,t} \in U_c, \\ \dfrac{d}{dt}(\boldsymbol{h}_c,\, \boldsymbol{h}_{c,t})_c + b_c(\boldsymbol{e}_c,\, \boldsymbol{h}_{c,t}) = 0, & \forall\, \boldsymbol{h}_{c,t} \in V_c, \\ \dfrac{d}{dt}(\boldsymbol{e}_f,\, \boldsymbol{e}_{f,t})_f - b_f(\boldsymbol{e}_{f,t},\, \boldsymbol{h}_f) + c_f(\jmath,\, \boldsymbol{e}_{f,t}) = 0, & \forall\, \boldsymbol{e}_{f,t} \in U_f, \\ \dfrac{d}{dt}(\boldsymbol{h}_f,\, \boldsymbol{h}_{f,t})_f + b_f(\boldsymbol{e}_f,\, \boldsymbol{h}_{f,t}) = 0, & \forall\, \boldsymbol{h}_{f,t} \in V_f, \\ c_c(\jmath_t,\, \boldsymbol{e}_c) = c_f(\jmath_t,\, \boldsymbol{e}_f), & \forall\, \jmath_t \in L. \end{cases} \tag{20}$$

The appropriate functional spaces are:

$$\begin{cases} U_{e,f} = H(\operatorname{rot}, \Omega_{e,f}), \quad V_{e,f} = L^2(\Omega_{e,f})^3, \\ L = \mathbf{H}_{\|}^{-1/2}(\operatorname{div}_\Sigma, \Sigma). \end{cases} \tag{21}$$

The various bilinear forms appearing in (20) are defined by:

$$\left\{\begin{array}{ll} (\boldsymbol{u}_c, \boldsymbol{v}_c)_c = \displaystyle\int_{\Omega_c} \boldsymbol{u}_c \cdot \boldsymbol{v}_c \, dx, & (\boldsymbol{u}_f, \boldsymbol{v}_f)_f = \displaystyle\int_{\Omega_f} \boldsymbol{u}_f \cdot \boldsymbol{v}_f \, dx, \\ b_c(\boldsymbol{e}_c, \boldsymbol{h}_c) = \displaystyle\int_{\Omega_c} \boldsymbol{h}_c \cdot \mathrm{rot}\, \boldsymbol{e}_c \, dx, & b_f(\boldsymbol{e}_f, \boldsymbol{h}_f) = \displaystyle\int_{\Omega_f} \boldsymbol{h}_f \cdot \mathrm{rot}\, \boldsymbol{e}_f \, dx, \\ c_c(\jmath, \boldsymbol{e}_c) \;=\; \displaystyle\int_{\Sigma} \jmath \cdot \pi_\tau \boldsymbol{e}_c \, d\sigma, & c_f(\jmath, e_f) \;=\; \displaystyle\int_{\Sigma} \jmath \cdot \pi_\tau \boldsymbol{e}_f \, d\sigma. \end{array}\right. \tag{22}$$

where $\pi_\tau$ denotes again the trace map: $\pi_\tau \boldsymbol{u} = n \wedge (\boldsymbol{u} \wedge n)_{|\Sigma}$.

### 3.3 Finite element approximation.

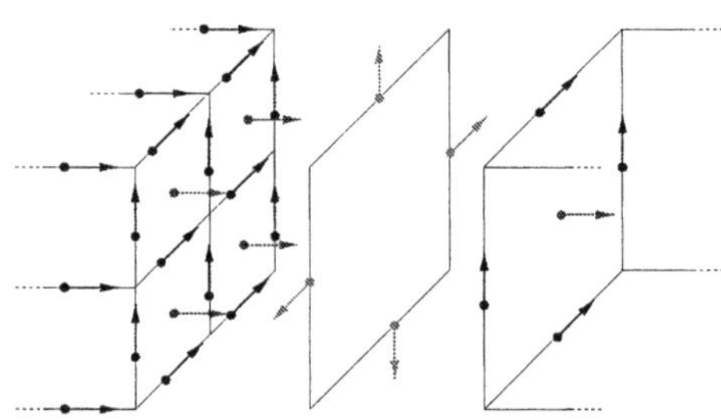

**Fig. 4.** The degrees of freedom: the electric field in blue, the magnetic field in red, the electric current in green

As in section 2.2, the space discretization of (20) is obtained by replacing the spaces $U_c, V_c, U_f, V_f$ and $L$ by finite dimensional approximation subspaces $U_{c,h}, V_{c,h}, U_{f,h}, V_{f,h}$ and $L_h \subset L$ . The spaces $U_{f,h}$ and $V_{f,h}$ ( respectively $U_{c,h}$ and $V_{c,h}$) are constructed as the spaces $U_h$ and $V_h$ in section 2.2, using a regular cubic mesh of stepsize $h$ in $\Omega_f$ (respectively of stepsize $2h$ in $\Omega_g$). For the space $L_h$, we have chosen as the mesh $\mathcal{T}_h(\Sigma)$ of the interface $\Sigma$, the trace of the coarse grid mesh in $\Omega_c$, i.e. a regular mesh made of squares $C$ of side $2h$. Then, we use standard (tangential) Raviart-Thomas 2D elements, which provides a conforming subspace of $L$:

$$L_h = \{ j_h \in L \,\|\, \forall\, K \in \mathcal{T}_h(\Sigma), j_h|_K \in \mathcal{Q}_{1,0} \times \mathcal{Q}_{1,0} \,\} \tag{23}$$

The degrees of freedom in $J_h$ are the fluxes (or equivalently the constant normal component) of the vector fields across (or on) each edge of the surface mesh. The global spacial discretization is summarized in figure 4. The semi-discrete problem can be rewritten as an algebraic-differential system (we use the same type of notation as in section 2.2):

$$\begin{cases} \mathcal{M}_f^E \dfrac{d\mathbf{e}_f}{dt} - \mathcal{B}_f\, \mathbf{h}_f - \mathcal{C}_f^*\, \mathbf{j} = 0, & \mathcal{M}_c^E \dfrac{d\mathbf{e}_c}{dt} - \mathcal{B}_c\, \mathbf{h}_c + \mathcal{C}_c^*\, \mathbf{j} = 0, \\ \mathcal{M}_f^H \dfrac{d\mathbf{h}_f}{dt} + \mathcal{B}_f^*\, \mathbf{e}_f = 0, & \mathcal{M}_c^H \dfrac{d\mathbf{h}_c}{dt} + \mathcal{B}_c^*\, \mathbf{e}_c = 0, \\ \mathcal{C}_c\, \mathbf{e}_c - \mathcal{C}_f\, \mathbf{e}_f = 0. & \end{cases} \tag{24}$$

### 3.4 Time discretization.

We look for a time discretization scheme of (24) that will use a time step $\Delta t$ in $\Omega_f$ and $2\Delta t$ in $\Omega_g$. More precisely, in $\Omega_f$, the discrete unknowns will be $e^n_{f,h}$, $h^{n+\frac{1}{2}}_{f,h}$ while in $\Omega_c$, the discrete unknowns will be $e^{2n}_{c,h}$, $h^{2n+1}_{c,h}$. The two computational grids only meet at "even instants" $t = t^{2n}$ at which the two electric fields (in $\Omega_c$ and $\Omega_f$) will be computed simultaneously.

Concerning the interface unknown, one will choose to discretize $j_h$ with the coarse time step (which is coherent with the choice of the coarse mesh for the interface mesh): in other words $\mathbf{j}$ is considered constant, equal to $\mathbf{j}^{2n+1}$ during the time interval $[t^{2n}, t^{2n+2}]$. As a consequence, the scheme that we shall apply to the first four equations of (24) is the following:

$$\begin{cases} \mathcal{M}_f^E \dfrac{\mathbf{e}_f^{2n+1} - \mathbf{e}_f^{2n}}{\Delta t} - \mathcal{B}_f\, \mathbf{h}_f^{2n+\frac{1}{2}} - \mathcal{C}_f^*\, \mathbf{j}^{2n+1} = 0, \\ \mathcal{M}_f^H \dfrac{\mathbf{h}_f^{2n+\frac{1}{2}} - \mathbf{h}_f^{2n-\frac{1}{2}}}{\Delta t} + \mathcal{B}_f^*\, \mathbf{e}_f^{2n} = 0, \\ \mathcal{M}_f^E \dfrac{\mathbf{e}_f^{2n+2} - \mathbf{e}_f^{2n+1}}{\Delta t} - \mathcal{B}_f\, \mathbf{h}_f^{2n+\frac{3}{2}} - \mathcal{C}_f^*\, \mathbf{j}^{2n+1} = 0, \\ \mathcal{M}_f^H \dfrac{\mathbf{h}_f^{2n+\frac{3}{2}} - \mathbf{h}_f^{2n+\frac{1}{2}}}{\Delta t} + \mathcal{B}_f^*\, \mathbf{e}_f^{2n+1} = 0, \end{cases} \tag{25}$$

$$\begin{cases} \mathcal{M}_c^E \dfrac{\mathbf{e}_c^{2n+2} - \mathbf{e}_c^{2n}}{2\Delta t} - \mathcal{B}_c\, \mathbf{h}_c^{2n+1} + \mathcal{C}_c^*\, \mathbf{j}^{2n+1} = 0, \\ \mathcal{M}_c^H \dfrac{\mathbf{h}_c^{2n+1} - \mathbf{h}_c^{2n-1}}{2\Delta t} + \mathcal{B}_c^*\, \mathbf{e}_c^{2n} = 0, \end{cases} \tag{26}$$

Solving these equations amounts to solving the Yee scheme in both the fine and coarse grids with an artificial source term $\mathbf{j}^{2n+1}$ located on the interface $\Sigma$. To determine this source term, we have to say how the last equation of (24), namely the continuity of the tangent electric field, will be discretized in time. This is the key point. This equation is taken into account in a (non trivial) mean sense:

$$\mathcal{C}_c \, \frac{\mathbf{e}_c^{2n+2} + \mathbf{e}_c^{2n}}{2} - \mathcal{C}_f \, \frac{\mathbf{e}_f^{2n+2} + 2\mathbf{e}_f^{2n+1} + \mathbf{e}_f^{2n}}{4} = 0. \tag{27}$$

The derivation of this equation is in fact driven, by the fact that, in order to ensure the stability of the scheme, one wants to guaranty the conservation of some discrete energy. Proceeding as in section 2.3, formula (13), we can define a fine grid energy $E_f^n$ at times $t^n$ and a coarse grid energy $E_c^{2n}$ at times $t^{2n}$. The total energy is only defined at the even instants:

$$E^{2n} = E_f^{2n} + E_c^{2n}. \tag{28}$$

If is not difficult to deduce from equations (25) and (26) the identity:

$$\left\{ \begin{aligned} \frac{E^{2n+2} - E^{2n}}{2\Delta t} &= \mathcal{C}_f^* \, j^{2n+1} \cdot \Big(\frac{\mathbf{e}_f^{2n+2} + 2\mathbf{e}_f^{2n+1} + \mathbf{e}_f^{2n}}{4}\Big) \\ &\qquad - \mathcal{C}_c^* \, j^{2n+1} \cdot \Big(\frac{\mathbf{e}_c^{2n+2} + \mathbf{e}_c^{2n}}{2}\Big) \end{aligned} \right. \tag{29}$$

Using the discrete transmission equation (27), one deduces that the total energy is a constant quantity. The stability results from the fact that this total energy is a norm under the strict CFL condition (namely (3) with strict inequality). For the computations, we remark that, assuming that all the unknowns have been computed up to time $t^{2n}$, it is easy to see that, once $\mathbf{j}^{2n+1}$ is known, equations (25) - (26) allow us to explicitly compute

$$\mathbf{h}_f^{2n+\frac{1}{2}}, \mathbf{e}_f^{2n+1}, \mathbf{h}_f^{2n+\frac{3}{2}}, \mathbf{e}_f^{2n+2}, \mathbf{h}_c^{2n+1} \text{ and } \mathbf{e}_c^{2n+2}.$$

To be complete, we must find an equation that permits to compute $\mathbf{j}^{2n+1}$. After some manipulations, one can show that $\mathbf{j}^{2n+1}$ is the solution of the following linear system ($T^{2n+1}$ is a known quantity from previous time steps):

$$\Big(\mathcal{C}_c \left(\mathcal{M}_c^E\right)^{-1} \mathcal{C}_c^* + \mathcal{C}_f \mathcal{Q}(\Delta t) \, \mathcal{C}_f^*\Big) \, \mathbf{j}^{2n+1} = T^{2n+1} \tag{30}$$

where the matrix $\mathcal{Q}(\Delta t)$ is defined by:

$$\mathcal{Q}(\Delta t) = \left(\mathcal{M}_f^E\right)^{-1} - \frac{\Delta t^2}{4} \left(\mathcal{M}_f^E\right)^{-1} \mathcal{B}_f \left(\mathcal{M}_f^H\right)^{-1} \mathcal{B}_f^* \left(\mathcal{M}_f^E\right)^{-1}. \tag{31}$$

Finally solving our scheme during each interval $[t^{2n}, t^{2n+2}]$ consists in:

- a succession of explicit steps if each grid separately,
- the resolution of the linear system (30).

The existence and uniqueness of the discrete solution is guaranteed by the non-singularity of the matrix of the symmetric system (30). It is easy to see that this matrix is positive definite as soon as the matrix $\mathcal{Q}(\Delta t)$ is positive definite. This last property is equivalent to the matrix inequality:

$$\frac{\Delta t^2}{4}\,\mathcal{B}_f\left(\mathcal{M}_f^H\right)^{-1}\mathcal{B}_f^* < \mathcal{M}_f^E$$

which is nothing but the (strict) CFL stability condition in the fine grid.

**Acknowledgements.** I would like to thank my collaborators Francis Collino, Thierry Fouquet, Sylvain Garcès, Florence Millot and Jeronimo Rodriguez for their collaboration and their essential contributions to some of the works presented in this article.

## References

1. A. Bendali. *Approximation par éléments finis de surface de problèmes de diffraction des ondes électromagnétiques.* PhD thesis, *Université Paris VI*, 1984.
2. F. Brezzi and M. Fortin. *Mixed and Hybrid Finite Element Methods.* Springer Series in Computational Mathematics (15). Springer Verlag, 1991.
3. A. Buffa and P. Ciarlet Jr. On traces for functional spaces related to Maxwell's equations. I. an integration by parts formula in Lipschitz polyhedra. *Mathematical Methods in the Applied Sciences*, 24(1):9–30, 2001.
4. M. Clemens, P. Thoma, T. Weiland, and U. van Rienen. Computational electromagnetic-field calculation with the finite-integration method. *Surveys Math. Indust.*, 8(3-4):213–232, 1999.
5. F. Collino, P. Joly, and F. Millot. Fictitious domain method for unsteady problems: Application to electromagnetic scattering. *Journal of Comp. Physics*, 138:907–938, 1997.
6. B. Engquist and A. Majda. Absorbing boundary conditions for the numerical simulation of waves. *Math. of Comp.*, 31:629–651, 1977.
7. S. Garcés. *Application des méthodes de domaines fictifs à la modélisation des structures rayonnantes tridimensionnelles Etude mathématique et numérique d'un modèle.* PhD thesis, *ENSAE (Toulouse)* , 1997.
8. V. Girault and R. Glowinski. Error analysis of a fict. domain method applied to a Dirichlet problem. *Japan J. Ind. Appl. Math.*, 12(3):487–514, 1995.
9. P. Joly and L. Rhaouti. Domaines fictifs, éléments finis $h(div)$ et condition de neumann: le problème de la condition inf-sup. *C. R. Acad. Sci. Paris, Série I*, 328:1225–1230, 1999.
10. J.C. Nedelec. Mixed finite elements in $\mathbb{R}^3$. *Num. Math.*, 35:315–341, 1980.
11. S. Piperno, M. Remaki, and L. Fezoui. A nondiffusive finite volume scheme for the three-dimensional Maxwell's equations on unstructured meshes. *SIAM J. Numer. Anal.*, 9(6):2089–2108, 2002.
12. A. Taflove. *Computational Electrodynamics, The Finite-Difference Time Domain method.* Artech House, London, 1995.

13. K.S. Yee. Numerical solution of initial boundary value problems involving Maxwell's equations in isotropic media. *IEEE Trans. on Antennas and propagation*, pages 302–307, 1966.

# Methods and Approaches for RF Circuit Simulation and Electromagnetic Modelling*

T.A.M. Kevenaar[1], E.J.W. ter Maten[2], H.H.J.M. Janssen[1], S.P. Onneweer[3]

[1] Philips Research Laboratories Eindhoven, Tom.Kevenaar@philips.com
[2] Philips Research Laboratories Eindhoven, Eindhoven University of Technology
[3] Philips Semiconductors, Sunnyvale, CA, USA

**Abstract.** RF circuits and systems are gaining importance because we are moving further into a society where information is very important and should be available any time and anywhere. In this paper we give an overview of RF circuit simulation with an emphasis on noise simulation which is important functionality for RF designers. Due to the high frequency signals, the standard circuit formulation using Kirchhoff and lumped elements is not sufficient anymore to accurately predict the behaviour of a design and Maxwell's equations should be used. We give several approximations of Maxwell's equations and scenarios how the results can be incorporated in RF circuit simulation.

## 1 Introduction

High frequency applications are becoming increasingly important. This is caused by the fact that we are moving further into the information society where (digital) information is becoming very important. The first consequence is that large amounts of data should be transported, routed and processed at very high speeds. As an example, a switching array for optical data transmission may route 20 input signals each at 10 Gbit/s to one or many specified outputs. These switching arrays are still mainly implemented in silicon (or GaAs) and although we are dealing with digital circuits, the high frequency issues in this kind of circuits are analogue in nature and they have to be treated and analysed from an analogue viewpoint.
A second consequence of the increasing importance of information is the notion that this information should be available any time and anywhere. Therefore there is a tremendous increase in wireless networks that allow flexible access to a wide variety of information. The increase in functionality of, for example, cell phones leads to larger amounts of data to be exchanged. The same holds for the emerging in-home wireless digital networks.
Wireless transmission in general uses high frequency (RF) carriers usually in the range of 1-10 GHz. RF circuits are analogue circuits and should be treated as such.
Finally, many high frequency designs are aimed at a consumer market. This has severe consequences for the design process: where in the past there was

* Invited paper at SCEE-2002

time to build and measure several prototypes, nowadays the demands on time-to-market, time-to-quality, price, production volume and production yield, etc. are very severe. Furthermore, the increasing complexity and the decreasing size of these systems makes measuring extremely difficult and time consuming.
Therefore designers must be provided with analogue design environments that help them to design as quickly as possible a working circuit at first silicon. At the heart of these environments are the simulators. In order to predict the behaviour of a design as accurately as possible they should be provided with accurate models, not only of the non-linear components such as transistors but also of the physical structures which are part of the design.
In this paper we will give an overview of RF circuit simulation. First we will describe two main characteristics of RF circuits and RF designs. Next we will look in more detail at an approach to RF simulation that is becoming increasingly popular. This is followed by an explanation on why and how the physical implementation is incorporated in RF circuit simulation.

## 2 Characteristics of RF circuits

Clearly the most obvious characteristic of RF circuits is that they work at high frequencies. A consequence is that one has to use Maxwell's equations rather than Kirchhoff's equations and lumped models. Although clearly Kirchhoff's equations can be used without problems up to a certain frequencies, at RF frequencies they no longer accurately predict the behaviour of a design. The physical structures required to implement a design, such as tracks on an IC, connectors, etc., begin to play an important role in the total behaviour of the design. Since these structures in general do not have a regular shape, one has to resort to numerical methods to solve Maxwell's equations and somehow incorporate the results in a simulation.
A second characteristic is that RF signals have a broad but sparse spectrum [10] with a dynamic range of more than 60dB ($10^6$). The weakest signals might be almost lost in noise while the strongest signals will introduce all kind of spurious intermodulation components due to non-linearities which are always present in a circuit. Noise and non-linear distortion translate to bit-error-rate in the transmitted data. Consequently, it is important that designers can predict the overall noise and distortion quickly and accurately.

## 3 RF circuit simulation

### 3.1 Overview of RF building blocks and specifications

Although RF systems can be quite complicated, they are typically built from a limited number of building blocks. When discussing RF simulation techniques it is important to first determine the special properties and characteristics of each building block and the information that should be obtained

in a simulation. The most important blocks are mixers (which perform a frequency shift of the input signal, but also add non-linear intermodulation products and noise), amplifiers and filters (which distort the signal and add noise to it; power amplifiers may be strongly non-linear), dividers (strongly non-linear; modify a frequency reference signal), oscillators (generate a signal that serves as a frequency reference, often of very high accuracy; are autonomous circuits). In all cases non-linearity and noise are important issues. In the case of oscillators noise present in the circuit manifests itself as phase noise.
In the following section we will explain why and how these properties are specified.

## 3.2 RF specifications

In RF circuits the non-linearities are of course no different from non-linearities in other analog circuits. However they are usually expressed as intercept points like IP2 and IP3. The term 'intercept' originates from a graphical construction which can be used to determine IP2 and IP3. The numbers '2' and '3' refer to the order of the intermodulation product which they define. Clearly, these specifications should be determined using a full-nonlinear transient-like or harmonic balance-like simulation after which the result can be post-processed to yield the IP numbers.
More interesting is the noise in RF circuits. Noise consists of (usually) small unwanted signals in the circuit and originates in the devices that make up the circuit. For input/output circuits, a designer is usually interested in how much noise will be present or added in a certain, relatively narrow, frequency band of interest (FBOI). It will be the task of the simulator to accurately determine what the final noise spectrum in the FBOI looks like. A special thing in RF circuits is that spectra of input and noise signals are shifted around in the frequency band due to wanted (e.g. mixers) and unwanted non-linearities in combination with large signals. These signals can be internal to the circuit or be part of the input signal. Due to this noise folding, the noise in the FBOI might originate at completely different frequencies. The simulator must be able to handle noise folding.
Where in input/output circuits a designer is usually interested in the noise spectrum in a certain FBOI, when designing oscillators he is interested in the phase noise. As mentioned, an oscillator generates a frequency reference in the form of a periodic signal. Noise in the circuit causes the output signal to become noisy too. It is very important to notice that the noise might change the frequency of the output signal. This has severe consequences for simulation algorithms: noise is usually seen as a small perturbation on the noiseless solution which means that we can linearise the circuit. With oscillators this is no longer true as will be explained in the next Section.

### 3.3 Basic RF simulation methods

Now that we have illustrated some important properties of RF circuits and signals it can be seen that conventional simulation methods like AC and transient simulation are not sufficient for simulating noise in RF circuits. AC simulation is fast but does not incorporate non-linearities nor frequency folding. Transient analysis could be used to simulate frequency folding but this would lead to prohibitively long simulation times if we want to obtain accurate information in the frequency domain. This is the reason why special RF simulation algorithms were developed.

The objective of RF circuit simulation is to obtain solutions for the network variables (the voltages, currents, charges and fluxes) in the circuit and to study the effects due to noise. All the newly developed RF simulation methods somehow use the Periodic Steady-State solution (PSS) as a starting point. Conceptually, the PSS solution can be seen as a generalisation of the DC solution: where the DC solution describes the voltages and currents after infinite time in a circuit containing only *DC sources*, the PSS solution describes the voltages and currents after infinite time in a circuit containing only *periodic sources* (which include DC sources). As with the DC solution, the PSS solution is useful in its own right but it is also used as a basis for other analyses (for example, periodic AC and periodic noise). This presents a basic two-step approach:

- Firstly a (noiseless) PSS solution is determined which deals with the non-linearities in the circuit [2,3,7,8,10,13,14,17–19],
- Secondly, a perturbation analysis is done around the PSS solution to analyse noise including frequency shifts [1,4–6,16].

It is important to notice that the PSS algorithms make a distinction between forced and oscillatory problems. In the former the period of the solution is known beforehand while in the latter, determining the (exact) period is part of the problem. In general the oscillatory problems are more difficult to solve [7,12,15]. Also the solution of perturbed oscillatory problems may not be periodic at all.

The noiseless PSS problems are defined as finding a solution $\mathbf{x}(t)$ for systems of DAEs of the form

$$\begin{cases} \frac{d}{dt}\mathbf{q}(t,\mathbf{x}) + \mathbf{j}(t,\mathbf{x}) = \mathbf{0} \in \mathbf{R}^{\mathbf{N}} \\ \mathbf{x}(0) = \mathbf{x}(T) \end{cases} \quad \text{or} \quad \begin{cases} \frac{d}{dt}\mathbf{q}(\mathbf{x}) + \mathbf{j}(\mathbf{x}) = \mathbf{0} \in \mathbf{R}^{\mathbf{N}} \\ \mathbf{x}(0) = \mathbf{x}(T) \end{cases} \tag{1}$$

where $T > 0$ is the period of the solution; $\mathbf{x}$ contains the node voltages and currents through voltage sources and through inductors; $\mathbf{q}$ describes the capacitor charges as well as fluxes through inductors; $\mathbf{j}$ describes currents, or voltages differences, as well as effects of sources. The explicit dependence on $t$, in the equations at the left, denotes that periodic sources are present in the circuit (forced problems) and therefore the period $T$ is known before hand. With oscillators (the autonomous equations at the right) this is not the case

and the period $T$ should be solved together with $\mathbf{x}(t)$.
The PSS solution can be obtained in the frequency domain by determining its Fourier coefficients by a Harmonic Balance approach. This has the advantage that one can easily deal with circuit components that are characterised in the frequency domain which is not uncommon in RF applications. The time domain methods more easily deal with strong non-linearities in the circuit and have good convergence properties. Of these methods we mention Poincaré-map based methods, where increase in speed is obtained by applying vector acceleration methods such as Minimal Polynomial Extrapolation. Typical points of attention are: restarting at consistent solutions (i.e. satisfying the DAE-manifold), as well as dealing with multiple oscillation frequencies. Alternative methods are provided by (multiple) shooting methods, or by applying a finite difference method. All methods can be enhanced to deal with oscillatory systems in which $T$ is an additional unknown (and a gauge equation is added to the system).
The PSS-solution is useful to, for example, determine the non-linearity of the circuit such as IP2, IP3 etc. (see Section 3.2). For studying effects due to noise, the PSS solution is a first step in RF noise analysis in the two-step approach described before. An important effect of noise in RF circuits (see also Section 3.2) is noise folding: noise components are moved around the frequency band when they interact with other signals in the circuit. We now illustrate how a two-step approach based on a PSS solution is capable of reproducing this effect.
Assume that we determined the (forced) PSS solution $\mathbf{x}_{PSS}(t)$ for (1). To incorporate noise we add a perturbation term $\mathbf{n}(t)$ and get

$$\frac{d}{dt}\mathbf{q}(t,\mathbf{x}) + \mathbf{j}(t,\mathbf{x}) + \mathbf{n}(t) = \mathbf{0} \in \mathbf{R}^{\mathbf{N}}. \tag{2}$$

A natural approach would be to assume that a small $\mathbf{n}(t)$ also introduces a small deviation to the large signal solution $\mathbf{x}_{PSS}(t)$ and to linearise (2) by choosing $\mathbf{x}(t) = \mathbf{x}_{PSS}(t) + \mathbf{x}_n(t)$ and we find the following system for $\mathbf{x}_n(t)$:

$$\frac{d}{dt}(\mathbf{C}(t)\mathbf{x}_n) + \mathbf{G}(t)\mathbf{x}_n + \mathbf{n}(t) = \mathbf{0} \in \mathbf{R}^{\mathbf{N}}, \tag{3}$$

$$\mathbf{C}(t) = \left.\frac{\partial \mathbf{q}(t,\mathbf{x})}{\partial \mathbf{x}}\right|_{\mathbf{x}_{\mathbf{PSS}}}, \qquad \mathbf{G}(t) = \left.\frac{\partial \mathbf{j}(t,\mathbf{x})}{\partial \mathbf{x}}\right|_{\mathbf{x}_{\mathbf{PSS}}}. \tag{4}$$

It turns out that for *forced* systems this is a good approach because the period $T$ of the solution is completely determined by the input sources, and the homogenous problem only has the trivial solution. If we consider a typical Fourier component of the noise source, $\mathbf{n}(t) = \mathbf{U}e^{j\nu t}$, one may consider $\mathbf{y}_n(t) = e^{-j\nu t}\mathbf{x}_n(t)$ that satisfies a $T$-periodic system of equations

$$\frac{d}{dt}(\mathbf{C}(t)\mathbf{y}_n) + [\mathbf{G}(t) + j\nu\mathbf{C}(t)]\mathbf{y}_n + \mathbf{U} = \mathbf{0} \in \mathbf{R}^{\mathbf{N}} \tag{5}$$

(which is parametrized by $\nu$). It is clear that, for a single input frequency $\nu$, the solution $\mathbf{x}_n(t)$ contains frequencies of the form $(\nu + \omega_k)$ (in which

$\omega_k = 2\pi k/T$), i.e. frequency folding occurs. If we allow for several input frequencies $\nu_i$, we can also say that a certain output frequency might originate from a large number of possible input frequencies. Hence, noise components at a certain frequency might end up in a different frequency band. This is why, for example, $1/f$ noise which has its main energy at low frequencies, still plays an important role in RF circuits.
It is important to note that we described a *linear* perturbation analysis and we will not find contributions containing for example $(\nu_1 + \nu_2 + \omega_k)$, $(\nu_1 + 2\nu_2 + \omega_k)$ etc. This assumption is in general not a severe limitation when simulating noise in RF circuits.
When dealing with perturbed *oscillatory* systems

$$\frac{d}{dt}\mathbf{q}(\mathbf{x}) + \mathbf{j}(\mathbf{x}) + \mathbf{n}(t) = \mathbf{0} \in \mathbf{R}^{\mathbf{N}} \tag{6}$$

it is no longer possible to assume that small perturbations $\mathbf{n}(t)$ lead to small deviations in $\mathbf{x}_{PSS}(t)$ [An instructive example is provided by considering $y'(t) + \cos(t)y(t) - 1 = 0$, of which the inhomogeneous solution is not periodic at all; however, note that $y(t+2\pi)$ still satisfies the differential equation]. The main reason is that the *period* of the large signal solution is influenced by $\mathbf{n}(t)$. This can lead to large (momentary) frequency deviations such that the difference between the noiseless and noisy solution can no longer be considered to be small.
In [5] a solution is given to deal with this problem by introducing an extra term which describes the frequency (or phase) shift of the solution due to $\mathbf{n}(t)$. Hence, rather than assuming $\mathbf{x}(t) = \mathbf{x}_{PSS}(t) + \mathbf{x}_n(t)$ as a solution for (6), we now assume $\mathbf{x}(t) = \mathbf{x}_{PSS}(t+\alpha(t)) + \mathbf{x}_n(t)$, where $\alpha(t)$ is a non-trivial scalar function that has to be determined as part of the solution process and leads to the *phase noise* of the system. $\mathbf{x}_n$ represents the *orbital deviation.*
In order to arrive at an expression for the phase- or time-shift function $\alpha(t)$ (assumed to be sufficiently smooth), we define $s = t + \alpha(t)$ and $\mathbf{y}(t) \equiv \mathbf{x}_{\mathrm{PSS}}(s) = \mathbf{x}_{\mathrm{PSS}}(t + \alpha(t))$. We observe that $\mathbf{y}(t)$ satisfies

$$\frac{\mathrm{d}}{\mathrm{d}t}\mathbf{q}(\mathbf{y}) + \mathbf{j}(\mathbf{y}) = \mathbf{C}(t + \alpha(t))\mathbf{u}_1(t + \alpha(t))\alpha'(t), \tag{7}$$

in which $\mathbf{u}_1(t) = \mathbf{x}'_{\mathrm{PSS}}(t)$, being the tangent to the orbit. Clearly $\mathbf{y}(t)$ itself satisfies a perturbed differential equation. We note that $\mathbf{u}_1(t)$ satisfies the homogeneous part of (6), linearised around the noiseless PSS solution

$$\frac{d}{dt}(\mathbf{C}(t)\mathbf{x}) + \mathbf{G}(t)\mathbf{x} = \mathbf{0} \in \mathbf{R}^{\mathbf{N}}, \tag{8}$$

with $\mathbf{C}(t)$ and $\mathbf{G}(t)$ as defined in (4). It follows from Floquet theory [5,11] that (8) has $N$ independent solutions (in which $\mathbf{u}_1$ coincides with our previously introduced one) $\mathbf{u}_1(t)e^{\mu_1 t}, \ldots, \mathbf{u}_m(t)e^{\mu_m t}, \mathbf{u}_{m+1}(t), \ldots, \mathbf{u}_N(t)$. In case of a stable index 1 problem we can assume $\mu_1 = 0$ and $\mathrm{Re}(\mu_i) < 0$ for $i \geq 2$.
The adjoint system of (8)

$$\mathbf{C}^T(t)\frac{d}{dt}(\mathbf{y}) - \mathbf{G}^T(t)\mathbf{y} = \mathbf{0}. \tag{9}$$

has similar properties: $\mathbf{v}_1(t)e^{-\mu_1 t}, \ldots, \mathbf{v}_m(t)e^{-\mu_m t}, \mathbf{v}_{m+1}(t), \ldots, \mathbf{v}_N(t)$ are $N$ independent solutions. It should be noted that the vectors $\mathbf{v}_i(t)$ and $\mathbf{u}_j(t)$ satisfy a special bi-orthogonality relation with respect to $\mathbf{C}(t)$ and $\mathbf{G}(t)$, namely

$$\mathbf{V}(t)\mathbf{C}(t)\mathbf{U}(t) = \begin{pmatrix} I_m & 0 \\ 0 & 0 \end{pmatrix}, \quad \mathbf{V}(t)\mathbf{G}(t)\mathbf{U}(t) = \begin{pmatrix} J_m^1 & 0 \\ J_m^2 & J_m^3 \end{pmatrix}. \tag{10}$$

In applications, the noise (perturbation) term $\mathbf{n}(t)$ in (6) has the form $\mathbf{n}(t) = B(\mathbf{x}(t))\mathbf{b}(t)$. It seems natural to decompose $B(\mathbf{x}_{\mathrm{PSS}}(t+\alpha(t)))\mathbf{b}(t)$ into components along a basis of which one basic function is $\mathbf{C}(t+\alpha(t))\mathbf{u}_1(t+\alpha(t))$ (see (7)). By multiplying (7) and $B(\mathbf{x}(t+\alpha(t)))\mathbf{b}(t)$ by $\mathbf{v}_1^T(t)$, the crucial bi-orthogonality implies a non-linear, scalar, differential equation for $\alpha(t)$

$$\alpha'(t) = -\mathbf{v}_1^T(t+\alpha(t))B(\mathbf{x}_{\mathrm{PSS}}(t+\alpha(t)))\mathbf{b}(t), \quad \alpha(0) = 0 \tag{11}$$

from which $\alpha(t)$ can be determined [The same bi-orthogonality also provides an elegant way to determine $\mathbf{v}_1(t)$, once $\mathbf{u}_1(t)$ is known]. Note that if $\mathbf{b}(t) = 0$, for $t \geq t_0$, then $\alpha$ becomes a constant phase shift, and the phase shifted function $\mathbf{y}(t)$ solves (6) exactly for $t \geq t_0$. In general, even for small $\mathbf{b}$, the phase shift function $\alpha(t)$ may increase with time. Because (11) is non-linear, phase shifts from individual sources do not add up to give a group phase shift.

In the above we assumed deterministic disturbances prescribed by the time function $\mathbf{b}(t)$. In noise analysis, however, the noise is usually not described by time functions but by statistical properties such as mean and standard deviation. In [5,6] the (stationary) autocorrelation of $\mathbf{y}(t)$ is studied more closely (here the $*$ denotes complex conjugation). One derives (assuming real $\alpha(t)$ and $\mathbf{X}_j$ being the $j$-th Fourier coefficient of $\mathbf{x}_{\mathrm{PSS}}(t)$)

$$R_{\mathbf{y}}(\tau) \equiv \lim_{t\to\infty} E[\mathbf{y}(t)\mathbf{y}^*(t+\tau)] = \sum_{j=-\infty}^{\infty} \mathbf{X}_j\mathbf{X}_j^* e^{-i\omega_j\tau} R_j(\tau)$$

with a corresponding relation between the spectral densities

$$S_{\tilde{\mathbf{x}}_{\mathrm{PSS}}}(\omega) = \sum_{j=-\infty}^{\infty} \mathbf{X}_j\mathbf{X}_j^* S_j(\omega+\omega_j), \text{ where } \int_{-\infty}^{\infty} \mathbf{X}_j\mathbf{X}_j^* S_j(2\pi f+\omega_j)\mathrm{d}f = \mathbf{X}_j\mathbf{X}_j^*$$

The interesting point is that the above formulas do not require the explicit evaluation of $\alpha(t)$! 'Only' the variance $\sigma^2(\tau)$ of $\alpha(t)$ is met, which can be related to the $\sigma^2$ of the individual source. This allows for deriving approximating expressions for $S_j(\omega)$, and also gives way to summing efficiently for getting group contributions.

In this section we summarised RF noise algorithms based on a two-step approach: a PSS analysis followed by a linear (for forced systems) or non-linear (oscillatory systems) perturbation analysis. Apart from accurate simulation algorithms, a circuit simulator needs accurate models. On the one hand there are the models of the non-linear components such as bipolar transistors, MOS transistors, etc. Although these models are very important they are outside the scope of this paper. On the other hand, for accurate RF simulation we also require models of physical structures which are somehow generated by solving Maxwell's equations. This will be the topic of the following section.

# 4 Electromagnetic modelling for circuit simulation

## 4.1 Concept of lumped elements

In the previous sections we described circuit analysis that is based on Kirchhoff's voltage and current law or KVL and KCL, respectively. However, Maxwell's equations are more fundamental and the electromagnetic (EM) field is the foundation of circuit theory and electronic modelling and simulation. In practice, a complete electromagnetic model of an electronic circuit is expensive to create and analyse and fortunately, electronic circuit theory has shown how to approximate many practical circuits by lumped element models: the energy-storage elements, (inductors and capacitors), and the dissipative elements (resistors) are connected to each-other and to sources or active elements within the circuit by conducting paths of negligible impedance. So apparently, the distributed effects, inherent to the solution of Maxwell's equations, in many real circuits can be represented by a few properly chosen lumped coupling elements. Circuit simulation is based on this concept of lumped elements, which ignores the electromagnetic interaction that is present within and between the physical circuit components and interconnections.

A first level of refinement to the lumped approach is to model the real physical interconnections (be it on-chip, in an IC package, a hybrid module or on a PCB) by means of lumped parasitic elements describing the conductor resistance and the capacitive and inductive coupling between the conductors. The ideal (lumped) circuit is extended with this parasitic network and can then be treated by the same network analysis and simulation tools. This approach can be used when the individual elements and the total circuit are small compared to the wavelength of the signals (quasi-static approach) [22]. For structures comparable in size to the wavelength there are two effects which will play a role which can not be taken into account using this approach: distributed effects (compared to lumped) and retardation effects from one part of the circuit to another.

Firstly, we consider the distributed effects. In general the lumped representation of an element is valid if the region it occupies is small compared to the wavelength and when only one type of energy storage, either electric or magnetic is important in that region. If the electric energy storage in parts of a primarily inductive element, or magnetic energy in a primarily capacitive element, becomes important, the approach through classic circuit theory is to divide each element into sub-elements that can be treated as one or the

other. A good example is the capacitive coupling between the turns of an inductor, which in a first approximation can be represented by adding a capacitive element across the terminals of the inductor. A further improvement is to add capacitive elements between each pair of adjacent turns.
Secondly, retardation effects, arising from the finite propagation time of electromagnetic effects across the circuit, can cause phase delays in the circuit. If there is an in-phase component of the induced electric field (due to changing magnetic fields) and magnetic field (due to the current), this represents an energy transfer, which is in fact the radiated energy. Another phenomenon that cannot easily be modelled with lumped parasitic elements is the presence of frequency dependent inhomogeneous current distributions in non-ideal conductors, e.g. due to skin effect.
Electromagnetic simulation is aimed at overcoming the limitations of the lumped element and lumped parasitics approach. Electromagnetic simulators build an accurate spatial model of the physical structures of the circuit. The spatial model is accompanied by the material properties of the structural elements: conductivity, permittivity and permeability. In order to incorporate EM effect in circuit simulation, ports (or pins) are attached to the physical structure, denoting the locations where e.g. lumped models of components or modules are to be attached. In many EM simulators, some approximation of Maxwell's equations is solved. Therefore, in the following section we will explain some often used approximations.

## 4.2 Maxwell's equations and the Kirchhoff approximation

Maxwell's equations describe the electromagnetic field and are given by:

$$\nabla \times \mathbf{E} = -\frac{\partial \mathbf{B}}{\partial t}, \qquad \text{(Faraday's law)} \tag{12}$$

$$\nabla \times \mathbf{H} = \mathbf{J} + \frac{\partial \mathbf{D}}{\partial t}, \qquad \text{(Ampère's law)} \tag{13}$$

$$\nabla \cdot \mathbf{B} = 0, \tag{14}$$

$$\nabla \cdot \mathbf{D} = \rho, \qquad \text{(Gauss' law)} \tag{15}$$

$$\mathbf{B} = \mu \mathbf{H}, \qquad \mathbf{D} = \epsilon \mathbf{E}, \qquad \text{(constitutive relations)} \tag{16}$$

$$\mathbf{J} = \sigma \mathbf{E}, \qquad \text{(Ohm's law).} \tag{17}$$

Here $\mathbf{E}$ and $\mathbf{H}$ are the electric and magnetic field, $\mathbf{B}$ and $\mathbf{D}$ are the magnetic and electric flux densities, $\mathbf{J}$ and $\rho$ are the current and charge density, and $\mu$, $\epsilon$ and $\sigma$ are the material parameters permeability, permittivity and conductivity, respectively. We can derive several approximations:

1. Assuming DC conditions ($\partial/\partial t = 0$) and taking the divergence of (13) we get $\nabla \cdot \mathbf{J} = 0$, and with Gauss' theorem over a closed surface $S$

$$\iiint_V \nabla \cdot \mathbf{J} dV = \oiint_S \mathbf{J} \cdot dS = 0. \tag{18}$$

Since the only current flowing out of the surface is in the wires, this gives Kirchhoff's current law (KCL), which simply states that the algebraic sum of currents flowing out of a circuit junction is zero.
From Faraday's law, we can introduce the potential (voltage) V according to $\mathbf{E} = -\nabla V$ and with Stoke's theorem over a closed loop $l$ we get:

$$\iint_A \nabla \times \mathbf{E} \cdot dA = \oint_l \mathbf{E} \cdot dl = 0. \tag{19}$$

This gives Kirchhoff's voltage law (KVL), which states that for any closed loop of a circuit, the algebraic sum of the voltages for the individual branches of the loop is zero. These two laws provide the basis for classical circuit theory.

2. If we only neglect the displacement current $\partial \mathbf{D}/\partial t$ in (13) we get the quasi-static approach. We still obtain the KCL by taking the divergence of (13). From (14) we can write $\mathbf{B} = \nabla \times \mathbf{A}$, where $\mathbf{A}$ is a magnetic vector potential. When we substitute this in (12), we get

$$\nabla \times (\mathbf{E} + \frac{\partial \mathbf{A}}{\partial t}) = 0 = -\nabla \times \nabla V, \tag{20}$$

which means that in this case we can also define a scalar electric potential $V$ by:

$$\mathbf{E} = -\nabla V - \frac{\partial \mathbf{A}}{\partial t}, \tag{21}$$

where V fulfils the KVL. Taking the rotation of (12) and substituting (13) and taking the rotation of (13), we get the equations for $\mathbf{E}$ and $\mathbf{H}$:

$$\Delta \mathbf{E} = \mu \frac{\partial \mathbf{E}}{\partial t} = \mu\sigma \frac{\partial \mathbf{J}}{\partial t}, \tag{22}$$

$$\Delta \mathbf{H} = -\nabla \times \mathbf{J}, \text{where} \quad \Delta = \nabla^2. \tag{23}$$

This gives the typical eddy current solutions with skin depth $\delta = 1/\sqrt{\pi\mu f\sigma}$ inside conductors with conductivity $\sigma$. Currents will run on the edge of the conductor within a depth $\delta$ and fields cannot penetrate any deeper in the conductor than this skin depth. This approach is required for frequencies $\omega > 1/\mu\sigma r^2$ where $r$ is the thickness of the interconnect. For frequencies high enough so that the current distribution is not uniform anymore (interconnect is thick compared to the skin depth), the resistance and internal reactance will become frequency dependent, since they will be determined by the skin depth. This approximation can also be seen as infinite wavelength approximation of the wave equation solution and is applicable when the wavelength is much larger than the dimension d of the problem: $\lambda > d$ or frequencies $\omega << c/d$, the so-called quasi-static approach. For higher frequencies, radiation losses become important and this is not taken into account by the quasi-static approach.

3. Clearly it is also possible to solve Maxwell's equations completely without any approximations which leads to the Helmholtz wave equations:

$$\Delta \mathbf{E} - \frac{1}{c^2}\frac{\partial^2 \mathbf{E}}{\partial t^2} = \mu \frac{\partial \mathbf{J}}{\partial t} + \nabla \frac{\rho}{\epsilon}, \quad \Delta \mathbf{H} + \frac{1}{c^2}\frac{\partial^2 \mathbf{H}}{\partial t^2} = -\nabla \times \mathbf{J} \tag{24}$$

The solution of this set of equations is called the full-wave solution.

According to the analysis above, electromagnetic simulation tools may be characterised as either quasi-static or full-wave. Quasi-static simulators only consider free charges and currents on electrical conductors. By contrast, full-wave simulators account for the propagation of electromagnetic waves in free space and dielectric materials. Radiation losses are not naturally incorporated in the quasi-static model and radiated electromagnetic fields can only be approximated by post-processing on the calculated charges and currents. These phenomena are naturally incorporated in full-wave tools. A common problem in full-wave electromagnetic simulation tools is their computational complexity: because of the very fine spatial discretisation, the size of the mathematical equations that have to be solved is enormous. Reduced order modelling (ROM), the construction of a simplified system to approximate the original system with reasonable accuracy, will be necessary to cope with this [23–25]. Also for large quasi-static problems, ROM will be necessary.

### 4.3 Electromagnetic simulation tools for RF circuits

To assess the need for on-chip EM simulation let us assume that RF signals of interest for present-day businesses are in the frequency range of 1-10 GHz. Digital signals may have pulse edges in the order of 100 *ps*. For a straight TEM transmission-line consisting of ideal conductors embedded in a homogeneous dielectric medium the propagation speed of the signal is $c/\sqrt{\epsilon_r}$, where $\epsilon_r \approx 2-4$ for state of the art RFIC processes. As a result, the wavelength of the 10 GHz RF signal is of the order of 15-20 mm. Likewise, the distance a digital pulse travels in 100 ps is some 15-20 mm. The die size is roughly 10 $mm^2$. The maximum geometrical length of digital signal lines on the chip is comparable to the figures shown above. On the other hand, on-chip RF interconnections are likely to be only a fraction of the die size and the largest components, spiral inductors, typically measure only 0.2-0.5 mm across. Consequently, from this perspective it is unlikely that it is necessary to apply full-wave electromagnetic simulation to on-chip components and interconnect structures. Long RF interconnections will be designed as coplanar wave-guides or similar structures and as long as the TEM propagation mode is dominant and the losses are small, the quasi-static approximation probably remains valid, even for very high frequencies.
Nevertheless, in order to tackle interconnect related problems Maxwell's equations (quasi-static and possibly full-wave) must be solved which can be done in the frequency domain or in the time domain. Within the frequency domain methods one assumes a harmonic time dependency, such that all time

derivatives can be replaced by $j\omega$. Some of the most popular frequency domain methods are the Finite Difference (FDM) and Finite Elements (FEM) Method, as well as the Boundary Elements Method (BEM), Method of Moments (MoM) and Spectral Domain Analysis (SDA) [26,27]. The first two methods are differential equation schemes, the latter ones integral equation schemes. Well-known time domain methods are the Finite Difference Time Domain algorithm (FDTD) and the Time Domain Transmission Line Method (TDTLM).

The FDM approximates the differential operators by finite differences. It is easy to implement and applicable to general configurations, however has difficulty to handle curved boundaries. It may also need a large mesh volume to implement the absorbing boundary conditions (ABC) for unbounded problems.

In the FEM, the solution domain is discretized into elements [28]. Making use of interpolation functions (shape functions), each element is mapped into a basic standard element. The unknown fields are locally expressed in terms of the interpolation functions over each individual element. By applying a variational or Galerkin procedure, a set of algebraic equations, described by sparse matrices, is obtained. The Galerkin method is in fact a weighted residual procedure with trial functions equal to the weighting functions and is one of the most widely used methods. This method is also the most general. It can handle curved boundaries and arbitrary inhomogeneous material distributions. As with the FDM, one has to take care with ABC for unbounded problems.

When the FEM is applied to the boundary integral equations, this results in the BEM [29]. However, instead of large sparse matrices, where iterative solvers can be applied, dense matrices are obtained, which in general are smaller. This method can handle arbitrary curved boundaries and small mesh volumes, due to the integral equation approach. However, inhomogeneous material distributions are more difficult to handle.

The MoM also employs the method of weighted residuals [30]. The method starts by establishing a set of trial functions with one or more variables. The residuals are a measure of the difference between trial and true solution. The variable parameters are determined in a manner that guarantees a best fit of the trial functions based on a minimisation of residuals.

In the SDA method, the integral equation is derived from the time-harmonic Maxwell's equations or the Helmholtz wave equations using Green's functions [31]. In this method only the surface of conductors needs to be discretized, resulting in a small dense matrix equation and thus a fast scheme. The major drawback is that it cannot easily be generalised, since Green's functions may not be available for a general configuration and inhomogeneous material distribution.

For integral equation methods, the so-called Fast Multi-pole Methods (FMM) can speed up the calculations and make large and complex geometries computable. The main idea behind these methods is the fact that the effect of

all points working on all points yields a method with $O(N^2)$ computations, whereas if the points are well separated, one can cluster the "far-away" points. This yields a method of at most $O(N \log(N))$ [32].
The Finite-Difference Time-Domain (FDTD) method is currently one of the most popular approaches. As first proposed by Yee in 1966 [33], it uses the differential form of Maxwell's equations. Yee used an electric field (**E**) grid which was offset both spatially and temporally from a magnetic field (**H**) grid to obtain update equations that yield the present fields throughout the computational domain in terms of the past fields. The initial lack of attention, in spite of the simplicity and elegance of Yee's approach, can be attributed to the high computational cost. However, recently many shortcomings of the original FDTD method were alleviated leading to a near exponential growth in publications in the past ten years [34].
Finally, the TDTLM method models the spatial electromagnetic field in terms of a distributed transmission line network after discretizing the solution domain [35]. Basically this method has the same limitations as the FDTM method.
Unfortunately, however, most of the currently available 2.5D electromagnetic simulators (such as those based on MoM or FDTD) are only suited for relatively small interconnect structures due to large computing times.

## 4.4 Coupling EM simulators to circuit simulators

The purpose of our explanation on EM simulation was to include the influence of physical structures such as interconnect in RF circuit simulation. This means that somehow we have to connect the EM simulator, or the results thereof, to the circuit simulator. In this section we give some possible approaches.
A possible solution would be to tightly couple a circuit simulator with an EM simulator (co-simulation). The idea is based on an iterative approach in the time domain and requires costly computations and hence in general it is not an attractive approach. However, when antennas are integrated on silicon [36], electromagnetic and circuit analysis must be combined. By its very nature (conversion of electromagnetic radiation into electrical energy and visa versa) an antenna cannot be treated as a parasitic component and co-simulation will be necessary.
Another approach is to use an EM simulator as a stand-alone tool to generate a compact model which can be incorporated in the circuit simulator in addition to the lumped circuit components connected to the physical structure. This requires a robust method to catch the behaviour of the EM model at the ports. Most of the present EM simulators can produce Y- or S-parameter models that can be incorporated in a circuit simulator and they are most easily used in (linear or non-linear) frequency domain analyses such as AC and Harmonic Balance. For non-linear time-domain methods, such as PSS, we need to reflect the behaviour as a system of DAEs. There are a number of pos-

sible ways to couple this to the simulator. One can generate these equations directly from the S-parameters or via a behavioural modelling step. Some simulators calculate the impulse response of an S-parameter component by the inverse Laplace transform and apply numerical convolution during transient simulation.
Other tools are capable of creating an equivalent circuit model i.e. an RLC network. This type of model is more versatile but more difficult to generate. The size of the model could be a problem as well as its passivity (non-stable). Instead of mapping onto a real network, the model could also be described in terms of the G- and C matrices which can be incorporated in the circuit simulator.
Interesting work was done in coupling an FDTD simulator to the circuit simulator where a behavioural model approach was used to generate a lumped element model [38].
A limitation of using an EM simulator as stand-alone tool is that the actual EM model is never exercised in its circuit context, which means that the current distribution in the physical structures is not explicitly known. However, some EM simulators can reconstruct the geometrical current distribution and EM radiation from the currents at the model ports and create a visual rendition of them. To use this feature it is necessary to make the circuit simulation results available to the EM simulator [37].

### 4.5 Future developments

Currently the most common applications, where an EM simulator generates some kind of model for the circuit simulator, are the simulation of printed circuit boards and micro-wave circuits. The physical structures in these applications consist of planar metallisation structures on (almost) loss-less substrates. The most successful methods to analyze these usually large structures (e.g. complete multi-layer PCBs) are based on 2.5D boundary element analysis. These methods produce either a rational approximation of the S-parameters or a reduced lumped element model by using a fit procedure (Momentum-RF, Fasterix). In case of a lumped circuit model, the resulting circuit is in general not passive. Reduced order modelling techniques that guarantee stability, such as PVL [25], will need to be used.
For modelling physical structures on an IC, one of the most promising solutions seems to be to adjust the tools and methods used for the planar structures. MoM and FDTD seem to be less favourable candidates for ICs, due to high computational cost. In order to be able to expand the above-mentioned tools for simulation of ICs, several problems need to be solved:

- 3D effects: due to the shape of IC interconnect, the effects of the side-walls have to be taken into account.
- Substrate effects: the actual IC interconnect behaviour can only be calculated when the effects of the silicon substrate are taken into account

(displacement and induced (eddy) currents and concomitant losses). The silicon substrate cannot be approximated by an ideal conductive plane. The resistance of the substrate is high, leading to dispersion and frequency dependent attenuation in signal lines and to reduced q-factors in passive components, such as spiral inductors. A number of models to account for substrate effects have already been developed but need to be coupled to the interconnect analysis tools.

– These 3D and substrate effects increase the complexity of the problem. Therefore robust methods should be developed to reduce the complexity of the final model which is used in the circuit simulator. The stability of these models should be a point of attention.

## References


1. W. Anzill, F.X. Kärtner, P. Russer: *Simulation of the phase noise of oscillators in the frequency domain*, AEÜ, Archive für Elektr. und Übertragungstechnik, Vol. 48-1, pp. 45-50, 1994.
2. H.G. Brachtendorf, G. Welsch, R. Laur, A. Bunse-Gerstner: *Numerical steady state analysis of electronic circuits driven by multi-tone signals*, Electrical Engineering 79, pp. 103-112, 1996.
3. H.G. Brachtendorf, R. Laur: *Analyse des transienten Verhaltens von Oszillatoren durch eine inverse Charakteristikenmethode*, ITG Workshop *Mikroelektronik für die Informationstechnik*, Darmstadt, Germany, 2000.
4. A. Demir, A. Sangiovanni-Vincentelli: *Analysis and Simulation of noise in nonlinear electronic circuits and systems*, Kluwer Academic Publ., Boston, USA, 1998.
5. A. Demir, A. Mehrotra, J. Roychowdhury: *Phase noise in oscillators: a unifying theory and numerical methods for characterisation*, DAC98, San Francisco, June 1998. Extended version IEEE Trans. on Circuits and Systems - I: Fund. Theory and Applics., Vol. 47-5, pp. 655-674, 2000.
6. A. Demir: *Phase noise in oscillators: DAEs and coloured noise sources*, Proc. ICCAD'98, Int. Conf. on Computer Aided Design, Nov. 8-12, 1998, San Jose, CA, USA, pp. 170-177, 1998.
7. S.H.M.J. Houben, E.J.W. ter Maten, J.M. Maubach, J.M.F. Peters: *Novel time-domain methods for free-running oscillators*, In: V. Porra, M. Valtonen, I. Hartimo, M. Ilmonen, O. Simula, T. Veijola: *ECCTD'01 - Proceedings of the 15TH European Conference on Circuit Theory and Design*, Helsinki University of Technology (ISBN 951-22-5571-5), Finland, pp III-393 – III-396, 2001.
8. T.A.M. Kevenaar: *Periodic Steady State Analysis using Shooting and Waveform-Newton*, Int. J. Circuit Theory and Applics., Vol 22, pp. 51-60, 1994.
9. T.A.M. Kevenaar, E.J.W. ter Maten: *RF IC Simulation: state-of-the-art and future trends*, Proc. SISPAD'99 (ISBN 4-930813-98-0), Kyoto, Japan, pp. 7-11, 1999.
10. K. Kundert: *Simulation methods for RF integrated circuits*, Proc. ICCAD'97, San Jose, CA, USA, 1997.
11. R. Lamour: *Floquet-Theory for differential-algebraic equations (DAE)*, ZAMM, Vol. 78-3, pp. S989-S990, 1998.

12. S. Lampe, H.G. Brachtendorf, E.J.W. ter Maten, S.P. Onneweer, R. Laur: *Robust limit cycle calculations of oscillators*, In: U. van Rienen, M. Günther, D. Hecht (Eds): *Scientific computing in electrical engineering*, Proc. SCEE-2000, Warnemünde, Germany, Lecture Notes in Computational Science and Engineering 18, Springer Verlag, Berlin, Germany, pp. 233-240, 2001.
13. E.J.W. ter Maten: *Numerical methods for frequency domain analysis of electronic circuits*, Survey on Mathematics for Industry, Vol. 8, pp. 171–185, 1999.
14. K. Mayaram, D.C. Lee, S. Moinian, D. Rich, J. Roychowdhury: *Overview of computer-aided analysis tools for RFIC: algorithms, features, and limitations*, IEEE 1997 Custom Integrated Circuit Conference, Santa Clara, pp. 505-512, 1997.
15. R. Neubert, A. Schwarz: *Efficient analysis of oscillatory circuits*, In: U. van Rienen, M. Günther, D. Hecht (Eds): *Scientific computing in electrical engineering*, Proc. SCEE-2000, Warnemünde, Germany, Lecture Notes in Computational Science and Engineering 18, Springer Verlag, Berlin, Germany, pp. 225-232, 2001.
16. M. Okumura, H. Tanimoto, T. Itakura, T. Sugawara: *Numerical noise analysis for nonlinear circuits with a periodic large signal excitation including cyclostationary noise sources*, IEEE Trans. on Circuits and Systems - I: Fund. Theory and Applics., Vol.40-9, pp. 581-590, 1993.
17. R. Pulch, M. Günther: *A method of characteristics for solving multirate partial differential equations in radio frequency application*, Preprint Nr. 00/07, IWRMM, Univ. Karlsruhe, subm. for publication, 2000.
18. M. Rösch, K. Antreich: *Schnelle stationäre Simulation nichtlinearer Schaltungen im Frequenzbereich*, AEÜ - Archive für Elektronik und Übertragungstechnik, 46, pp. 1-10 (1992).
19. J. Roychowdhury: *Multi-time PDEs for dynamical system analysis*, In: U. van Rienen, M. Günther, D. Hecht (Eds): *Scientific computing in electrical engineering*, Proc. SCEE-2000, Warnemünde, Germany, Lecture Notes in Computational Science and Engineering 18, Springer Verlag, Berlin, Germany, pp. 3-14, 2001.
20. D.A. Smith, W.F. Ford, and A. Sidi: *Extrapolation methods for vector sequences*, SIAM Review, 29(2), pp. 199–233, 1987.
21. R. Telichevesky, K. Kundert: *Efficient AC and noise analysis of two-tone RF circuits*, Proc. DAC'96, Las Vegas, 1996.
22. S. Ramo, J.R. Whinnery and T. Van Duzer, *"Fields and Waves in Communication Electronics"*, Wiley, New York, 1996.
23. A. Odabasioglu and M. Celik, *"PRIMA: Passive Reduced-Order Interconnect Macromodeling Algorithm"*, IEEE. Trans. Computer-Aided Design, Vol. 17, pp. 645-654, 1999.
24. L. Knockaert and D. De Zutter, *"Passive Reduced Order Multi-port Modeling: The Pade-Arnoldi-SVD Connection"*, Int. J. Electronics and Communications, no. 53, pp. 254-260, 1999.
25. P. Feldman and R.W. Freund, *"Efficient Linear Circuit Analysis by Padé Approximation via the Lanczos Process"*, IEEE Trans. Computer-Aided Design, Vol. 14, pp. 639-649, May 1995.
26. G. Wang, *"A Survey on Simulations of Metal-Insulator-Semiconductor Interconnects"*, Stanford internal report, Nov. 1998, http://www-tcad.stanford.edu/~gaofeng/research/mis_review.htm

27. T.H. Hubing, *"Survey of Numerical Electromagnetic Modeling Techniques"*, Univ. of Missouri-Rolla internal report, Sep. 1991, http://www.emclab.umr.edu/pdf/TR91-1-001.pdf
28. P.P. Silvester and R.L. Ferrari, *"Finite Elements for Electrical Engineers"*, Cambridge Univ. Press, 1996.
29. C.A. Brebbia, J.C. Telles and L.C Wrobel, *"Boundary Element Techniques"*, Berlin: Springer-Verlag, 1984.
30. R.F. Harrington, *"Field Computation by Moment Methods"*, IEEE Press Series on EM Waves, 1993.
31. T.G. Livernois and P.B. Katehi, *A Generalized Method for Deriving the Space-domain Green's Function in a Shielded, Multi-layer Substrate Structure with Applications to MIS Transmission Lines"*, IEEE Trans. Microwave Theory Tech., Vol. MTT-37, pp. 1761-67, 1989.
32. W.C. Chew, J.M. Jin, E. Michielssen and J.M.Song (Editors), *"Fast and Efficient Algorithms in Computational Electromagnetics"*, Artech House, Boston, MA, 2001.
33. K.S. Yee, *"Numerical solution of initial boundary value problems involving Maxwell's equations in isotropic media"*, IEEE Trans. Antennas and Propagation, Vol. 14, pp. 302-307, 1966.
34. A. Taflove (Editor), *"Advances in Computational Electrodynamics: The Finite-Difference Time-Domain Method"*, Artech House, Boston, MA, 1998.
35. P.B. Johns, *"A Symmetrical Condensed Node for the TLM Method"*, IEEE Trans. Microwave Theory Tech., Vol. MTT-35, pp. 370-377, 1987.
36. D. Leenaerts, G. Gielen and R.A. Rutenbar, *"CAD Solutions and Outstanding Challenges for Mixed-Signal and RF IC-Design"*, Proc. ICCAD 2001.
37. R. du Cloux, G.P.J.F.M. Maas, A.J.H. Wachters, R.F. Milsom and K.J. Scott, *"Fasterix, an environment for PCB simulation"*, Proc. 11th Int. Conf. On EMC, Zürich, Switzerland, 1993.
38. P.B.L. Meijer, *"Generalized Neural Networks for Modeling Non-linear Dynamic Multi-demensional Systems in Time and Frequency Domain"*, Philips Research Internal Report 6726, 1993.

# Recent Advances in Finite Integration Technique for High Frequency Applications*

Rolf Schuhmann and Thomas Weiland

Technische Universität Darmstadt,
Computational Electromagnetics Laboratory (TEMF), Schloßgartenstraße 8,
64289 Darmstadt, Germany, schuhmann@temf.tu-darmstadt.de

**Abstract.** We review some recent extensions of the Finite Integration Technique (FIT), which is known to be a generalization of the Finite Difference Time Domain (FDTD) method. Some shortcomings of the standard formulation are discussed which limit the applicability or at least the efficiency of the time domain variant of FIT. The novel developments which are proposed in this paper cover both the basic geometrical modeling in space and time and advanced methods to solve the algebraic problems in time and frequency domain. A numerical application is presented to demonstrate the performance of the algorithms in the high frequency regime.

## 1 Introduction

For many high frequency electromagnetic applications — especially whenever broadband results are to be calculated — time domain simulation methods can be considered as the most suited approach. This is reflected by the great popularity of the Finite Difference Time Domain algorithm (FDTD, [1]) and the vast number of publications in this area.

An alternative formulation that leads to a computationally equivalent update scheme in time domain is supplied by the Finite Integration Technique (FIT [2,3]). Due to its general approach, FIT can also be applied to static or time-harmonic problems or be combined with alternative time integration schemes, e.g. for low-frequency problems.

However, there are also several shortcomings of the standard FIT (or FDTD) formulation. Two of them are the poor modeling quality for arbitrarily shaped geometric objects, if Cartesian grids are used ('staircase'-problem), and the reduced efficiency in the simulation of resonant structures, where the long transients lead to a long simulation time.

In the following we first give a short introduction into the notation and some basic properties of the FIT. In the main part we discuss some recent developments facing the problems mentioned above: Techniques to increase the modeling accuracy of the FIT for curved boundaries (section 2), and to simulate highly resonant structures (section 3). Finally we present an application (section 4) where several of the proposed techniques are applied to obtain accurate simulation results with moderate computing resources.

* Invited paper at SCEE-2002

### 1.1 Maxwell Grid Equations

Similar to the FDTD method, the Finite Integration Technique uses a pair of staggered computational grids, the primary grid $G$ and the dual grid $\widetilde{G}$, which however can have a more general structure as the standard 'Yee cell' of FDTD. The state variables of FIT are so-called grid voltages and grid fluxes which are defined as integrals of the electric and magnetic fields on edges $L_i$, $\widetilde{L}_i$, or faces $A_i$, $\widetilde{A}_i$ of $G$ and $\widetilde{G}$, respectively:

$$\widehat{e}_i = \int_{L_i} \boldsymbol{E} \cdot d\boldsymbol{s}, \qquad \widehat{\widehat{b}}_i = \int_{A_i} \boldsymbol{B} \cdot d\boldsymbol{A}, \tag{1a}$$

$$\widehat{\widehat{d}}_i = \int_{\widetilde{A}_i} \boldsymbol{D} \cdot d\boldsymbol{A}, \qquad \widehat{h}_i = \int_{\widetilde{L}_i} \boldsymbol{H} \cdot d\boldsymbol{s}, \qquad \widehat{\widehat{j}}_i = \int_{\widetilde{A}_i} \boldsymbol{J} \cdot d\boldsymbol{A}. \tag{1b}$$

Using this kind of state variables, an *exact* representation of the integral form of Faraday's and Ampere's law applied to facets of the grids can be found. In matrix-vector form, the so-called Maxwell Grid Equations [3] read

$$\mathbf{C}\widehat{\mathbf{e}} = -\frac{d}{dt}\widehat{\widehat{\mathbf{b}}}, \qquad \widetilde{\mathbf{C}}\widehat{\mathbf{h}} = \frac{d}{dt}\widehat{\widehat{\mathbf{d}}} + \widehat{\widehat{\mathbf{j}}}, \tag{2}$$

with the curl-matrices $\mathbf{C}$ and $\widetilde{\mathbf{C}} = \mathbf{C}^T$, and the vectors of voltages and fluxes $\widehat{\mathbf{e}}$, $\widehat{\mathbf{h}}$ and $\widehat{\widehat{\mathbf{d}}}$, $\widehat{\widehat{\mathbf{b}}}$, $\widehat{\widehat{\mathbf{j}}}$, respectively. The approximations of the method finally take place in the material matrices (the discrete constitutive relations), given here for the linear case:

$$\widehat{\widehat{\mathbf{d}}} = \mathbf{M}_\varepsilon \widehat{\mathbf{e}}, \quad \widehat{\widehat{\mathbf{b}}} = \mathbf{M}_\mu \widehat{\mathbf{h}}, \quad \widehat{\widehat{\mathbf{j}}} = \mathbf{M}_\kappa \widehat{\mathbf{e}} + \widehat{\widehat{\mathbf{j}}}_S \tag{3}$$

($\widehat{\widehat{\mathbf{j}}}_S$ = source currents). For *dual orthogonal* grid systems, where edges of $G$ and facets of $\widetilde{G}$ (and vice versa) intersect by 90°, $\mathbf{M}_\varepsilon$, $\mathbf{M}_\mu$, and $\mathbf{M}_\kappa$ are diagonal matrices which can be easily inverted.

Equations (2) can be symmetrized and rewritten in one large system:

$$\frac{d}{dt}\mathbf{x} = \underbrace{\begin{pmatrix} -\mathbf{M}_\varepsilon^{-1/2}\mathbf{M}_\kappa\mathbf{M}_\varepsilon^{-1/2} & \mathbf{M}_\varepsilon^{-1/2}\mathbf{C}^T\mathbf{M}_\mu^{\;1/2} \\ -\mathbf{M}_\mu^{-1/2}\mathbf{C}\mathbf{M}_\varepsilon^{-1/2} & \mathbf{0} \end{pmatrix}}_{\mathbf{A}} \mathbf{x} + \mathbf{b} \tag{4}$$

with the state vector $\boldsymbol{x}$ and the excitation $\mathbf{b}$:

$$\mathbf{x} = \begin{pmatrix} \mathbf{M}_\varepsilon^{1/2}\widehat{\mathbf{e}} \\ \mathbf{M}_\mu^{1/2}\widehat{\mathbf{h}} \end{pmatrix}, \quad \mathbf{b} = \begin{pmatrix} -\mathbf{M}_\varepsilon^{-1/2}\widehat{\widehat{\mathbf{j}}}_S \\ \mathbf{0} \end{pmatrix}. \tag{5}$$

### 1.2 Formulations for the High-Frequency Regime

*Frequency Domain.* In frequency domain we can eliminate the magnetic voltages $\widehat{\mathbf{h}}$ in (2) and obtain the discrete *curl-curl-equation* which reads (for the lossless case with $\mathbf{M}_\kappa = \mathbf{0}$):

$$\Big(\underbrace{\mathbf{M}_\varepsilon^{-1}\mathbf{C}^T\mathbf{M}_\mu^{-1}\mathbf{C}}_{\mathbf{A}_{CC}} - \omega^2\mathbf{I}\Big)\,\widehat{\mathbf{e}} = -j\omega\mathbf{M}_\varepsilon^{-1}\widehat{\widehat{\mathbf{j}}}_S \tag{6}$$

It can be easily shown that the *curl-curl system matrix* $\mathbf{A}_{CC}$ has only real eigenfrequencies $\omega_i$, if the material operators $\mathbf{M}_\varepsilon^{-1}$ and $\mathbf{M}_\mu^{-1}$ are symmetric and positive definite. This important property which is referred to as *spatial stability* [3] is responsible for the long-time stability of the FIT system in time domain. In frequency domain these eigensolutions correspond to undamped oscillating resonant modes in lossless structures. It is a standard task of many electromagnetic simulation codes to calculate the dominant eigensolutions of $\mathbf{A}_{CC}$, i.e. the resonant modes with lowest frequencies $\omega_i > 0$. To solve the inhomogeneous curl-curl equation (6) for current-driven devices at a fixed frequency $\omega_0$ is often a quite expensive task, and the development of robust algebraic solvers including appropriate preconditioners is still a topic of ongoing research [4].

*Time Domain.* In time-domain, we can approximate the time derivatives in (2) by central differences on a staggered time grid $t^{(m)} = t_0 + m \cdot \Delta t$ and obtain the update formulas of the leapfrog scheme:

$$\widehat{\mathbf{h}}^{(m+1)} = \widehat{\mathbf{h}}^{(m)} \qquad - \Delta t \mathbf{M}_\mu^{-1} \mathbf{C} \widehat{\mathbf{e}}^{(m+1/2)}, \tag{7a}$$

$$\widehat{\mathbf{e}}^{(m+3/2)} = \widehat{\mathbf{e}}^{(m+1/2)} + \Delta t \mathbf{M}_\varepsilon^{-1} \left( \mathbf{C}^T \widehat{\mathbf{h}}^{(m+1)} - \widehat{\widehat{\mathbf{j}}}_S^{(m+1)} \right). \tag{7b}$$

The size of the time-step width $\Delta t$ is limited by the well-known Courant-criterion due to stability reasons. Combined with the spatial operator matrices for a Cartesian grid (the 'Yee-cell'), these formulas and the FDTD-scheme can be shown to be computationally equivalent [5].

*Multiport Devices.* Many electrodynamic devices can be modeled as $n$-port systems, where only the behaviour at distinct input and output ports is of interest. So-called generalized ports include the standard transmission line ports of microwave devices, but also some more sophisticated input/output concepts like filament currents or the interaction of electromagnetic fields with charged particle beams in accelerating cavities [6].

In the most common case, the ports are waveguide-type apertures with one propagating mode per port[1]. If the signals considered are the amplitudes of incoming ($a_i$) and outgoing ($b_i$) waves, we can define the *scattering matrix* $\mathbf{S}$ of the structure by $\mathbf{b} = \mathbf{S}\mathbf{a}$.

Alternatively the *impedance matrix* $\mathbf{Z}$ is given by the relation $\mathbf{u} = \mathbf{Z}\mathbf{i}$ of the generalized voltage and current quantities

$$u_i = Z_i^{\frac{1}{2}} \, (a_i + b_i) \qquad i_i = Z_i^{-\frac{1}{2}} \, (a_i - b_i), \tag{8}$$

with the line impedance $Z_i$ of each port. The waveguide modes are normalized such that $u_i \cdot i_i = a_i^2 - b_i^2$ describes the power flow through a port.

[1] Apertures where more than one waveguide mode shall be considered are conceptually identified as separate ports.

There are two possibilities to simulate such multiport devices with FIT: Either the ports are treated with a special absorbing boundary condition in time or frequency domain [7], where the incoming wave amplitudes $a_i$ are directly realized as excitation of the structure. This approach yields the **S**-matrix of the structure and is the most efficient approach in time domain.

In frequency domain it is often advantageous to treat the ports with a closed (perfectly magnetic conducting) boundary condition and to excite the structure with equivalent input currents $i_i$ in the port planes, respectively. This technique leads (in the lossless case) to real-valued system matrices which are much easier to solve. It directly yields the impedance matrix **Z** which can easily transformed into the scattering matrix **S** if necessary.

Following here the second approach, we can define coupling matrices which map the $n$ input currents $i_i$ onto the discrete source current vector $\widehat{\widehat{\mathbf{j}}}_S$, and the electric field vector $\widehat{\mathbf{e}}$ on the output voltages $u_i$:

$$\mathbf{b} = \mathbf{B}\mathbf{i} \qquad \mathbf{u} = \mathbf{F}\mathbf{x}. \tag{9}$$

**B** and $\mathbf{F}=\mathbf{B}^T$ contain the discrete field patterns of the waveguide modes, which can be obtained by the solution of a related 2D eigenvalue problem for each port. With (4) we obtain the state-space formulation

$$\frac{d}{dt}\mathbf{x} = \mathbf{A}\mathbf{x} + \mathbf{B}\mathbf{i} \quad , \quad \mathbf{u} = \mathbf{F}\mathbf{x}. \tag{10}$$

In frequency domain the impedance matrix finally is given by

$$\mathbf{Z} = \mathbf{F}\,(j\omega\mathbf{I} - \mathbf{A})^{-1}\,\mathbf{B}. \tag{11}$$

## 2 Advanced Geometric Modeling Techniques

The restriction of the original FDTD algorithm to 2D and 3D equidistant Cartesian grids is often considered to be the major disadvantage of this method. While guaranteeing the simplicity as well as the computational efficiency of FDTD, Cartesian grids suffer from their poor modeling quality for applications with complex geometries ('staircase'-approximations).

In the following, we will concentrate on two approaches to improve the geometric modeling capabilities of FIT: The application of triangular grids in 2D and an enhanced technique to derive the related material matrices, and recent results for so-called sub-cell methods, where the geometrical information below the level of the cell-resolution is taken into account.

### 2.1 Extended Algorithm for Triangular Grids in 2D

If conformal triangular grids are used in two-dimensional problems, the generation of these grids seems to be a well-understood task with a large variety of efficient algorithms available. For the application of the FIT also a dual grid is needed, and in the classic implementation [8] this dual grid has to fulfill the orthogonality property mentioned above. Such a dual orthogonal grid, however, can only be found, if all primary grid angles are below 90°, and this demand on the grid generation process can not always be met [8].

Recently, a more flexible scheme with an extended material operator has been proposed [9]. For the new material operator we introduce special interpolating functions, so-called *Whitney forms*, which are well-known from Finite Element (FE)-approaches [10]. They define a map from the degrees of freedom of the discrete formulation (here: the electric grid voltages $\widehat{e}_i$) to field values at arbitrary points inside the triangular cells:

$$\boldsymbol{E}(\boldsymbol{r}) = \sum_i \widehat{e}_i \boldsymbol{w}_i^{(1)}(\boldsymbol{r}). \tag{12}$$

Note, that this approach is consistent with the integral-type definition of the electric voltages in (1) due to the relations

$$\int_{L_i} \boldsymbol{w}_j^{(1)} \cdot d\boldsymbol{s} = \begin{cases} 1 \ (i = j) \\ 0 \ (i \neq j) \end{cases} \quad \Rightarrow \quad \int_{L_i} \boldsymbol{E}(\boldsymbol{r}) = \widehat{e}_i. \tag{13}$$

Knowing the electric field at any point inside the cells, the electric flux components $\widehat{\widehat{d}}_i$ can be 'exactly' (referring to the interpolation function) calculated by an integration of (12) over the corresponding dual facets:

$$\widehat{\widehat{d}}_i = \int_{\widetilde{A}_i} \varepsilon \boldsymbol{E}(\boldsymbol{r}) \cdot d\boldsymbol{A} = \sum_j \widehat{e}_j \int_{\widetilde{A}_i} \varepsilon \boldsymbol{w}_j^{(1)}(\boldsymbol{r}) \cdot d\boldsymbol{A}. \tag{14}$$

Due to the local support of the basis functions, this approach also yields a sparse (but non-diagonal) material matrix $\mathbf{M}_\varepsilon$. Its off-diagonal entries

$$M_{\varepsilon,ij} = \int_{\widetilde{A}_i} \varepsilon \boldsymbol{w}_j^{(1)}(\boldsymbol{r}) \cdot d\boldsymbol{A}. \tag{15}$$

$(i \neq j)$ are generally non-zero also for orthogonal grid systems.

A thorough analysis of this new matrix reveals that it is non-symmetric in the general case, which deteriorates the spatial stability of the discretization scheme. However, it can be symmetrized by varying the position of the dual point in each cell and thus the shape of the integration areas for the electric grid fluxes in (14). It can be shown that there is one unique position for the dual point $\widetilde{P}$, which results in a symmetric matrix for both the electric and the magnetic material operator [9].

Numerical evaluations in [9] show that these new matrices lead as expected to a 2nd order accurate scheme, where the grid angles of the primary (triangular) can now be extended to a maximum of about 120°. As an additional result, the new material operator is found to be different from an FE-type matrix based on Whitney-forms as well as from the classic FIT-matrix (even in the case of dual-orthogonal grids). Thus, this derivation supplies a comparison — and some important hints on the differences between FIT and FE discretization schemes on triangular grids.

## 2.2 Extended Conformal Algorithm for Cartesian Grids

The most efficient choice to tackle the staircase problems of FIT in three dimensions is probably to use so-called sub-cell methods, which are able to model arbitrary partial material fillings within Cartesian grids cells.

In the *partially filled cells* (PFC) methods proposed in [11] and [12] this is realized for interfaces to perfect electric conducting (PEC) materials by considering reduced cell length and areas in the derivation of the material matrices. As shown in Fig. 1, left, this technique may lead to very small dimensions of the 'remaining' cell. This is no problem for the accuracy of the PFC method, but it may induce a considerable overhead in the time domain algorithm, as such small cell lengths require to reduce the maximum stable time step according to the Courant criterion.

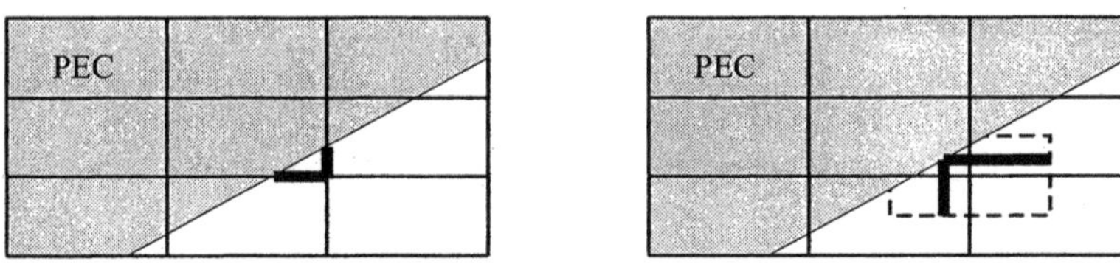


**Fig. 1.** PFC-approach (left) with reduced cell edges, and new USC-scheme applying a larger virtual cell to avoid the reduction of the maximum stable time step (right).

As a remedy, an extended version of these schemes, named *Uniformly Stable Conformal (USC)* scheme has been proposed in [13]. The idea of the new method is to use also the information from *adjacent* cells to complete the local curl-operation. Thus, the effective cell size is enlarged instead of reduced, and the same time step can be used as in the standard scheme.

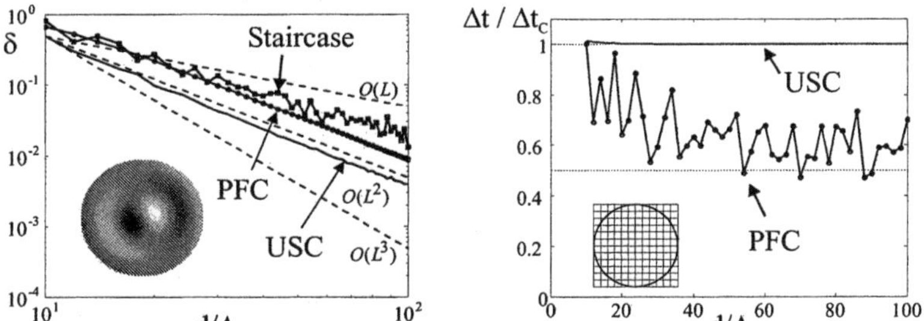


**Fig. 2.** Performance of the USC scheme: Relative field error $\delta$ of eigensolutions in a cylindrical cavity with diameter $d$ (left), and maximum stable time step (normalized to the Courant limit, right) for a varying grid step size $\Delta$.

The additional computational cost of this approach is usually negligible, as only for a small number of field components (at the PEC interfaces) off-diagonal elements in the material matrices have to be processed. Some numerical results for the USC scheme compared to the standard PFC approach are shown in Fig. 2. The accuracy of the USC method is typically as good as with PFC, whereas from the larger time step a speed-up factor in time domain of about two can be achieved.

# 3 Highly Resonant Structures

In standard time domain methods, the long settling times in *highly resonant* devices such as cavities and microwave filters require a large number of time steps to reach steady state. If the goal of the simulation is a precise prediction of the system's response near a resonance, the use of approximative techniques, such as introducing artificial losses or non-conservative time integration schemes, is out of the question. On the other hand, a conventional eigenvalue analysis (in frequency domain) of the 'closed' structure, with short circuits at the input/output ports, only gives a hint on the resonance frequencies, but no quantitative results for the behavior of the original (open) device around that frequency.

## 3.1 Model Order Reduction

An alternative approach to analyze highly resonant devices arises from the idea that the number of degrees of freedom in the original system (2) (the number of $\widehat{e}$ and $\widehat{h}$-components in the grid) is typically much higher than the number of poles and zeros which determine its system response within a certain frequency range. Thus, the goal of Model Order Reduction (MOR)-approaches is to approximate the system by a smaller one *before* actually solving it.

*Lanczos-based Techniques.* The Lanczos algorithm generates two Krylov-spaces, spanned by vectors $\mathbf{v}_i$ and $\mathbf{w}_i$ $(i = 1 \ldots p)$ with

$$\mathbf{W}^T\mathbf{V} = \mathbf{I} \qquad \text{and} \quad \mathbf{W}^T\mathbf{A}\mathbf{V} = \mathbf{T}, \tag{16}$$

where $\mathbf{T}$ is a banded matrix. Using the (approximative) projection $\mathbf{x} = \mathbf{V}\mathbf{z}$ leads to

$$\begin{aligned}\mathbf{Z}' &= (\mathbf{F}\mathbf{V})\,(j\omega\mathbf{I} - \mathbf{T})^{-1}\,(\mathbf{W}^T\mathbf{B}) \\ &= \mathbf{F}'\,(j\omega\mathbf{I} - \mathbf{T})^{-1}\,\mathbf{B}'.\end{aligned} \tag{17}$$

The dimension of this reduced system is only $p \times p$, and if the projection takes into account all dominant eigenvectors of the original system, $\mathbf{Z}'$ will be a good approximation of the impedance matrix $\mathbf{Z}$.

If the Lanczos algorithm is applied to the shifted and inverse matrix $(\mathbf{A}-s_0\mathbf{I})^{-1}$ we obtain a PVL-approximation which needs a considerably lower number of basis vectors. However, inverting the matrix is computationally very expensive and can only be performed for small systems. More details and an algorithm to combine both ideas can be found in [15].

*Modal Approach.* A very similar technique is obtained, if instead of the Lanczos basis vectors $\mathbf{v_i}$ several eigenvectors $\mathbf{x}_i$ of the system are used in the projection above. The accuracy of this *modal approach* [6] depends on the eigenmodes which are not considered in the expansion, and their coupling to the input quantities $\mathbf{b}_n$. For eigenfrequencies well above the frequency range of interest, $\omega_i \gg \omega$ $(i > p)$, this contributions can be approximated to depend only weakly on $\omega$. Thus, a *constant correction vector* can be added which is based on the solution of the inhomogeneous equation (6) for one frequency $\omega_0$ [16]. This yields for one entry in the impedance matrix:

$$Z_{mn}(\omega) \approx \sum_{i=1}^{p} \frac{(\mathbf{b}_m^T \mathbf{x}_i)\,(\mathbf{x}_i^T \mathbf{b}_n)}{j\omega - j\omega_i} + Z_{mn}^{corr}(\omega_0) \tag{18}$$

For the required solution of the inhomogeneous curl-curl equation a good start solution for iterative solvers is supplied by the (truncated) modal expression.

An important advantage of this modal approach in combination with the correction term is that only a moderate number of modes is needed for accurate results. In contrast to the Lanczos scheme, they can typically be completely held in memory and be used for the construction of field solutions at arbitrary frequencies.

## 4 Numerical Application

As a numerical example we present the calculation of effective material coefficients in so-called *metamaterials* [18–20]. Metamaterials consist of a lattice of conducting, non-magnetic elements that can be described by an effective magnetic permeability $\mu_{eff}$ and an effective electrical permittivity $\varepsilon_{eff}$. Both of these parameters can exhibit values not found in naturally occurring materials — e.g. simultaneously $\mu_{eff} < 0$ *and* $\varepsilon_{eff} < 0$ — with some very interesting consequences for wave propagation [18]. Typically each cell of the lattice represents a resonant structure with spatial dimensions much smaller than the incident wavelength.

Fig. 3 (taken from [20]) shows a metamaterial consisting of an array of split ring resonators (SRRs) and accompanying wires, which has been subject to measurements as well as simulations before ([20], and in a simplified form in [19]). The task for the numerical methods is to perform field simulation in this structure and to extract the material parameters from the electromagnetic fields over a certain frequency range.

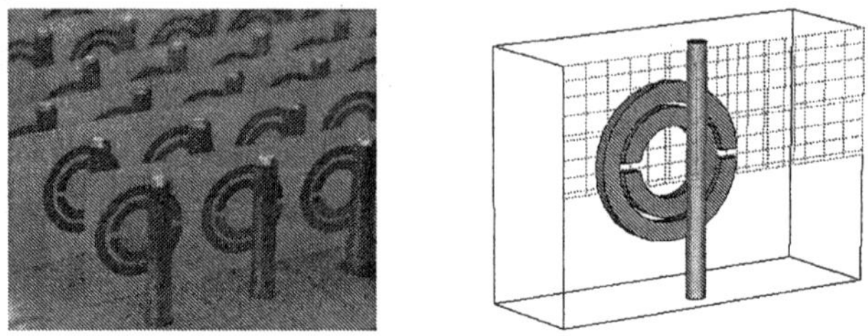

**Fig. 3.** SRR-type metamaterial and one cell in the simulation model.

The effective material parameters in metamaterials describe an equivalent homogeneous medium which has the same propagation properties as (macroscopic) planar waves in the lattice. According to [18], they can be calculated from averaged field quantities in one cell of the lattice, e.g. for the permeability:

$$H_{x,av}(\omega) = \frac{1}{L_x}\int_{L_x} \boldsymbol{H}(\omega,\boldsymbol{r})\cdot d\boldsymbol{r}, \qquad B_{x,av}(\omega) = \frac{1}{S_x}\int_{S_x} \boldsymbol{B}(\omega,\boldsymbol{r})\cdot d\boldsymbol{A}, \tag{19}$$

where $L_x$ is the length of the cell, $S_x$ is its normal facet, and

$$B_{x,av}(\omega) = \mu_0 \mu_{x,\mathit{eff}}(\omega) H_{x,av}(\omega) \tag{20}$$

It can be shown [16] that for waves in coordinate direction only one cell of the lattice has to be simulated. This cell can then be considered as a 2-port structure, and for the extraction of the effective material parameters the complete $2 \times 2$ scattering matrix **S** or the impedance matrix **Z** has to determined.

Our first discrete model consists of a rather coarse mesh with only $22 \times 10 \times 9 = 1,980$ grid points. Due to the small dimensions of the SRR, this is still equivalent to more than 70 lines per wavelength. However, to be able to model the geometric details of the SRR in this mesh, a sub-cell method as described above has to be applied. In a second simulation, a finer mesh with $38 \times 18 \times 16 = 10,944$ grid points (130 lines per wavelength) is used.

As a result, Fig. 4 shows the scattering parameter (reflection $S_{11}$) of one cell, as well as the real part of the effective permeability for the frequency range from 4GHz to 5.5GHz. Obviously the coarse discretization gives already qualitatively good results, but exhibits a frequency shift of about 200 MHz compared to the simulation with the finer mesh. The extracted permeability quantity is as expected negative near resonance.

For the simulation of this resonant structure different approaches have been applied (with indistinguishable result curves): A standard time domain simulation (TD), the Lanczos-based model-order-reduction technique (MOR) according to [15], and the modal approach with correction term (M+C). The computational cost is summarized in Table 1. Note that one time step in TD is equivalent to one matrix-vector operation in the other approaches, and

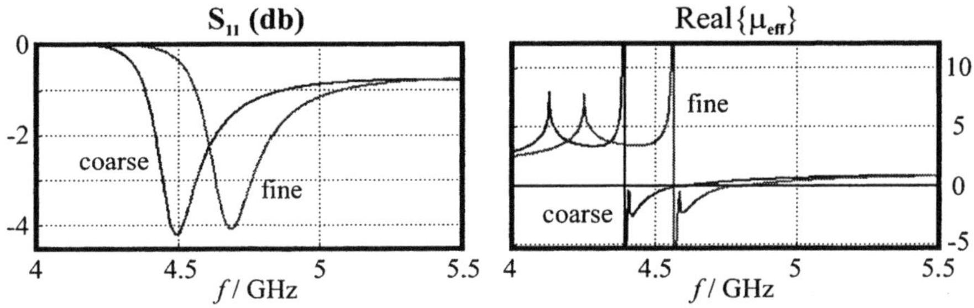


**Fig. 4.** Scattering parameter (reflection coefficient $S_{11}$, left) of one cell, and real part of the effective permeability in the SRR medium (right). The results of the fine and the coarse grid resolution agree qualitatively, but exhibit a slight frequency shift by about 200MHz.

that in TD, FD, and for the correction term in the modal approach two runs are needed to obtain the full **S** or **Z**-matrix. For comparison, also a direct solution of the curl-curl equation (6) for a single frequency $\omega_0$ (frequency domain, FD) is included.

**Table 1.** Computational cost (number of matrix-vector operations) for different methods: Time domain (TD), frequency domain (FD, single frequency point), model order reduction (MOR), modal approach with correction (M+C).

| grid points | TD | FD (single freq.) | MOR (no fields) | M+C 10 Modes | M+C Correction |
|---|---|---|---|---|---|
| 1,980 | $2 \times 50,000$ | $2 \times 300$ | 260 | $\sim 1,100$ | $2 \times 70$ |
| 10,944 | $2 \times 85,000$ | $2 \times 600$ | 520 | $\sim 2,300$ | $2 \times 180$ |

Due to the small geometric details compared to the wave length (lattice parameter $L_x = 1.46$ cm compared to $\lambda \approx 5\ cm$ @ 6 GHz, grid step sizes $\Delta < 0.07$ cm), the model is spatially 'oversampled'. In time domain with the Courant-limit for a stable time integration this leads to a large number of time steps per period ($1/(f \cdot \Delta t) = 700$, or 1000 for the coarse and the fine grid, respectively). Thus, together with the long settling times, the time domain approach is not competitive in this application.

As long as only the S-parameters are needed, the most efficient approach is the MOR method. However, since the complete Krylov-space generated in MOR can usually not be held in memory, no field solutions are available for the extraction of the effective permeability.

The computational cost of the modal approach depends mainly on the eigenmode-solver applied (and often also on its internal settings). Although in this case 2 or 3 eigenmodes in the expansion are sufficient for accurate results, we recommend to compute a set of at least 7 to 10 modes: The computation time in this range is only slightly higher than for 3 or 5 modes, and the equation solvers in the correction part benefit from the improved start solution. Furthermore, the number of modes actually used can be varied to supply for a simple convergence check in the post-processing.

# 5 Conclusion

Together with the recently developed sub-cell extensions for the FIT (or FDTD) discretization scheme, this technique can be considered to represent the state-of-the-art method for a large area of microwave applications. If the flexibility of a Cartesian computational grid is not sufficient, the FIT approach allows to switch to generalized grid types without changing the algebraic notation or loosing its important conservation and stability properties. Whereas in 3D Cartesian grids seem to be superior to most other grid types in terms of their easy and robust data structure (grid generation, data format and access), triangular grids can be a competitive alternative for many 2D applications.

The efficiency of FIT's time domain approach, is not satisfying in some cases, especially if highly resonant devices are to be handled. Such an application is the simulation of so-called metamaterials, where the electromagnetic fields can be described by negative effective material parameters in a small frequency band near the resonance.

The proposed model order reduction techniques, based on a projection of the electromagnetic fields on a Krylov subspace or the subspace defined by some eigensolutions of the model, turn out to be very efficient and robust tools for such applications. Depending on the required output data — transfer matrices or also field solutions — considerable speed-ups in computation time can be achieved.

# 6 Acknowledgements

This paper summarizes recent results of the research group *Discretization methods & RF* at TEMF, with its current members Monika Balk, Frank Mayer, Rolf Schuhmann (spokesman), Anastassios Skarlatos, Holger Spachmann, Bernd Trapp, and Tilmann Wittig. The USC scheme has been developed by our colleague Igor Zagorodnov.

# References

1. Yee, K.S.: Numerical Solution of Initial Boundary Value Problems Involving Maxwell's Equations in Isotropic Media. IEEE Trans. AP **14** (1966) 302–307
2. Weiland, T.: A Discretization Method for the Solution of Maxwell's Equations for Six-Component Fields. Electr. Commun. (AEÜ) **31** (1977) 116

3. Weiland, T.: Time Domain Electromagnetic Field Computation with Finite Difference Methods. Int. J. of Num. Modelling **9** (1996) 295–319
4. Clemens, M., Schuhmann, R., v. Rienen, U., Weiland, T.: Modern Krylov Subspace Methods in Electromagnetic Field Computation Using the Finite Integration Theory. ACES Journal, Spec. Issue on Appl. Math.: Meeting the Challenges Presented by Comp. Electromagn. **11**,1 (1996) 70–84
5. Weiland, T.: Advances in FIT/FDTD Modeling. Proc. of the ACES Conference, Monterey, CA (2002) 1,1–1,14
6. Dohlus, M., Schuhmann, R., Weiland, T.: Calculation of Frequency Domain Parameters Using 3D Eigensolutions. Int. J. of Num. Modelling **12** (Special Issue) (1999) 41–68
7. Schuhmann, R., Clemens, M., Thoma, P., Weiland, T.: Frequency and Time Domain Computations of S-Parameters Using the Finite Integration Technique. Proc. of the ACES Conference, Monterey, CA (1996) 1295–1302
8. V. Rienen, U., Weiland, T.: Triangular Discretization Method for the Evaluation of RF-Fields in Cylindrically Symmetric Cavities. IEEE Trans. Magn. **21** (1985) 2317–2320
9. Schuhmann, R., Schmidt, P., Weiland, T.: A New Whitney-based Material Operator for the Finite-Integration Technique on Triangular Grids. IEEE Trans. Magn. **38**,2 (2002) 409–412
10. Bossavit, A.: Computational Electromagnetism. Academic Press, San Diego (1997)
11. Thoma, P.: Zur numerischen Lösung der Maxwellschen Gleichungen im Zeitbereich. PhD Thesis, TU Darmstadt (1997)
12. Dey, S., Mittra, R.: A Locally Conformal Finite-Difference Time-Domain (FDTD) Algorithm for Modeling Three-Dimensional Perfectly Conducting Objects. IEEE Microwave and Guided Wave Letters **7**,9 (1997) 273–275
13. Zagorodnov, I., Schuhmann, R., Weiland, T.: A Uniformly Stable Conformal FDTD-Method in Cartesian Grids. Submitted to: Int. J. of Num. Modelling
14. Schuhmann, R., Wittig, T., Munteanu, I., Trapp, B., Weiland, T.: Time and Frequency Domain Simulations of Highly Resonant RF-Devices. Proc. of the ICEAA 2001, Turin (2001) 125–128
15. Wittig, T., Munteanu, I., Schuhmann, R., Weiland, T.: Two-Step Lanczos Algorithm for Model Order Reduction. IEEE Trans. Magn. **38**,2,1 (2002) 673–676
16. Schuhmann, R., Weiland, T.: Efficient Calculation of Effective Material Parameters in Metamaterials Using FDTD and a Modal Approach. Proc. of the IEEE Int. Microwave Symp. (2002) 2037–2040
17. Schuhmann, R., Weiland, T.: Rigorous Analysis of Trapped Modes in Accelerating Cavities. Phys. Rev. Special Topics - Accelerators and Beams **3**,122002 (2000), http://publish.aps.org/abstract/PRSTAB/v3/e122002
18. Pendry, J.B., Holden, A.J., Robbins, D.J., Stewart, W.J.: Magnetism From Conductors and Enhanced Nonlinear Phenomena. IEEE Trans. MTT **47** (1999) 2075–2084
19. Smith, D.R., Vier, D.C., Kroll, N., Schultz, S.: Direct Calculation of Permeability and Permittivity for a Left-Handed Metamaterial. Appl. Phys. Lett. **77**,14 (2000) 2246–2248
20. Weiland, T., Schuhmann, R., Greegor, R.B., Parazzoli, C.G., Vetter, A.M., Smith, D.R., Vier, D.C., Schultz, S.: *Ab Initio* Numerical Simulation of Left-Handed Metamaterials: Comparison of Calculations and Measurements. J. Appl. Phys. **90**, 10 (2001) 5419–5424

# Toward Generalized Finite Element Difference Methods for Electro- and Magnetostatics*

Igor Tsukerman

Department of Electrical and Computer Engineering, The University of Akron, Akron, OH 44325-3904, USA
igor@uakron.edu

**Abstract.** The paper explores a class of "Finite Element Difference" (FED) schemes with Finite Difference–type data structures but based on Finite Element – variational principles. Curved material boundaries are approximated *algebraically* on relatively coarse regular rectangular or hexahedral grids by a judicious choice of local approximating functions, rather than geometrically on conforming meshes. The grids do not have to resolve small geometric details. The proposed approach combines the ideas of the Generalized Finite Element – Partition of Unity methods, Discontinuous Galerkin Methods and Finite Difference / Finite Volume / Finite Integration Techniques.

## 1 Introduction: Finite Differences and Finite Elements for Complex Geometries

The major complicating factors in numerical simulation of fields are the presence of curved interface boundaries between materials with different parameters and point / edge / corner singularities. The standard Finite Difference (FD) and Finite Element (FE) Methods do not handle singularities well. Curved boundaries can be accurately rendered in conventional FE analysis by geometrically conforming meshes, whereas in FD a 'staircase' approximation, with related numerical artifacts and loss of accuracy, is typical. The success and popularity of FEM in the engineering community can be attributed, to a large extent, to its ability to handle complex geometries – although on a more fundamental level it can be argued that the advantages of FEM stem from its underlying basic variational principles.

However, the geometric and physical flexibility of FEM comes with a price. Mesh generation can be complicated, especially in 3D, and the associated data structures are complex as well. Consequently, FEM is not competitive with FD methods for simple geometries. Our aspiration is to explore a class of "Finite Element Difference" (FED) methods loosely defined as FD-style schemes with FE-type capabilities, with as many of the following features as possible:

- Finite difference data structures and schemes based on Finite Element variational principles.

* Invited paper at SCEE-2002

- Ability to handle curved interface boundaries on relatively coarse regular rectangular or hexahedral grids with high accuracy and without having to resolve small geometric details.
- Flexibility in the choice of local approximating functions.
- Few or none additional unknowns other than the nodal values of the potential.
- For elliptic problems, symmetric positive definite discrete systems.
- Feasibility of $h$- and $p$-type refinement.
- Optimal convergence rate.

The following computational techniques have some, or many, of these desired features and are relevant to our analysis:

1. Generalized FEM by Partition of Unity [27], [3], [20], [39].
2. Heuristic homogenization schemes by volume averaging [18] or edge-length averaging [46].
3. Homogenization schemes based on variational principles [28].
4. "Finite Integration Techniques" (FIT) and its extensions and enhancements developed by researchers in Darmstadt [24], [13].
5. Discontinuous Galerkin (DG) methods [1], [8], [11], [14].
6. Methods rooted deeply in differential geometry and algebraic topology [7], [22], [25], [40], [41].
7. Miscellaneous other techniques, such as special elements with holes or inclusions [26], [37], and "immersed surface" methodology [44]. The latter modifies the Taylor expansions to account for derivative jumps at material boundaries but leads to rather unwieldy expressions.

Obviously, FD-FV-FIT methods have always been oriented toward regular meshes, and the recent advancements in their application to inhomogeneous problems are stimulating (see Acknowledgment). Nevertheless variational approaches remain valuable, as they tend to automatically preserve important physical and mathematical characteristics of the problem. Section 2 reviews some of the existing techniques that are most closely related to the material of this paper. A FED-type method of Section 3 is a natural extension of the variational-difference scheme proposed by Moskow *et al.* [28] and can be viewed as a nonconforming FEM. Numerical experiments are described in Section 4.

The new and promising framework of "multivalued approximation" outlined in Section 5 offers a simple and general way of constructing FED-type schemes. A discussion of the results and future directions appears in 6.

## 2 Numerical Methods for Inhomogeneous Media

### 2.1 Generalized FEM by Partition of Unity

A detailed explanation and analysis of this method proposed originally by Babuška and Melenk [27], [3], is widely available (e.g. [39]). Of all interesting features of GFEM, the most salient for our present goals is the ability to employ a variety of special nonpolynomial approximating functions. In particular, functions representing jumps of the normal derivatives of the potential at interface boundaries can be added to the basis. Strouboulis *et al.* [39] present an extensive set of application examples with special functions for material inclusions in stress analysis. Babuška *et al.* [2] apply GFEM to problems with material interfaces. Plaks *et al.* [31] implemented GFEM for problems with magnetized nanoparticles.

In GFEM the computational domain is considered to be covered with overlapping subdomains ('patches'), and different local approximations are merged by Partition of Unity (PU) $\{\Omega_i\}_{i=1}^{n_{patches}}$ constructed on this system of patches.

$$\sum_{i=1}^{n_{patches}} \varphi_i \equiv 1 \quad \text{in } \Omega, \quad \text{supp}(\varphi_i) = \Omega_i \tag{1}$$

That is, each function $\varphi_i$ is associated with the respective patch $\Omega_i$ and vanishes outside that patch. Then the global solution $u$ can be decomposed into its 'patch components' $u_i$:

$$u = u \sum_i \varphi_i = \sum_i u\varphi_i = \sum_i u_i, \quad \text{with } u_i \equiv u\varphi_i \tag{2}$$

Decomposition (2) can be used in a natural way for assembling the global approximation of the solution from the local ones. Suppose that locally, within each patch $\Omega_i$, the exact solution $u$ can be approximated by a linear combination $u_h^{(i)}$ of some approximating functions $\left\{ g_m^{(i)} \right\}_{m=1}^{M}$:

$$u_h^{(i)} = \sum_m c_m^{(i)} g_m^{(i)} \tag{3}$$

$c_m^{(i)}$ being some (real- or complex-valued) coefficients. The final system of approximating functions $\left\{ \psi_m^{(i)} \right\}$ is built with $\varphi_i$ as weight functions:

$$\psi_m^{(i)} = g_m^{(i)} \varphi_i \tag{4}$$

Multiplication (4) by $\varphi_i$ guarantees seamless merging of patch-wise approximations, with rigorously provable estimates of the global error in terms of local errors and the norms of the PU functions $\varphi$ [3]. On the negative side, however, this multiplication complicates the set of approximating functions and tends to make it more ill-conditioned (in some cases even linearly dependent, see [3]). The computation of gradients, implementation of the Dirichlet conditions and numerical Galerkin quadratures also gets more complicated. In addition, GFEM-PU may lead to a combinatorial increase in the number of degrees of freedom. For illustration, consider a regular hexahedral mesh where a 'patch' is defined as a set of eight hexahedra around a common node. In the presence of material boundaries, it is sensible to replace the usual eight trilinear basis functions with eight special functions satisfying the derivative jump condition at the interface, as described in the following subsection. In GFEM-PU, each of these special functions gets multiplied by the 'shape function' $\varphi$ of the patch. As each hexahedral element of the mesh is an intersection of eight patches (centered at its eight respective nodes) and each of these patches contributes eight approximating functions, the stiffness matrix for elements close to material interfaces is $64 \times 64$ instead of the usual $8 \times 8$. For all of the above reasons, alternative approaches may be worth exploring. One such approach is discussed in Section 3 and another promising one [43] is outlined in Section 5.

## 2.2 Generalized Special FEM for 'Rough' Coefficients

Babuška *et al.* [2] proposed special approximating functions satisfying the boundary condition on material interfaces for an elliptic problem where the coefficient can vary sharply in at most one coordinate direction:

$$\nabla \cdot a(x) \nabla u(x,y) = f(x,y) \tag{5}$$

in a rectangular domain with homogeneous Dirichlet boundary conditions. The coordinate transformation

$$\tilde{x}(x) = \int_0^x \frac{ds}{a(s)}, \qquad \tilde{y}(y) = y \tag{6}$$

"removes" the derivative jumps at material discontinuities. Therefore the conventional FEM on a standard (say, triangular) mesh can be considered in the transformed domain. Corresponding to that, there is an FE method on a curvilinear mesh in the original domain. The approximating functions are piecewise- linear with respect to the transformed coordinates $\tilde{x}$, $\tilde{y}$ but are more complex in the original coordinates.

This coordinate mapping forms a common basis for three different variational schemes in [2]. The first approach is to use the special approximating functions on curvilinear triangles and the same Galerkin test functions. The method is conforming and is equivalent to the standard FEM in the transformed domain.

In the second approach, regular (rather than curvilinear) triangles are employed in the *original* domain, and the approximating functions are the same as in the previous method. The standard piecewise-linear continuous test functions are applied. Since the spaces of approximating and test functions are in this case different, nonsymmetric systems result and the stability constant is higher. The approximation space is not a subspace of $H^1(\Omega)$, and hence this version of the method is nonconforming. The third approach of [2] fits in the framework of GFEM-PU [3] (historically, though, GFEM was formally introduced a few years *after* [2]). An ordinary triangular mesh is used, and the standard nodal basis functions form a PU. The local approximating functions, while the same as in the previous two approaches, get multiplied by the PU functions in accordance with the GFEM-PU procedure.

The method is conforming, standard meshes are used and matrix symmetry is preserved - thus this approach is arguably the best of the three. However, the multiplication by the PU functions in GFEM does add some complexity and redundancy, as discussed above.

The authors of [2] extend their first and third approaches to a general case where the material coefficient can vary sharply locally in at most one direction. The extension of the first method also relies on curvilinear meshes, and the extension of the third method is again a GFEM-PU algorithm. The description, analysis and implementation of these methods in this general setting are rather complex and for the first method may be impractical.

### 2.3 Homogenization Schemes Based on Variational Principles

This interesting approach uses the special approximating functions of [2] as a starting point for the construction of a difference scheme. Within each grid cell, the authors seek a (generally anisotropic) material parameter such that the discrete and continuous energy inner products are the same over the chosen discrete space. The overall construction in [28] relies on a special partitioning of the grid ("red-black" numbering, or the "Lebedev grid") and on a specific, central difference, representation of the gradient. A rigorous proof of discrete weak convergence for the proposed scheme is given, and the numerical experiments show that the method can give accurate results even for curved boundaries and high contrasts of material parameters between different regions.

The method proposed in Section 3 can be viewed as a generalization of the variational-difference approach of [28]. The special 'Lebedev' grids and the specific approximation of gradients by central differences adopted in [28] turn out not to be really critical for the algorithm.

### 2.4 Discontinuous Galerkin Methods

The idea to relax the interelement continuity requirements of the standard FEM and to use nonconforming elements appeared at the early stages of FE research ([17], [38]). For example, in the Crouzeix-Raviart elements [17], the continuity of piecewise-linear functions is imposed only at midpoints of the edges.

Over recent years, a substantial amount of work has been devoted to Discontinuous Galerkin Methods DGM) [1], [14]; a consolidated view with extensive bibliography is presented in [1]. Many of the approaches start with the "mixed" formulation that includes additional unknown functions for the fluxes on element edges (2D) or faces (3D). However, these additional unknowns can be replaced with their numerical approximations, thereby producing a "primal" variational formulation in terms of the scalar potential alone. In DGM, the interelement continuity is ensured, at least in the weak sense, by retaining the surface integrals of the jumps, generally leading to saddle-point problems even if the original equation is elliptic.

### 2.5 Heuristic Homogenization Schemes

Finite Difference Time Domain methods [45] of applied electromagnetics require enormous computational work due to a large number of time steps for numerical wave propagation. Therefore regular orthogonal grids are strongly preferred and the need to avoid 'staircase' approximations of curved boundaries has been quite acute. To this end, Yu, Dey and Mittra advocate very simple homogenization techniques where the equivalent material parameter for grid cells containing two materials is obtained by either volume averaging [18] or edge-length averaging [46]. Numerical experiments have shown good results but the method remains heuristic, and its accuracy and possible limitations need further theoretical and practical investigation.

### 2.6 Finite Integration Techniques

Another class of methods for applied electromagnetics is Finite Integration Techniques (FIT) developed primarily by researchers in Darmstadt [12], [24], [35], [36]. Various extensions and enhancements of FIT, such as "the Perfect Boundary Approximation technique" and "Conformal FIT", approximate curved material interfaces on regular orthogonal grids. Although a detailed description of PBA has not yet been published, PBA is apparently related to some of the homogenization techniques mentioned in the previous subsection. FIT utilizes the integral form of Maxwell's equations imposed on a pair of grids. In recent years, this model has been gaining popularity quite rapidly (see e.g. [13], [24], [35], [36]).

## 3 A Finite Element Difference Method

### 3.1 The General Setup

For simplicity and clarity, this paper focuses on a model linear elliptic boundary value problem with homogeneous Dirichlet conditions:

$$\nabla \cdot \sigma \nabla u \;=\; f \quad \text{in } \Omega \subset R^n \;\; (n = 2, 3); \;\; u_{|\partial\Omega} \;=\; 0 \tag{7}$$

Here $\sigma$ is a generic material parameter that can be discontinuous across material boundaries and can depend on coordinates but not, in the linear case under consideration, on the potential $u$. The physical meaning of $\sigma$ depends on the nature of the problem, and we shall keep in mind electrostatics and magnetostatics as practical examples.

Due to special choices of approximating functions, regular hexahedral or rectangular grids will in many cases suffice and will therefore be used exclusively in the remainder of the paper – even though the approach applies to general FE meshes equally well. Consequently, the domain will be assumed hexahedral / rectangular. The computational problem is nontrivial due to the presence of nonconforming material interface boundaries – straight or curved - with the standard interface conditions

$$u_1 = u_2 \quad \text{on } \Gamma \tag{8}$$

$$\sigma_1 \frac{\partial u_1}{\partial n} = \sigma_2 \frac{\partial u_2}{\partial n} \quad \text{on } \Gamma \tag{9}$$

where the subscripts refer to the two subdomains $\Omega_1$ and $\Omega_2$ sharing the material boundary $\Gamma$, and $n$ is the normal direction to $\Gamma$. A regular mesh is obtained by subdividing each edge of $\Omega$ into a number of subintervals, not necessarily equal. Let $\Omega_i$ be a brick element of the mesh, so that $\Omega = \cup \Omega_j$, $j = 1, 2, \ldots, m$. The energy inner product is defined as

$$(u, v)_{\sigma,\omega} = \int_\omega \sigma \nabla u \cdot \nabla v \, d\omega \quad \text{in } H^1(\omega) \times H^1(\omega) \tag{10}$$

for any domain $\omega \subset \Omega$ with a Lipschitz boundary.

### 3.2 Special Approximating Functions

This section reviews the construction of special approximating functions proposed by Babuška et al. [2] and later used by Moskow et al. [28]. (The construction is presented here in a somewhat modified form convenient for our purposes.) The main observation is that the normal derivative of a potential $u$ satisfying the jump conditions (9) can be rendered continuous by an appropriate coordinate mapping $S$:

$$S: \quad (\tau_1, \tau_2, n) \to (\tilde{\tau}_1, \tilde{\tau}_2, \tilde{n}); \qquad \tilde{\tau}_1 = \tau_1, \ \tilde{\tau}_2 = \tau_2, \ \tilde{n} = n/\sigma \tag{11}$$

where $(\tau_1, \tau_2, n)$ is an orthogonal system, $\tau_1, \tau_2$ being tangential to the (sufficiently smooth) interface boundary $\Gamma$ and $n$ normal to that boundary, with $n = 0$ corresponding to $\Gamma$. Consider two specific but most practical cases:

*Case 1.* A plane interface boundary. Let $R$ be the (constant) orthogonal rotation matrix transforming $(x, y, z)$ to $(\tau_1, \tau_2, n)$, i.e.

$$(\tau_1, \tau_2, n)^T = R \cdot (x, y, z)^T$$

Coordinate stretching (11) in the $n$-direction can be associated with the diagonal matrix $S = \text{diag}(1, 1, \sigma^-)$ and the overall coordinate transformation – with a symmetric positive definite matrix

$$T = R^T S R \tag{12}$$

Note that $T$ depends on coordinates through the dependence of $S$ on $\tau_1$, $\tau_2$ and, most importantly, on $n$. For any function $\psi$: $R^3 \to R$, one can define the transformed function $\tilde{\psi}$ as

$$\tilde{\psi}(\zeta) \equiv \psi(T\zeta), \quad \forall \zeta \equiv (x, y, z)^T \in R^3 \tag{13}$$

It is straightforward to verify that if $\psi$ has continuous derivatives, then $\tilde{\psi}$ satisfies the derivative jump condition (9) at the interface.

*Case 2.* A spherical interface boundary with a radius $r_0$. For simplicity, let the material parameter $\sigma$ be piecewise-constant and equal to $\sigma_{in}$ inside the sphere and $\sigma_{out}$ outside. Consider the coordinate 'stretching' in the radial direction for points outside the sphere:

$$\tilde{r} = \begin{cases} r, & r \le r_0 \\ r_0 + \sigma_{in}/\sigma_{out}(r - r_0), & r > r_0 \end{cases} \tag{14}$$

If a function $\psi(r,\theta,\varphi)$ has continuous derivatives, the transformed function $\tilde{\psi}(r,\theta,\varphi) \equiv \psi(\tilde{r},\theta,\varphi)$ satisfies the interface condition (9). In Cartesian coordinates, $\tilde{\psi}$ can be written as

$$\tilde{\psi}(x,y,z) \equiv \psi(\tilde{x},\tilde{y},\tilde{z}), \quad \text{with } \tilde{x} = x\tilde{r}/r, \quad \tilde{y} = y\tilde{r}/r, \quad \tilde{z} = z\tilde{r}/r \tag{15}$$

In a hexahedral cell $[x_1,x_2] \times [y_1,y_2] \times [z_1,z_2]$ containing the spherical interface, one may choose, for example, a set of eight basis functions trilinear with respect to the *transformed* coordinates $\tilde{x}$, $\tilde{y}$, $\tilde{z}$:

$$\tilde{\psi}_1 = (\tilde{x}_2 - \tilde{x})(\tilde{y}_2 - \tilde{y})(\tilde{z}_2 - \tilde{z}), \quad \text{etc.} \tag{16}$$

where the transform of parameters $x_2$, $y_2$, $z_2$ is theoretically unimportant but tends to make the basis better conditioned. The $\tilde{\psi}$'s, being smooth functions of $\tilde{x}$, $\tilde{y}$, $\tilde{z}$, have the necessary derivative jumps across the spherical interface in the original coordinates.

This construction is not affected by the relative size or position of the spherical boundary with respect to the grid cell. In particular, the approximating functions are applicable even to spherical particles contained entirely within one cell – quite a desirable feature for problems with small particles [31]. One numerical example of this kind is given in Section 4.3.

### 3.3 Variational Formulations

Over each hexahedral element $\Omega_i$, one can introduce the approximation space $V_{hi} = \text{span}\{\tilde{\psi}_j^{(i)}\}$, $j = 1,2,\ldots 8$. A conforming FE method could be obtained by partition of unity, as was done in [31] for a different choice of special functions $\tilde{\psi}$. GFEM-PU does introduce additional complexity, and a much simpler nonconforming method is therefore worth exploring. The FE space is defined as

$$V_h = \{v_h : v_{h|\Omega_i} \in V_{hi}; \;\; v_h \text{ continuous at the nodes of the grid}\} \tag{17}$$

This space is nonconforming because the continuity is imposed only at the nodes and not over all interelement boundaries. $V_h$ is a finite-dimensional subspace of the broken Sobolev space [29]

$$V_h \subset H^1_{\cup}(\cup\Omega_i) \equiv \{v \in L_2(\Omega): \;\; v_{|\Omega_i} \in H^1(\Omega_i), \; \forall i\,(\{1,2,\frac{1}{4},m\}\,\}$$

with the energy inner product

$$(u, v)_{\sigma\cup} = \sum_{i=1}^{m} \int_{\Omega_i} \sigma \nabla u \cdot \nabla v \, d\Omega \quad \text{in} \quad H^1_{\cup}(\cup\Omega_i) \times H^1_{\cup}(\cup\Omega_i) \tag{18}$$

where the union symbol in the subscripts is indicative of $\cup\Omega_i$. The energy inner product over the whole domain $\Omega$ is, as before,

$$(u, v)_{\sigma} = \int_{\Omega} \sigma \nabla u \cdot \nabla v \, d\Omega \quad \text{in} \quad H^1(\Omega) \times H^1(\Omega) \tag{19}$$

For any pair of functions that are actually in $H^1(\Omega)$, the $\sigma$-product and the partitioned $\sigma\cup$-product are obviously identical; but for functions discontinuous across interelement boundaries the $\sigma$-product is undefined. It is known that $(u, u)_{\sigma\cup}$ not only is strictly positive (for nonzero $u$) and hence defines a valid inner product, but also satisfies the discrete Friedrichs inequality with a constant independent of the mesh size [19], [23], [9]. The continuous variational problem is formulated as

$$(u, u')_{\sigma} = (f, u'), \quad u, \forall u' \in H^1_0(\Omega) \tag{20}$$

Since $(u, u')_{\sigma} = (u, u')_{\sigma\cup}$ in $H^1_0(\Omega)$, (20) can also be written as

$$(u, u')_{\sigma\cup} = (f, u'), \quad u, \forall u' \in H^1_0(\Omega) \tag{21}$$

The discrete problem is

$$(u_h, u'_h)_{\sigma\cup} = (f, u'_h), \quad u_h, \forall u'_h \in V_h \subset H^1_{\cup}(\cup\Omega_i) \tag{22}$$

Note that there are no interface surface integrals or other penalty factors for conformity involved. The main difference with the standard FEM is in the energy inner product in the broken Sobolev space – this product is computed element-wise due to possible discontinuities of functions across the interelement boundaries. It is also important that $H^1_{\cup}(\cup\Omega_i) \not\subset H^1_0(\Omega)$ and therefore the test function $u'_h$ in (22) in general does not lie in the Sobolev space of the original continuous problem.

### 3.4 Stiffness Matrices and the Discrete System

Unlike conventional FE nodal bases, the approximating functions $\tilde{\psi}$ do not generally satisfy the Kronecker-delta conditions at the eight nodes. Let $N$ be the matrix of the nodal values

$$N = \begin{pmatrix} \tilde{\psi}_1(r_1) & \tilde{\psi}_2(r_1) & \dots & \tilde{\psi}_8(r_1) \\ \tilde{\psi}_1(r_2) & \tilde{\psi}_2(r_2) & \dots & \tilde{\psi}_8(r_2) \\ \dots & \dots & \dots & \dots \\ \tilde{\psi}_1(r_8) & \tilde{\psi}_2(r_8) & \dots & \tilde{\psi}_8(r_8) \end{pmatrix} \tag{23}$$

where $r_j$ is the position vector of the $j$-th node. Then for any function $u_h = \sum_{j=1}^{8} \underline{u}_{\psi_j} \tilde{\psi}_j$, where $\underline{u}_\psi \in R^8$ is a coefficient vector, the respective vector $\underline{u}_{nodal} \in R^8$ of the nodal values of $u_h$ is

$$\underline{u}_{nodal} = N\underline{u}_\psi \tag{24}$$

Assuming that $N$ is nonsingular, it follows immediately from (24) that each column of $N^{-1}$ constitutes a coefficient vector transforming the $\{\tilde{\psi}\}$ basis into the Kronecker-delta nodal basis $\{\psi_{1-8}\}$. If both bases are arranged as *column* matrices $\psi$ and $\varphi$, then

$$\varphi = N^{-T}\tilde{\psi} \quad \text{with} \quad \varphi_j(r_k) = \delta_{jk} \tag{25}$$

The transform between the Gram matrices of the two bases is

$$\varphi\varphi^T = N^{-T}\tilde{\psi}\,\tilde{\psi}^T N^{-1} \tag{26}$$

To each node $k$ of the mesh there corresponds, in an obvious way, a basis function $\varphi_k$ that is equal to $\varphi_k^{(i)}$ within each element $i$ adjacent to node $k$. Since no matching conditions between the bases in the element-wise spaces have been introduced, $\varphi_k$ will in general be discontinuous across interelement boundaries.

The stiffness matrix corresponding to the left hand side of (22) is formed in the usual way. First, one notes a one-to-one correspondence between Euclidean vectors of nodal values $\underline{u} \in R^n$ and the respective functions $u_h \in V_h$:

$$u_h = \sum_{l=1}^{n} u_k \varphi_k \tag{27}$$

Further, let $\mathcal{I} : R^n \to V_h$ be the interpolation operator defined by (27). The energy inner products in $V_h \times V_h$ have their counterparts in $R^n \times R^n$:

$$\begin{aligned} (u_h, u_h')_{\sigma\cup} &= (\mathcal{I}\underline{u}, \mathcal{I}\underline{u}')_{\sigma\cup} = (\textstyle\sum_{k=1}^{n} u_k\varphi_k, \sum_{l=1}^{n} v_k\varphi_k)_{\sigma\cup} \\ &= (Lu, u') \equiv (\underline{u}, \underline{u}')_L \end{aligned} \tag{28}$$

where $\underline{u}$, $\underline{u}'$ are Euclidean vectors in $R^n$ and $L$ is the stiffness matrix with the elements

$$L_{kl} = (\varphi_k, \varphi_l)_{\sigma\cup}$$

Consequently, the discrete formulation (22) can be rewritten in an equivalent form

$$(\underline{u}, \underline{u}')_L = (f, \mathcal{I}\underline{u}') \ \forall \underline{u}' \, ( R^n \tag{29}$$

The *stiffness* matrices in the $\psi$– and nodal–bases are related, due to (24), by the similarity transform

$$L_N = N^{-T} L_\psi N^{-1} \tag{30}$$

where $L_N$ stands for the nodal stiffness matrix with the entries $(\sigma \nabla\varphi_i, \nabla\varphi_j)$ and the entries of $L_\psi$ are

$$L_{\psi ij} = (\sigma \nabla\psi_i, \nabla\psi_j) \tag{31}$$

If the material interface boundary is plane, the entries of $L_\psi$, and consequently $L_N$, can be found analytically. For a curved boundary and/or higher order approximation, numerical quadratures may be unavoidable.

Convergence analysis can be adapted, with some changes, from [28] and is outlined in Appendix I. Please see the Discussion section for further remarks.

# 4 Numerical Results

## 4.1 2D Problem: Cylinder in an External Field

As the first and simplest illustrative example, let us consider a dielectric cylinder placed in an external field (Fig. 4.1). The domain is the unit square [0,1] × [0,1], with the inhomogeneous Dirichlet condition $u = x$ imposed on its boundary (this condition would correspond to a uniform applied field if the domain were much larger). The particular results below are for the material parameter $\sigma = 10$ of the cylinder and for $\sigma = 1$ outside the cylinder. The cylinder axis is at (0.52, 0.49), and its radius is 0.15.

Fig. 4.1 shows the potential distribution as a function of $x$ along the line $y = 0.5$ for several grids. Even for a very crude $4 \times 4$ grid, in no way capable of capturing the cylindrical boundary, the computed values of nodal potentials are quite accurate. Problems with small particles (e.g. for nanotechnological

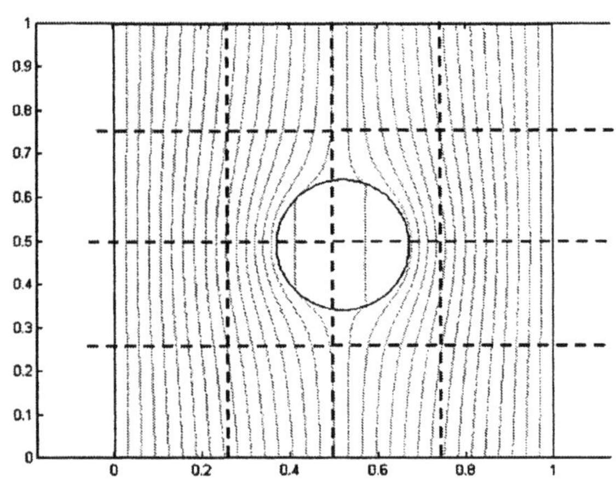


**Fig. 1.** Field lines for a cylindrical particle (a 2D problem). Good results obtained with FED even on a very coarse $4 \times 4$ grid shown.

applications [31]) would require a high level of *hp* mesh refinement around the particles in standard FEM or FD methods. In FED, a coarse grid suffices because the special approximating functions with jumps represent the

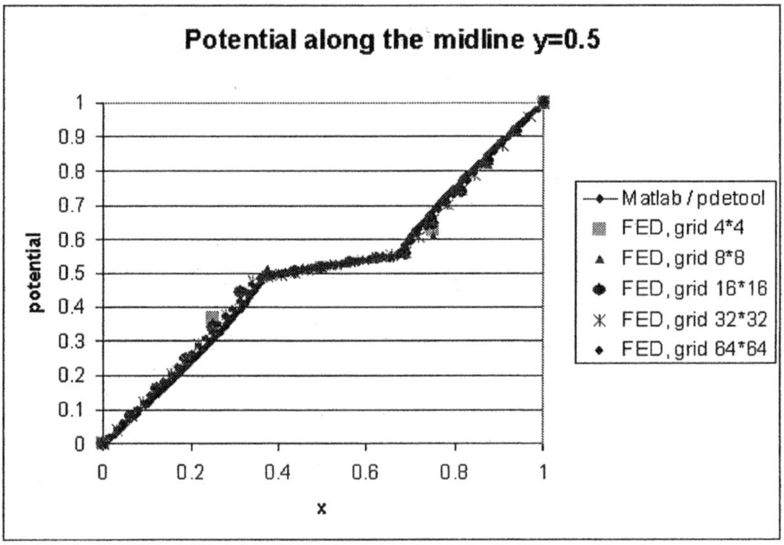


**Fig. 2.** Potential as a function of $x$ for one cylindrical particle, at $y = 0.5$ (a 2D problem).

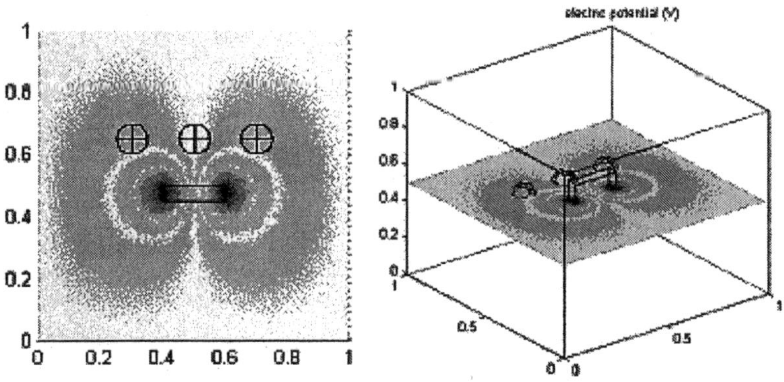


**Fig. 3.** Potential distribution for the test problem with three spherical particles near a magnetized island.

behavior of the potential well. A 3D example of this nature is given in the following sections.

### 4.2 A 3D Problem: Magnetic Particles Near a Magnetized Substrate

This test problem is related to multiparticle assembly processes in nanotechnology and was previously solved using Generalized FEM [31], albeit with different approximating functions. As an additional point of comparison, a FEMLAB™[1] simulation was also performed.

Three spherical particles are located near a magnetized substrate. The computational domain is a unit cube, with homogeneous Dirichlet condition at its surface. The radii of all particles are equal to $r_0 = 0.05$; the particles are centered at (0.3, 0.65, 0.5), (0.5, 0.65, 0.5) and (0.7, 0.65, 0.5), respectively. The relative permeabilities of the particles are assumed to be $\mu_r = 10$. The substrate is a parallelepiped, $[0.4, 0.6] \times [0.45, 0.5] \times [0.4, 0.6]$, with unit magnetization in the $x$-direction. The magnetic scalar potential distribution is illustrated in Fig. 4.1 produced by FEMLAB.

The results obtained by FED and FEMLAB are in a good agreement (Fig. 4.2). Note that the grid size in FED is equal to the radius of the particles, i.e. not at all sufficient to resolve the geometry of the spherical boundary. This is possible because the behavior of the solution at particle boundaries is analytically incorporated into the FED approximation. The popular but more basic schemes that rely on simple averaging of the material parameter in the volume of the cell or along its edges (see Section 2.5) cannot produce a comparable result; for example, the volume-averaging scheme yields a smoothed potential distribution (Fig. 4.2).

### 4.3 3D Particles Smaller Than the Grid Cell

Special approximating functions representing derivative jumps at interfaces make the numerical solution adequate even if fine geometric details are not resolved on the grid. It is even feasible to solve problems with particles much smaller than a grid cell. For illustration, several problems with a varying radius of a spherical particle with $\sigma = 10$ in the unit cube were solved on a fixed $5 \times 5 \times 5$ mesh (i.e. $h = 0.2$). The particle was placed a little off-center of the domain to eliminate possible symmetry artifacts. To make an exact solution available for validation, Dirichlet boundary conditions were set to corresponded to the well-known potential of a sphere in a uniform field (the induced potential of a sphere has only the dipole component).

For the radius of the particle $r_0 = 0.05$, i.e. *four times smaller* than the mesh size, the error in the nodal values of the potential in the cross-section $z = 0.5$ is plotted in Fig. 4.3a. For comparison, a similar plot of the

[1] FEMLAB is a registered trademark of Comsol, Inc.

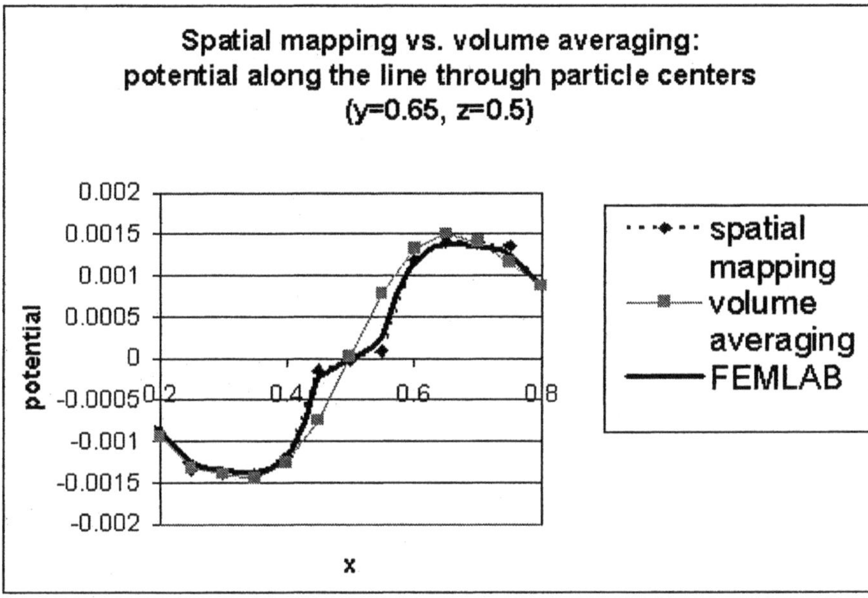


**Fig. 4.** Three spherical particles near a magnetized island: potential as a function of $x$ at $y = 0.65$, $z = 0.5$.

potential obtained by volume-averaging of the material parameter is shown in Fig. 4.3b. The error for volume-averaging is an order of magnitude higher than for FED (note the scaling factors in the figures); this is not surprising because for volume-averaging the solution obviously does not even depend on the position of the particle within the cell.

# 5 Future Development: the Framework of Multivalued Approximation

The method described above provides very reasonable accuracy even for fairly coarse regular grids that do not have to resolve the geometric details. Nevertheless the approach has two main drawbacks. First, as in GFEM, the computation of matrix entries in (31) requires numerical quadratures over 3D regions of complex shape (such as, for example, the intersection of a spherical particle with a hexahedral grid cell). Second, the optimal convergence rate with respect to the mesh size cannot generally be expected.

Of these two drawbacks, the second one is alleviated by the fact that we are interested in obtaining reasonable accuracy on relatively *coarse* grids, when the asymptotic rate may not be relevant. The first drawback is more serious, as numerical quadratutes in 3D carry a large computaitonal overhead.

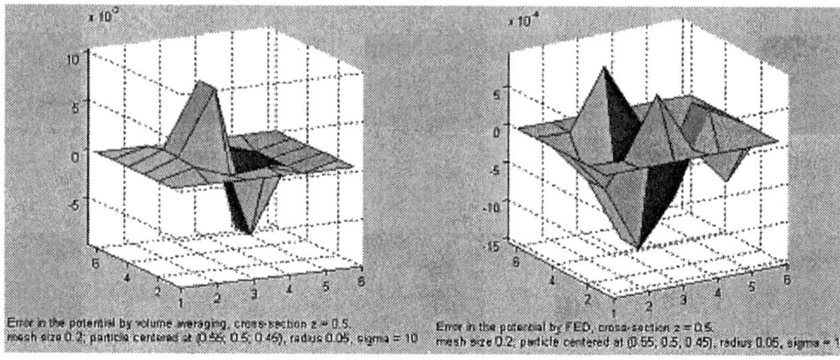


**Fig. 5.** Numerical errors for the potential of a spherical particle in the unit cube (plane z = 0.5). Left: volume averaging, scaling factor $10^{-3}$. Right: FED with a spatial mapping, scaling factor $10^{-4}$.

We now briefly describe a new framework that eliminates the unpleasant volume integration while preserving the best features of the previous approach. The main ideas are listed below, and the details will be given in [43].

A New Framework: Multivalued Approximaiton

- A system of overlapping patches is introduced as before. The nodal values on the grid are single-valued and provide the necessary "information transfer" between the overlapping patches. The transformation between the nodal values the numerical solution and its expansion in terms of the approximating functions is given by the same matrix N as before (23).
- Any relevant approximating functions can be used within each patch, independently of other patches. (Examples: singular Coulomb potentials; functions generated by spatial mappings to satisfy the jump conditions on material interfaces.)
- Simple regular grids can be used.
- When patches overlap, the approximation is generally multivalued. Not only is this not problematic, as long as all approximations converge to the exact solution, but can be turned into an advantage: the discrepancy between multiple numerical values available in the intersection of patches may serve as an *a posteriori* error estimator. (The well-known Bank-Weiser error indicator based on the jumps of normal derivatives across element boundaries is a direct analogy.)
- Since a unique globally continuous interpolant is not defined, the Galerkin method in $H^1(\Omega)$ is generally not applicable. However, within each patch there is a sufficiently smooth local approximation (3), and a general moment (weighted residual) method can be applied, provided that the support of the test function is contained entirely within the patch.

– In particular, by introducing the standard 'control volume' centered at a given node of the grid and setting the test function equal to one within that control volume and zero elsewhere, one arrives at a flux balance scheme. This is a generalization of the standard 'control volume' technique to any set of suitably defined local approximating functions. Only surface intergals, rather than volume quadratures, are needed, which greatly reduces the computational overhead.

## 6 Discussion

Inhomogeneous static problems can be solved either on geometrically conforming meshes, possibly with *hp*-refinement that has now become conventional in FE methodology, or, alternatively, on regular meshes with additional approximating functions representing derivative jumps on material interfaces. The trade-off is between the complexity of mesh generation and adaption on the one hand, and the burden of numerical quadratures on the other.

The paper reviewed the most promising ways to construct "Finite Element Difference" algorithms that are based on variational FE principles but produce FD-like schemes on regular grids. One such algorithm was developed as a generalization of the variational-difference homogenization scheme proposed in [28]. Analysis then revealed that this algorithm is equivalent to a nonconforming FE method in broken Sobolev spaces on regular grids. Although the convergence rate of this method in the energy norm is suboptimal with respect to $h$, practical simulation results are quite reasonable because in many cases, and in particular for problems with small particles, the mesh size is far from the asymptotic regime.

An even more promising alternative is the new framework of multivalud approximation that has been briefly described in the previous section and will be further explored elsewhere (e.g. [43]). The most interesting application areas for this class of schemes include macromolecular modeling in solvents [4], [16], where the interface boundaries are geometrically extremely complex [4] but the solute region admits a rather simple analytical definition as a combination of spheres. The proposed approach provides a way to solve this type of problems accurately on relatively coarse regular grids.

In the context of purely finite difference methods, a similar approach gives rise, for example, to numerical models of point charge singularities in electrostatics with virtually no loss of accuracy and no extra computational cost, as explained in [43]. This type of problems is critical for molecular dynamics simulations of macromolecules [4], [16], [21].

**Acknowledgment**

My communication with the research group from Technische Universität Darmstadt was a major stimulating factor for pursuing the line of research reflected in this paper. I would like to thank Markus Clemens, Rolf Schuhmann and Thomas Weiland for very informative discussions. I am also grateful to Prof. Susanne Brenner of the University of South Carolina for answering my questions.

## Appendix I. Convergence

Let $u^*$ be the exact solution of the continuous problem (20) or (21) and $u_h \in V_h$ be the solution to (22). We assume that $u^*$ belongs to a space $W(\Omega) \subset H_0^1(\Omega)$ of potentials $u$ that admit an approximation $\mathcal{A}u \in V_h$ such that

$$\| u - \mathcal{A}u \|_{\sigma\cup} \le c \| u \|_W h^p \tag{32}$$

$$\| u - \mathcal{A}u \| \le c \| u \|_W h^q, \quad u \in W(\Omega) \tag{33}$$

where the orders of approximation $p > 0$, $q > 0$ and the $L_2$-norm appears without a subscript. In general terms, $W$ comprises functions that are rendered sufficiently smooth by the spatial mapping of Section 3.2. An elaborate mathematical analysis of approximation (32) in 2D is presented in [2], and its technical details are quite involved. To avoid these mathematical complications, we shall treat estimate (32) as an additional assumption.

Even though we have not assumed a special grid construction or a specific approximation of gradients as in Moskow *et al* [28], weak convergence analysis can be adapted, with some changes, from [28]. Beware that in the transformations below $\sigma-$ and $\sigma\cup-$products are used interchangeably for functions in $H_0^1(\Omega)$.

For any function $u' \in W(\Omega)$, consider the energy inner product

$$\begin{aligned} (\mathcal{A}u^* - u_h, \mathcal{A}u')_{\sigma\cup} &= (\mathcal{A}u^*, \mathcal{A}u')_{\sigma\cup} - (u_h, \mathcal{A}u')_{\sigma\cup} \\ &= \{(\mathcal{A}u^*, \mathcal{A}u')_{\sigma\cup} - (u^*, u')_{\sigma\cup}\} + \{(u^*, u')_\sigma - (u_h, \mathcal{A}u')_{\sigma\cup}\} \ \ldots \end{aligned} \tag{34}$$

[the terms in the second curly brackets correspond to the discrete and continuous variational problems (20, 22)]

$$\ldots = \{(\mathcal{A}u^*, \mathcal{A}u')_{\sigma\cup} - (u^*, u')_{\sigma\cup}\} + \{(f, u') - (f, \mathcal{A}u')\} \tag{35}$$

Due to the approximation property of $\mathcal{A}$ for both $u^*$ and $u'$, i.e. $u^* \approx \mathcal{A}u^*$ and $u' \approx \mathcal{A}u'$ in the sense of (32), the last expression (33) can be rewritten as the following estimate of weak convergence:

$$|(\mathcal{A}u^* - u_h, \mathcal{A}u')_{\sigma\cup}| \le \left[c(u^*)h^p + c(u') \| f \| h^q\right] (\| u^* \|_W + \| u' \|_W) \tag{36}$$

A more detailed analysis shows that the convergence rate in the energy norm is suboptimal.

## References


1. Arnold, Douglas N., Brezzi, F., Cockburn, B. and Marini, L. D.: Unified analysis of discontinuous Galerkin methods for elliptic problems, SIAM J. Numer.Analysis **39**, No.5 (2002), 1749–1779.
2. Babuška I, Caloz G., Osborn J.E.: Special finite-element methods for a class of 2nd-order elliptic problems with rough coefficients, SIAM Journal on Numerical Analysis, **31**, No. 4 (1994), 945–981.
3. Babuška I., Melenk, J.M.: The partition of unity method, International Journal for Numerical Methods in Eng., **40** , No. 4, (1997) 727–758.
4. Baker, N.A., Sept, D., Simpson, J., Holst, M.J., and McCammon, J. A.: Electrostatics of nanosystems: Application to microtubules and the ribosome, PNAS, **98**, No. 18, (2001), 10037–10041, www.pnas.org/cgi/doi/10.1073/pnas.181342398
5. Belytschko, T., Krongauz, Y., Organ, D., Fleming, M., Krysl, P.: Meshless methods: an overview and recent developments, Computer Methods in Applied Mechanics and Engineering, **139**, No. 1-4, (1996), 3–47.
6. Bossavit, Alain: Computational Electromagnetism: Variational Formulations, Complementarity, Edge Elements, San Diego: Academic Press, 1998.
7. Bossavit, A., Kettunen, L.: Yee-like schemes on staggered cellular grids: A synthesis between FIT and FEM approaches, IEEE Trans. Magn. **36**, (2000), 861–867.
8. Bottasso C.L., Micheletti S, Sacco R, The discontinuous Petrov-Galerkin method for elliptic problems, Computer Methods in Applied Mechanics and Engineering, **191**, No. 31, 3391–3409, 2002.
9. Brenner, S.C.: Poincare-Friedrichs inequalities for piecewise H1 functions, Research Report 2002:01, Department of Mathematics, University of South Carolina (to appear in SIAM Journal on Numerical Analysis).
10. Briggs, E. L., Sullivan, D. J., and Bernholc, J. Real-space multigrid-based approach to large-scale electronic structure calcula-tions, Physical Review B, 54 (1996), No. 20, 14362–14375.
11. Castillo P., Cockburn, B., Perugia, I., and Schöotzau, D.: An a priori error analysis of the local discontinuous galerkin method for elliptic problems, SIAM J. Numer.Analysis 38, No.5, (2000), 1676–1706.
12. Ciarlet, P. G., The finite element method for elliptic problems, Amsterdam; New York: North-Holland Pub. Co. 1980.
13. Clemens, M, Weiland, T.: Magnetic field simulation using Conformal FIT formulations, IEEE Trans Magn. **38**, No. 2 (2002), 389-392.
14. Cockburn, B., Karniadakis, G.E., and Shu, C.-W., The development of discontinuous Galerkin methods, in Discontinuous Galerkin Methods. Theory, Computation and Applications, B. Cockburn, G.E.Karniadakis, and C.-W.Shu, eds., Lecture Notes in Comput. Sci. Engrg. **11**, Springer-Verlag, New York (2000), 3–50.
15. Collatz, Lothar, The numerical treatment of differential equations, New York: Springer, 1966.
16. Cortis, C.M., Friesner, R.A.: Numerical solution of the Poisson-Boltzmann equation using tetrahedral finite-element meshes, Journal of Computational Chemistry, **18**, No. 13, (1997), 1591–1608.

17. Crouzeix, M., Raviart, P. A.: Conforming and nonconforming finite element methods for solving the stationary Stokes equation, RAIRO Anal. Numer. **7**, R-3 (1973), 33-76. MR 49:8401.
18. Dey, S., Mittra, R.: A conformal finite-difference time-domain technique for modeling cylindrical dielectric resonators, IEEE Transactions on Microwave Theory and Techniques, **47** (1999), No. 9, 1737–1739.
19. Dolejsi, V, Feistauer, M, Felcman, J.: On the discrete Friedrichs inequality for nonconforming finite elements, Numerical Functional Analysis and Optimization, **20** (1999), No. 5–6, 437-447.
20. Duarte, C.A., Babuška, I., Oden, J.T.: Generalized finite element methods for three-dimensional structural mechanics problems, Computers & Structures, **77** (2000), No. 2, 215–232.
21. Fogolari, F., Esposito, G., Viglino, P., Molinari, H.: Molecular mechanics and dynamics of biomolecules using a solvent continuum model, Journal of Computational Chemistry, **22** (2001), No.15, 1830–1842.
22. Hiptmair, R.: Discrete Hodge operators, Numer. Math. **90** (2001), 265–289.
23. Knobloch, P.: Uniform validity of discrete Friedrichs' inequality for general nonconforming finite element spaces, Numerical Functional Analysis and Optimization, **22** (2001), No. 1, 107–126.
24. Krietenstein, B., Schuhmann, R., Thoma, P., Weiland T.: The perfect boundary approximation technique facing the big challenge of high precision field computation, Proceedings of the XIX International Linear Accelerator Conference (LINAC 98), Chicago, USA (1998), 860–862.
25. Mattiussi, C.: An analysis of finite volume, finite element, and finite difference methods using some concepts from algebraic topology, Journal of Computational Physics **133** (1997), No. 2, 289–309.
26. Meguid, S.A., Zhu, Z.H.: A novel fnite element for treating inhomogeneous solids. International Journal for Numerical Methods in Engineering, **38** (1995), 1579–1592.
27. Melenk, J.M., Babuška, I.: The partition of unity finite element method: Basic theory and applications, Comput. Methods Appl. Mech. Engrg., **139** (1996), 289–314.
28. Moskow, S., Druskin, V., Habashy, T., Lee, P., Davydycheva, S.: A finite difference scheme for elliptic equations with rough coefficients using a Cartesian grid nonconforming to interfaces, SIAM J. on Numerical Analysis, **36** (1999), No. 2, 442–464.
29. Oden, J. T, Babuška, I., and Baumann, C.E.: A discontinuous hp finite element method for diffusion problems, Journal of Com-putational Physics, **146** (1998), 491–519.
30. Plaks, A., Tsukerman, I., Painchaud, S., and Tabarovsky, L.: Multigrid methods for open boundary problems in geophysics, IEEE Trans. Magn., **36** (2000), No. 4, p.633–636.
31. Plaks, A., Tsukerman, I, Friedman, G., Yellen, B.: Generalized Finite Element Method for magnetized nanoparticles, to appear in IEEE Trans. Magn., May 2003.
32. Proekt, L., Tsukerman, I.: Method of overlapping patches for electromagnetic computation, IEEE Trans. Magn., **38** (2002), No. 2, 741–744.
33. Sagui, C. and Darden, T.A.: Molecular dynamics simulations of biomolecules: long-range electrostatic effects, Annu. Rev. Bio-phys. Biomol. Struct. **28** (1999), 155–79.

34. Sagui, C. and Darden, T.: Multigrid methods for classical molecular dynamics simulations of biomolecules, Journal of Chemical Physics, **114** (2001), No. 15.
35. Schuhmann, R. and Weiland, T.: A stable interpolation technique for FDTD on non-orthogonal grids, International Journal of Numerical Modelling: Electronic Networks, Devices and Fields, **11** (1998), 299–306.
36. Schuhmann, R. and Weiland, T.: Recent advances in finite integration technique for high frequency applications, invited paper, Proceedings of SCEE-2002, Eindhoven, June 2002.
37. Soh, A.K., Long, Z.F.: Development of two-dimensional elements with a central circular hole, Comput. Methods Appl. Mech. Engrg., **188** (2000), 431–440.
38. Strang, G., Variational crimes in the ?nite element method, in: The Mathematical Foundations of the Finite Element Method with Applications to Partial Differential Equations, A.R. Aziz, ed., New York: Academic Press, 1972, 689–710.
39. Strouboulis T., Babuška, I. Copps, K.L.: The design and analysis of the Generalized Finite Element Method, Computer Methods in Applied Mechanics and Engineering, **181**, (2000), No. 1–3, 43-69.
40. Tarhasaari, T, Kettunen, L, Bossavit, A.: Some realizations of a discrete Hodge operator: A reinterpretation of finite element techniques, IEEE Trans. Magn. **35**: (1999) No. 3, 1494-1497.
41. Tonti, E.: Finite formulation of electromagnetic field, IEEE Trans. Magn. **38** (2002), No. 2, 333–336.
42. Tsukerman, I.: Spurious solutions, paradoxes and misconceptions in computational electromagnetics, to appear in IEEE Trans. Magn., May 2003.
43. Tsukerman, I.: Finite Element Difference schemes for electro- and magnetostatics, Proceedings of Compumag'2003, Saratoga Springs.
44. Wiegmann, A., and Bube, K.P., The explicit-jump immersed interface method: Finite difference methods for PDEs with piece-wise smooth solutions, SIAM J. Numer. Analysis **37** (2000), No. 3, 827–862.
45. http://www.fdtd.org/
46. Yu, W. and Mittra, R.: A conformal finite difference time domain technique for modeling curved dielectric surfaces, IEEE Mi-crowave Wireless Comp. Lett., **11** (2001), 25–27.

# Electrical RLC Networks and Semiconductor Devices

Giuseppe Alì*

IAC-CNR, via P. Castellino 111, I-80131 Naples, Italy

**Abstract.** In refined network analysis, a compact network model is combined with drift-diffusion models for the semiconductor devices which are part of the network, in a multiphysics approach. For linear RLC networks containing diodes as distributed devices, we construct a mathematical model that combines the differential-algebraic network equations of the circuit with elliptic boundary value problems modelling the diodes. For this mixed initial-boundary value problem of partial differential-algebraic equations a first existence result is given, based on a non-standard application of Schauder's fixed point theorem.

## 1 Introduction

In semiconductor technology, each device is part of a more complex electrical circuit. The interaction between the single device and the whole circuit can be neglected whenever the scale of the device is larger than the scale of interconnects. As spatial dimensions of semiconductor technology shrink steadily, accurate models of device/interconnect interactions are needed [1,3].

In this article, we concentrate on RLC networks with bipolar semiconductor devices, such as diodes. Due to the very different time scales related to the relaxation of diodes to equilibrium and to the electric current in the network, it is appropriate to model the devices by stationary drift-diffusion equations [7,8]. For the sake of simplicity, we derive appropriate coupling conditions for one-dimensional diodes and linear RLC networks, set up by Modified Nodal Analysis (MNA) [5,9]. This multiphysics approach yields a coupled system of (elliptic) partial differential equations (PDEs) and differential-algebraic equations (DAEs), for short, a system of partial differential-algebraic equations (PDAEs): the node potentials of the network define boundary conditions for the diode model and this causes each diode to produce a current flow, such that each diode acts as a voltage-defined current source for the electric network.

This paper is mainly devoted to a short presentation of an existence result for this PDAE. Further details can be found in [2]. The issue of uniqueness is more delicate and will be addressed in a forthcoming paper. We recall that non-uniqueness is a general feature of drift-diffusion equations, for input data far from equilibrium (see [6–8,10] and the references therein). Nevertheless,

* e-mail: ali@iam.na.cnr.it

from a physical point of view, we expect uniqueness when the circuit equations enter into play. At the moment it does not seem conceivable to prove such a result for general data, but only when the boundary conditions for the diodes remain close to equilibrium, following the analysis of the stationary drift-diffusion model [7].

To the author's knowledge, a mathematical analysis of coupled systems of DAEs and PDEs has been performed only for linear PDEs. Specifically, an existence result has been obtained for an RLC networks containing uniform lossy transmission lines [3,4], which are described by the telegrapher's equations. This model leads to a system of *linear* hyperbolic PDAEs.

The paper is organized as follows. In Section 2 we present the model, which will be analyzed in the subsequent section. Finally, we draw some conclusions and discuss several open problems.

## 2 Network Models for Electrical Circuits

### 2.1 Electrical RLC Networks

An RLC network is an electrical circuit whose basic components are resistors, inductors and capacitors. In addition, we assume that the network contains semiconductor devices. Using Modified Nodal Analysis (MNA), such an RLC network is described by certain given capacitance, inductance and conductance matrices, $\tilde{\mathbf{C}} \in \mathbb{R}^{n_C \times n_C}$, $\tilde{\mathbf{L}} \in \mathbb{R}^{n_L \times n_L}$, and $\tilde{\mathbf{G}} \in \mathbb{R}^{n_G \times n_G}$, which are positive-definite and symmetric. The indices $n_C$, $n_L$ and $n_G$ denote the number of capacitors, inductors and resistors, respectively.

If the network has $n$ nodes, the unknowns to be determined are the node potentials $\mathbf{u} \in \mathbb{R}^n$, the currents through inductive branches $\jmath_L \in \mathbb{R}^{n_L}$, and the currents through branches with voltage sources $\jmath_V \in \mathbb{R}^{n_V}$. The input data of the network are the independent voltage sources $\boldsymbol{v} \in \mathbb{R}^{n_V}$ and the independent current sources $\imath \in \mathbb{R}^{n_I}$. The currents through the Ohmic contacts of the semiconductor devices, which depend on the applied potentials, are $\boldsymbol{\lambda} \in \mathbb{R}^{2n_d}$, where $n_d$ is the number of devices. Finally, the network topology is described by the incidence matrices $\mathbf{A}_C \in \mathbb{R}^{n \times n_C}$, $\mathbf{A}_L \in \mathbb{R}^{n \times n_L}$, $\mathbf{A}_R \in \mathbb{R}^{n \times n_G}$, $\mathbf{A}_V \in \mathbb{R}^{n \times n_V}$, $\mathbf{A}_I \in \mathbb{R}^{n \times n_I}$, and $\mathbf{A}_\lambda \in \mathbb{R}^{n \times 2n_d}$, for capacitors, inductors, resistors, voltage sources, current sources and Ohmic contacts of devices, respectively.

Using Kirchhoff's laws, it is possible to derive the network equations for the circuit part,

$$
\begin{aligned}
&\mathbf{A}_C \tilde{\mathbf{C}} \mathbf{A}_C^\top \frac{d\mathbf{u}}{dt} + \mathbf{A}_R \tilde{\mathbf{G}} \mathbf{A}_R^\top \mathbf{u} + \mathbf{A}_L \jmath_L + \mathbf{A}_V \jmath_V + \mathbf{A}_\lambda \boldsymbol{\lambda} + \mathbf{A}_I \imath(t) = 0, \\
&\tilde{\mathbf{L}} \frac{d\jmath_L}{dt} - \mathbf{A}_L^\top \mathbf{u} = 0, \\
&-\mathbf{A}_V^\top \mathbf{u} + \boldsymbol{v}(t) = 0.
\end{aligned}
\tag{1}
$$

The unknowns of the RLC network equations (1) can be decomposed into differential and algebraic components, after introducing the projector $\mathbf{Q}_C$ onto the kernel of $\mathbf{A}_C^\top$, and the orthogonal projector $\mathbf{P}_C = \mathbf{I} - \mathbf{Q}_C$. Then, the differential component is given by $\mathbf{y} \equiv (\mathbf{y}_1, \mathbf{y}_2)^\top \stackrel{\text{def}}{=} (\mathbf{P}_C\mathbf{u}, \boldsymbol{\jmath}_L)^\top$, and the algebraic component by $\mathbf{z} \equiv (\mathbf{z}_1, \mathbf{z}_2)^\top \stackrel{\text{def}}{=} (\mathbf{Q}_C\mathbf{u}, \boldsymbol{\jmath}_V)^\top$. The RLC network equations can be rewritten as

$$
\begin{aligned}
&\mathbf{A}\frac{d\mathbf{y}}{dt} + \mathbf{B}_P\mathbf{y} + \mathbf{C}_P\mathbf{z} + \mathbf{F}_P(\boldsymbol{\lambda}, \boldsymbol{\imath}(t)) = 0, \\
&\mathbf{B}_Q\mathbf{z} + \mathbf{C}_Q\mathbf{y} + \mathbf{F}_Q(\boldsymbol{\lambda}, \boldsymbol{\imath}(t), \boldsymbol{v}(t)) = 0,
\end{aligned}
\tag{2}
$$

where

$$
\mathbf{A} = \begin{pmatrix} \tilde{\mathbf{H}} & 0 \\ 0 & \tilde{\mathbf{L}} \end{pmatrix}, \qquad \mathbf{B}_Q = \begin{pmatrix} \mathbf{Q}_C^\top\mathbf{A}_R\tilde{\mathbf{G}}\mathbf{A}_R^\top\mathbf{Q}_C & \mathbf{Q}_C^\top\mathbf{A}_V \\ \mathbf{A}_V^\top\mathbf{Q}_C & 0 \end{pmatrix},
$$

$$
\mathbf{F}_P = \begin{pmatrix} \mathbf{P}_C^\top\mathbf{A}_\lambda\boldsymbol{\lambda} + \mathbf{P}_C^\top\mathbf{A}_I\boldsymbol{\imath} \\ 0 \end{pmatrix}, \qquad \mathbf{F}_Q = \begin{pmatrix} \mathbf{Q}_C^\top\mathbf{A}_\lambda\boldsymbol{\lambda} + \mathbf{Q}_C^\top\mathbf{A}_I\boldsymbol{\imath} \\ -\boldsymbol{v} \end{pmatrix}.
$$

The matrix $\mathbf{A}$ is positive-definite and symmetric, since both $\tilde{\mathbf{H}} = \mathbf{A}_C\tilde{\mathbf{C}}\mathbf{A}_C^\top + \mathbf{Q}_C^\top\mathbf{Q}_C$ and $\tilde{\mathbf{L}}$ are positive-definite, symmetric matrices. The system (2) is supplemented with initial data $\mathbf{y}(t_0) = \mathbf{y}_0$.

The topological conditions

$$
\ker(\mathbf{A}_C, \mathbf{A}_R, \mathbf{A}_V)^\top = \{0\}, \tag{3}
$$

$$
\ker \mathbf{Q}_C^\top\mathbf{A}_V = \{0\} \tag{4}
$$

$$
\ker \mathbf{P}_C^\top\mathbf{A}_\lambda = \{0\}, \qquad (\mathbf{Q}_C^\top\mathbf{A}_\lambda = 0). \tag{5}
$$

imply that the matrix $\mathbf{B}_Q$ is invertible, and $\mathbf{z}$ can be expressed as a function of $\mathbf{y}$, so that the system (2) has algebraic index 1. The topological conditions (3)–(5) are physically reasonable. In particular, condition (5) requires that any device's terminal is connected to ground through a path of capacitors.

## 2.2 Semiconductor devices and electrical networks

In this subsection we deal with the device equations and with their coupling to the network equations. We assume that an electrical circuit contains $n_d$ bipolar devices. We model the $a$-th device as a segment of length $l$, characterized by a doping profile $N_a(x)$, $x \in (0, l)$. Then, each device is described by the drift-diffusion equations

$$
\begin{aligned}
&-\partial_x\left(q\mu_n n_a \partial_x \phi_{n,a}\right) = qR_a, \\
&-\partial_x\left(q\mu_p p_a \partial_x \phi_{p,a}\right) = -qR_a, \\
&-\partial_x\left(\epsilon\partial_x V_a\right) = q(N_a + p_a - n_a),
\end{aligned}
\tag{6}
$$

where $\phi_{n,a}$, $\phi_{p,a}$ are the quasi-Fermi potentials, $V_a$ is the electric potential, and $n_a, p_a$ are the electron and hole densities. The electron and hole mobilities

$\mu_n$, $\mu_p$ are assumed to be given functions of the densities, the electric field, and the space variable. The generation-recombination term, $R_a$, is modelled by

$$R_a(n,p,V) = F_a(n,p,V)\left(\frac{np}{n_i^2} - 1\right).$$

This expression includes both Shockley-Read-Hall and Auger generation-recombination terms [6–8,10].

In (6), the densities $n_a$ and $p_a$ will be expressed in terms of the quasi-Fermi potentials $\phi_{n,a}$, $\phi_{p,a}$, and of the electric potentials $V_a$, by

$$n_a = n_i \exp\left(\frac{V_a - \phi_{n,a}}{U_T}\right), \qquad p_a = n_i \exp\left(-\frac{V_a - \phi_{p,a}}{U_T}\right),$$

where $U_T = k_B T/q$ is the thermal potential. In particular, the generation-recombination term becomes

$$R_a = \tilde{R}_a(\phi_n, \phi_p, V) = \tilde{F}_a(\phi_n, \phi_p, V)\left(\exp\left(\frac{\phi_p - \phi_n}{U_T}\right) - 1\right).$$

The end points of each device are Ohmic contacts, for which the following conditions have to be imposed [7,8],

$$\phi_{n,a}(0,t) \equiv \phi_{p,a}(0,t) = u_{1,a}(t), \qquad V_a(0,t) = V_{\mathrm{bi},a}(0) + u_{1,a}(t), \tag{7}$$

$$\phi_{n,a}(l,t) \equiv \phi_{p,a}(l,t) = u_{2,a}(t), \qquad V_a(l,t) = V_{\mathrm{bi},a}(l) + u_{2,a}(t), \tag{8}$$

where $u_{1,a}$, $u_{2,a}$ are the applied potentials, and the built-in potential $V_{\mathrm{bi},a}$ is defined by

$$V_{\mathrm{bi},a} = U_T \log\left(\frac{N_a}{2n_i} + \sqrt{\left(\frac{N_a}{2n_i}\right)^2 + 1}\right).$$

The applied voltages $\mathbf{u}_1 = (u_{1,1}, \dots u_{1,d})^\top$ and $\mathbf{u}_2 = (u_{2,1}, \dots u_{2,d})^\top$ are part of the unknowns of the network equations. Using the incidence matrix $\mathbf{A}_\lambda$, we can write the following coupling condition to the electrical network,

$$\begin{pmatrix} \mathbf{u}_1(t) \\ \mathbf{u}_2(t) \end{pmatrix} = \mathbf{A}_\lambda^\top \mathbf{u}. \tag{9}$$

Finally, the electrical currents $\boldsymbol{\lambda}_1 = (\lambda_{1,1}, \dots, \lambda_{1,n_d})^\top$ and $\boldsymbol{\lambda}_2 = (\lambda_{2,1}, \dots, \lambda_{2,n_d})^\top$ at the Ohmic contacts are explicitly given by

$$\lambda_{1,a}(t) = j_{n,a}(0,t) + j_{p,a}(0,t), \;\; \lambda_{2,a}(t) = -j_{n,a}(l,t) - j_{p,a}(l,t), \tag{10}$$

$$j_{n,a} = -q\mu_n n_a \partial_x \phi_{n,a}, \;\; j_{p,a} = -q\mu_p p_a \partial_x \phi_{p,a}.$$

The plus and minus sign in the definition of $\lambda_{1,a}$ and $\lambda_{2,a}$ take care of the incoming directions to the device. For later reference, it is convenient to introduce the vector $\boldsymbol{\lambda} = (\boldsymbol{\lambda}_1, \boldsymbol{\lambda}_2)^\top$.

## 3 Existence of solutions

In this section, we prove an existence result for the coupled system of the network equations and the drift-diffusion equations. The main difficulty resides in the term $\boldsymbol{\lambda}$, which is a nonlinear, differential functional of $\mathbf{A}_\lambda^\top \mathbf{u}$. In general, this functional cannot be expected to be continuous, due to the lack of uniqueness of the drift-diffusion equations [7]. Therefore, the existence of a solution to the DAE describing the electrical network is not ensured.

For simplicity, we assume that the circuit contains only one device. All the results of this section can be easily extended to arbitrary $n_d$ devices.

Before stating the main result (Theorem 2), Lemma 1 collects some a priori estimates which will be needed later.

**Lemma 1.** *Let the given electric network provide symmetric, positive definite matrices $\tilde{\mathbf{C}}$, $\tilde{\mathbf{L}}$, $\tilde{\mathbf{G}}$, and let the network's incidence matrices satisfy the topological conditions (3)–(5). Furthermore, let $(\mathbf{y}, \mathbf{z}) \in (C([t_0, t_1]))^{n+n_L+n_V}$ be a solution to the corresponding network equation (2), with initial value $\mathbf{y}_0$. We assume that $\boldsymbol{\imath} \in \left(L^2([t_0, t_1])\right)^{n_I}$, $\boldsymbol{v} \in \left(L^2([t_0, t_1])\right)^{n_V}$, and that $\boldsymbol{\lambda}$ satisfies the condition*

$$(\mathbf{A}_\lambda^\top \mathbf{u})^\top \boldsymbol{\lambda} \geq 0. \tag{11}$$

*Then, for all $t \in [t_0, t_1]$, the solution satisfies the estimates*

$$|\mathbf{y}|^2(t) \leq C_y e^{c_1(t-t_0)} \left( |\mathbf{y}_0|^2 + \|\boldsymbol{\imath}\|^2_{(L^2([t_0,t_1]))^{n_I}} + \|\boldsymbol{v}\|^2_{(L^2([t_0,t_1]))^{n_V}} \right), \tag{12}$$

$$|\mathbf{z}|^2(t) \leq C_z \left( |\mathbf{y}|^2(t) + |\boldsymbol{\imath}|^2(t) + |\boldsymbol{v}|^2(t) \right), \tag{13}$$

*for some positive constants $C_y$, $C_z$ and $c_1$.*

*Proof.* The lemma follows from equation (2), after multiplying by $\mathbf{y}^\top$, using Schwarz inequality and Gronwall's lemma (see [2]). □

We remark that no assumption is made on the functional structure of $\boldsymbol{\lambda}$. It is possible to check that the natural power condition (11) is satisfied by $\boldsymbol{\lambda} = (\boldsymbol{\lambda}_1, \boldsymbol{\lambda}_2)^\top$, defined by (10).

Guided by Lemma 1, we introduce the Banach space

$$X = A \times A \times B, \quad A = C\left([t_0, t_1], L^2([0, l])\right), \quad B = C\left([t_0, t_1], \mathbb{R}^{n_y}\right),$$

and the subset $M \subset X$ containing all $(\phi_n, \phi_p, \mathbf{y}) \in X$ which satisfy for all $x \in [0, l]$, $t \in [t_0, t_1]$ the inequalities (12) and

$$u_1(t) \wedge u_2(t) \leq \phi_n(x, t), \phi_p(x, t) \leq u_1(t) \vee u_2(t), \tag{14}$$

and the coupling conditions (9), with $\mathbf{y} = (\mathbf{y}_1, \mathbf{y}_2)^\top$, $\mathbf{y}_2(t) \in \mathbb{R}^{n_L}$. In (14), the composition of two functions by the symbols $\wedge$ and $\vee$ denotes their minimum and maximum function, respectively. It is not hard to see that $M$ is a bounded, convex subset of $X$. We are ready to state the main result of this paper. The remainder of this section is devoted to its proof.

**Theorem 2.** *Existence. Let the source functions $\imath$ and $v$ be continuous, the network matrices be symmetric, positive definite and the topological conditions (3)–(5) be fulfilled. Then there exists a solution in $M$ to the coupled system (2), (6), (7), (8), (9).*

*Proof.* The proof is based on Schauder's fixed point theorem and requires the next Lemmas 3 and 4. In order to define an iteration operator $T$ which maps $M$ into itself, we consider a triplet $(\phi_n', \phi_p', \mathbf{y}') \in M$. The operator will be defined in two steps. First we consider the following problem,

$$\begin{aligned}
&\frac{\epsilon}{q}\partial_x^2 V = n(V, \phi_n') - p(V, \phi_p') - N, \\
&V(0,\cdot) = V_{\mathrm{bi}}(0) + u_1'(\cdot), \qquad V(l,\cdot) = V_{\mathrm{bi}}(l) + u_2'(\cdot), \\
&\begin{pmatrix} u_1' \\ u_2' \end{pmatrix} = \mathbf{A}_\lambda^\top \mathbf{y}_1'.
\end{aligned} \tag{15}$$

We denote by $V'$ the solution to (15) (see Lemma 3). Next, we consider the following linearized drift-diffusion equations,

$$\begin{aligned}
&\partial_x\left(\mu_n' n' \partial_x \phi_n\right) = -\bar{F}' \cdot \left(\exp\left(\frac{\phi_p - \phi_n}{U_T}\right) - 1\right), \\
&\partial_x\left(\mu_p' p' \partial_x \phi_p\right) = \bar{F}' \cdot \left(\exp\left(\frac{\phi_p - \phi_n}{U_T}\right) - 1\right), \\
&\phi_n(0,\cdot) = \phi_p(0,\cdot) = u_1(\cdot), \qquad \phi_n(l,\cdot) = \phi_p(l,\cdot) = u_2(\cdot),
\end{aligned} \tag{16}$$

where the prime denotes evaluation at $(\phi_n', \phi_p', V')$. The system (16) is coupled to the DAE

$$\begin{aligned}
\mathbf{A}_P \frac{d\mathbf{y}}{dt} + \mathbf{B}_P \mathbf{y} + \mathbf{C}_P \mathbf{z} + \mathbf{F}_P(\imath(t), \tilde{\boldsymbol{\lambda}}) &= 0, \\
\mathbf{B}_Q \mathbf{z} + \mathbf{C}_Q \mathbf{y} + \mathbf{F}_Q(\imath(t), v(t)) &= 0, \\
\mathbf{y}(t_0) &= \mathbf{y}'(t_0),
\end{aligned} \tag{17}$$

by the coupling conditions

$$\begin{pmatrix} u_1 \\ u_2 \end{pmatrix} = \mathbf{A}_\lambda^\top \mathbf{y}_1, \qquad \tilde{\boldsymbol{\lambda}} = \begin{pmatrix} -q\mu_n' n' \partial_x \phi_n - q\mu_p' p' \partial_x \phi_p \\ q\mu_n' n' \partial_x \phi_n + q\mu_p' p' \partial_x \phi_p \end{pmatrix}. \tag{18}$$

We denote by $(\phi_n'', \phi_p'', \mathbf{y}'')$ the solution to (16)–(18) (see Lemma 4). In this way, we have constructed a map $T(\phi_n', \phi_p', \mathbf{y}') = (\phi_n'', \phi_p'', \mathbf{y}'')$. Using Lemmas 3 and 4 below, it is immediate to see that $T$ is continuous and maps $M$ into itself. In order to apply Schauder's fixed point theorem, we need to show that $T(M)$ is relatively compact in $M \subset X$. We introduce the decomposition $T(M) = M_1 \times M_2$, with

$$\begin{aligned}
M_1 &= \left\{(\phi_n'', \phi_p'') \in C\left([t_0, t_1], (L^2([0,l]))^2\right) \,\middle|\, (\phi_n'', \phi_p'', \mathbf{y}'') \in T(M)\right\}, \\
M_2 &= \left\{\mathbf{y}'' \in C\left([t_0, t_1], \mathbb{R}^{n_y}\right) \mid (\phi_n'', \phi_p'', \mathbf{y}'') \in T(M)\right\}.
\end{aligned}$$

Recalling the estimate (12), and using the network equations to express $\frac{d\mathbf{y}''}{dt}$ in term of $\mathbf{y}''$, it is immediate to see that $\mathbf{y}''$ belongs to a bounded subset of $C^1([t_0,t_1],\mathbb{R}^{n_y})$. Therefore, $M_2$ is relatively compact in $C([t_0,t_1],\mathbb{R}^{n_y})$.

Next, the Rellich-Kondrachov theorem implies that the set

$$\left\{(\psi_n,\psi_p)\in\left(L^2([0,l])\right)^2 \,\middle|\, (\psi_n,\psi_p)(x)=\mathbf{w}(x,t),\ \mathbf{w}\in M_1\right\}$$

is relatively compact in $L^2([0,l])^2$, for all $t\in[t_0,t_1]$. Since $(\phi_n'',\phi_p'')$ is Lipschitzian with respect to $(\mathbf{u}_1,\mathbf{u}_2)$, for any $t\in[t_0,t_1]$ and for any $\epsilon>0$ there exist $\delta(\epsilon,t)>0$ such that

$$\max_{[t_0,t_1]}\|\mathbf{w}(\cdot,t)-\mathbf{w}(\cdot,s)\|_{L^2}<\epsilon$$

for all $s\in[t_0,t_1]$ with $|s-t|<\delta(\epsilon,t)$. Then, by invoking Arzelà-Ascoli theorem, $M_1$ is relatively compact in $C\left([t_0,t_1],L^2([0,l])^2\right)$.

Finally, after applying Tichonov's theorem, we can conclude that $T(M)=M_1\times M_2$ is relatively compact in $X$, and Schauder's theorem implies the existence of a fixed point of the map T. □

As we have mentioned, the proof of Theorem 2 resides on the following two lemmas.

**Lemma 3.** *If $\phi_n'$ and $\phi_p'$ satisfy the inequality (14), the problem (15) has for any time $t\in[t_0,t_1]$ a unique solution $V=V'(\cdot,t)$ in $H^1([0,l])$. This solution satisfies the estimate*

$$\inf_{[0,l]}V_{\text{bi}}+u_1(t)\wedge u_2(t)\le V'(x,t)\le\sup_{[0,l]}V_{\text{bi}}+u_1(t)\vee u_2(t). \tag{19}$$

The proof of this lemma can be found in [7] and [8].

**Lemma 4.** *The problem (16)–(18) has a unique solution*

$$(\phi_n,\phi_p,\mathbf{y})=(\phi_n''(x,\cdot),\phi_p''(x,\cdot),\mathbf{y}''(\cdot))$$

*in $C\left([t_0,t_1],H^1([0,l])\right)^2\times C([t_0,t_1],\mathbb{R}^{n_y})$. Moreover, the solution satisfies the estimates (12), (14) and, therefore, belongs to $M$.*

*Proof.* We give a sketch of the proof. The details can be found in [2]. First, it is possible to see that the quasilinear problem (16) is solvable for any boundary data $u_1$, $u_2$, and that any solution satisfies (14). This statement can be assessed by a standard application of Schauder's fixed point theorem, following the guideline of Lemma 3. Next, it can be shown that both the solution and the current $\tilde{\boldsymbol{\lambda}}$ are Lipschitz continuous with respect to the boundary data. The Lipschitz continuity of the solution implies uniqueness and well-posedness of the problem (16). Then, the solution is continuous with respect to time. The Lipschitz continuity of $\tilde{\boldsymbol{\lambda}}$ implies that the problem (17) has a unique solution, as implied by Picard-Lindelöf theorem. Finally, recalling Lemma 1, the solution satisfies (12), since it is possible to verify the power condition $(\mathbf{A}_\lambda^\top\mathbf{u})^\top\tilde{\boldsymbol{\lambda}}\ge 0$. □

## 4 Open problems

We have proposed an existence result for a system of elliptic PDEs, describing a set of semiconductor devices, nonlinearly coupled to a system of DAEs, describing a linear RLC network. The proof is based on a fixed point map, defined in an appropriate function space. In order to translate the idea of the proof into an efficient numerical scheme, we need to define a contractive iteration map, rather than a generic fixed point map. Therefore, the issue of uniqueness of the solution becomes particularly relevant, and it is strictly related to the stability of any scheme based on such ideas. This is still an open problem. A second problem concerns the extension of the existence proof to evolutionary drift-diffusion and, possibly, hydrodynamical models for semiconductor devices, involving time derivatives too. A third problem is related to the inclusion of thermal effects. Finally, we plan to extend the theory presented in these pages to multidimensional devices (rather than the strip model used here) coupled to circuits, and to similar coupling problems arising when parts of a complex system are replaced by spatially lower dimensional averaged models (system simulation, mixed-level simulation, etc.).

## References


1. SIA Semiconductor Industry Association, The National Technology Roadmap for Semiconductors, 1997, (http://www.oslo.sintef.no/fora/bilder/roadmap5.pdf)
2. G. Alì, A. Bartel, M. Günther, C. Tischendorf: Elliptic Partial Differential-Algebraic Multiphysics Models in Electrical Network Design. Universität Karlsruhe - IWRMM Preprint Nr. 02/05 (2002)
3. M. Günther: Partielle differential-algebraische Systeme in der numerischen Zeitbereichsanalyse elektrischer Schaltungen. (VDI, 2001)
4. M. Günther: A PDAE model for interconnected linear RLC networks. Mathematical and Computer Modelling of Dynamical Systems **7** No. 2 (2001) 189–203
5. C. W. Ho, A. E. Ruehli, P. A.Brennan: The modified nodal approach to network analysis. IEEE Trans. Circuits and Systems, CAS **22** (1975) 505–509
6. J. W. Jerome: Analysis of charge transport. A mathematical study of semiconductor devices. (Springer, 1995)
7. P. A. Markowich: The Stationary Semiconductor Device Equations. (Springer, 1986)
8. P. A. Markowich, C. A. Ringhofer and C. Schmeiser: Semiconductor Equations. (Springer, 1990)
9. W. J. McCalla: Fundamentals of Computer-Aided Circuit Simulation. (Kluwer Academic Publisher, 1988)
10. S. Selberherr: Analysis and Simulation of Semiconductor Devices. (Springer, 1984)

# Modified Extended BDF Time-Integration Methods, Applied to Circuit Equations

Sandra Allaart-Bruin[1], Jan ter Maten[2], and Sjoerd Verduyn Lunel[3]

[1] Eindhoven University of Technology, sbruin@win.tue.nl
[2] Eindhoven University of Technology, Philips Research Laboratories
[3] Leiden University

**Abstract.** Electric circuits designers are frequently interested in the transient behaviour of the designed circuit. A common method for time integration of the Differential Algebraic circuit Equations (DAE) is the Backward Differentiation Formula (BDF) method. In 1983, J. Cash proposed the Modified Extended BDF (MEBDF) method, which combines better stability properties and higher order of convergence than BDF, but requires more computations per step. We prove reduction of convergence order for MEBDF when applied to DAE's with higher DAE-index. However, because in practice, in circuit analysis, the DAE-index does not exceed 2, the reduction is quite moderate and it equals the BDF-order in that case. One gains better, or even unconditional, stability. One also obtains consistent solutions.

## 1 Introduction

In Circuit Design, Transient Analysis is most heavily used. There is constant interest in methods that offer better performance with respect to robustness as well as to reduction of CPU-time. Because the underlying circuit equations are Differential-Algebraic Equations (DAE), a point for robustness is how well the time integrator behaves when problems of higher DAE-index have to be treated: does reduction of order of convergence happen, and does one obtain consistent solutions. Other points of interest are stability conditions (see [8]), and the damping behavior along the imaginary axis.
Backward Differentiation Formula (BDF) Methods do not suffer from reduction of order of convergence and generate consistent solutions [1,9,12], which made them very popular for circuit simulation, however at the cost of being conditionally stable when the order exceeds 2. Improvements were looked for in combining BDF with Trapezoidal Rule [10] (less damping), or in new methods, like Implicit Runge-Kutta methods like Radau-methods [11] (3rd order L-stable Implicit Runge-Kutta methods with options for parallelism), or CHORAL (embedded method of order (2)3, stiffly accurate and L-stable)[7].
In this paper, we focus on Modified Extended BDF (MEBDF) methods [3] and we report on newly proven convergence order reduction when applied to DAE's of higher index and on the generation of consistent solutions. The MEBDF-methods offer better stability conditions than BDF. For instance, when applied to index-1 DAE's, the 3-step method has order 4 and is A-stable (see [9]). When applied to index-2 DAE's, the order of convergence reduces

to 3, but the method's stability property remains, which compares favorably to the 3-step BDF method. The MEBDF-methods easily fit a datastructure that was designed for an implementation of BDF-methods. Variants also allow parallelism [6].
We start with an explanation of the MEBDF method. After that we formulate two new theorems on the order of the MEBDF method. We continue with some numerical results of an example circuit and present results for a general test problem.

## 2 Modified Extended BDF

One timestep with the MEBDF method consists of three BDF steps and an evaluation step. This results in more work compared to BDF, but the order of convergence increases with one for most circuits. This implies that for convergence order 3 we normally apply the 3-step BDF method, while with the MEBDF method a 2-step method suffices. Well-posed circuits lead to an index-1 DAE. For this type of DAE's the k-step MEBDF has order $k+1$, while the k-step BDF has order $k$ (see Section 3 and [12]). Checking the circuit topology enables one roughly to distinguish between index-1 and index-2 problems [12]. Circuits that contain controlled elements require a much deeper analysis [5]. Several numerical methods suffer from convergence order reduction when applied to DAE's with higher index. For this reason, one is interested in time integration methods of which the order of convergence is robust with respect to the DAE-index quantity. A second property that one likes to see fulfilled is that the numerical solution satisfies the consistency property for DAE's: if one starts on the manifold defined by the algebraic constraints one wants to stay on it. An appreciated property would be that the method is able to generate a consistent solution even if one starts from an inconsistent initial value.
We consider the following quasi-linear DAE

$$A\mathbf{x}'(t) + \mathbf{g}(\mathbf{x}(t), t) = \mathbf{0}, \tag{1}$$

where $\mathbf{x}(t) \in D \subseteq \mathbb{R}^m, t \in I \subseteq \mathbb{R}$ and $A$ is a constant $m \times m$ matrix.
MEBDF is defined by the following steps for integrating one timestep[1].

- **First BDF step**: Solve for $\overline{\mathbf{x}}_\ell^{(\ell-k)}$[2] at $t_\ell$ by the $k$-step BDF-method

$$A \sum_{i=0}^{k} \hat{\alpha}_i \mathbf{x}_{\ell-i} = -h\hat{\beta}_k \mathbf{g}(\mathbf{x}_\ell, t_\ell).$$

[1] For occurring coefficients $\hat{\alpha}_0, \ldots \hat{\alpha}_k, \hat{\beta}_k, \alpha_0, \ldots, \alpha_k, \beta_k, \beta_{k+1}$ see [2,3]

[2] $\cdot_\ell$, is to denote that it is an approximation on time $t_\ell$; $\bar{\cdot}$ is to denote that it is a BDF solution; $\cdot^{(\ell-k)}$ is to denote that it is computed in the $(\ell-k)$-th step.

- **Second BDF step**: Compute an "extrapolated" solution at $t_{\ell+1}$ by BDF

$$A\sum_{i=0}^{k}\hat{\alpha}_i\mathbf{x}_{\ell+1-i} = -h\hat{\beta}_k\mathbf{g}(\mathbf{x}_{\ell+1},t_{\ell+1}),$$

where $\mathbf{x}_\ell = \overline{\mathbf{x}}_\ell^{(\ell-k)}$. This solution is called $\overline{\mathbf{x}}_{\ell+1}^{(\ell-k)}$.
- **Evaluation step**: Compute

$$\overline{\mathbf{g}}_\ell = \mathbf{g}(\overline{\mathbf{x}}_\ell^{(\ell-k)},t_\ell), \qquad \overline{\mathbf{g}}_{\ell+1} = \mathbf{g}(\overline{\mathbf{x}}_{\ell+1}^{(\ell-k)},t_{\ell+1}).$$

- **MEBDF step**: Solve for a new solution $\mathbf{x}_\ell^{(\ell-k)}$ at $t_\ell$:

$$A\sum_{i=0}^{k}\alpha_i\mathbf{x}_{\ell-i} = -h\hat{\beta}_k\mathbf{g}(\mathbf{x}_\ell,t_\ell) - h(\beta_k-\hat{\beta}_k)\overline{\mathbf{g}}_\ell - h\beta_{k+1}\overline{\mathbf{g}}_{\ell+1}.$$

For solving all equations we use a quasi-Newton method. This means that we don't compute a new Jacobian every iteration, but apply the same Jacobian as for the $k$-step BDF during the several iteration steps.

## 3 Theoretical Results

In this section we prove new results on the order of the MEBDF applied to DAE's. Here we restrict ourselves to a constant stepsize $h$, i.e. $t_{\ell+1} = t_\ell + h$ for all $\ell$. For $\ell \geq k$, the numerical solution $\{x_\ell\}$ of the $k$-step MEBDF applied to (1), satisfies

$$A\frac{1}{h}\sum_{i=0}^{k}\alpha_i\mathbf{x}_{\ell-i} + \hat{\beta}_k\mathbf{g}(\mathbf{x}_\ell,t_\ell) + (\beta_k-\hat{\beta}_k)\overline{\mathbf{g}}_\ell + \beta_{k+1}\overline{\mathbf{g}}_{\ell+1} = \delta_\ell. \tag{2}$$

Here, $\delta_\ell$ describes the perturbations in the $\ell$-th step for $\ell \geq k$, which are introduced by numerical computations including the errors arising from solving the nonlinear equations. Moreover, we introduce

$$\begin{aligned}
\mathbf{x}_*(t_\ell) &:= \text{exact solution of (1)}\\
\tilde{\mathbf{x}}_\ell &:= \mathbf{x}_\ell - \mathbf{x}_*(t_\ell), \quad \ell \geq 0,\\
\tau_\ell &:= A\frac{1}{h}\sum_{i=0}^{k}\alpha_i\mathbf{x}_*(t_{\ell-i}) + \beta_k\mathbf{g}(\mathbf{x}_*(t_\ell),t_\ell) + \beta_{k+1}\mathbf{g}(\mathbf{x}_*(t_{\ell+1}),t_{\ell+1}), \quad \ell \geq k.
\end{aligned}$$

Here $\tau_\ell$ represents the local discretization error of the $k$-step MEBDF in the $\ell$-th step ($\ell \geq k$). Now (2) may be written as

$$A\frac{1}{h}\sum_{i=0}^{k}\alpha_i\tilde{\mathbf{x}}_{\ell-i} + B_\ell\tilde{\mathbf{x}}_\ell + \mathbf{h}(\tilde{\mathbf{x}}_\ell,t_\ell) + \tau_\ell - \delta_\ell - \nu_\ell - \tilde{\nu}_\ell = \mathbf{0}, \tag{3}$$

where

$$\begin{aligned} B_\ell &:= \hat{\beta}_k \mathbf{g}_\mathbf{x}(\mathbf{x}_*(t_\ell), t_\ell), \\ \mathbf{h}(\mathbf{x}, t) &:= \hat{\beta}_k \left[\mathbf{g}(\mathbf{x} + \mathbf{x}_*(t), t) - \mathbf{g}(\mathbf{x}_*(t), t) - \mathbf{g}_\mathbf{x}(\mathbf{x}_*(t), t)\mathbf{x}\right], \\ \nu_\ell &:= \left(\beta_k - \hat{\beta}_k\right)\left[\mathbf{g}(\mathbf{x}_*(t_\ell), t_\ell) - \overline{\mathbf{g}}_\ell\right], \\ \tilde{\nu}_\ell &:= \beta_{k+1}\left[\mathbf{g}(\mathbf{x}_*(t_{\ell+1}), t_{\ell+1}) - \overline{\mathbf{g}}_{\ell+1}\right]. \end{aligned}$$

### 3.1 Index-1 DAE

In this section, we assume that (1) is an index-1 DAE. This implies that the pair $\{A, B_\ell\}$ is index-1 tractable [12]. Let $Q$ be a projector onto Ker$[A]$ and $P := I - Q$. We define $G_{1\ell} := A + B_\ell Q$ and $\hat{\delta}_\ell$ by

$$\hat{\delta}_\ell := \begin{cases} QG_{1\ell}^{-1}(\delta_\ell + \nu_\ell + \tilde{\nu}_\ell), & \ell \geq k \\ QG_{1\ell}^{-1}\mathbf{h}(\tilde{\mathbf{x}}_\ell, t_\ell) + Q\tilde{\mathbf{x}}_\ell + QG_{1\ell}^{-1}B_\ell P\tilde{\mathbf{x}}_\ell, & \ell < k \end{cases}$$

$\hat{\delta}_\ell$ represents the defect in the algebraic part for $\ell \geq k$. For the starting values, the corresponding defects are also described by $\hat{\delta}_\ell$.

**Theorem 1.** *Suppose a constant $C_1 > 0$ exists such that the starting values satisfy the relation*

$$\|P\mathbf{x}_\ell - P\mathbf{x}_*(t_\ell)\| \leq C_1 h^{k+1}, \quad \ell < k, \tag{4}$$

*and suppose constants $C_2$ and $C_3$ exist such that*

$$\|\delta_\ell\| \leq C_2 h^k, \quad \|\hat{\delta}_\ell\| \leq C_3 h^{k+1}, \quad \ell \geq k, \tag{5}$$

*Then a constant $C > 0$ exists such that*

$$\|\mathbf{x}_*(t_\ell) - \mathbf{x}_\ell\| \leq C h^{k+1}, \quad \ell \geq 0. \tag{6}$$

**Outline of the proof** (for full details see [2]): We use the projectors $P$ and $Q$ to split the DAE into an algebraic part and a differential part. Next, we show the existence of a solution of the algebraic part. We use the inverse function theorem to show that this solution is locally unique. Substituting this solution into the differential part of the DAE results in an equation involving only index-1 variables. This equation can also be solved uniquely with the inverse function theorem. By applying the assumptions of the theorem we get the final result.

**Remarks.** Most well-posed circuits lead to an index-1 DAE [12]. For this type of DAE we conclude that the $k$-step MEBDF has order $k$+1 (cf. (6)), while the $k$-step BDF has order $k$. Note that the $k$-step MEBDF-methods are A-stable for $k \leq 3$, while for BDF this is restricted to the case $k \leq 2$ [3].

### 3.2 Index-2 DAE

Next, we assume that (1) is an index-2 DAE. Let $Q_{1\ell}$ be a projector onto $\mathrm{Ker}[G_{1\ell}]$ and $P_{1\ell} = I - Q_{1\ell}$.

**Theorem 2.** *Suppose a constant $C_1 > 0$ exists such that the starting values satisfy the relation*

$$\|PP_{1\ell}\mathbf{x}_\ell - PP_{1\ell}\mathbf{x}_*(t_\ell)\| \leq C_1 h^{k+1}, \quad \ell < k$$

*and suppose constants $C_2$ and $C_3$ exits such that*

$$\|\delta_\ell\| \leq C_2 h^k, \qquad \ell \geq k, \tag{7}$$

$$\|\hat{\delta}_\ell\| \leq C_3 h^{k+1}, \quad \ell \geq 0. \tag{8}$$

*Then a constant $C > 0$ exists such that*

$$\|\mathbf{x}_*(t_\ell) - \mathbf{x}_\ell\| \leq Ch^k, \qquad \ell \geq k. \tag{9}$$

**Outline of the proof** (again, for full details see [2]). The structure of the proof of this theorem is similar to the proof of the index-1 theorem, but more lengthly. With the use of projectors, the DAE is split into an algebraic, an index-1 and an index-2 part. We first solve the algebraic equations and use the solution for solving the other parts [2].

**Remarks.** Circuits which contain L-I cutsets or C-V loops lead to an index-2 DAE [12]. For this case, we proved that order reduction for MEBDF occurs (which does not happen for BDF [12]). However the reduction for MEBDF is moderate: the $k$-step MEBDF and the $k$-step BDF, applied to index-2 DAE's, have the same order, namely $k$. However, MEBDF has much better stability properties than BDF [3].

## 4 Numerical Tests

We performed some tests in Matlab with a variable step size fixed order BDF and MEBDF method. We made Work-Precision Diagrams, to compare the work needed for both methods. For more details we refer to [2].

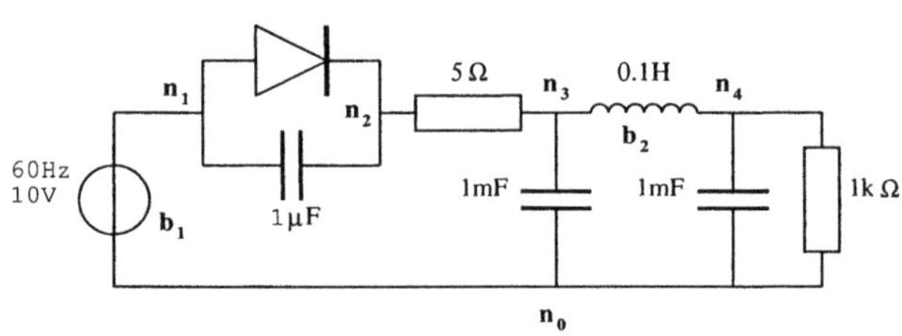


**Fig. 1.** Rectifier Circuit

### 4.1 Rectifier Circuit

Fig. 1 shows a rectifier circuit that serves the AC to DC conversion. It is designed in such a way that it damps the incoming sine-wave. We computed a reference solution $\mathbf{x}_{ref}$ at $t = 0.16$ using the BDF method with a very small time step. In Fig. 2(a) we see a Work-Precision Diagram in which we compare the 1-step MEBDF method with the 2-step BDF method. We see that the 1-step MEBDF method even performs slightly better than the 2-step BDF. Fig. 2(b) presents a comparison for the 3-step BDF and the 2-step MEBDF, in which the BDF method performs better. Clearly the methods are comparable in these cases, however the MEBDF has the potential of better stability.

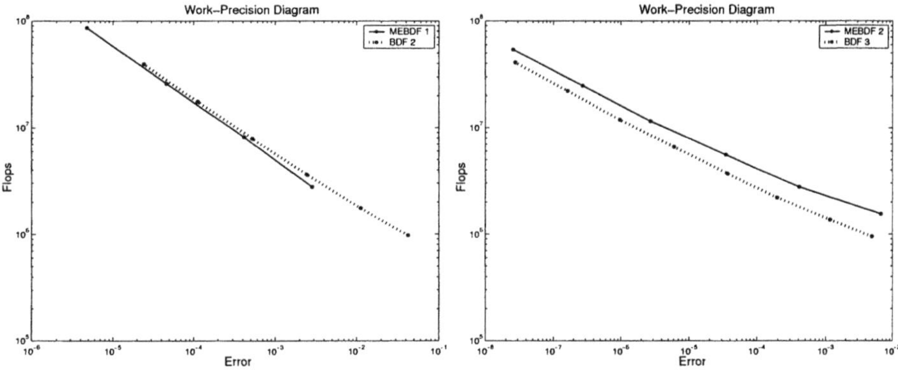

(a) (b)

**Fig. 2.** Work-Precision Diagram for second order (a) and third order (b) methods

### 4.2 Oscillatory Test Problem

To show how the MEBDF deals with oscillatory problems, we consider an ODE example problem of the form

$$x' = y + h(r)x, \tag{10}$$

$$y' = -x + h(r)y, \tag{11}$$

where $r = \sqrt{x^2 + y^2}$. We assume that $0 < x_0 < 1$, $0 < y_0 < 1$. The equations can be re-written in the following form:

$$r' = h(r)r,$$
$$\theta' = 1.$$

For $h(r) = \frac{1-r}{10}$, $r = 1$ is the only stable limit, $\lim_{t\to\infty} r(t) = 1$. We apply the BDF and MEBDF method to equations (10)-(11), which results in approximate solutions $(x_n, y_n)$ $(n = 0, 1, \ldots)$. We compute $r_n = \sqrt{x_n^2 + y_n^2}$ and plot this as function of $t$. In Fig. 3(a) we did this for the 1-step MEBDF and the 2-step BDF. We see that the MEBDF solution is less damped than the BDF solution of the same order. In Fig.3(b) we see that the 3-step BDF does not perform very well on this problem. This is due to the fact that the 3-step BDF is not A-stable and for oscillatory problems only A-stable methods perform well. Clearly, the MEBDF method is more suited for oscillatory

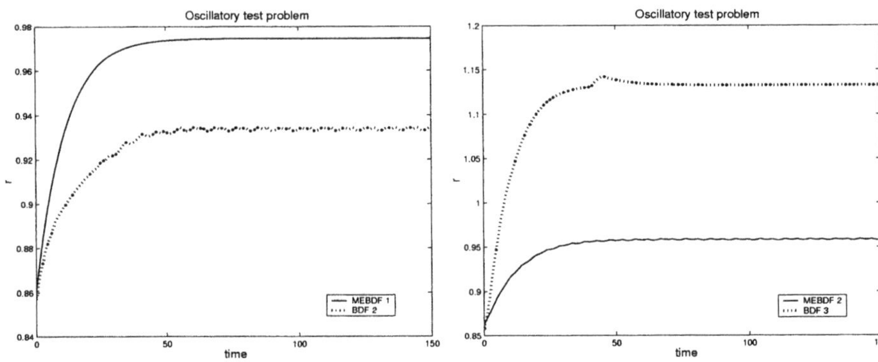


(a) (b)

**Fig. 3.** Plot of $r_n = \sqrt{x_n^2 + y_n^2}$ by second order (a) and third order (b) methods.

problems than BDF. Here one can also fully exploit the better stability and convergence order properties of MEBDF.

## 5 Consistent Solutions

BDF generates consistent solutions, even if one starts with not consistent initial values [2,12]. Because BDF serves in predicting for the extrapolation point, MEBDF inherits this property from BDF. This is in contrast to for instance solutions obtained by using the Trapezoidal Rule. This property is important when visualizing results for problems with discontinuities. It is known that computing consistent initial values from the DC operating point only requires solving an additional linear system [5]. Another way of getting consistent initial values is integrating forward, till there is no influence of the (inconsistent) initial values and then integrating backward.

## 6 Conclusion

MEBDF methods offer an alternative to BDF methods in the case of DAE's with index not exceeding 2. The MEBDF methods combine higher convergence order with better stability properties and better numerical damping:

Hence, more rigorous benchmark tests for circuits occurring in practice are worth to be done.

Finally, we gratefully acknowledge discussions, including assistance in providing implementation, with Jeff Cash (Imperial College, London), Jason Frank (CWI, Amsterdam) and Caren Tischendorf (Von Humboldt Universität, Berlin).

## References


1. Brenan, K.E., Campbell, S.L., Petzold, L.R.: Numerical solution of initial-value problems in differential-algebraic equations, SIAM - Society for Industrial and Applied Mathematics, Philadelphia, PA, USA (1996).
2. Bruin, S.M.A.: Modified Extended BDF applied to circuit equations. Philips Research Laboratories, Nat.Lab. Unclassified Report **2001/826**, MSc-Thesis Free University Amsterdam (2001).
http://www.extra.research.philips.com/publ/rep/nl-ur/ur2001-826.pdf
3. Cash, J.R.: The integration of stiff initial value problems in ODEs using Modified Extended Backward Differentiation Formulae, Comp. & Maths. with Appls, Vol **9**, No **5**, (1983) pp. 645–657.
4. Cash, J.R.: Modified extended backward differentiation formulae for the numerical solution of stiff initial value problems in ODEs and DAEs, Journal of Computational and Applied Mathematics, **125**, (2000) pp. 117–130.
5. Estévez Schwarz, D., Tischendorf, C.: Structural analysis of electric circuits and consequences for MNA, Int. J. of Circuit Theory and Applications **28**, (2000) pp. 131-162.
6. Frank, J.E., van der Houwen, P.J.: Diagonalizable extended backward differential formulas, BIT, Vol 40, No 3, (2000) pp. 497–512.
7. Guenther, M., Rentrop, P., Feldmann, U.: CHORAL - A One Step Method as Numerical Low Pass Filter in Electrical Network Analysis. In: van Rienen, U., Günther M., Hecht D. (Eds.): Scientific Computing in Electrical Engineering, Lecture Notes in Computational Science and Engineering, No **18**, (SCEE-2000, Warnemünde), Springer Verlag (2001) pp. 199-215.
8. Hairer, E., Nørsett, S.P., Wanner, G.: Solving Ordinary Differential Equations I, Springer Berlin (1987).
9. Hairer, E., Wanner, G.: Solving Ordinary Differential Equations II, Springer Berlin (1996).
10. Hosea, M.E., Shampine, L.F.: Analysis and implementation of TR-BDF2, Applied Numerical Mathematics 20, (1996) pp. 21-37.
11. de Swart, J.J.B.: Parallel software for implicit differential equations, PhD-Thesis Univ. of Amsterdam (1997).
12. Tischendorf, C.: Solution of index-2 differential algebraic equations and its application in circuit simulation, Logos Verlag Berlin, PhD-Thesis Humboldt Univ. zu Berlin (1996).

# Non-parabolic Tail Electron Hydrodynamical Model for Silicon Semiconductors

Angelo Marcello Anile and Giovanni Mascali

Università di Catania, Viale A. Doria 6, I-95125 Catania, Italia

**Abstract.** In this paper we present a theoretical foundation for tail electron hydrodynamical models (TEHM) in semiconductors with application to bulk silicon.

## 1 Introduction

The description of the functioning of modern electron devices requires increasingly accurate physical models of carrier transport in semiconductors in order to deal with high-field phenomena such as impact ionization, thermal self-heating, etc. High-energy electron phenomena are of particular interest in the accurate evaluation of the degradation and breakdown of devices.
Monte Carlo methods (MC) are extremely CPU intensive and therefore not practical for routine design applications. Traditional hydrodynamical models for carrier transport in semiconductors cannot describe high energy electrons since they deal only with average values over the whole carrier population.
On the other hand several considerations, [1], support the existence of two thermal distributions at different temperatures for electrons having energies respectively lower and higher than a suitable threshold energy, the so-called *cold* and *hot* electrons. This is the reason why several authors, [2,3] have introduced new fluid dynamical models in which two well-defined subpopulations of electrons are considered. Recentely [4], a theoretical foundation for these models has been given by utilizing the moment method and a closure technique by which one can obtain both the constitutive fluxes and the production terms, appearing in the moment equations, as functions of the fundamental hydrodynamical variables, without resorting to MC.
However, for the sake of simplicity, a parabolic band approximation has been used also for the hot electrons. Here, we propose an improvement of the model in which the Kane dispersion relation is employed for approximating the electron energy in the conduction band. Furthermore we assume as threshold energy that corresponding to the energy gap in Si, so to have direct information about the electrons which can give rise to impact ionization.
In order to test the model we apply it to bulk Si and compare the results with those obtained by an usual hydrodynamical model. Comparisons with Monte Carlo results are also present relatively to hot electron variables.

## 2 Boltzmann transport equation and Moment equations

We treat the case of silicon unipolar devices for which the electrons contributing to charge transport are those in the six equivalent valleys around the six minima of the conduction band [5]. We assume that, for those electrons, the relation between the energy, $\mathcal{E}$, and the quasi-wave vector, $\mathbf{k}$, both measured from the bottom of the conduction band, is given by the Kane dispersion relation

$$\mathcal{E}(k)\,[1+\alpha\,\mathcal{E}(k)] = \frac{\hbar^2\,k^2}{2m^*}, \quad \mathbf{k} \in \Re^3, \tag{1}$$

which involves a parameter $\alpha = 0.5\,eV^{-1}$, called the non-parabolicity factor, while $m^* = 0.32\,m_e$ is the electron effective mass, $m_e$ being the free electron mass.

At a kinetic level, electrons in a semiconductor are described by a one-particle distribution function, $f(\mathbf{x}, t, \mathbf{k})$, whose evolution is governed by the semiclassical Boltzmann equation (BTE) coupled to the Poisson equation for the electric field $\mathbf{E}$

$$\begin{aligned} &\frac{\partial f}{\partial t} + \mathbf{v}(\mathbf{k})\cdot\frac{\partial f}{\partial \mathbf{x}} - \frac{q\mathbf{E}}{\hbar}\cdot\frac{\partial f}{\partial \mathbf{k}} = \mathcal{C}[f], \\ &\nabla_{\mathbf{x}}\cdot(\epsilon\mathbf{E}) = q\,[N_+(\mathbf{x}) - N_-(\mathbf{x}) - n(\mathbf{x})]\,, \end{aligned} \tag{2}$$

here, $q$ represents the absolute value of the electron charge, $\hbar$ the reduced Planck constant, $\epsilon$ the dielectric constant, $N_+$ and $N_-$ the donor and acceptor concentrations respectively, and $n$ the total electron number density. $\mathbf{v}$, the electron group velocity, depends on the energy $\mathcal{E}$ through the relation

$$\mathbf{v}(\mathbf{k}) = \frac{1}{\hbar}\nabla_{\mathbf{k}}\mathcal{E}(\mathbf{k}).$$

$\mathcal{C}[f]$ is the collision term, which takes account of the various scattering mechanisms the electrons undergo in a semiconductor. In the non-degenerate case, its form is

$$\mathcal{C}[f] = \int_{\Re^3} [w(\mathbf{k}',\mathbf{k})\,f(\mathbf{k}') - w(\mathbf{k},\mathbf{k}')f(\mathbf{k})]\;d\mathbf{k}'\,,$$

where $w(\mathbf{k},\mathbf{k}')$ represents the sum of the various electron scattering rates from a state with wave vector $\mathbf{k}$ to one with wave vector $\mathbf{k}'$. We will take into account the following scattering mechanisms for silicon:

- electron - acoustical phonon intravalley scattering,
- electron - phonon intervalley scattering, for which there are six contributions: the three $g_1, g_2, g_3$ and the three $f_1, f_2, f_3$ optical and acoustical intervalley scatterings [5],
- electron-impurity scattering.

Introducing a threshold energy, $\mathcal{E}_{thr}$, for the electrons, (here we use the threshold energy of impact ionization $\mathcal{E}_{thr} = 1.12\,eV$), and the kinetic quantities: $1, \mathbf{v}, \mathcal{E}, \mathcal{E}\,\mathbf{v}$, we describe the electron flow by means of the fundamental variables: number density, average velocity, energy and energy flux of electrons having energy less than and greater than $\mathcal{E}_{thr}$,

$$\begin{cases} n_H = \int_{\Delta_H} f_H \, d\mathbf{k}, & n_H \, \mathbf{V_H} = \int_{\Delta_H} \mathbf{v} f_H \, d\mathbf{k}, \\ n_H \, W_H = \int_{\Delta_H} \mathcal{E} f_H \, d\mathbf{k}, & n_H \, \mathbf{S_H} = \int_{\Delta_H} \mathcal{E} \mathbf{v} f_H \, d\mathbf{k}, \end{cases} \tag{3}$$

$$\begin{cases} n_C = \int_{\Delta_C} f_C \, d\mathbf{k}, & n_C \, \mathbf{V_C} = \int_{\Delta_C} \mathbf{v} f_C \, d\mathbf{k}, \\ n_C \, W_C = \int_{\Delta_C} \mathcal{E} f_C \, d\mathbf{k}, & n_C \, \mathbf{S_C} = \int_{\Delta_C} \mathcal{E} \mathbf{v} f_C \, d\mathbf{k}, \end{cases} \tag{4}$$

with $\Delta_H = \{\mathbf{k} : \mathcal{E}(\mathbf{k}) \geq \mathcal{E}_{thr}\}$ and $\Delta_C = \Re^3 - \Delta_H$, henceforth the subscripts $H$ and $C$ will indicate quantities referring to hot and cold electrons respectively.
From the BTE one can obtain the following evolution equations for these macroscopic quantities:

$$\begin{cases} \frac{\partial n_H}{\partial t} + \frac{\partial n_H V_H^i}{\partial x^i} - q \, E_i \, \mathcal{N}^i = C_{n_H}, \\ \frac{\partial n_H P_H^i}{\partial t} + \frac{\partial n_H U_H^{ji}}{\partial x^j} - q \, E_j \, \mathcal{T}^{ij} + q \, E^i \, n_H = C_{P_H^i}, \\ \frac{\partial n_H W_H}{\partial t} + \frac{\partial n_H S_H^i}{\partial x^i} - q \, \mathcal{E}_{thr} E^i \, \mathcal{N}_i + q \, E_i \, n_H \, V_H^i = C_{W_H}, \\ \frac{\partial n_H S_H^i}{\partial t} + \frac{\partial n_H F_H^{ij}}{\partial x^j} + q \, E_j \, n_H \, G_H^{ij} - q \, \mathcal{E}_{thr} E_j \, \mathcal{T}^{ij} = C_{S_H^i}, \end{cases} \tag{5}$$

$$\begin{cases} \frac{\partial n_C}{\partial t} + \frac{\partial n_C V_C^i}{\partial x^i} = C_{n_C} - q \, E_i \, \mathcal{N}^i, \\ \frac{\partial n_C P_C^i}{\partial t} + \frac{\partial n_C U_C^{ji}}{\partial x^j} + q \, E^i \, n_C = C_{P_C^i} - q \, E_j \, \mathcal{T}^{ij}, \\ \frac{\partial n_C W_C}{\partial t} + \frac{\partial n_C S_C^i}{\partial x^i} + q \, E_i \, n_C \, V_C^i = C_{W_C} - q \, \mathcal{E}_{thr} E^i \, \mathcal{N}_i, \\ \frac{\partial n_C S_C^i}{\partial t} + \frac{\partial n_C F_C^{ij}}{\partial x^j} + q \, E_j \, n_C \, G_C^{ij} = C_{S_C^i} - q \, \mathcal{E}_{thr} E_j \, \mathcal{T}^{ij}, \end{cases} \tag{6}$$

where summation over repeated lowercase letters is understood and

$$\begin{aligned} P_A^i &= \frac{1}{n_A} \int_{\Delta_A} \hbar k^i f_A d\mathbf{k} = m^* (V_A^i + 2\alpha S_A^i) \text{ (average crystal momentum)}, \\ U_A^{ij} &= \frac{1}{n_A} \int_{\Delta_A} v^i \hbar k^j \, f_A d\mathbf{k} \quad \text{(average crystal momentum flux)}, \\ G_A^{ij} &= \frac{1}{n_A} \int_{\Delta_A} [\frac{\partial}{\partial k^j} (\frac{1}{\hbar} v^i \, \mathcal{E}(k))] f_A d\mathbf{k}, \\ F_A^{ij} &= \frac{1}{n_A} \int_{\Delta_A} v^i v^j \mathcal{E}(k) f_A d\mathbf{k} \quad \text{(average flux of energy flux)}, \\ C_{n_A} &= \int_{\Delta_A} \mathcal{C}[f] d\mathbf{k} \quad \text{(density production)}, \end{aligned} \tag{7}$$

$$C_{P_A^i} = \int_{\Delta_A} \hbar k^i \mathcal{C}[f] d\mathbf{k} \text{ (crystal momentum production)},$$

$$C_{W_A} = \int_{\Delta_A} \mathcal{E}(k)\mathcal{C}[f] d\mathbf{k} \quad \text{(the energy production)},$$

$$C_{S_A^i} = \int_{\Delta_A} v^i \mathcal{E}(k)\mathcal{C}[f] d\mathbf{k} \quad \text{(energy flux production)},$$

all these quantities referring to electrons in zone $A$, with $A = H, C$. The first set of equations is found multiplying the BTE by the kinetic quantities $1, \hbar\mathbf{k}, \mathcal{E}, \mathcal{E}\,\mathbf{v}$ and integrating over $\Delta_H$, while the second set is derived by subtracting the first one to the corresponding usual moment equations.
The surface terms

$$\mathcal{N}^i = \frac{1}{\hbar}\int_{\Sigma} f_H\, \nu^i d\sigma \qquad \mathcal{T}^{ij} = \int_{\Sigma} k^i f_H\, \nu^j d\sigma, \tag{8}$$

with $\Sigma = \{\mathbf{k} : \mathcal{E}(\mathbf{k}) = \mathcal{E}_{thr}\}$, and $\nu^i$ inner normal to $\Sigma$, represent the increasing rate of the corresponding macroscopic quantities due to the net migration of carriers from one energy zone to the other owing to the driving electric field.
In equations (5), (6), in addition to the fundamental variables (3), (4), the extra-unknowns (7), (8) appear, so that, in order to have a closed system of equations, it is necessary to express the latter variables in terms of the former ones. A way to get constitutive relations, which lies on sound physical bases, is to use the maximum entropy principle, for more details the reader is referred to [6–8,10]. This principle furnishes the form of the distribution functions that make the best use of the knowledge of a finite number of moments.
In particular, having assumed as fundamental variables $n_A$,$\mathbf{V}_A$, $W_A$ and $\mathbf{S}_A$, the maximum entropy distributions $f_A$, $A = H, C$, are those which make the electron entropy extremal under the constraints of fixed values of those variables.
The electron entropy can be written as

$$s\,[f_C, f_H] = -k_B \left[\int_{\Delta_C} (f_C \log f_C - f_C)\, d\mathbf{k} + \int_{\Delta_H} (f_H \log f_H - f_H)\, d\mathbf{k}\right],$$

with $k_B$ Boltzmann constant. Therefore the distribution functions $f_A$, $A = H, C$, that maximize it under the constraints $n_A$,$\mathbf{V}_A$, $W_A$ and $\mathbf{S}_A$ are given by

$$f_A^{ME} = \exp\left[-\left(\frac{1}{k_B}\lambda_A + \lambda_A^W\, \mathcal{E}_A + \lambda_A^{V^i}\, v_A^i + \lambda_A^{S^i}\, v_A^i\, \mathcal{E}_A\right)\right], \quad A = H, C,$$

where the $\lambda$'s are Lagrange multipliers that take care of the constraints.
In order to determine the Lagrange multipliers in terms of $n_A$, $\mathbf{V}_A$, $W_A$, $\mathbf{S}_A$, $A = H$, $C$, one has to insert the expressions of the maximum-entropy-distribution functions into (3)-(4) and solve the resulting system. After that

the closure relations can be obtained by substituting the distribution functions, expressed in terms of the fundamental variables, into (7) and (8). However, on account of the algebraic difficulties, it is possible to get only approximate expressions for the Lagrange multipliers under physically reasonable assumptions on the distribution functions.
In particular, on the basis of MC results, we assume that the anisotropy of the $f_A^{ME}$, $A = H\,,C$ remains small even out of equilibrium. We formally introduce a *small* anisotropy parameter $\delta$, assume that the Lagrange multipliers are analytic in $\delta$ and expand them around $\delta = 0$ up to the first order. The representation theorems for isotropic functions imply that $\lambda_A$ and $\lambda_A^W$ are of order zero in $\delta$, while $\lambda_A^{V^i}$ and $\lambda_A^{S^i}$ are of the first order. Therefore the approximate expressions of the maximum entropy functions

$$f_A^{ME} = \exp\left(-\frac{\lambda_A}{K_B} - \lambda_A^W \mathcal{E}\right)\left[1 - \delta\left(\lambda_A^{V_i} v^i + \lambda_A^{S_i} v^i \mathcal{E}\right)\right], \quad A = H,C \quad (9)$$

are used to obtain the following results for the Lagrange multipliers

$$\begin{aligned}
\lambda_H &= -K_B \log\left(\frac{\hbar^3 n_H}{4\,\pi\, m^*\,\sqrt{2\,m^*} d_0^H}\right), & \lambda_C &= -K_B \log\left(\frac{\hbar^3 n_C}{4\,\pi\, m^*\,\sqrt{2\,m^*} d_0^C}\right),\\
\lambda_H^W &= g_H^{-1}(W_H), & \lambda_C^W &= g_C^{-1}(W_C),\\
\lambda_H^{V_i} &= b_{11}^H\, V_H^i + b_{12}^H\, S_H^i, & \lambda_C^{V_i} &= b_{11}^C\, V_C^i + b_{12}^C\, S_C^i,\\
\lambda_H^{S_i} &= b_{12}^H\, V_H^i + b_{22}^H\, S_H^i, & \lambda_C^{S_i} &= b_{12}^C\, V_C^i + b_{22}^C\, S_C^i,
\end{aligned} \quad (10)$$

where

$$d_k^A\left(\lambda_A^W\right) = \int_{\Delta\mathcal{E}_A} \mathcal{E}^k \sqrt{\mathcal{E}\,(1+\alpha\,\mathcal{E})}\,(1 + 2\,\alpha\,\mathcal{E}) \exp(-\lambda_A^W\,\mathcal{E})\, d\,\mathcal{E},$$

with $\Delta\mathcal{E}_H = (\mathcal{E}_{thr}, +\infty)$ and $\Delta\mathcal{E}_C = (0, \mathcal{E}_{thr})$. $g_A^{-1}$ are the inverse functions of

$$g_A\left(\lambda_A^W\right) = \frac{d_1^A\left(\lambda_A^W\right)}{d_0^A\left(\lambda_A^W\right)}, \quad A = H, C, \quad (11)$$

and the coefficients $b_{ij}^A$, $A = H\,,C$, are given by

$$b_{11}^A = \frac{a_{22}^A}{D^A}, \qquad b_{12}^A = -\frac{a_{12}^A}{D^A}, \qquad b_{22}^A = \frac{a_{11}^A}{D^A}, \quad (12)$$

with

$$a_{11}^A = -\frac{2\,p_0^A}{3\,m_A^*\,d_0^A}, \qquad a_{12}^A = -\frac{2\,p_1^A}{3\,m_A^*\,d_0^A}, \qquad a_{22}^A = -\frac{2\,p_2^A}{3\,m_A^*\,d_0^A},$$

and

$$D^A = a_{11}^A\, a_{22}^A - \left(a_{12}^A\right)^2, \qquad A = H\,,C\,,$$

$p_k^A$ being

$$p_k^A = \int_{\Delta_A} \frac{\mathcal{E}^k \left[\mathcal{E}\,(1+\alpha\,\mathcal{E})\right]^{3/2}}{1+2\,\alpha\,\mathcal{E}} \exp(-\lambda_A^W\,\mathcal{E})\,d\,\mathcal{E}\,, \quad A = H\,,C\,.$$

For the inversion of the functions (11) we have resorted to a numerical approach.

## 3 Application to the case of bulk Si

In this section we test the model in the case of bulk Si. The physical situation is represented by a Si semiconductor with uniform doping, which we take to be $N_+ = 10^{17}\mathrm{cm}^{-3}$. All the above-mentioned scattering mechanisms are considered. As regards the balance equations, taking into account the symmetry with respect to space translations, we can drop the spatial dependence. Moreover, in cases when a constant bias voltage is applied to the semiconductor, the Poisson equation $(2)_2$ is satisfied with $n$ equal to the value of the doping concentration and $\mathbf{E}$ constant. Being the motion along the direction of the electric field, we can take this as the x-direction, and therefore the balance equations reduce to the following set of ordinary differential equations

$$\frac{d}{dt} n_H = n_C\, G_n^H\,(W_C) - n_H\, L_n^H\,(W_H) + qE\, D_n(W_H, V_H, S_H), \tag{13}$$

$$m^* \frac{d}{dt} n_H V_H = -q n_H E(1 - 2\,\alpha\, m^*\, G_H) + n_H[(c_{11}^H - 2\,\alpha\, m^* c_{21}^H) V_H + \\ +(c_{12}^H - 2\,\alpha\, m^* c_{22}^H) S_H] - n_H\, q\, \frac{D_W(W_H)}{1+2\,\alpha\,\mathcal{E}_{thr}}\, E, \tag{14}$$

$$\frac{d}{dt} n_H W_H = -n_H\, q\, V_H\, E + n_C\, G_W^H(W_C) - n_H\, L_W^H(W_H) + n_H\, C_W^{HH}(W_H) + \\ +q\,\mathcal{E}_{thr}\, n_H E\, D_n(W_H, V_H, S_H), \tag{15}$$

$$\frac{d n_H S_H}{dt} = -q n_H E G_H + n_H V_H (c_{21}^H V_H + c_{22}^H S_H) - n_H q \frac{\mathcal{E}_{thr} D_W(W_H)}{1+2\,\alpha\,\mathcal{E}_{thr}} E, \tag{16}$$

$$\frac{d}{dt} n_C = -\frac{d}{dt} n_H, \tag{17}$$

$$m^* \frac{d}{dt} n_C V_C = -q n_C E(1 - 2\,\alpha\, m^*\, G_C) + n_C[(c_{11}^C - 2\,\alpha\, m^* c_{21}^C) V_C + \\ +(c_{12}^C - 2\,\alpha\, m^* c_{22}^C) S_C] + n_H\, q\, \frac{D_W(W_H)}{1+2\,\alpha\,\mathcal{E}_{thr}}\, E, \tag{18}$$

$$\frac{d}{dt} n_C W_C = -n_C\, q\, V_C\, E + n_H\, G_W^C(W_H) - n_C\, L_W^C(W_C) + n_C\, C_W^{CC}(W_C) + \\ -q\,\mathcal{E}_{thr}\, n_H E\, D_n(W_H, V_H, S_H), \tag{19}$$

$$\frac{d n_C S_C}{dt} = -q n_C E G_C + n_C V_C (c_{21}^C V_C + c_{22}^C S_C) + n_H q \frac{\mathcal{E}_{thr} D_W(W_H)}{1+2\alpha\mathcal{E}_{thr}} E, \tag{20}$$

where $V_A$ and $S_A$ are the x-components of $\mathbf{V}_A$ and $\mathbf{S}_A$ and $G_A$ is the xx-component of $G_A^{ij}$, $A = H,\, C$, while the the production terms $G_n^H$, $L_n^H$, $G_W^A$, $L_W^A$, $c_{ij}^A$, $A = H, C$, $i, j = 1, 2$ and the surface terms $D_n$ and $D_W$ are tabled functions of their arguments.

The solutions of (13)-(20) for electric fields respectively equal to $1, 2, 5$, and $10\frac{V}{\mu m}$, are reported in Fig.s 1, 2, and 3. We notice that, in agreement with the results in [2], the stationary value of $W_H$ slowly varies with the electric field, due to the Kane approximation to the structure of the silicon conduction band.
We also compare the results concerning the hot electron variables with those obtained by MC, see Table 1 and Table 2. The different qualitative behaviour in the velocity of the hot electrons is due to the fact that in the two fluid model $V_H$ starts decreasing at electric fields higher than 10 $V/\mu m$, so as the total electron mean velocity in the one fluid model. In Figs. 4 and

| | TEHM | MC |
|---|---|---|
| $n_H$ | 0.14 % | 0.13 % |
| $W_H$ | 1.1640 eV | 1.1814 eV |
| $V_H$ | 2.1305 $10^7$ $cm/s$ | 2.7377 $10^7$ $cm/s$ |

**Table 1.** Comparison between the results obtained by the tail electron hydrodynamical model and those derived by MC simulation, E=7.5$V/\mu m$ .

| | TEHM | MC |
|---|---|---|
| $n_H$ | 0.97 % | 0.87 % |
| $W_H$ | 1.1751 eV | 1.2133 eV |
| $V_H$ | 2.2637 $10^7$ $cm/s$ | 2.6390 $10^7$ $cm/s$ |

**Table 2.** Comparison between the results obtained by the tail electron hydrodynamical model and those derived by MC simulation, E=10$V/\mu m$ .

5 we also compare the results relative to the average quantities over the whole electron population with those derived by a standard hydrodynamical model for semiconductors based on the maximum entropy principle [9]. The agreement is very good so that it is possible to state that, while the model gives direct information on the hot electrons, no accuracy is lost regarding the total electron population.

## Acknowledgments

We acknowledge support from MURST, ex fondi 60%, and from CNR grants n. 99.01714.01, 00.0128.ST74.

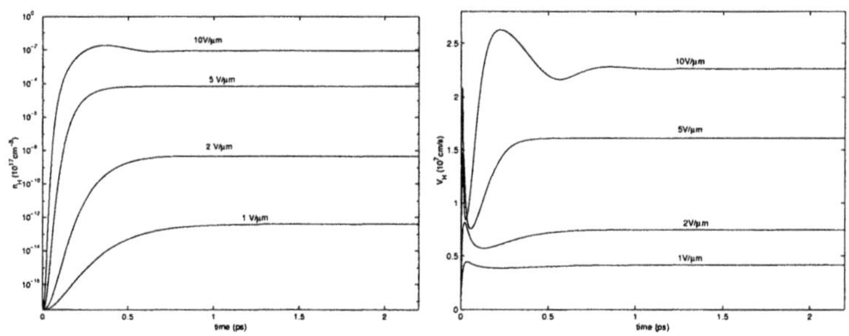


**Fig. 1.** The time evolution of the hot electron density and average velocity for different values of the electric field and for $N_{+} = 10^{17}/cm^3$.

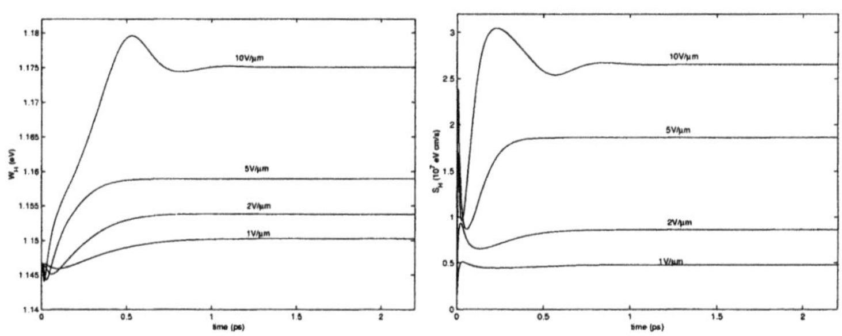


**Fig. 2.** The time evolution of the hot electron energy and energy flux for different values of the electric field and for $N_{+} = 10^{17}/cm^3$.

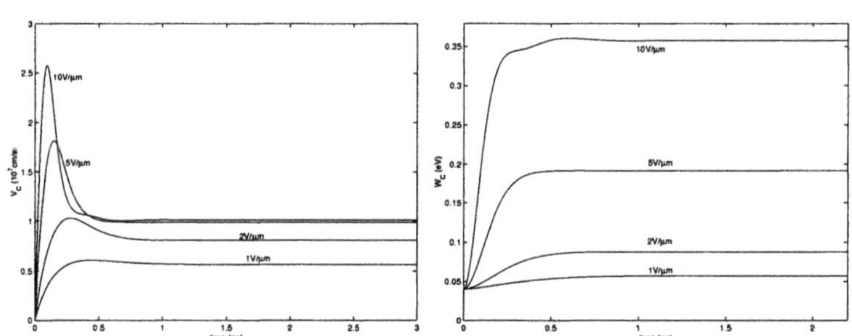


**Fig. 3.** The time evolution of the cold electron velocity and energy for different values of the electric field and for $N_{+} = 10^{17}/cm^3$.

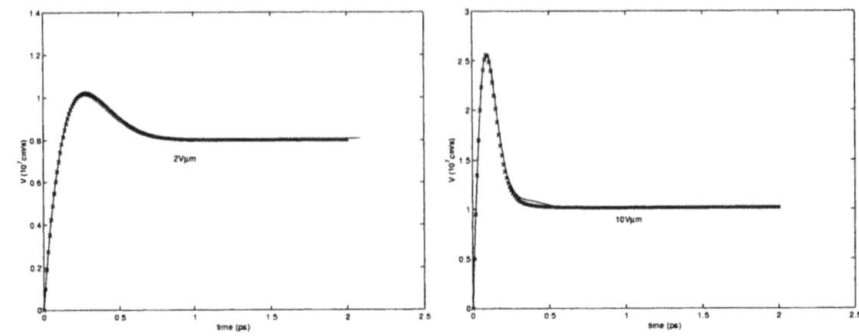


**Fig. 4.** Comparison of the time evolution of the average electron velocity for $E = 2, 10\,\frac{V}{\mu m}$.Continuous line: two population model, crosses: single population model

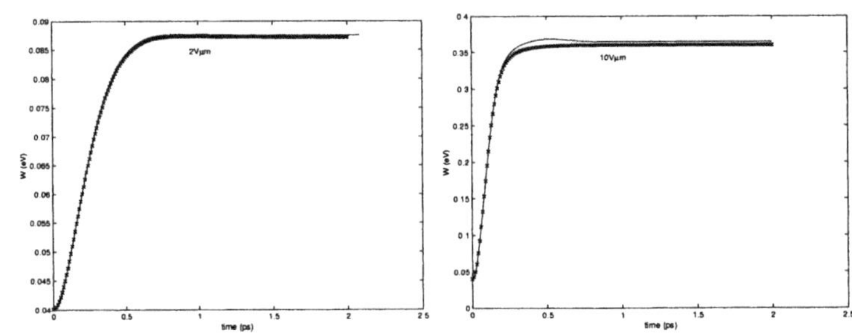


**Fig. 5.** Comparison of the time evolution of the average electron energy for $E = 2, 10\,\frac{V}{\mu m}$.Continuous line: two population model, crosses: single population model

## References

1. S. Laux, M. Fischetti, Transport Models for Advanced Device Simulation-Truth or Consequences, Proceedings of the Bipolar-BiCMOS Technology Meeting, October 1-3, 1995, Minneapolis.
2. C. -S. Yao, J. -G. Ahn, Y. -J. Park, H. -S. Min and R. W. Dutton, Formulation of a Tail electron Hydrodynamic model based on Monte Carlo Simulations, IEEE Electron Dev. Lett.,**16**, 26-29 (1995).
3. P. Scrobohaci,T. -W. Tang, Modeling of the hot electron subpopulation and its application to the impact ionization in silicon submicron devices: part I, IEEE Trans. Electron Devices,**41** 1197-1205 (1996).
4. A. M. Anile, G. Mascali, Theoretical Foundations for Tail Electron Hydrodynamical models in semiconductors, Appl. Math. Letters,**14** 245-252 (2001).
5. C. Jacoboni, L. Reggiani, The Monte Carlo method for the solution of charge transport in semiconductors with application to covalent materials *Rev. Mod. Phys.*, **55**, 645-705, (1983).
6. I. Müller, T. Ruggeri, *Rational Extended Thermodynamics*, second edition, Springer, Berlin, 1998.

7. C. D. Levermore, Moment Closure Hierarchies for Kinetic Theories, J. Stat. Phys., **83**, 1021-1065 (1996).
8. A .M. Anile, V. Romano, Non-parabolic band transport in semiconductors: Closure of the moment equations, Cont. Mech. Thermodyn,**1**, 307-325(1999)
9. V. Romano, Non-parabolic band hydrodynamical model of silicon semiconductors and simulation of electron devices,Math. Meth in Appl. Sciences,**24**, 439-471(2001)
10. G. Mascali, V. Romano, Hydrodynamical model of charge transport in seiconductors based on the maximum entropy principle, Cont. Mech. Thermodyn,**14**, 405-423(2002)

# Multirate Co-simulation of First Order Thermal Models in Electric Circuit Design

A. Bartel, M. Günther

Institut für Wissenschaftliches Rechnen und Mathematische Modellbildung, Universität Karlsruhe (TH), Engesser Straße 6, D-76128 Karlsruhe, Germany.

**Abstract.** Thermal effects influence the electrical behaviour of circuits more and more. Therefore it is necessary to take power dissipation and temperature evolution into account. In order to analyize large systems of integrated circuits, this has to be realized very efficiently. Thus we introduce a thermal network model consisting of 0D and 1D thermal elements approximating the full heat aspect, but keeping the system relatively small. After semi-discretization, this approach yields a coupled DAE system. According to the largely differing time scales, we outline the basics of a multirate co-simulation algorithm, which bases on an averaging technique. Its potential and feasibility is demonstrated on a simple, however, instructive test-circuit. As an outlook we discuss the application to thermal models of SOI-chips.

## 1 Introduction – Task

According to growing package densities, former secondary effects like self heating get important for the general behaviour of a chip. Industry predicts up to 100 Watt/cm$^2$ in the near future [7]. Therefore the temperature rise has to be in the scope of circuit simulation in an adequate way.

Commonly circuit simulators use more or less independent networks for the thermal aspect: Thermally modelled elements like transistors are equipped with an thermal network, leaving mostly out some efficient interaction of these networks.

Thus modelling takes place in the following circumstances: On the one hand, circuits are described by zero dimensional electric objects equipped with an incidence matrix conveying the topology. On the other hand, powers are dissipated and stored *locally*. Moreover, temperature slowly levels out according to the temperature gradient. And the thermal problem is a problem in real space, with a spatial heat distribution. That is, coming from ordinary circuits, the needed spatial coordinate is not at hand. Thus first in this paper, we address the modeling of the electro-thermal problem

Furthermore, the time scales of electric systems and thermal aspects differ by several orders of magnitude for a heat conduction over a centimeter. Thus we have a multirate setting, which has to be addressed [2]. In this multiphysical setup, an energy coupling enables the application of a multirate co-simulation, which is addressed in the later part of this work and applied to a test circuit.

## 2 First order thermal modelling

The modelling constraints are:

(1) feasibility in circuit simulators, that is, no large thermal networks than electric ones, with respect to the number of equations;
(2) usage of the available information as far as possible – only a few additional topological informations shall be supplied;
(3) facilitation of a fast simulation technique.

A thermal network with interacting domains has to have a spatial dimension and we have to deal at least with 1D approximations. Regarding the computational effort, lower dimensions are preferable. On that basis we propose a thermal model, the *accompanied thermal network*, which consists of 0D and 1D thermal elements – thus it is a modelling-cost trade-off. However, that network can be regarded as a step on the hierarchy of thermal modelling towards a full 3D evolution.

Now, let us render modelling precisely. Let $\mathbf{A}$ be the circuit topology (obtained by modified nodal analysis – see [5]), then $\mathbf{A}_{\mathrm{lp}} \subset \mathbf{A}$ (in terms of branches) shall describe the thermally relevant zero dimensional elements; and $\mathbf{A}_{\mathrm{tr}}$ the 1D elements, which could be branches of $\mathbf{A}$ as well as additional thermal-'interconnects'. For simplicity we normalize the spatial extension of each element: $x \in [0,1]$. Furthermore, we separate left ($x=0$) and right-hand ($x=1$) ends in the incidence matrix $\mathbf{A}_{\mathrm{tr}} := \mathbf{B} - \mathbf{D}$, where the matrix entries of $\mathbf{B}, \mathbf{D}$ belong to $\{0,1\}$ — that is, a ordering with respect to the ends.
Since coupled 0D elements have a common heat capacity, we form connected units – this involves two identifications: $\mathbf{P}$ and $\mathbf{S}$ identify thermal branches and electric network nodes with the corresponding 0D units, respectively. In that way, heat mass for each unit is given as: $\widehat{\mathbf{M}} = \mathbf{P}\mathbf{M}\mathbf{P}^{\top}$, where $\mathbf{M}$ is the diagonal matrix of heat mass with entries: $m = \rho_d \, c \, V$, for each 0D element using its density $\rho_d$, heat capacity $c$ and volume $V$.
For a simplified setup, we define *artificial* 0D units for the junctions of 1D elements only, which allow to treat boundaries equally. The unit's temperatures are denoted by $\widehat{\mathbf{T}}$.
In total, there shall be a number of $m_{\mathrm{tr}}$ 1D elements and $m_{\mathrm{lp}}$ 0D elements.

The networks, and 0D, 1D elements have to interact, therefore the *coupling* structure has to be described: If we split the dissipated power in the 1D and 0D contribution, $\mathbf{E}_{\mathrm{tr}}$ and $\mathbf{E}_{\mathrm{lp}}$, respectively, then we have:

- Circuit-to-Heat: The power dissipation of 0D elements is a global source for its thermal unit. The net unit power transition is given by $\mathbf{P}\mathbf{E}_{\mathrm{lp}}$.
  A 1D element dissipates power as a 0D quantity, only if the branch belongs to $\mathbf{A}$. We obtain a local dissipate power via a specified distribution function: $\rho_i(x,T)/(R_i(T)a_i(x)) \cdot \mathbf{E}_{tr,i}$, where $\nabla_x R = \rho$ and $a_i(x)$ denotes the cross-section.

- Heat-to-Heat: In the accompanied thermal network, every 1D element is connected to a 0D unit and vice versa. The heat passing from thermal element to element is given by a corresponding 1D-to-0D flux. Thus we sum up these fluxes: for each connected unit the sum runs over all attached 1D elements: inward flow minus outward flow gives

$$\begin{aligned} &\mathbf{B}_{\text{fac}}\Lambda(0)\nabla_x\mathbf{T}(0,t) - \mathbf{D}_{\text{fac}}\Lambda(1)\nabla_x\mathbf{T}(1,t), \\ &= (\mathbf{B}_{\text{fac}}, -\mathbf{D}_{\text{fac}})\begin{pmatrix}\Lambda(0)\nabla_x\mathbf{T}(0,t)\\ \Lambda(1)\nabla_x\mathbf{T}(1,t)\end{pmatrix} = \mathbf{S}(\mathbf{B},-\mathbf{D})\begin{pmatrix}\Lambda(0)\nabla_x\mathbf{T}(0,t)\\ \Lambda(1)\nabla_x\mathbf{T}(1,t)\end{pmatrix}, \end{aligned}$$

  where $\mathbf{B}_{\text{fac}} = \mathbf{S}\mathbf{B}$, i.e., a projection of $\mathbf{B}$ into the accompanied thermal network ($\mathbf{D}_{\text{fac}}$ is analog), and $\Lambda$ the diagonal matrix of heat fluxes times the corresponding local cross-sections.
- Heat-to-Circuit: The 0D branch temperatures are simply obtained by: $\mathbf{P}^\top\widehat{\mathbf{T}}$. Moreover, it is possible to have heat distribution dependent circuit parameters for the 1D thermal elements. In that case, there is a kind of density, which describes the ratio of local electric effect to the total effect (cf. [3]).

Now we can formulate the corresponding mathematical description given in Box 1 (boxed appear the coupling variables). First, using modified nodal analysis, the electric network is described by a set of differential algebraic network equations in the node potentials $\mathbf{u}$ and branch currents $\jmath_V, \jmath_L$ (through voltage sources and inductors) and the system is closed by a consistent initial value, (DAE-IVP). Secondly, the thermal network is formed: for each 1D-element there is a heat equation in temperature variables $\mathbf{T}: [0,1]\times[0,\infty) \to \mathbb{R}^{m_{\text{tr}}}$; for the connected units, temperatures $\widehat{\mathbf{T}}: [0,\infty) \to \mathbb{R}^{m_{\text{lp}}}$ are determined solely from temporal heat change. Additionally, we model cooling to an ambient temperature $T_{\text{env}}$ as first order effect, i.e., it is described by Newton's cooling and depends on the surface to volume ratio, roughly, $F$ and a transition factor $\gamma$. The thermal network is accomplished by the identification for the boundaries of 1D elements with 0D units temperatures (BC) and initial values (IV). In the overall, thus we have parabolic PDEs coupled with DAEs via right-hand sides and source terms.

## 3 Simulation algorithm

A large potential of savings in the computation of thermal-electric systems is found in comparing the time scales of the subsystems: circuit signal time $\tau_{\text{circuit}}$ (e.g. input) and thermal relaxation time $\tau_{\text{heat}}$:

$$\begin{aligned} &\tau_{\text{circuit}} &&\approx 10^{-9}\dots 10^{-6}\ \text{sec} \\ &\tau_{\text{heat}} &&= l^2C/\lambda \approx 1\dots 10\ \text{sec} \qquad (l = \text{length} \approx \text{cm}). \end{aligned}$$

Here we have an extra-ordinary multirate behaviour. A straight forward and popular *idea* of tackling coupled problems is co-simulation, where each physical subsystem is addressed separately by some adapted algorithm or available

**Box 1:** COUPLED THERMO-ELECTRIC PROBLEM.

electric network: (DAE-IVP) $\mathbf{A} = (\mathbf{A}_C,\ \mathbf{A}_G,\ \mathbf{A}_L, \mathbf{A}_I,\ \mathbf{A}_V)$

$$\mathbf{0} = \mathbf{A}_C \dot{\mathbf{q}}(\mathbf{A}_C^\top \mathbf{u}(t), \boxed{\mathbf{T}_C^{\text{br}}}) + \mathbf{A}_G \mathbf{r}(\mathbf{A}_G^\top \mathbf{u}(t), \boxed{\mathbf{T}_G^{\text{br}}, \mathbf{R}}) + \mathbf{A}_L \jmath_L(t) + \mathbf{A}_I \imath(t) + \mathbf{A}_V \jmath_V(t)$$

$$\mathbf{0} = \dot{\phi}(\jmath_L(t), \boxed{\mathbf{T}_L^{\text{br}}}) - \mathbf{A}_L^\top u(t)$$

$$\mathbf{0} = \mathbf{A}_V^\top \mathbf{u}(t) - \mathbf{v}(t)$$

$$\text{(IV)} \quad \mathbf{x}(t_0) = (\mathbf{u}_0,\ \jmath_{L,0},\ \jmath_{V,0})^\top$$

coupling interface:

$$(\mathbf{E}_{\text{tr}}^\top(t), \mathbf{E}_{\text{lp}}^\top(t))^\top = \mathbf{E}(t) = \text{diag}(\jmath_E)\mathbf{A}_{\text{tp}}^\top \mathbf{u}, \qquad \mathbf{T}^{\text{br}} = \mathbf{P}^\top \widehat{\mathbf{T}}, \qquad \mathbf{R}$$

thermal network: (PDAE-BIVP) $x \in \Omega = [0,1]^{m_{\text{tr}}}$

$$\text{(1D)} \quad \mathbf{C}_i \dot{T}_i(x,t) = \nabla_x(\lambda_i \nabla_x T_i(x,t)) - \gamma F_i \cdot (T_i(x,t) - T_{\text{env}}) + \boxed{E_{\text{tr},i}(t)} \cdot \frac{\widetilde{\rho}_i(x,T_i)}{R_i(t,T_i)\, a_i}$$

$$0 = \nabla_x R_i - \widetilde{\rho}_i(x, T_i(x,t)) \qquad (i = 1, \ldots, m_{\text{tr}})$$

$$\text{(0D)} \quad \widehat{\mathbf{M}}\dot{\widehat{\mathbf{T}}}(t) = (\mathbf{B}_{\text{fac}},\ -\mathbf{D}_{\text{fac}}) \begin{pmatrix} \Lambda(0)\nabla_x \mathbf{T}(0,t) \\ \Lambda(1)\nabla_x \mathbf{T}(1,t) \end{pmatrix} - \gamma \widehat{\mathbf{F}}(\widehat{\mathbf{T}} - T_{\text{env}} \mathbb{1}_k) + \mathbf{P}\boxed{\mathbf{E}_{\text{lp}}(t)}$$

$$\text{(BC)} \quad \mathbf{T}(0,t) = \mathbf{B}_{\text{fac}}^\top \widehat{\mathbf{T}}(t), \qquad \mathbf{T}(1,t) = \mathbf{D}_{\text{fac}}^\top \widehat{\mathbf{T}}(t)$$

$$\text{(IC)} \quad \mathbf{T}(x,0) = T_{\text{env}} \mathbb{1}_{m_{\text{tr}}}$$

Notation: $C$ heat capacity per volume; $\lambda$ heat conductivity; $\widetilde{\rho}/R$ local distribution of electric parameter; $a$ cross-section.

tool. Therefore time-stepping and coupling, which is needed to create an efficient and robust scheme, is slightly more delicate.

Our proposal is to use *multirate-co-simulation*, that is, to perform no iteration (as long as possible). To enable that, following [4] we pay some additional effort in computing the coupling variables (dissipated powers) carefully and use a averaging technique. The outline of that algorithm is given in Box 2.

To obtain a full waveform relaxation scheme, we have to add to the algorithm between position 2) and 3) a convergence check, and need to iterate steps 1) and 2). The thermal network in step 2) is preferably semi-discretized in space and then time integration is performed (method of lines), since we think of common SPICE-like simulators. That means a coupled system of DAEs is integrated over time. By introducing the dissipated energies (integral over the powers), we couple the subsystems by differential variables, and thus waveform relaxation is suitable [1].

**Box 2:** SIMULATION ALGORITHM.

0) define communication step size $H$ and window $[t_0, t_0 + H]$
1) compute electric network and energies of desired accuracy over $[t_0, t_0 + H]$; (keep temperature constant)
   → total 'local' dissipated energies
   (local = elementwise, using available quantities in the simulator)
2) compute thermal part for $[t_0, t_0 + H]$, add energy linearly!
3) go to one: $t_0 := t_0 + H$

Multirate is very naturally achieved by the 'decoupled' computation of electric and thermal network. But this has a price: the solution of the full coupled system is only of order 1. However, we are mostly interested in the precise computation of the electric network. A splitting technique can restore the order.

Next we want to study the performance of that algorithm for a test-circuit.

## 4 Simulation of a test-circuit

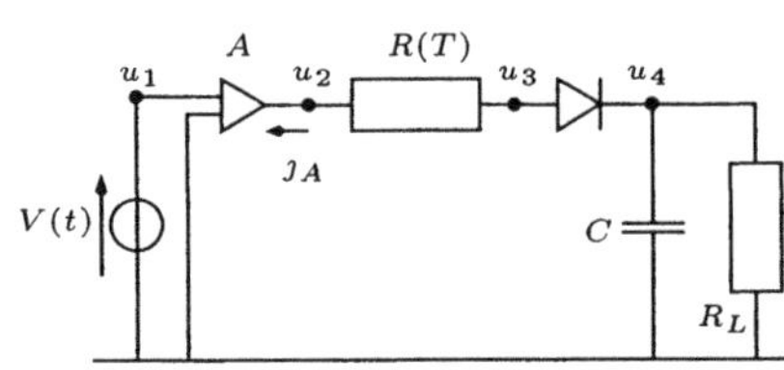


**Fig. 1.** Test-Circuit.

For testing, we use a simple, but instructive, circuit (see Fig. 1), which is capable of all effects. It consists of an operational amplifier, two resistors, a diode, and some capacitor. Its general electrical behaviour is as follows: the capacitor $C$ is charged via the diode, and later discharged via the load resistor $R_L$.

The accompanied thermal network consists of two 0D elements: the amplifier as heat source, and the diode being thermally dependent. Last, the resistor $R(T)$ is modelled as 1D thermal element. Thus we cover the full capabilities of the accompanied thermal network and address so all thermal effects.

Using network analysis the circuit equations can be set up. Here we model $R(T)$ as a piece of very thin copper with the following local resistance over $[0, l] = [0, 4\,\text{cm}]$:

$$\rho(x, T) = r_0(10 \cdot (l - x)x + c_0)(1 + \alpha(T(t, x) - T_{\text{meas}}) + \beta(T(t, x) - T_{\text{meas}})^2)/a$$

(with $T_{\text{meas}}$ reference temperature, basic material resistivity $r_0 \cdot c_0 = \rho_{\text{Cu}, 291}$, and cross-section $a$; the specific heat shall be equally distributed). Furthermore, we need a diode current:

$$I_D(u_{\text{di}}, \widehat{T}_{\text{di}}) = \widehat{I}_S(\widehat{T}_{\text{di}})[\exp(u_{\text{di}}/v_T) - 1]$$

(with thermal voltage $v_T$ and saturation current $\widehat{I}_S$, see [6]). For the thermal network, some dimensions of the concurring physical objects have to be assigned: volume, surface, texture.

To run the above simulation algorithm, we have to equip the network with two additional equations

$$\frac{d}{dt}E_{\text{tot,op}} = (u_{\text{op}}) \cdot \jmath_A, \qquad \frac{d}{dt}E_{\text{tot,tr}} = (u_{R(T)}) \cdot \jmath_A$$

for the total energy over the communication window $[t_0, t_0 + H]$. Now, we excite $V(t)$ as single sine wave extending over $[0\text{ sec}, 25\mu\text{ sec}]$, and integrate until $t = 0.5$ sec. The solution of the thermally modelled resistor is depicted in Fig. 2 (left) and the single-rate and multirate solution are compared at time $t = 0.5$ sec. Here an absolute error of less than $5 \cdot 10^{-4}$ K occurs in the multirate method.

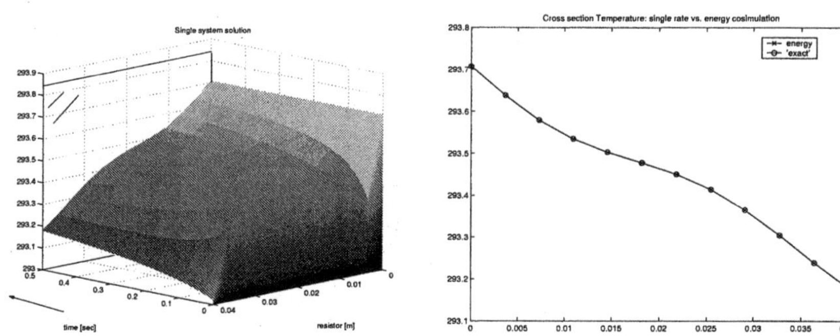


**Fig. 2.** Heat evolution of dragged on resistor (left). Final heat distribution at $t =$ 0.5sec (right).

We compare single-rate results versus multirate co-simulation in terms of time steps:

| | steps | = signal time + [0, 26.$\mu$ sec] | heat evol [26.$\mu$ sec, 0.5 sec] | comm. steps |
|---|---|---|---|---|
| single-rate | 406 | 348 | 58 | – |
| multi-rate | network | 339 | 60 | 26 |
| co-simulation | heat | 1 | 25 | |

We observe that the electric network conducts the number of steps in the single rate algorithm (since it is the faster component) and these steps are reproduced by the multirate algorithm. Furthermore, in any communication step the thermal network is computed in a single step. Thus the simulation result is very close to the optimal: the electric circuit takes as many steps it needs to resolve the diode's switching (this number depends also on the chosen parameters), while temperature has to do almost nothing. This of course justifies the use of a decoupled thermal network, but only for time integration over small scales.

Of course, these multirate scales can be found in chip technology, too. Therefore we are going to point out the arising thermal 1D-structures in the recent SOI chip technology.

## 5 Application and Outlook

Especially silicon-on-insulator (SOI) MOSFET-chip technology suffers from self-heating: An additional oxide layer causes the MOSFET body to float (smaller power consumption). Unfortunately, this layer behaves as thermal insulation. The temperature rise effects the particle mobility and through that the switching behaviour of the transistor. Arrays of SOI transistors (arrays of gates) can be regarded as 1D elements: They are lined up on the silicon waver and power supply and ground lines separate these chip portions. Mainly, the push-pull amplifier dissipates power and heats up the chip. The local gate activity causes a small additional heating. Thus the above modelling can be applied (only the source term of (1D) in Box 1 gets a sum). We obtain a simple thermal network, consisting of 1D-elements for each thermally modelled line of transistors and a single 0D element for the amplifier.

The further step in the modelling hierarchy would involve 2D approximations: A simple strategy is to use 1D elements in a grid formation. But, also 2D elements can be integrated in the above approach by specification of the boundary conditions.
Of course, if the space dimensions shrink the multirate potential decreases: the thermal time constant is proportional to the square of the extension and so the application of this algorithm is tailored to dimensions of a few micrometers for gigahertz signals and circuits (depending on the actual parameters). The less seperated the time scales are the more a co-simulation version of the algorithm seems to be adequate, i.e., the application of iterations. In the region of few differing time scales a full co-simulation is still valid and an alternative to the all at once solution. Then the energy coupling can be dropped as additional effort. – So if a small scale electric technology exhibits macrostructures (1D) of some orders of magnitude larger, a fruitful application is expected.

Summing up, we have pointed out the relevance of thermal-electric models in chip design and the construction of fast simulation techniques. Our proposal is a first order thermal approximation called *accompanied thermal network*. This can be solved using a multirate co-simulation algorithm, which is based on an averaging technique. A test-circuit shows feasibility.

This basic mode of an algorithm has to be elaborated in the future; there are various refinements and developments conceivable, for instance, how behave refined coupling techniques, or interpolated extrapolation; how can we predict the communication step size $H$; and is it possible to use Newton-type techniques for exploitation of the available derivative information? Parallel computation of thermal 1D elements? Moreover stronger couplings and smaller spatial scales have to be investigated. Also the analytical side of the coupled PDAE system poses interesting questions of index and solvability.

## Acknowledgements

Thanks go to U. Feldmann (Infineon Technologies) and P. Rentrop (TU Munich) for many fruitful discussions. Moreover, this work is part of the project "Numerische Simulation von elektrischen Netzwerken mit Wärmeeffekten" (No. 03REM3KA), which is founded by the BMBF programme "Neue mathematische Verfahren in Industrie und Dienstleisungen".

## References

1. M. Arnold, M. Günther. Preconditioned dynamic iteration for coupled differential algebraic equations. *Bit*, **41** No. 1 (2001), pp. 1-25.
2. A. Bartel, M. Günther, R. Pulch, P. Rentrop. Numerical Techniques for Different Time Scales in Electric Circuit Simulation. In: M. Breuer, F. Durst, Ch. Zenger: *High Performance Scientific and Engineering Computing*, Lecture Notes in Computational Science and Engineering, Springer 2002, pp. 343-360.
3. L. Codecasa, D. D'Amore, P. Maffezzoni. A Rigorous Approach to Electro-Thermal Network Modeling. *ECCTD'01, Espoo*, (2001) II-161–164.
4. Ch. Deml, P. Türkes. Fast Simulation Technique for Power Electronic Circuits with Widely Different Time Constants. *IEEE Transactions on Industry Applications.* **35** No. 3 (1999), pp. 657-662.
5. M. Günther, U. Feldmann: *CAD-based electric-circuit modeling in industry.* Surv. Math. Ind. **8** (1999), pp. 97–157.
6. G. Massobrio, P. Antognetti: *Semiconductor Device Modeling with SPICE.* 2nd. ed., McGraw-Hill (1993).
7. The national technology roadmap for semiconductors, 1997.

# Parallel Solution Techniques for Sparse Linear Systems in Circuit Simulation

Achim Basermann[1], Fabienne Cortial-Goutaudier[1], Uwe Jaekel[1], and Koutaro Hachiya[2]

[1] C&C Research Laboratories, NEC Europe Ltd., Rathausallee 10, D-53757 Sankt Augustin, Germany; E-mail: {basermann, cortial, jaekel}@ccrl-nece.de
[2] System LSI Design Engineering Division, NEC Electron Devices, Kawasaki, Kanagawa 211-1888, Japan; E-mail: k-hachiya@ax.jp.nec.com

**Abstract.** For solving sparse linear systems from circuit simulation whose coefficient matrices include a few dense rows and columns, a parallel Bi-CGSTAB algorithm with distributed Schur complement (DSC) preconditioning is presented. The parallel efficiency of the solver is increased by transforming the equation system into a problem without dense rows and columns as well as by exploitation of parallel graph partitioning methods. The costs of local, incomplete *LU* decompositions are decreased by fill-in reducing reordering methods of the matrix and a threshold strategy for the factorization. The efficiency of the parallel solver is demonstrated with real circuit simulation problems on a PC cluster.

## 1 Introduction

The simulation of large, highly integrated circuits leads to non-linear differential algebraic equations. For integration of these equations, accurate solution methods for sparse linear systems are required within the non-linear iterations. The corresponding matrices are real, non-symmetric, very ill-conditioned, have an irregular sparsity pattern, and include a few dense rows and columns.

When the systems become large, iterative solvers are very likely to outperform direct methods. For convergence acceleration of iterative solvers, parallelization and appropriate preconditioning are suited techniques to reduce the execution time.

We present a parallel Bi-CGSTAB algorithm with distributed Schur complement (DSC) preconditioning [4] which achieves an accuracy of the solution similar to a direct solver but usually is distinctly faster for large problems. The parallel efficiency of the method is increased by transforming the equation system into a problem without dense rows and columns as well as by exploitation of parallel graph partitioning methods. The costs of local, incomplete *LU* decompositions are decreased by fill-in reducing reordering methods of the matrix and a threshold strategy for the factorization.

## 2 Problem of Dense Rows and Columns

Matrices from circuit simulation problems are usually very sparse but include a few (nearly) dense rows and columns. In the parallel case, dense rows and columns are difficult to handle for partitioning methods since they result in couplings between all equations. In addition, good load balance is hard to achieve if the matrix is distributed row-wise. Fill-in reducing ordering methods may become very costly due to a few dense rows and columns, and the matrices may get very ill-conditioned.

Fortunately, dense rows and columns are usually easy to remove from circuit simulation matrices since the corresponding columns or rows as a rule have only one non-zero entry on the diagonal. Such equations normally include voltage sources (constraints). A dense column whose corresponding row has only one diagonal entry can be removed since the corresponding unknown can be determined from the row equation and substituted in all other equations. On the other hand, a dense row (equation) whose corresponding column has merely one diagonal entry is only responsible for the corresponding unknown. All other equations can be solved independently. If the corresponding columns or rows of dense rows and columns do not have one diagonal entry only such rows and columns can be handled by using the Woodbury formula [2]. This case is rare for circuit simulation problems and does not occur for the matrices investigated here.

## 3 Distributed Schur Complement Techniques

### 3.1 Definitions

Fig. 1 (left) schematically displays the row-wise distribution of a matrix $A$ to two processors. Each processor owns its local row block. The square matrices $A_i$ are the local diagonal blocks of $A$. We assume that the local rows are arranged in such a way that the rows without couplings to the other processor(s) come first and then the rows with couplings. The former are called *internal rows*, have only entries in the $A_i$ part of the local rows and are not coupled with rows of other processors. The latter additionally have entries outside the $A_i$ part or are coupled with rows of other processors. These local rows are named *local interface rows.* The part outside $A_i$ which represents couplings between the processors is called *local interface matrix* $X_i$. From the view of processor 2 in Fig. 1 (left), the local interface rows of processor 1 with entries at column positions in the area of $X_2$ are *external interface rows.* Since the sparsity pattern of circuit simulation matrices usually is non-symmetric local interface rows of processor $i$ may have entries in $A_i$ only but are uni-directionally coupled with rows of other processors. These rows are external interface rows from the view of the other processors. This can not be determined locally on processor $i$, communication is necessary. Since each row of the matrix corresponds to a specific unknown of the equation system

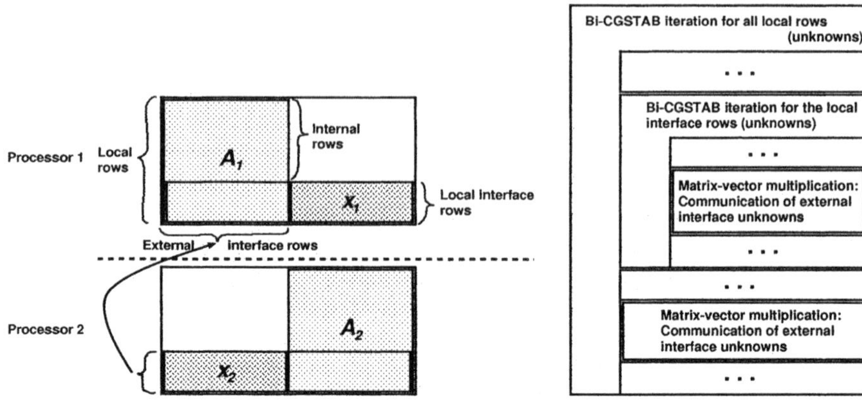


**Fig. 1.** Left: DSC definitions: Matrix distributed to two processors. Right: Schematic view of the DSC algorithm on each processor.

(row 1 to solution vector component 1 and to right hand side component 1, e.g.) *internal unknowns*, *local interface unknowns*, and *external interface unknowns* can be defined correspondingly.

### 3.2 Algorithm

Fig. 1 (right) gives a schematic survey of the DSC algorithm per processor. On each processor an outer Bi-CGSTAB iteration [5] is performed for all local rows (unknowns). As basic iterative method, a flexible variant of GMRES, FGMRES [3,4], is also well suited for the DSC algorithm but is not considered here because of its higher storage requirements. The outer iteration contains a partial matrix-vector multiplication which requires communication since each processor only owns its local segment of the vector. It is necessary to exchange components of non-local vector segments which correspond to external interface unknowns (rows).

Within the outer Bi-CGSTAB iteration, an inner Bi-CGSTAB iteration for the local interface rows (unknowns) only is performed. This includes a partial matrix-vector multiplication of the interface system but the communication scheme is the same as for the outer matrix-vector multiplication and thus has to be implemented only once.

From the mathematical point of view, each processer $i$ solves the following equation:

$$A_i\, x_i + X_i\, y_{i,\text{ext}} = b_i\,, \quad x_i = \begin{pmatrix} u_i \\ y_i \end{pmatrix}, \quad b_i = \begin{pmatrix} f_i \\ g_i \end{pmatrix}\,. \tag{1}$$

$x_i$ are the local vector components, $y_{i,\text{ext}}$ the external interface vector components, and $b_i$ is the local segment of the right hand side vector. $x_i$ is split into the internal vector components $u_i$ and the local interface vector components $y_i$, $b_i$ accordingly.

$A_i$ is then split (see [4] for details), and (1) is reformulated:

$$A_i = \begin{pmatrix} B_i & F_i \\ E_i & C_i \end{pmatrix} \rightarrow \begin{pmatrix} B_i & F_i \\ E_i & C_i \end{pmatrix} \begin{pmatrix} u_i \\ y_i \end{pmatrix} + \begin{pmatrix} 0 \\ \sum_{\text{Neighbours } j} E_{ij}\, y_j \end{pmatrix} = \begin{pmatrix} f_i \\ g_i \end{pmatrix} . \quad (2)$$

The result of the sum over all neighbouring processors $j$ with couplings to processor $i$ in (2) is the same as that of $X_i\, y_{i,\text{ext}}$ in (1). $E_{ij}\, y_j$ is the part of $X_i\, y_{i,\text{ext}}$ which reflects the contribution to the local equation from the neighbouring processor $j$.

The matrix equation (2) represents two equations. From the first, we derive an expression for $u_i$, substitute $u_i$ in the second equation and get

$$u_i = B_i^{-1}(f_i - F_i\, y_i) \;\rightarrow\; S_i\, y_i + \sum_{\text{Neighbours } j} E_{ij}\, y_j = g_i - E_i\, B_i^{-1} f_i \;. \quad (3)$$

$S_i = C_i - E_i\, B_i^{-1} F_i$ is the *local Schur complement.* Note that (3) is an equation for the interface vector components only.

(3) can be rewritten as a block-Jacobi preconditioned Schur complement system [4]:

$$y_i + S_i^{-1} \sum_{\text{Neighbours } j} E_{ij}\, y_j = S_i^{-1}(g_i - E_i\, B_i^{-1} f_i) \;. \quad (4)$$

### 3.3 Preconditioning

Fig. 2 illustrates the principle of preconditioning within the DSC algorithm per processor. The outer iteration from Fig. 1 (right) is preconditioned per processor by a block incomplete *LU* decomposition with threshold (ILUT) [3] of the local diagonal block ($L_1\, U_1$ in Fig. 2). For preconditioning the inner iteration, a block ILUT for the local interface rows only is exploited. This factorization need not be computed but can be used from the lower right part of the decomposition for the outer iteration ($L_{1,S}\, U_{1,S}$ in Fig. 2).

Mathematically speaking, we perform a block factorization of $A_i$ on processor $i$ using the splitting from (2):

$$A_i = \begin{pmatrix} B_i & F_i \\ E_i & C_i \end{pmatrix} = \begin{pmatrix} B_i & 0 \\ E_i & S_i \end{pmatrix} \begin{pmatrix} I & B_i^{-1} F_i \\ 0 & I \end{pmatrix} . \quad (5)$$

We then assume that we have the *LU* decomposition $S_i = L_{i,S}\, U_{i,S}$ of the local Schur complement. With this, we formulate the *LU* factorization

$$L_i\, U_i = \begin{pmatrix} L_{i,B} & 0 \\ E_i\, U_{i,B}^{-1} & L_{i,S} \end{pmatrix} \begin{pmatrix} U_{i,B} & L_{i,B}^{-1} F_i \\ 0 & U_{i,S} \end{pmatrix} \quad (6)$$

with $B_i = L_{i,B}\, U_{i,B}$ the *LU* decomposition of $B_i$. By transforming the right hand side of (6) into

$$\left[ \begin{pmatrix} L_{i,B} & 0 \\ E_i\, U_{i,B}^{-1} & L_{i,S} \end{pmatrix} \begin{pmatrix} U_{i,B} & 0 \\ 0 & U_{i,S} \end{pmatrix} \right] \begin{pmatrix} I & U_{i,B}^{-1} L_{i,B}^{-1} F_i \\ 0 & I \end{pmatrix} = \begin{pmatrix} B_i & 0 \\ E_i & S_i \end{pmatrix} \begin{pmatrix} I & B_i^{-1} F_i \\ 0 & I \end{pmatrix}$$

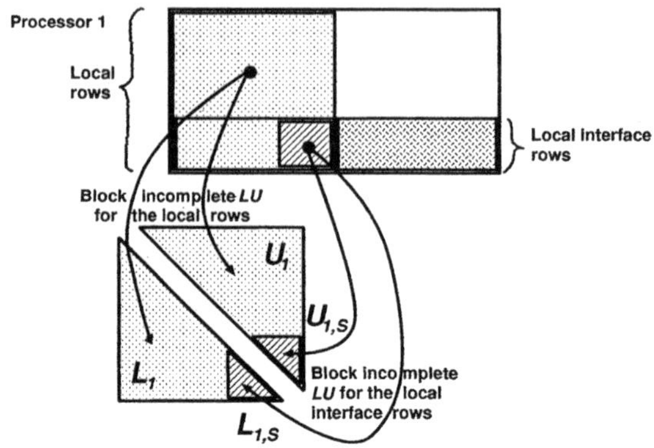


**Fig. 2.** Principle of preconditioning within the DSC algorithm.

we find after comparison with (5) that $L_i U_i$ is an $LU$ factorization of $A_i$. The other way round, we also see the practical advantage from (6) that the $LU$ factorization $S_i = L_{i,S} U_{i,S}$ of the local Schur complement has not to be computed explicitly if we already have an $LU$ factorization of the local diagonal block $A_i$.

If we perform incomplete decompositions we get an approximate, preconditioned Schur complement system with the approximation $\widetilde{S}_i$ of the local Schur complement $S_i$ (compare with (4)):

$$y_i + \widetilde{S}_i^{-1} \sum_{\text{Neighbours } j} E_{ij}\, y_j = \widetilde{S}_i^{-1}(g_i - E_i\, B_i^{-1} f_i)\ . \tag{7}$$

### 3.4 Repartitioning and Reordering

The distributed sparsity pattern of the matrix can be represented as a distributed graph with nodes and edges. Graph repartitioning can then be used to reduce the number of couplings between the distributed matrix row blocks. In graph theory formulation, the reduction is done by a minimization of the number of edges cut in the graph. This goal of graph partitioning corresponds to a minimization of the number of interface unknowns in the DSC algorithm, and thus problem (7) is made very small. For graph partitioning, we use the ParMETIS software from the University of Minnesota [1]. Since ParMETIS requires an undirected graph as input the non-symmetric pattern of the matrix has to be symmetrized for the matrix graph construction.

For local, incomplete decompositions, we use METIS nested dissection reordering to reduce fill-in into the factors [1]. Nested dissection reordering usually generates a similar sparsity pattern for the local diagonal blocks $A_i$ on each processor $i$. This results in similar fill-in for each ILUT and thus supports load balancing.

# 4 Results

The following experiments were performed on an SGI $O^2$ workstation (300 MHz, 1 GB main memory), NEC CCRLE's PC cluster SUCCESS (six 4-way SMP nodes with Pentium III Xeon CPUs, 500 MHz, 1 GB main memory per node, Giganet interconnection network between the nodes), and NEC CCRLE's PC cluster GRISU (32 2-way SMP nodes with AMD Athlon MP 1900+ CPUs, 1.6 GHz, 1 GB main memory per node, Myrinet2000 interconnection network between the nodes).

In all tests with the DSC algorithm, the outer iteration was stopped if the residual norm of the total equation system divided by the initial norm was less than $10^{-11}$. The inner iteration was stopped if the residual norm of the interface system divided by the initial norm was less than $10^{-13}$. These criteria resulted in high accuracy of the solution in all our experiments.

## 4.1 Sequential Results: Iterative versus Direct Solver

Table 1 compares sequential execution times of a direct method (software from Saad's SPARSKIT [3], threshold for ILUT: 0) and of the iterative DSC solver for five equation systems from the simulation of NEC circuits [1] on an SGI $O^2$ workstation. The ILUT thresholds in the DSC case for the matrices `row2m`, `256md`, `Simys`, `ccp`, and `circ2a` are $10^{-6}$, $10^{-6}$, $10^{-4}$, $10^{-4}$, and $10^{-7}$, respectively. In all following experiments, these thresholds are applied. METIS nested dissection is used for reordering. For the reduced matrices, dense rows and columns are treated by direct substitution (see 2). The results in table 1 show that this cancellation of dense rows and columns accelerates the direct solver significantly. The times for the iterative DSC solver are distinctly shorter or comparable with the times of the direct solver for the reduced matrices. In the case of the matrices `256md` and `circ2a`, fill-in in the complete $LU$ factors is exceptionally small.

## 4.2 Parallel Results

**Original versus Reduced System** Table 2 presents execution times of the DSC algorithm on eight processors of SUCCESS for the original and the reduced system. The number of interface variables and of outer iterations (see Fig. 1, right) are given in addition. Repartitioning and reordering are applied. The results in table 2 show that the times for the reduced systems are significantly shorter than the times for the original systems. This is mainly due to a distinctly smaller number of interface variables in the case of the reduced systems. Repartitioning is effective in this case and results in very low costs for the inner iteration from Fig. 1 (right). Therefore, all following experiments are performed with the reduced systems.

[1] The matrices can be made available on request.

**Table 1.** Sequential results on an SGI $O^2$: Direct versus iterative DSC solver.

| | Original matrix | | Reduced matrix | | Direct, time/s | | DSC, time/s |
|---|---|---|---|---|---|---|---|
| Matrix | Order | Non-zeros | Order | Non-zeros | Original | Reduced | Reduced |
| `row2m` | 16570 | 127130 | 16502 | 109731 | 4.4 | 3.5 | 1.1 |
| `256md` | 26114 | 160079 | 26076 | 141011 | 2.3 | 1.5 | 1.9 |
| `Simys` | 24705 | 183713 | 24585 | 127695 | 13.4 | 5.4 | 1.0 |
| `ccp` | 89556 | 760630 | 89378 | 603753 | 51.1 | 27.9 | 5.6 |
| `circ2a` | 482969 | 3912413 | 482963 | 2750390 | 7169.5 | 15.5 | 20.0 |

**Table 2.** DSC times on 8 SUCCESS processors: Original versus reduced system.

| | Original system | | | Reduced system | | |
|---|---|---|---|---|---|---|
| Matrix | #interface vars | Time/s | #iter. | #interface vars | Time/s | #iter. |
| `row2m` | 9123 | 21.69 | 3 | 1792 | 0.80 | 2 |
| `256md` | 14263 | 1.53 | 2 | 8160 | 0.69 | 2 |
| `Simys` | 22950 | 1.33 | 2 | 1504 | 0.12 | 2 |
| `ccp` | 35484 | 23.70 | 2 | 2102 | 1.01 | 2 |
| `circ2a` | 482874 | 2944.90 | 2 | 481 | 1.85 | 1 |

**Effect of Reordering and Repartitioning** In table 3, times of the DSC method on eight SUCCESS processors with both repartitioning and reordering, with repartitioning only, and without repartitioning and reordering are displayed. The number of interface variables is given for the first and last scenario; for the second, the number is the same as for the first one. Additionally, the fill-in deviation, the difference of the maximum fill-in and the mean fill-in per processor divided by the mean fill-in, is specified. This is a measure for the degree of load imbalance during the construction and application of local ILUT factorizations.

The shortest times by far in table 3 are achieved for the DSC method with repartitioning and reordering. Without reordering, load imbalance increases significantly (see column *Deviation* and 3.4). In addition, total fill-in for ILUT without reordering usually is distinctly higher than with reordering. For matrix `circ2a`, e.g., total fill-in is 5245267 in the first scenario and 17561396 in the second. For the third scenario, graph partitioning is not applied, and thus the significantly increased number of interface variables leads to further loss of performance.

**Scalability** Table 4 shows times of the DSC method on up to 64 GRISU processors for the two largest test cases. The speedup for `ccp` on eight processors is 2.3, the corresponding speedup for `circ2a` is 4.0. This shows that the scaling behaviour improves significantly with increasing problem size. On 24 processors, a speedup of 6.7 is achieved for `circ2a` while the speedup on 64 processor is only moderately increased to 7.4. The use of higher processor numbers only makes sense for larger problems since the matrices are extremely sparse.

**Table 3.** DSC times on 8 SUCCESS processors: Effect of ordering and partitioning.

| | Partitioning + ordering | | | Partitioning only | | No permutation | | |
|---|---|---|---|---|---|---|---|---|
| Matrix | #If. vars | Deviation | Time/s | Deviation | Time/s | #If. vars | Deviation | Time/s |
| row2m | 1792 | 4.0% | 0.80 | 139.0% | 2.96 | 10823 | 63.5% | 22.15 |
| 256md | 8160 | 10.4% | 0.69 | 251.3% | 2.38 | 25182 | 102.7% | 6.30 |
| Simys | 1504 | 9.0% | 0.12 | 16.2% | 0.31 | 17996 | 95.5% | 2.68 |
| ccp | 2102 | 10.7% | 1.01 | 30.5% | 2.01 | 65856 | 133.2% | 76.10 |
| circ2a | 481 | 0.6% | 1.85 | 2.2% | 6.62 | 8752 | 1.1% | 37.86 |

**Table 4.** DSC times on GRISU: Scalability.

| | Time/s on $p$ processors | | | | | | | |
|---|---|---|---|---|---|---|---|---|
| Matrix | $p = 1$ | $p = 2$ | $p = 4$ | $p = 8$ | $p = 12$ | $p = 16$ | $p = 24$ | $p = 64$ |
| ccp | 0.78 | 0.64 | 0.38 | 0.34 | | | | |
| circ2a | 2.11 | 1.70 | 0.90 | 0.53 | 0.41 | 0.37 | 0.32 | 0.29 |

## 5 Conclusions

For equation systems from real circuit simulation runs, we demonstrated that the removal of dense rows and columns is crucial for direct and iterative solvers, in particular in the parallel case. For large problems, the iterative DSC algorithm presented usually is superior to a direct solver and shows a favourable scaling behaviour. To achieve the latter, graph partitioning and local ordering methods are necessary since partitioning keeps the interface system small and local ordering improves load balance besides fill-in reduction for local factorizations. Combined with these techniques, the DSC method presented is a well suited iterative solver for circuit simulation.

## References

1. Karypis, G., Kumar, V.: ParMETIS: Parallel graph partitioning and sparse matrix ordering library. Tech. rep. # 97-060, University of Minnesota (1997)
2. Press, W.H., Teukolsky, S.A., Vetterling, W.T., Flannery, B.P.: Numerical Recipes in C. 2nd ed., Cambridge University Press (1994)
3. Saad, Y.: Iterative Methods for Sparse Linear Systems. PWS, Boston (1996)
4. Saad, Y., Sosonkina, M.: Distributed Schur complement techniques for general sparse linear systems. SISC **21** (1999) 1337–1356
5. Van der Vorst, H.: Bi-CGSTAB: A fast and smoothly converging variant of Bi-CG for the solution of nonsymmetric linear systems. SIAM J. Sci. Statist. Comput. **13** (1992) 631–644

# Finitely Large Phased Arrays of Microstrip Antennas – Analysis and Design

Dave J. Bekers[1], Stef J.L. van Eijndhoven, Alphons A.F. van de Ven, Peter-Paul Borsboom[2], and Evert W. Kolk[3]

[1] Eindhoven University of Technology, Department of Mathematics and Computer Science, P.O. Box 513, 5600 MB Eindhoven, The Netherlands, e-mail: d.j.bekers@tue.nl
[2] Thales Nederland B.V., Department JRS, P.O. Box 42, 7550 GD Hengelo, The Netherlands
[3] Thales Air Defense, Haneau de Roussigny, 91470 Limours, France

**Abstract.** This paper focuses on the development of a model to obtain qualitative insight in the behaviour of large, but finite, phased arrays of microstrip antennas. This model concerns a finite array of simple elements, namely perfectly conducting, infinitely thin, narrow rings, excited by voltage gaps and positioned in free or half space. The currents on the rings, and from that the electromagnetic field, are calculated by a moment method. Dimension analysis is carried out to reduce numerical effort and to acquire insight in the behaviour of the array. The qualitative analysis shows promising results and although numerically a brute force method has been applied, CPU times are still acceptable.

## 1 Problem Description

Currently, Thales Nederland is realizing new radar systems consisting of large phased arrays of microstrip antennas. These arrays consist of about 1000 antenna elements positioned on an antenna face of about $16\,\mathrm{m}^2$. The systems scan in azimuth by rotation and in elevation by phase shifts. A narrow main lobe and low side lobe level, an impedance match with the feeding network of the array, and a low cross polarization are design goals.

To analyse such arrays, either a finite array model (or element-by-element approach) or an infinite array model is used at Thales Nederland. The infinite array model requires much less computation time and data storage demand than the finite array model. However, since it cannot account for edge effects and differences between the antenna elements, the need for a finite array model still exists.

Since the actual geometry of the antenna elements is complicated, simulation of a finite array taking into account in detail these elements will require too much computing resources to be realistically feasible. Therefore, we have decided to develop a model based on simple elements that will enable us to find the characteristics that describe the qualitative behaviour of large phased arrays.

## 2 Modelling

The main requirements on our model are the following. Firstly, the arrays are finite such that boundary effects can be incorporated. Secondly, computation times should be in the order of hours. Thirdly, the algorithm should be based on analytical expressions to provide insight in the characteristics or characteristic parameters of an array.

The qualitative model concerns a finite planar array of simple elements in free space or above an infinitely wide ground plate. Considering the radiating part of the actual antenna elements, i.e. a rectangular microstrip loop, we have chosen perfectly conducting, infinitely thin, narrow ring-shaped microstrips, shortly rings, as elements; see Fig. 1. The reasons for this choice are twofold. A ring is the most simple loop geometry and the modes on this geometry can be described analytically.

The rings are excited by voltage gaps at a certain frequency with corresponding wavelength $\lambda$ and wave number $k$. On each ring, the gap is uniform with respect to the width and can be positioned arbitrarily. The widths $2b_q$ of the rings are all of the same order, but much smaller than the wavelength, the radii $a_q$, and the distances between the rings. In other words, $kb_q \ll 1$, $\beta_q = b_q/a_q \ll 1$, and $b_p/(L_{pq} - a_p - a_q) \ll 1$, where $L_{pq}$ is the distance between the centers $\mathbf{c}_p$ and $\mathbf{c}_q$ of ring $p$ and $q$; see Fig. 1. If a ground plate is present, the rings are situated above this plate at height $h$.

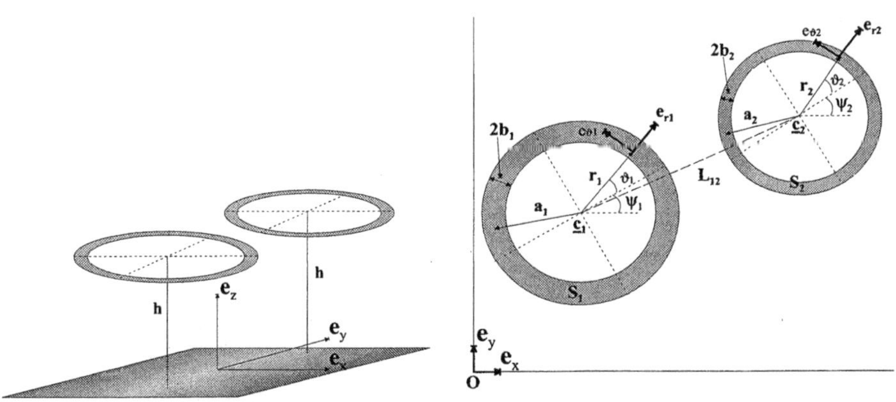


**Fig. 1.** Geometry of an array of two rings.

Since the excitation is time harmonic, the electromagnetic field is so also, and therefore, a (spatial) time-harmonic representation of this field is used. The time dependence is suppressed. The total electric field is written as the sum of a scattered electric field $\mathbf{E}^{scat}$ and an excitation field $\mathbf{E}^{ex}$. The scattered electric field is expressed in terms of the unknown current density $\mathbf{J}$ on the rings by an integro-differential operator acting on the current,

$$\mathbf{E}^{scat} = \mathcal{L}\mathbf{J}. \tag{1}$$

The operator $\mathcal{L}$ can be factorized as $\mathcal{L} = \mathcal{D}\mathcal{T}$, where $\mathcal{D}$ is the differential operator $-j\omega\mu(\mathcal{I} + k^{-2}\nabla\nabla\cdot)$ and $\mathcal{T}$ is the integral operator described by a Green's kernel, where $\mathcal{T}\mathbf{J}$ is the magnetic vector potential. The condition that the tangential component of the total electric field is zero at the surface $S$ of the rings yields an equation for the current,

$$(\mathcal{L}\mathbf{J})_{tan} = -(\mathbf{E}^{ex})_{tan}, \qquad \text{on } S. \tag{2}$$

Here, $(\,\cdot\,)_{tan}$ is a trace operator that restricts a vector field to the surface $S$ and projects this field tangentially on $S$. Equation (2) is solved by a moment method expanding the unknown current into a finite number of expansion functions with unknown coefficients. Once the current is known, the electromagnetic far field can be calculated analytically.

## 3 Analytical Aspects

In this section, we describe globally all essential steps of the calculation of the current $\mathbf{J}$. A publication with more details is in preparation and will appear in 2003 as pre-publication in [1].

Let the surface of the q-th ring be $S_q$. On each $S_q$, a polar parameter representation is chosen, the orientation of which is described by the angle $\psi_q$; see Figure 1. The voltage gap on $S_q$ is positioned in $\vartheta_q = \pi$. Hence, $\psi_q$ determines not only the orientation of the parameter representation, but also the position of the gap. Two tangent vectors $\mathbf{e}_{r_q}$ and $\mathbf{e}_{\vartheta_q}$ correspond in the usual way to this representation. Together with the corresponding normal, they form a local coordinate system on $S_q$, which is extended straightforwardly to global coordinate system.

It is assumed that the current $\mathbf{J}_q = \mathbf{J}|_{S_q}$ is directed along the centerlines of the rings and that it is uniform with respect to the width $b_q$,

$$\mathbf{J}_q(r_q, \vartheta_q) = w(\vartheta_q)\mathbf{e}_{\vartheta_q}. \tag{3}$$

The basis of this assumption is that the wavelength is much larger than the widths of the rings. Expressing $\mathcal{L}\mathbf{J}_q$, i.e. the scattered electric field induced by ring $q$, into the system of $S_p$, we calculate $(\mathcal{L}\mathbf{J}_q)_{tan}$ on $S_p$ straightforwardly by putting the axial coordinate equal to zero and omitting the axial component. Then, a differential operator $\mathcal{D}_{r_p\vartheta_p}$ is determined such that

$$(\mathcal{L}\mathbf{J}_q)_{tan}|_{S_p} = \mathcal{D}_{r_p\vartheta_p}(\mathcal{T}\mathbf{J}_q|_{S_p}), \tag{4}$$

where $|_{S_p}$ denotes the restriction to $S_p$. Hence, the projection of the trace operator is incorporated in $\mathcal{D}_{r_p\vartheta_p}$.

The impedance matrix component for a test function $\mathbf{v}_p$ and an expansion function $\mathbf{J}_q$ is given by

$$<\mathbf{v}_p, (\mathcal{L}\mathbf{J}_q)_{tan}|_{S_p}>, \tag{5}$$

where $<\cdot,\cdot>$ is the inner product on functions defined from $S$ to the tangent space of $S$. It is shown that the variation of $(\mathcal{L}\mathbf{J}_q)_{tan}|_{S_p}$ in $r_p$ is of order $\beta_p$ with respect to its variation in $\vartheta_p$. This implies that $(\mathcal{L}\mathbf{J}_q)_{tan}|_{S_p}$ depends only weakly on $r_p$. Therefore, the test function $\mathbf{v}_p$ is chosen uniform with respect to $r_p$ and tangentially directed only. This means

$$\mathbf{v}_p(r_p, \vartheta_p) = v(\vartheta_p)\mathbf{e}_{\vartheta_p}. \tag{6}$$

Then, the impedance matrix component (5) turns into

$$<\mathbf{v}_p, (\mathcal{L}\mathbf{J}_q)_{tan}|_{S_p}> = \int_{\vartheta_p} v_p(\vartheta_p) \int_{r_p} (\mathcal{D}_{r_p\vartheta_p}[\mathcal{T}\mathbf{J}_q|_{S_p}])_{\vartheta_p}\, d\sigma(r_p, \vartheta_p). \tag{7}$$

Neglecting terms of order $\beta_p$, we reverse the differential operator $\mathcal{D}_{r_p\vartheta_p}$ and the integral with respect to $r_p$. This leads to averaging of the Green's kernel with respect to the radial source and observation coordinates. In case $p = q$, the averaged kernel has a logarithmic singularity, otherwise it is regular. Requiring that the test function $v_p$ and the expansion function $w_q$ have square integrable second and first generalized derivatives, respectively, we transfer the reversed differential operator to $v_p$. The resulting differential operator incorporates the Helmholtz operator. Together with the periodic boundary conditions, this operator induces a Sturm-Liouville problem for $v_p$, the eigen functions of which are chosen as test functions. Then, the expansion functions are chosen equal to the test functions. The resulting impedance matrix component is a double integral, which can be reduced to a single integral in case $p = q$ by use of properties of inner product and convolution.

Choosing a finite number of test and expansion functions on each $S_p$, we obtain an impedance matrix built up of blocks, which describe the self and mutual coupling of the rings. The blocks on the diagonal are diagonal matrices describing the self coupling of each ring, while the other blocks are dense matrices describing the mutual coupling between each pair of rings. The expansion coefficients are calculated by an LU-factorization of the impedance matrix.

## 4 Numerical Results

The first result we show is used for validation of the implementation. We show the real and imaginary part of the current through one ring in free space, excited at a frequency of 3 GHz; see Fig. 2. We can compare the result with known results from literature: the current through a wire loop excited by a voltage gap at the same frequency; see [2, Fig. 2 and 3] and [3]. Here, we use a rule of thumb found by Kraus [4, p. 238], which states that the results for a thin strip of width $w$ and a wire with cross-sectional radius $w/4$ are equivalent. Two expansion functions show already an accurate result for the real part of the current. Four expansion functions show a quite accurate result for the imaginary part of the current, except near the voltage gap in $\vartheta = \pi$. The reason for the latter is the following. It can be shown that since the current has a square integrable generalized derivative, the excitation field should be square integrable. However, the delta functions that describe the voltage gaps are not square integrable. Furthermore, the expansion functions do not only have square integrable generalized derivatives, but are even continuously differentiable.

Figure 3 shows the effect of a ground plate. Figure 3.(a) shows the current amplitude for a ring in free space, and at $h = \lambda/4$, $h = \lambda/2$, and $h = \lambda$ above a ground plate. The current is normalized on the maximum amplitude in free space. Due to interference, the amplitude for $h = \lambda/4$ is lower than for free space, and for $h = \lambda/2$ and $h = \lambda$ higher. Due to space attenuation, the amplitude for $h = \lambda$ is lower than for $h = \lambda/2$. Figure 3.(b) shows the far field components in the $yz$-plane for free space and for $h = \lambda/4$. Here, a spherical coordinate is chosen that is related to the cartesian coordinate system in Fig. 1 in the usual way. The influence of the ground plate can be observed from the behaviour of the $\phi$-component, i.e. the cross-polarization, that vanishes at $\vartheta = 90^o$ (endfire) for the ground plate, but not for free space.

Figure 4 shows results for two line arrays of 7 identical rings with spacings $7\lambda/15$ and $3\lambda/5$. Here, 8 expansion functions per ring is a suitable choice. The orientation of the local coordinate systems on the rings is such that $\psi_q = 0$. As aforementioned, the voltage gaps, all of equal amplitude, are positioned in $\vartheta_q = \pi$. The centers of the rings are positioned on the positive $x$-axis, where the center of the first ring is in the origin; see Fig. 1. The CPU time of a Matlab implementation on a HP PC with Windows NT, an Intel Pentium IV processor at 1.0 GHz, and 256 Mb of RAM is 69 sec. Figures 4.(a)-(b) show the normalized radiation intensities in the $xz$ and $yz$-plane ($H$ and $E$-plane) together with the intensity of one ring. In the $xz$-plane, one main lobe and several side lobes are observed for both spacings, where the number of lobes is related to the spacing. In the $E$-plane, only one lobe is observed. These results are in qualitative correspondence with results from literature; see [5, Chapter 3]. Besides that the array with larger spacing has more side lobes, also its maximal radiation intensity is higher. This effect is due to the degree of mutual coupling. Influence of mutual coupling on the current is shown in Fig. 4.(c)-(f). For smaller spacing, the amplitudes of the currents differ significantly from the amplitude on one ring, while for larger spacing, they differ only slightly. In particular, the maxima of the amplitudes for the spacing $7\lambda/15$ are significantly lower than the maximum for one ring. For the spacing $3\lambda/5$, these maxima are both slightly higher and lower than the maximum for one ring. The phases differ for both spacings from the phase on one ring.

## 5 Conclusions

We have developed a tool for analysing finite arrays of rings that is easy to handle. The algorithm is based on analytic expressions as required. The validation has been successful; results have shown to be in qualitative correspondence with literature and practice. Although numerically, a brute force method has been applied, the CPU times are acceptable. However, to analyse large arrays of about 100 elements or more, they should be reduced. The accuracy is sufficient for qualitative analysis.

## 6 Prospects

Research on characteristics of arrays and essential aspects of the antenna elements will be topic of further research. A transparent relation should be established between excitation, geometry, and scattered field. Finally, feedback should be provided to the hardware designers of Thales Nederland.

## References


1. Reports on Applied and Numerical Analysis, Department of Mathematics and Computing Science, Eindhoven University of Technology, P.O. Box 513, 5600 MB Eindhoven, The Netherlands (http://www.win.tue.nl/analysis/preprints/).
2. Poljak, D.: Finite Element Integral Equation Modelling of a Thin Wire Loop Antenna. Commun. Numer. Meth. Engng. **14** (1998) 347-354.
3. Champagne, N.J., Williams, J.T., Wilton, D.R.: The use of curved segments for numerically modeling thin wire antennas and scatterers. IEEE Trans. Ant. Prop. **40** (1992) 682-688.
4. Kraus, J.: Antennas. McGraw-Hill, New York (1950).
5. Stutzman, W.L., Thiele, G.A.: Antenna Theory and Design. Wiley, New York (1998, 2nd, ed.).

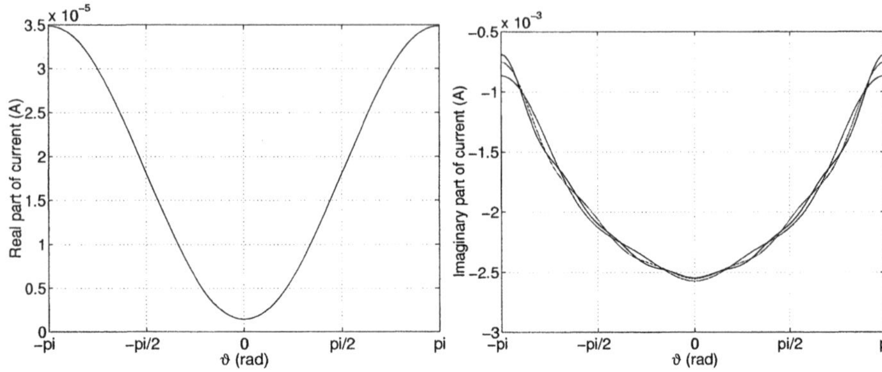


(a) (b)

**Fig. 2.** The total current on a ring excited by a voltage gap of 1 V at a frequency of 3 GHz.; $a = 0.0637\,\lambda$, $b = 0.0027\,\lambda$, $\mathbf{c} = (0,0,0)$, and $\psi = 0$. (a) Real part; 2 expansion functions of cosine type. (b) Imaginary part; blue, red, black: 4, 6, and 8 expansion functions of cosine type.

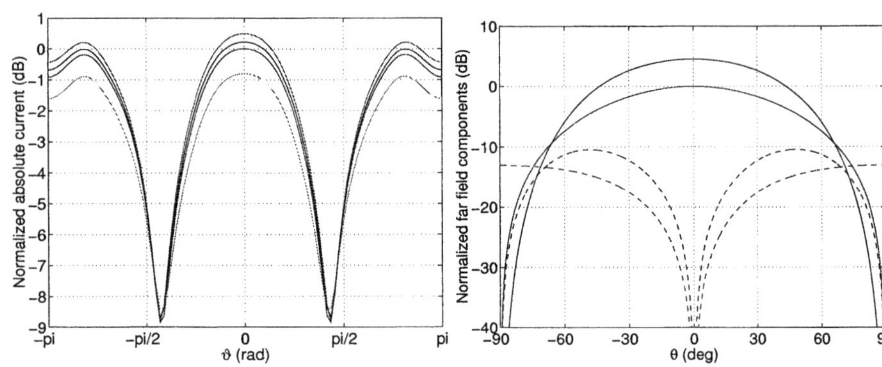


(a) (b)

**Fig. 3.** A ring excited by a voltage gap of 1 V with 8 expansion functions of cosine type; $ka = 2\pi/5$, $\beta = 1/40$, $\mathbf{c} = (0,0,0)$, $\psi = 0$. (a) Current amplitudes normalized on the maximum amplitude in free space; blue: free space; green: $h = \lambda/4$; red: $h = \lambda/2$; black: $h = \lambda$. (b) Far field components in the $yz$-plane normalized on the maximum of $|E_\vartheta|$; blue: free space; black: $h = \lambda/4$; solid lines: $E_\vartheta$-components (co-polarization); dashed lines: $E_\phi$-components (cross-polarization).

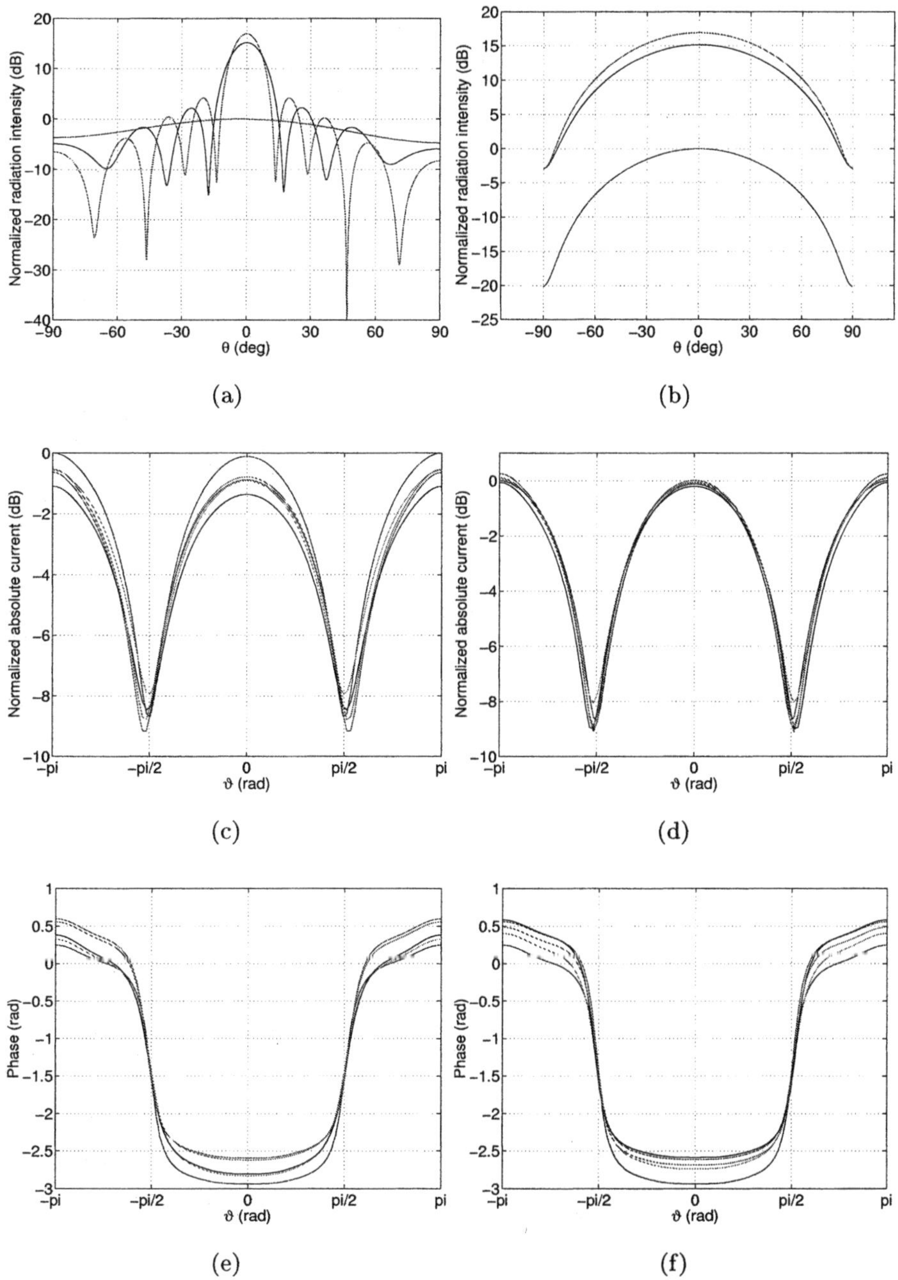


**Fig. 4.** Results for line arrays of 7 rings of equal radii ($ka_q = 2\pi/6$) and widths ($\beta_q = 3/50$), with voltage gaps of equal amplitude (1 V), and positioned in free space.
(a)-(b): Radiation intensities in the $xz$ and $yz$-plane, resp., normalized on the maximum intensity of one ring; blue: one ring; black: spacing $7\lambda/15$ ($\mathbf{c}_q = 7\lambda(q-1)/15$); red: spacing $3\lambda/5$ ($\mathbf{c}_q = 3\lambda(q-1)/5$).
(c)-(f): Current amplitudes and phases for spacings $7\lambda/15$ ((c), (e)) and $3\lambda/5$ ((d), (f)); blue: one ring; green/red/black/purple: from first ring to center ring.

# The Sommerville Mesh in Yee-like Schemes

Alain Bossavit

LGEP, 11 Rue Joliot-Curie, 91192 Gif-sur-Yvette, France

**Abstract.** The idea of modelling space as two interacting equivalent networks, one for currents, one for magnetic fluxes, pervades computational electromagnetics since its beginnings. The Yee scheme, the TLM method, can thus be interpreted. But this is also true of finite element- or finite volume-inspired more recent proposals, as we show, so the idea is not incompatible with "unstructured" meshes. Yet, meshes with some rotational and translational symmetry (locally, at least) are desirable on many accounts. The tetrahedral Sommerville mesh we describe here, able to fit curved boundaries and yet regular, looks like an interesting compromise.

## 1 Introduction

Sommerville [3], working in the Twenties on the problem of space-filling tetrahedra (the story is well told in [2]), found one which is especially interesting in the context of modern FDTD-like methods.

Such Yee-like methods [4][5] retain the essential features of FDTD, i.e., they can be interpreted as the formation of two interlocked equivalent networks, one magnetic, one electric, but the underlying pavings need not be the two staggered cubic lattices of Yee's scheme. They can be "cell-complexes in duality", meaning there is a primal mesh, made of nodes, edges, etc., of a paving by convex cells of any shape, and a dual mesh, made of dual volumes, dual facets, etc., in one-to-one correspondence with the primal ones. Degrees of freedom are magnetic fluxes, emf's, relative to primal cells, and mmf's, currents, relative to dual ones. "Network" equations, which are nothing else than a generalization of Kirchhoff's laws, are then set up, in a way which is essentially unique. But they must be supplemented by "network constitutive laws", in order to couple emf's with currents, and mmf's with fluxes. These laws, which are encoded in square symmetric matrices $\boldsymbol{\nu}$ and $\boldsymbol{\epsilon}$, the size of which is the number of primal facets and edges respectively, must be constructed in analogy with the constitutive laws $H = \nu B$, $D = \epsilon E$ (or $J = \sigma E$).

But such analogies are not that compelling, and in contrast to network equations, network constitutive laws come in many flavors, almost as many as investigators. For instance [1], the use of edge elements in a Galerkin variational approach does result in such laws, but with the drawback of producing non-diagonal $\boldsymbol{\nu}$ and $\boldsymbol{\epsilon}$ matrices, while efficiency in the simulation requires diagonal matrices, in order to have an explicit time evolution scheme. "Diagonal lumping" procedures can help in this respect, but their implementation still raises many problems.

Another approach (which on the face of it doesn't seem to require finite elements, but holding this view would be wrong) consists in building "mutually orthogonal" meshes: each primal edge pierces its associate dual facet at right angle, etc. It's then straightforward [4] [5] to get diagonal $\mu$ and $\epsilon$ matrices. But whereas the Galerkin approach is available on any reasonably behaved primal mesh, very few primal meshes will allow construction of an orthogonal dual. This condition seems to impose a kind of rigidity on the primal mesh which severely constrains its topology and its metric. The cubic lattice of Yee's scheme satisfies these constraints, and still does after some distortion. But the notorious "staircasing" problem plays against it. Hence the search for more flexible substitutes. The Sommerville mesh, based on the first Sommerville space-paving tetrahedron, promises to be one.

## 2 Yee-like methods

The geometric approach to electromagnetism stems from the remark that all its observables are integrals over some manifold of dimension $p$ (a line if $p = 1$, a surface if $p = 2$, etc.). For instance, the electric field is known via electromotive forces (emf's), which are line integrals such as $\int_c \tau \cdot E$, where $c$ is a curve with unit tangent vector $\tau$, and $E$ the electric vector field. Let's denote by $e(c)$, or for better mnemonic value, by $\int_c e$, this emf. This enhances the status of the electric field as a *mapping*, from curves to real numbers (with obvious properties of additivity and continuity), assuming the value $\int_c e$ when applied to $c$, a map we denote by the symbol $e$. What counts, physically, is this mapping $e$, not the "proxy" vector field $E$ by which it can be expressed, once a metric-defining scalar product has been introduced. A similar argument shows that magnetic induction is a map, denoted $b$, from surfaces to reals, whose values $b(S)$, again better denoted by $\int_S b$, are induction fluxes. Mappings of this kind are called "$p$ forms" in differential geometry, where $p$ refers to the dimension.

In this spirit, Maxwell's equations become differential relations between forms, as follows:

$$d_t \int_S b + \int_{\partial S} e = 0, \tag{1}$$

$$-d_t \int_\Sigma d + \int_{\partial \Sigma} h = \int_\Sigma j, \tag{2}$$

where $\partial$ denotes the boundary, for all surfaces $S$ and $\Sigma$. (The $S$ vs $\Sigma$ notation is meant to stress a difference in the way these surfaces are oriented: each $S$ has "inner" orientation [4], i.e., a specified clockwise gyration sense, which matches the orientation of the boundary $\partial S$, whereas $\Sigma$ has "outer" orientation, i.e., a crossing direction through it, which matches the way to "turn around" $\partial \Sigma$. See [1] for more details.)

The basic discretization move, then, is to satisfy (1) and (2), not on *all* surfaces $S$ and $\Sigma$, but on all those (in finite number, if one works in a bounded region) built from facets of suitable meshes.

So let $\mathfrak{m}$ be a meshing of the computational domain $D$, made of volumes $v$ (indexed over the set $\mathcal{V}$), which have in common facets $f$ (indexed over $\mathcal{F}$), which hinge along edges $e$ (indexed over $\mathcal{E}$), which meet at nodes $n$ (indexed over $\mathcal{N}$). Each of these $p$-cells is given an orientation of its own, hence the definition of incidence numbers (e.g., $\mathbf{R}^e_f$ between edge $e$ and facet $f$), which say whether two cells of dimension $p$ and $p+1$ meet (otherwise, the number is 0), and if they do, whether their orientations match or not (e.g., $\mathbf{R}^e_f = \pm 1$ depending on whether $e$ goes along the gyratory sense attributed to $f$ or counter to it). Other incidence numbers $\mathbf{G}^n_e$ and $\mathbf{D}^f_v$ are similarly defined and form (rectangular) matrices $\mathbf{G}$, $\mathbf{R}$, $\mathbf{D}$. It's easy to see that $\mathbf{RG} = 0$ and $\mathbf{DR} = 0$.

Let's now decide to represent $b$ and $e$ by (time-dependent) arrays $\mathbf{b} = \{\mathbf{b}_f : f \in \mathcal{F}\}$ and $\{\mathbf{e}_e : e \in \mathcal{E}\}$ of "degrees of freedom" (DoF), interpreted as fluxes and emf's relative to the individual facets and edges of $\mathfrak{m}$. It's an approximate representation, in the sense that it tells about fluxes [resp. emf's] relative to "$\mathfrak{m}$-surfaces" only, i.e., those made of facets of $\mathfrak{m}$ [resp. to "$\mathfrak{m}$-lines"]. To satisfy (1) for facet $f$ means, as one will easily realize,

$$d_t \mathbf{b}_f + \sum_{e \in \mathcal{E}} \mathbf{R}^e_f \mathbf{e}_e = 0,$$

and this, if true for all $f$, enforces (1) for all $\mathfrak{m}$-surfaces, by additivity. Hence, in matrix form, a first group of equations between DoF arrays:

$$d_t \mathbf{b} + \mathbf{Re} = 0, \tag{3}$$

a spatial discretization of (1). (Time-discretization will be straightforward, and we gloss over it.) In search for an analogous group in similar relation to (2), we notice that DoF's for $h$ and $d$ (or $j$) should "sit at the same place" as those for $b$ and $e$, owing to the local character of the constitutive laws $b = \mu h$ and $d = \epsilon e$ (or $j = \sigma e$). Hence the decision to consider a *dual* mesh $\tilde{\mathfrak{m}}$, whose respective nodes, edges, etc., are in one-to-one correspondence with the volumes, facets, etc., of $\mathfrak{m}$: a dual node inside each primal volume, a dual edge piercing each primal facet, and so forth. (Notice this is exactly the situation with the staggered grids in FDTD.) This way, the incidence matrices of the dual mesh appear to be the transposes $\mathbf{D}^t, \mathbf{R}^t, \mathbf{G}^t$ of the primal ones. Then, again, DoF arrays $\mathbf{h} = \{\mathbf{h}_f : f \in \mathcal{F}\}$ and $\mathbf{d} = \{\mathbf{d}_e : e \in \mathcal{E}\}$ approximately represent $h$ and $d$, data about $j$ translate (by integration over the dual facets) into a known time-dependent array $\mathbf{j} = \{\mathbf{j}_e : e \in \mathcal{E}\}$, and a second group of equations is obtained:

$$-d_t \mathbf{d} + \mathbf{R}^t \mathbf{h} = \mathbf{j}. \tag{4}$$

Note the almost compulsory nature of this discretization process: All we have done is, enforce (1) and (2) for all $\mathfrak{m}$-surfaces $S$ and all $\tilde{\mathfrak{m}}$-surfaces $\Sigma$ in the only possible manner.

By duality of the two meshes, the form of the discrete constitutive laws is $\mathbf{h} = \boldsymbol{\nu}\mathbf{b}$ and $\mathbf{d} = \boldsymbol{\epsilon}\mathbf{e}$, where $\boldsymbol{\nu}$ and $\boldsymbol{\epsilon}$ are square matrices. How to construct them is the core of the discretization problem. It can be solved in essentially two ways, the "orthogonal construction" and the "barycentric construction", which give birth to a lot of variants.

First, one may in some cases have two "mutually orthogonal" meshes. By this it is meant that not only all cells, primal and dual, are straight (i.e., for $p = 1$ or 2, contained in a line or a plane), a requirement not made so far, but that a pair of mated cells are orthogonal: the dual edge $\widetilde{f}$ is orthogonal to facet $f$, etc. (Fig. 1), and that their orientations (inner for one, outer for the other) match, too. Then a very natural way to build $\boldsymbol{\nu}$ and $\boldsymbol{\epsilon}$ is to have them *diagonal* with entries

$$\boldsymbol{\nu}^{ff} = \nu_f \,\text{length}(\widetilde{f})/\text{area}(f), \quad \boldsymbol{\epsilon}^{ee} = \epsilon_e \,\text{area}(\widetilde{e})/\text{length}(e), \tag{5}$$

where $\nu_f$ and $\epsilon_e$ are the local values of $\nu$ and $\epsilon$, if these are smooth functions of position. (The case of discontinuous ones is hardly more difficult, see [1].) With such matrices and a leap-frog scheme with time step $\delta t$ small enough (for stability), (3)(4) plus the "network constitutive laws",

$$\mathbf{h} = \boldsymbol{\nu}\mathbf{b}, \tag{6}$$

$$\mathbf{d} = \boldsymbol{\epsilon}\mathbf{e}, \tag{7}$$

give birth to an explicit forward-marching integration scheme, which is nothing else than FDTD in the case of a cubic lattice-like mesh and its dual.

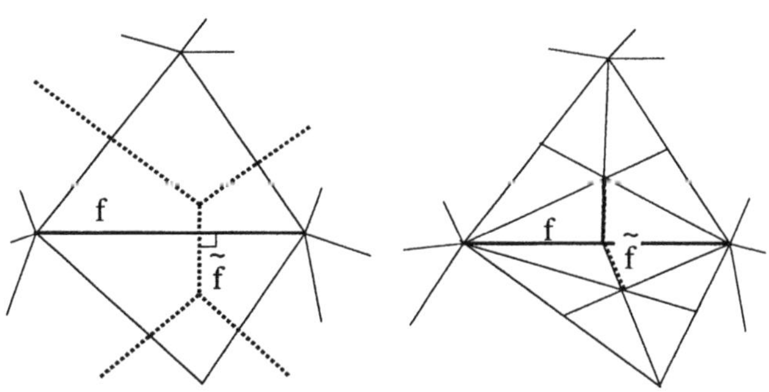


**Fig. 1.** (one dimension suppressed). Dual edges in the orthogonal construction (left) and the barycentric one (right)

Actually, (5) is a way to enforce, in the orthogonal case, a "consistency criterion" of wider scope, namely

$$\sum_{f' \in \mathcal{F}} \boldsymbol{\nu}^{ff'} f' = \nu_f \widetilde{f}, \qquad \sum_{e' \in \mathcal{E}} \boldsymbol{\epsilon}^{ee'} \widetilde{e'} = \epsilon_e e, \tag{8}$$

where the on-the-line $f'$, $\widetilde{e}'$ and $\widetilde{f}$, $e'$, stand for vectors: the vectorial area of facets $f'$ and $\widetilde{e}'$, the vector joining the end points of edge $\widetilde{f}$, the vector along $e'$. It can be shown (this is where finite elements are necessary, as a way to reconstruct fields from DoF arrays, and thus make a comparison between the exact and the approximate possible) that (8), which mixes the entries of $\boldsymbol{\nu}$ and *metric* information about the mesh, is a necessary—and, under reasonable complementary assumptions, sufficient—condition for the convergence of the above scheme.

Designing numerical schemes for the Maxwell equations, in this approach, thus reduces to building a pair of meshes in duality and matrices $\boldsymbol{\nu}$ and $\boldsymbol{\epsilon}$ which satisfy (8). (Given the pair of meshes, such matrices always exist. The difficulty is to have them *symmetric*.) Remarkably—and this is the second way to solve the discretization problem—the so-called "mass matrices" of facet elements and edge elements on a *tetrahedral* mesh, when taken as $\boldsymbol{\nu}$ and $\boldsymbol{\epsilon}$ respectively, happen to satisfy (8) when the dual mesh is the barycentric one. The Galerkin approach with Whitney elements, therefore, does not essentially differ from the Yee scheme and its modern avatars [4][5]: it just trades the inconvenience of having non-diagonal matrices $\boldsymbol{\epsilon}$ and $\boldsymbol{\mu}$ (this can be alleviated by "mass lumping" techniques) for easiness in building the dual mesh.

This is a genuine trade-off, for the "orthogonal dual and diagonal matrices" approach is not that straightforward: As Fig. 2 should make it plain, only a very limited family of primal meshes will admit suitable duals in the required mutual orthogonality relationship. On the other hand, mutual orthogonality, because of diagonality that comes with it owing to (8), is a very desirable feature, which offers an attractive interpretation of (3)(4)(6)(7). We now turn to that.

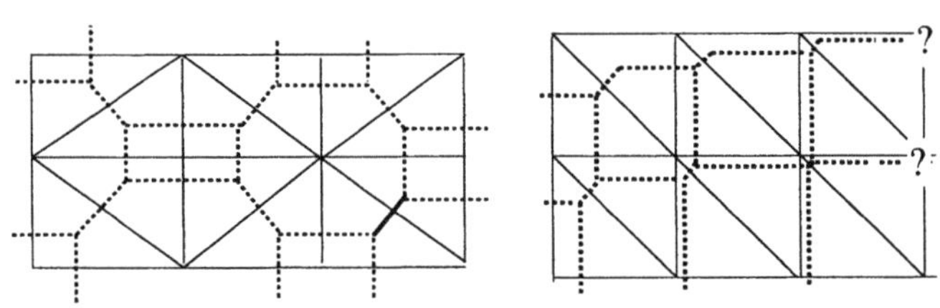


**Fig. 2.** Mutual orthogonality is easily achieved with some meshes (left), but regularity is not enough in this respect, even in dimension 2 (right). Odds are rather small that an arbitrary primal mesh, even good-looking, admit an orthogonal dual.

## 3 Interlocked equivalent networks

Suppose $\mathbf{b} = 0$ and $\mathbf{e} = 0$ at time $t = 0$, and let $\mathbf{q}$ stand for the node-based array of electric charges defined by $\mathbf{q}(t) = -\int_0^t \mathbf{G}^t\mathbf{j}(s)\,ds$. Then (3) and (4) imply $\mathbf{Db} = 0$ and $-\mathbf{G}^t\mathbf{d} = \mathbf{q}$. Displaying the equations like this ($\boldsymbol{\mu}$ is the inverse of $\boldsymbol{\nu}$):

$$\mathbf{G}^t\mathbf{d} = -\mathbf{q}, \quad \mathbf{d} = \boldsymbol{\epsilon}\mathbf{e}, \quad \mathbf{Re} = -d_t\mathbf{b}, \tag{9}$$

$$\mathbf{Db} = 0, \quad \mathbf{b} = \boldsymbol{\mu}\mathbf{h}, \quad \mathbf{R}^t\mathbf{h} = \mathbf{j} + d_t\mathbf{d}, \tag{10}$$

with a symmetry which would be even more striking if we added (non-physical) magnetic currents and charges, we see that (9)(10) describe two networks, subject to Kirchhoff's laws, each one exciting the other. Equations (9) rule the primal, or *electric* network: $\mathbf{G}^t\mathbf{d} = -\mathbf{q}$ expresses charge conservation at nodes, $\mathbf{Re} = -d_t\mathbf{b}$ is the loop law, with loop emf's due to flux variations at the right-hand side, and $\mathbf{d} = \boldsymbol{\epsilon}\mathbf{e}$ links branch emf's with branch (displacement) currents. Sources are $\mathbf{q}$, a given, and $d_t\mathbf{b}$, broadcast by the other network. The latter, magnetic, also has its node law, $\mathbf{Db} = 0$, its branch permeances (the entries of $\boldsymbol{\mu}$), and its loop law, with a source mmf provided, concurrently, by the given currents $\mathbf{j}$ and the displacement currents communicated by the first network. A lot of variations are possible on this basic theme, including the coupling with outside circuits, lumped elements, etc., which come very naturally.

Of course, the suggested interpretation of $\boldsymbol{\mu}$ and $\boldsymbol{\epsilon}$ as branch permeabilities and permittivities is a bit blurred when these matrices are not diagonal: It all goes as if, for instance, flux in "branch" $f$ (which one can construe as going through facet $f$) depended not only on the mmf $\mathbf{h}_f$ in this very branch, but also on mmf's associated with the nearby dual edges. Hence the interest for mesh-design methods that would give the best of both worlds: Easy construction of an orthogonal dual, on the one hand, and better fitting of curved boundaries, which is what tetrahedral volumes are felt useful for.

## 4 The Sommerville mesh

The Sommerville mesh is very promising in this respect. Its generating volume, here referred to as "the sommerville" (Fig. 3), is a tetrahedron with two edges of length $a$ and four of length $b$, with $3a^2 = 4b^2$. It paves space (Figs 4 and 5), forming a primal mesh which happens to be the Voronoi–Delaunay one associated with a body-centered cubic lattice (hence the orthogonality of the dual mesh). Note the octahedra are not regular, here. They are slightly squashed along the vertical direction. So this is not the standard "octet truss" tessellation, with, in two-to-one proportion, regular tetrahedra and octahedra (the latter, cut into four parts). But remarkably, it is an even more symmetrical paving. Although there seems to be, at first glance, a privileged direction (vertical), along which center masts would align, one soon realizes that so is not the case. Let's call "beams" the edges parallel to the natural Cartesian axes, and "struts" the slanted edges. *Each* beam, not only each vertical one, is surrounded by four sommervilles forming a squashed tetrahedron—but now squashed along the central beam's direction. The network, therefore, is as "isotropic" as can be achieved, with a correlative improvement of numerical dispersion relations with respect to FDTD.

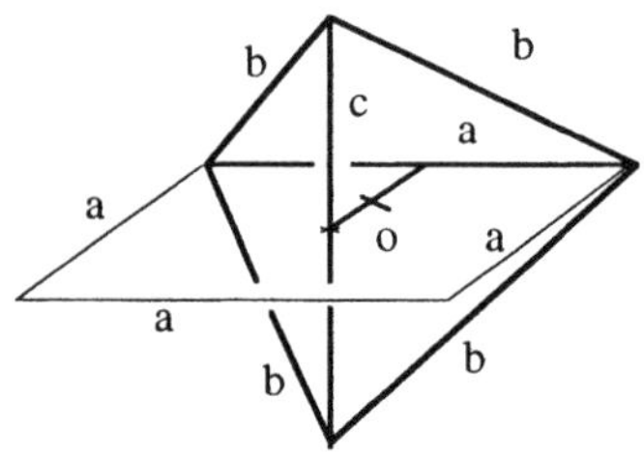


**Fig. 3.** Four tetrahedra like the one in the rear can be assembled around the vertical pillar to form an octahedron (not a regular one). If $c = a$, i.e., if $3a^2 = 4b^2$, we have a Sommerville tetrahedron there. Point $o$ marks the position of the center of the circumscribed sphere, well inside the tetrahedron (even if the latter is slightly deformed).

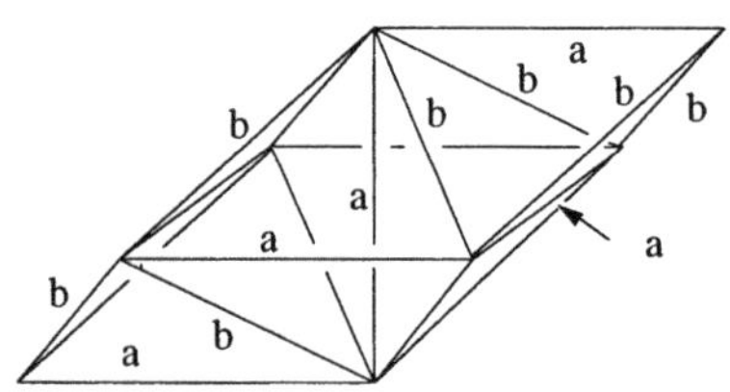


**Fig. 4.** Thanks to the $3a^2 = 4b^2$ relation, two additional Sommerville tetrahedra (up right and bottom left) complete the octahedron to make a space-filling hexahedron. The sommerville is therefore a space-filler.

The dual mesh is provided by the orthogonal construction: Take the circumcenters, and join them. Dual edges obtained this way are automatically orthogonal to primal facets, and the other way round. (Moreover, they do meet each other, which is not warranted in the more general Voronoi–Delaunay construction.) Like the primal mesh, this one is also a paving by a single space-filler, namely, the truncated octahedron (Fig. 7), one of the so-called Archimedean polyhedra (also known as "tetrakaidecahedron", which would already by itself justify the visit). But the staircasing problem, though

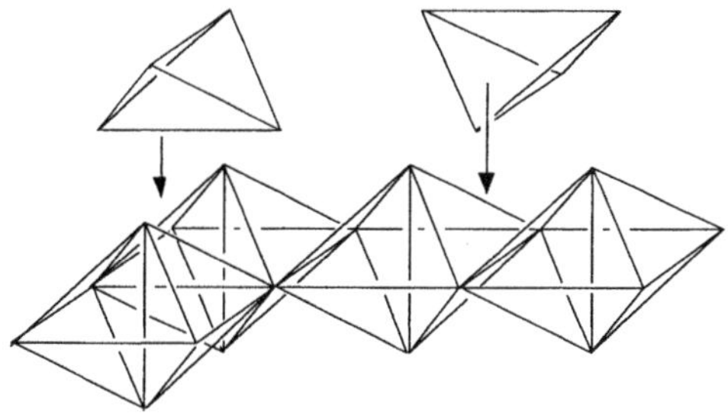

**Fig. 5.** Indeed, one may stack the hexahedra thus obtained, which amounts to combine octahedra and tetrahedra in the familiar "octet truss" pattern: First lay the octahedra side by side, then add sommervilles, two for each octahedron, thus obtaining a horizontal egg-crate shaped slab, with pyramidal holes, ready to be filled by a similar slab, superposed, and so on.

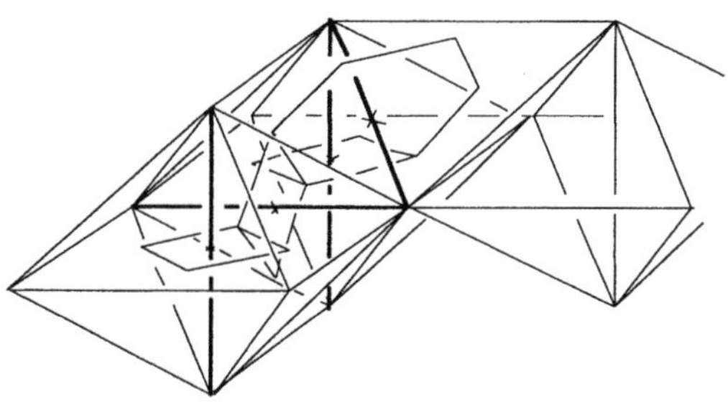

**Fig. 6.** A few dual facets (bounded by dual edges, themselves obtained by connecting the circumcenters of adjacent tetrahedra while turning around a given primal edge). Those around beams are squares, those around struts are hexagons.

still present, is much less acute. The mesh tolerates distortion, like the cubic one. Refinement is easy, as a sommerville splits into eight smaller ones.

Implementation is not difficult. One first sets up a cubic lattice, nodes of which are numbered with even integers, the generic node thus being $\{2p, 2q, 2r\}$. A second, staggered lattice with odd nodal coordinates is then added. Edges of both grids are our beams. Now join each "even" node to the eight "odd" nodes around it: here are the struts. It's convenient to label edges (and for that matter, all cells) by the coordinates of their mid-points. This way, $\{2p, 2q+1, 2r\}$ points to a beam, $\{2p \pm 1/2, 2q \pm 1/2, 2r \pm 1/2\}$ to a strut, etc. This makes it easy to span the set of edges which "interact", in the discrete analogue of the curl-curl equation, with a given one, i.e., those edges which share a facet with it. Two struts interact, in this sense, if only one of their "coordinates" differ, by $\pm 1$. A strut and a beam interact if the three coordinates differ by $\pm 1/2$, etc. Thanks to such rules, the network equations can be formed "by hand", in easy finite difference fashion, as with the Yee scheme.

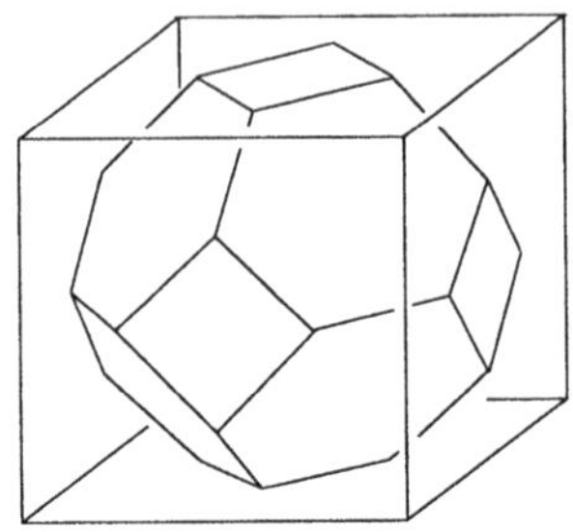

**Fig. 7.** The dual volume is a truncated octahedron, a known space filler. There is one around each primal node.

## References

1. Bossavit, A.: 'Generalized Finite Differences' in computational electromagnetics, PIER 32 (F.L. Teixeira, ed.), EMW (Cambridge, Ma), 2001, pp. 45-64. `http://ceta-mac1.mit.edu/pier/pier32/02.bossavit.pdf`
2. Senechal, M.: Which Tetrahedra Fill Space? Math. Magazine, **4,** 5 (1981), pp. 227-43.
3. Sommerville, D.M.Y.: Space-filling Tetrahedra in Euclidean Space. Proc. Edinburgh Math. Soc., **41** (1923), pp. 49-57.
4. Tonti, E.: A Direct Formulation of Field Laws: The Cell Method. CMES, **2,** 2 (2001), pp. 237-58.
5. Weiland, T.: Time domain electromagnetic field computation with finite difference methods. Int. J. Numer. Modelling, **9** (1996), pp. 295-319.

# Hierarchical Simulation of Substrate Coupling in BiCMOS Structures Using the Boundary Element Method

Thomas Brandtner[1] and Robert Weigel[2]

[1] Infineon Technologies, Development Center Graz, Austria
[2] University Erlangen-Nuremberg, Erlangen, Germany

**Abstract.** A novel substrate coupling simulation tool named *SubCALM* is presented. It is well suited to floorplanning of large mixed-signal designs since it exploits the boundary element method and contains a Poisson solver based on a hierarchical $O(n)$ conjugate gradient algorithm. Sophisticated preconditioners are applied, which further increase the computation speed by a factor of about 10. The approach is verified by experimental results in a 0.25 $\mu m$ BiCMOS technology.

## 1 Introduction

Modern mixed-signal IC designs consist of an increasing number of analogue and digital subcircuits due to the ongoing trend to higher integration. This enables digital interference to couple to sensitive analogue nodes through the substrate [1]. Typical substrate structures of a BiCMOS technology are shown in Fig. 1a. Most of the substrate noise on chip level is coupled via low-ohmic substrate contacts and the power supply network [2]. Hence floorplanning software tools should be provided in order to estimate how a certain placement of subcircuits and a power supply strategy influence substrate coupling effects in a given mixed-signal design. Traditional substrate coupling simulation tools do not meet this demand since they simply perform an impedance extraction of a flattened layout [3].

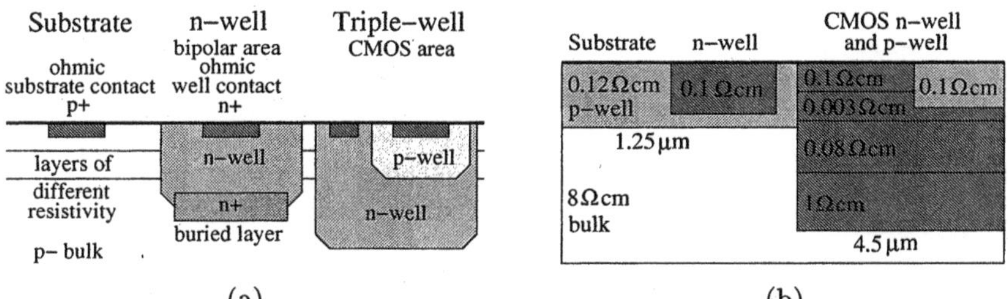


**Fig. 1.** **(a)** Typical substrate structures in BiCMOS integrated circuits. **(b)** Definition of layers in a BiCMOS technology.

Therefore, a novel simulation tool named *SubCALM* has been developed. It is well suited to floorplanning of mixed-signal designs due to its hierarchical approach and capability to process large layouts. It exploits a substrate model based on the boundary element method (BEM) as explained in Sect. 2. The subsequent section presents a new approach how to incorporate wells in this BEM model. The proposed algorithm may be accelerated by preconditioners (Sect. 4). Simulation and measurement results are shown in Sect. 5 for a $0.25\,\mu m$ BiCMOS technology.

## 2 Substrate Resistance Extraction

The substrate of a CMOS, BiCMOS or bipolar technology may be seen as a semi-infinite halfspace of silicon with different layers of resistivity depending on the doping density level, like channel stoppers, epitaxial layers and the bulk as shown in Fig. 1b. In most cases time-variant magnetic fields and displacement currents can be neglected for frequencies below 10 GHz leading to a simple electrostatic current flow problem between $m$ contacts [4]. The admittance of an $m$-contact geometry can be summarised by a symmetric $m \times m$ admittance matrix $\boldsymbol{Y}$.

### 2.1 Boundary Element Description

*SubCALM* exploits the boundary element method [5] for solving such an electrostatic problem. Choosing the BEM leads to a smaller number of variables, which is more suitable for huge problems, although it is clear that BEM does not reach the accuracy of finite element methods [4].

If a current with a certain density distribution $J(\boldsymbol{r_0})$ is injected into contact $j$, the potential $\phi$ at a certain point $\boldsymbol{r}$ inside the substrate may be calculated as

$$\phi(\boldsymbol{r}) = \int_{\Gamma_j} G(\boldsymbol{r}, \boldsymbol{r_0})\, J(\boldsymbol{r_0})\, d\boldsymbol{r_0} \tag{1}$$

with $\Gamma_j$ equals to the surface of contact $j$. $G(\boldsymbol{r}, \boldsymbol{r_0})$ is the modified Green's function which accommodates the special problem of a semi-infinite layered halfspace. G simply expresses the potential at a certain point $\boldsymbol{r}$, if a unit current is sourced into an infinitely small point $\boldsymbol{r_0}$.

### 2.2 Green's Function for Layered Media

Inside the substrate the Green's function $G$ has to satisfy

$$\nabla \left( \frac{1}{\rho(\boldsymbol{r})} \nabla\, G(\boldsymbol{r}, \boldsymbol{r_0}) \right) = -\delta(\boldsymbol{r} - \boldsymbol{r_0}) \tag{2}$$

where $\rho(\boldsymbol{r})$ denotes the resistivity at point $\boldsymbol{r}$. Due to radial invariance separation of variables in cylindrical coordinates may be used to simplify (2) to

$$G(r, z, z_0) = \frac{1}{2\pi} \int_0^\infty Z(k, z, z_0)\, J_0(kr)\, dk \tag{3}$$

$J_0()$ is the Bessel function of the first kind of order 0. $G$ has to meet several boundary conditions that can be used to determine $Z$:

1. Homogeneous Neumann boundary at $z = 0$
2. $Z(k, z, z_0)$ and $\frac{1}{\rho}\frac{\partial Z(k,z,z_0)}{\partial z}$ continuous at layer interfaces
3. $\lim_{z\to\infty} Z(k, z, z_0) = 0$
4. At injection point $z_0$: $\frac{\partial Z(k,z,z_0)}{\partial z}|_{z\to z_0+} - \frac{\partial Z(k,z,z_0)}{\partial z}|_{z\to z_0-} = -\rho k$

Considering these properties $G$ may be calculated according to (3). This is a one-dimensional Hankel transform, which can be evaluated by using a Fast Hankel transform algorithm [6] as $Z$ is smooth in $k$-domain.

Figure 2 presents a Green's function $G$ of the given arbitrary layer structure between contacts located at the surface $z = 0$. This example reveals some important properties of $G$. The electric field near the injection point has a simple $1/r$-relationship which is similar to the field of a single layer halfspace with $\rho = \rho_1$. However, if the distance gets bigger, the current does not penetrate the thin high-ohmic layer 2 underneath. Hence $G$ changes to a $\log(r)$-relationship up to certain distance $r$, where the interface area $r^2\pi$ to the underlying layer 3 becomes that huge that it is now inevitable for the current to flow along the shortest possible path through layer 2 into the low-ohmic layer 3. If distance is increased further, the $1/r$-relationship returns with $\rho = \rho_3$.

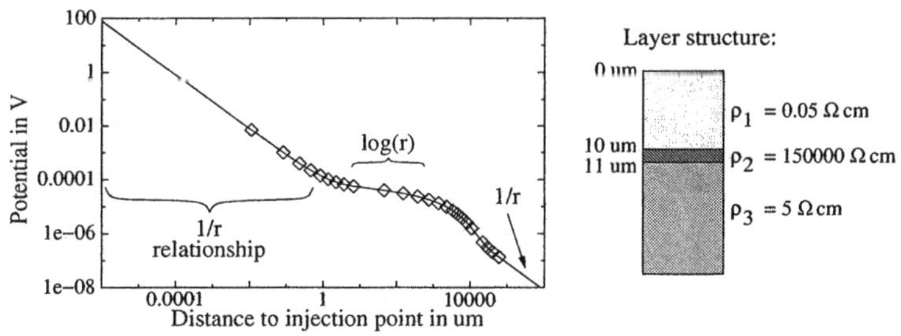


**Fig. 2.** Green's function for an semi-infinite layered halfspace.

The same arguments may be used in order to derive some major properties of the substrate coupling behaviour of the BiCMOS technology shown in Fig. 1b. Guard rings will be quite effective in this technology since the resistivity between adjacent substrate contacts is low, whereas the coupling between distant ones is smaller due to the less doped bulk.

### 2.3 Accelerated Resistance Calculation

We use Galerkin's approach in order to solve (1) numerically for $J$. Each contact is subdivided into smaller rectangles assuming constant potential and constant current density across the rectangle. For each rectangle an equation is derived which relates the known potential $\phi_i$ of rectangle $i$ to the contribution of injected currents $I_j$ into all rectangles $j$ of all contacts. If there are $n$ such rectangles, this results in a dense linear system $\boldsymbol{Z}\boldsymbol{i} = \boldsymbol{v}$ with an impedance matrix $\boldsymbol{Z} \in \mathbb{C}^{n\times n}$, a vector $\boldsymbol{i} \in \mathbb{C}^n$ containing all currents $I_j$ and a vector $\boldsymbol{v} \in \mathbb{C}^n$ consisting of potentials $\phi_i$. Each entry of $\boldsymbol{Z}$ yields

$$Z_{ij} = \frac{1}{A_i\, A_j} \int_{\Gamma_i} \int_{\Gamma_j} G(\boldsymbol{r_i}, \boldsymbol{r_j})\, d\boldsymbol{r_j}\, d\boldsymbol{r_i} \tag{4}$$

where $\Gamma_i$ is the surface of rectangle $i$ with area $A_i$ and $\Gamma_j$ denotes rectangle $j$ with area $A_j$. The admittance matrix $\boldsymbol{Y} \in \mathbb{C}^{m\times m}$ may be subsequently determined with $\boldsymbol{Y} = \boldsymbol{P^T}\, \boldsymbol{Z}^{-1}\, \boldsymbol{P}$ where $\boldsymbol{P} \in \{0,1\}^{n\times m}$ with $P_{ij} = 1$ if rectangle $i$ is part of contact $j$. Direct methods such as Gaussian elimination require $\mathrm{O}\left(n^3\right)$ operations solving this equation. Iterative algorithms like conjugate gradient (CG) usually require $\mathrm{O}\left(n^2\right)$ per iteration. Several different algorithms have been proposed in the literature, which offer a subquadratic runtime [7,8]. *SubCALM* is based on the hierarchical algorithm by Appel [9], with $\mathrm{O}\left(n\right)$ time consumption [10].

If the kernel of (4) is $1/r$, two rectangles may be combined if $R/r < \epsilon$ [5], where $R$ is the length of the longest side of the area covered by these two rectangles, $\epsilon$ is a certain error and $r$ is the distance to the injection point. If the kernel is arbitrary as in our case, this expression has to be generalised to $L\left(r, \epsilon\right) > R$. $L$ is derived from $G$ as shown in Fig. 3a. Two rectangles may be combined together, if $G$ can be linearised within the whole area. If $G$ were totally linear, an evaluation of $G$ at the centre of gravity would minimise the average integration error over $\Gamma_i$ and $\Gamma_j$. An appropriate maximum value for the linearity interval $L\left(r, \epsilon\right)$ with respect to distance is determined in such a way that the approximation error is limited to a given relative tolerance $\epsilon$. The application of such a simplification to a layout is visualised in Fig. 3b.

## 3 Modelling of Wells

Well structures cannot be modelled by the BEM easily. However, it becomes possible by modifying the calculation of $\boldsymbol{Z}$. A simplified well structure is shown in Fig. 4a. The pn-junction between wells is replaced by special 'well contacts'. Since this pn-junction is reversely biased, its junction capacitance $C_\mathrm{j}$ has to be considered as well. Two different potentials $V_\mathrm{w}$ and $V_\mathrm{w}'$ are dedicated to each well contact depending from which side it is looked at.

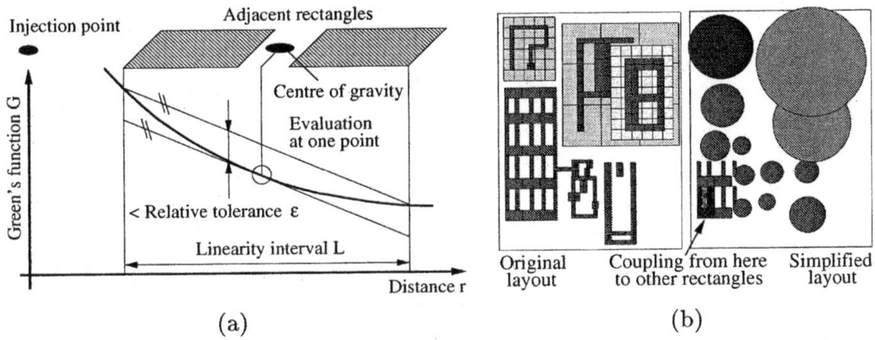


**Fig. 3.** **(a)** Local linear approximation for integration and simplification of $\boldsymbol{Z}$. **(b)** Combination of adjacent rectangles to bigger circular contacts.

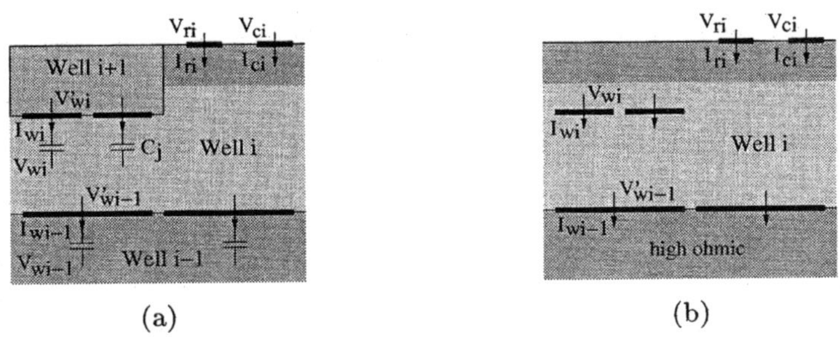


**Fig. 4.** Modelling of wells in the BEM algorithm.

Considering only well $i$, all subwells $i+1$ are removed first. In addition, a high ohmic layer is introduced instead of parent well $i-1$ as shown in Fig. 4b. Now $\boldsymbol{Z}'_{\mathbf{well}}$ of this modified structure is calculated using (4):

$$\begin{pmatrix} V'_{\mathrm{w},i-1} \\ V_{\mathrm{r},i} \\ V_{\mathrm{c},i} \\ V_{\mathrm{w},i} \end{pmatrix} \begin{pmatrix} Z'_{11} & Z'_{12} & Z'_{13} & Z'_{14} \\ Z'_{21} & Z'_{22} & Z'_{23} & Z'_{24} \\ Z'_{31} & Z'_{32} & Z'_{33} & Z'_{34} \\ Z'_{41} & Z'_{42} & Z'_{43} & Z'_{44} \end{pmatrix} \cdot \begin{pmatrix} -I_{\mathrm{w},i-1} \\ I_{\mathrm{r},i} \\ I_{\mathrm{c},i} \\ I_{\mathrm{w},i} \end{pmatrix} \tag{5}$$

Since a well is a closed volume, $I_{\mathrm{r},i} = I_{\mathrm{w},i-1} - I_{\mathrm{c},i} - I_{\mathrm{w},i}$ has to hold. Hence $\boldsymbol{Z}'_{\mathbf{well}}$ is over-determined. The variables dedicated to the reference contact $V_{\mathrm{r},i}$ and $I_{\mathrm{r},i}$ are removed leading to $\boldsymbol{Z}_{\mathbf{well}}$ with entries $Z_{\mathrm{well},jk}$. The influence of the junction capacitances can be considered as follows.

$$V_{\mathrm{w},i-1} = V'_{\mathrm{w},i-1} - \frac{1}{j\,\omega\,C_{\mathrm{j}}} I_{\mathrm{w},i-1} \tag{6}$$

This results in an impedance matrix $\boldsymbol{Z}_{\mathbf{well},i}$ for every well.

$$\begin{pmatrix} V_{\mathrm{w},i-1} - V_{\mathrm{r},i} \\ V_{\mathrm{c},i} - V_{\mathrm{r},i} \\ V_{\mathrm{w},i} - V_{\mathrm{r},i} \end{pmatrix} = \begin{pmatrix} Z_{\mathrm{well},11} + \mathrm{diag}\left(\frac{1}{j\,\omega\, C_{\mathrm{j}}}\right) & \cdots \\ Z_{\mathrm{well},21} & \ddots \\ Z_{\mathrm{well},31} & \cdots \end{pmatrix} \cdot \begin{pmatrix} -I_{\mathrm{w},i-1} \\ I_{\mathrm{c},i} \\ I_{\mathrm{w},i} \end{pmatrix} \tag{7}$$

All these matrices can be summed up in a symmetric impedance matrix $\boldsymbol{Z}$ for the whole system.

$$\begin{pmatrix} V_{\mathrm{c}0} - V_{\mathrm{r}0} \\ V_{\mathrm{r}1} - V_{\mathrm{r}0} \\ -\left(V_{\mathrm{c}1} - V_{\mathrm{r}1}\right) \\ -\left(V_{\mathrm{r}2} - V_{\mathrm{r}1}\right) \\ V_{\mathrm{c}2} - V_{\mathrm{r}2} \end{pmatrix} = \begin{pmatrix} Z_{\mathrm{well},0} & & 0 & 0 & 0 \\ & + & & & 0 \\ 0 & & Z_{\mathrm{well},1} & & 0 \\ 0 & & & + & \\ 0 & 0 & 0 & & Z_{\mathrm{well},2} \end{pmatrix} \cdot \begin{pmatrix} I_{\mathrm{c}0} \\ I_{\mathrm{w}0} \\ -I_{\mathrm{c}1} \\ -I_{\mathrm{w}1} \\ I_{\mathrm{c}2} \end{pmatrix} \tag{8}$$

The above approach can be easily extended to more contacts per well. All entries in $\boldsymbol{Z}'_{\mathbf{well}}$ and $\boldsymbol{Z}_{\mathbf{well}}$ except those for reference contacts will convert to submatrices. A modified CG-algorithm [11] has to be used in order to invert the complex symmetric, but not hermitian system impedance matrix $\boldsymbol{Z}$.

## 4 Preconditioner for CG-Algorithm

The convergence rate of the CG-algorithm depends on the ratio between the maximum and the minimum eigenvalue of $\boldsymbol{Z}$. If a $\boldsymbol{Z}$ with well structures is evaluated for low frequencies, it can become ill-conditioned. Hence the usage of a preconditioner is recommended. Although $\boldsymbol{Z}$ is dense, several different possibilities for a preconditioning matrix $\boldsymbol{M}$ still exist:

1. DIAG: Only diagonal elements of $\boldsymbol{Z}$ are considered. $\boldsymbol{M} = \mathrm{diag}\,(1/Z_{ii})$
2. BLOCK-DIAG: Internally, adjacent rectangles are stored inside one rectangle object in a binary tree structure [12]. Now only entries $Z_{ij}$ are considered, if rectangles $i$ and $j$ are located in the same rectangle object. Hence diagonal blocks $\boldsymbol{Z_k}$ of adjacent entries inside $\boldsymbol{Z}$ are chosen to form $\boldsymbol{M_k} = \mathrm{diag}\left(\boldsymbol{Z_k^{-1}}\right)$
3. BLOCK-BAND: In addition to BLOCK-DIAG, parts of $\boldsymbol{Z}$ defining the coupling between adjacent rectangle objects are taken into account as well. Let $\boldsymbol{Z_{k1}}$ and $\boldsymbol{Z_{k2}}$ be the diagonal block matrices for rectangle objects $k1$ and $k2$ and $\boldsymbol{Z_{k1,k2}}$ the coupling block matrix between the rectangles of these two objects.

$$\begin{pmatrix} \boldsymbol{Y_{11}} & \boldsymbol{Y_{12}} \\ \boldsymbol{Y_{12}^T} & \boldsymbol{Y_{22}} \end{pmatrix} = \begin{pmatrix} \boldsymbol{Z_{k1}} & \boldsymbol{Z_{k1,k2}} \\ \boldsymbol{Z_{k1,k2}}^T & \boldsymbol{Z_{k2}} \end{pmatrix}^{-1} \tag{9}$$

$\boldsymbol{Y_{12}}$ will be used as $\boldsymbol{M_{k1,k2}}$. The diagonal elements must be modified to $M_{ii} - \sum_{j=1,j\neq i}^{n} M_{ij}$ to make sure that $M$ is positive definite. Table 1 presents acceleration results due these preconditioning measures.

**Table 1.** Number of iterations (It) and runtime (t in seconds) of the CG-algorithm with different preconditioners applied to problems with different number of variables (NoV) at certain frequencies f, carried out on a Sun Fire 4800 with UltraSparcIII 750 MHz processors and 24 GB memory.

| NoV | f | Pure CG | | DIAG | | BLOCK-DIAG | | BLOCK-BAND | |
|---|---|---|---|---|---|---|---|---|---|
| | | It | t | It | t | It | t | It | t |
| 936 | 1 Hz | 268 | 4.94 | 18 | 0.39 | 19 | 0.41 | 12 | 0.30 |
| | 1 MHz | 194 | 3.66 | 21 | 0.48 | 22 | 0.48 | 14 | 0.35 |
| | 1 GHz | 134 | 2.54 | 24 | 0.53 | 24 | 0.52 | 18 | 0.45 |
| 8424 | 1 Hz | 527 | 98.80 | 37 | 8.23 | 28 | 6.33 | 19 | 4.77 |
| | 1 MHz | 396 | 78.26 | 38 | 8.54 | 32 | 7.40 | 19 | 5.33 |
| | 1 GHz | 227 | 43.71 | 42 | 9.49 | 34 | 7.83 | 22 | 6.06 |

## 5 Simulation and Measurement Results

A testchip in a 0.25 $\mu m$ BiCMOS technology with a layer structure shown in Fig. 1b has been designed and manufactured. It contains various different guard rings and well structures both for DC and RF measurements. Results of the DC part have already been published in [12]. Structures for RF were measured between 100 MHz and 20 GHz with a network analyser HP8510C. The LRRM method was used for calibration. Measurement results of an n-well structure are presented in Fig. 5. The influence of the voltage dependent pn-junction capacitance may be easily seen. The difference between measurement and simulation is less than 5 dB, which is sufficient for a coupling estimation.

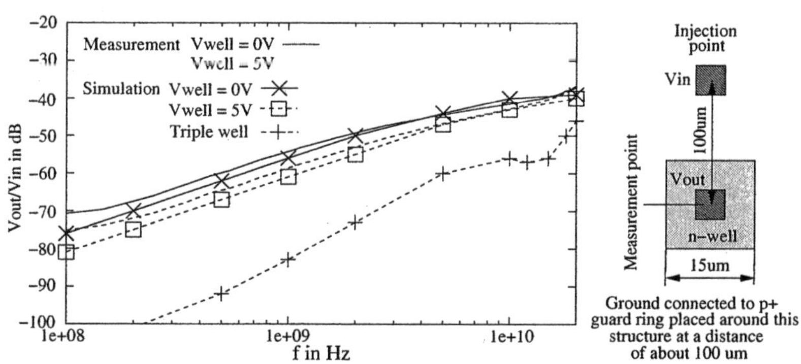


**Fig. 5.** RF measurement and simulation for a well structure.

In addition, simulated data for coupling from substrate into a p-well of a triple well structure is shown in Fig. 5. It was impossible to measure this small coupling on the testchip. However, simulation results for every pn-junction being involved (substrate to n-well and n-well to p-well) were verified by measurement. Hence one can assume that coupling into a triple well is at least another 20 dB smaller than into a single well.

## 6 Conclusion

A hierarchical substrate simulation tool has been presented, which offers the opportunity of substrate coupling estimation during the process of floorplanning. A novel approach of a well structure model for boundary element description has been proposed. An accelerated conjugate gradient based algorithm is used which is capable to deal with large mixed-signal design layouts. The application to a real-world mixed-signal design will be part of investigations in the near future.

## Acknowledgement

This work has been performed within the MEDEA+ T102 project ASDESE and has been funded by the German Bundesministerium für Bildung und Forschung (BMBF) under funding No. 01M3053.

## References

1. Singh, R. (Ed): Signal Integrity Effects in Custom IC and ASIC Designs. Wiley Interscience (2002)
2. Van Heijningen, M., Compiet, J., Wambacq, P., Donnay, S., Engels, M., Bolsens, I.: Analysis and Experimental Verification of Digital Substrate Noise Generation for Epi-Type Substrates. IEEE J. Solid-State Circuits **35(7)** (2000) 1002–1008
3. Cadence Design Systems Inc. (former Simplex Solutions): SubstrateStorm, *www.cadence.com (www.simplex.com)*
4. Verghese, N. K., Schmerbeck, T. J., Allstot, D. J.: Simulation Techniques and Solutions for Mixed-Signal Coupling in Integrated Circuits. Kluwer Academic Publishers (1995)
5. Shi, W., Liu, J., Kakani, N., Yu, T.: A Fast Hierarchical Algorithm for Three-Dimensional Capacitance Extraction. IEEE Trans. on CAD of Integrated Circuits and Systems **21(3)** (2002) 330–336
6. Anderson, W. L.: Fast Hankel Transforms Using Related and Lagged Convolutions. ACM Trans. Math. Software. **8(4)** (1982) 344–368
7. Nabors, K., White, J. K.: Fastcap: A Multipole Accelerated 3-D Capacitance Extraction Program. IEEE Trans. on CAD **10(11)** (1991) 1447–1459
8. Kapur, S., Long, D. E.: IES3: A Fast Integral Equation Solver for Efficient 3-Dimensional Extraction. Int. Conf. on CAD (ICCAD) (1997) 448–455
9. Appel, A. W.: An Efficient Program for Many-Body Simulation. SIAM J. Sci. Stat. Comp. **6(1)** (1985) 85–103

10. Esselink, K.: The Order of Appel's Algorithm. Information Processing Letters **41** (1992) 141–147
11. Freund, R. W.: Conjugate Gradient-Type Methods for Linear Systems with Complex Symmetric Coefficient Matrices. SIAM J. Sci. Stat. Comp. **13(1)** (1992) 425–448
12. Brandtner, T., Weigel, R.: Hierarchical Simulation of Substrate Coupling in Mixed-Signal ICs Considering the Power Supply Network. DATE Conf. (2002) 1028–1032

# An Embedding Method for High Frequency Circuits

Barbara Lang[1], Angelika Bunse-Gerstner[1], Henning Lemanczyk[1], Hans Georg Brachtendorf[2], and Rainer Laur[3]

[1] Zentrum für Technomathematik, Universität Bremen, Germany
[2] Fraunhofer IIS, Erlangen, Germany
[3] Institut für Theoretische Elektrotechnik und Mikroelektronik, Universität Bremen, Germany

**Abstract.** Widely seperated time scales appear in many electronic circuits, making analysis with the usual numerical methods very difficult and costly. In this article we present a quasilinear system of partial differential equations (PDE) of first order, where the time scales are treated seperately. The PDE corresponds to the system of differential-algebraic equations (DAE) describing the electronic circuit in the sense that the solution of the PDE restricted to one of its characteristics is the solution of the DAE. This embedding method is described in a general setting. Hence it can be used for various applications in circuit simulation.

Since generalized quasiperiodic functions, which are presented here, conceptualize physical properties, they have a basic significance for the embedding method.

Theoretical investigations are presented as well as new approaches for numerical methods based on the connection between the PDE and the DAE.

## 1 Introduction and Motivation

An electronic circuit is simulated by solving a system of differential algebraic equations (DAEs), which is derived via the modified nodal analysis from Kirchhoff's laws. The system is of the form

$$\frac{d}{dt}q(x(t)) + f(x(t)) + b(t) \quad = \quad 0, \qquad \text{(DAE)} \tag{1}$$

with node voltages and branch currents of impedant elements $x : \mathbb{R} \to \mathbb{R}^n$, charge and flux terms of impedant elements $q : \mathbb{R}^n \to \mathbb{R}^n$, static currents $f : \mathbb{R}^n \to \mathbb{R}^n$ and input signals $b : \mathbb{R} \to \mathbb{R}^n$, where the stimulus $b(t)$ is constant for autonomous circuits and time-dependent otherwise. Because the derivative of q is in general not invertible, there are not only dynamical but also algebraic conditions and the system is a proper system of DAEs.

Electronical circuits often have solutions consisting of oscillations with widely separated time scales. Mixers for example have solutions which are superpositions of fast and slow oscillations requiring very tiny stepsizes over a long time interval when calculating the solution numerically. Switching to a multivariate function we can seperate the time scales. If e.g. for the ring modulator described in [7] we choose artificially extreme input signals $U_{\text{in}1}(t) = 0.5\sin(2000\pi t)$ and $U_{\text{in}2}(t) = 2\sin(200000\pi t)$, the frequencies differ by a factor of 100. The solution for one of the node voltages is shown in Fig.1 a). Assuming that this is a quasiperiodic function of the form $x(t) = \sum_{z\in\mathbb{Z}^2} X_z\ e^{i(z_1\omega_1 t + z_2\omega_2 t)}$, with $X_z \in \mathbb{R}$ and $\omega_1 = 2000\pi, \omega_2 = 200000\pi$, we can associate a function in two variables of the form $\widehat{x}(\tau_1,\tau_2) = \sum_{z\in\mathbb{Z}^2} X_z\ e^{i(z_1\tau_2 + z_2\tau_2)}$. This function is $2\pi-$periodic in both dimensions. If we know $\widehat{x}(\tau_1,\tau_2)$ on $[0,2\pi]\times[0,2\pi]$ we know it on all of $\mathbb{R}^2$. Given the path $\gamma(t) := (\omega_1 t, \omega_2 t)$ in $\mathbb{R}^2$ we see that $\widehat{x}(\gamma(t)) = x(t)$. Note that $\gamma(t)$ has a very steep slope and is close to the $\tau_2-$ axis. If we had $\widehat{x}(\tau_1,\tau_2)$ on $[0,2\pi]\times[0,2\pi]$ we could easily retrieve $x(t)$.

We could try to compute such a bi-periodic function as a solution of a suitable system of partial differential equations (PDE) associated with the initial system of DAEs. The PDE has to be such that from its solution we can retrieve the solution of the DAE by following a specific path. Solving the DAE via a numerical solution of the associated PDE can decrease the costs drastically, because we have to compute the solution of the PDE only on $[0,2\pi]\times[0,2\pi]$, where it is typically a slowly varying function and can be approximated using normal or large stepsizes. We call this procedure "embedding the system of DAEs into a system of PDEs". For the above mentioned ring modulator it leads to a system of PDEs which gives the bi-periodic solution on $[0,2\pi]\times[0,2\pi]$ in Fig.1 b) corresponding to the node voltage considered on the left.

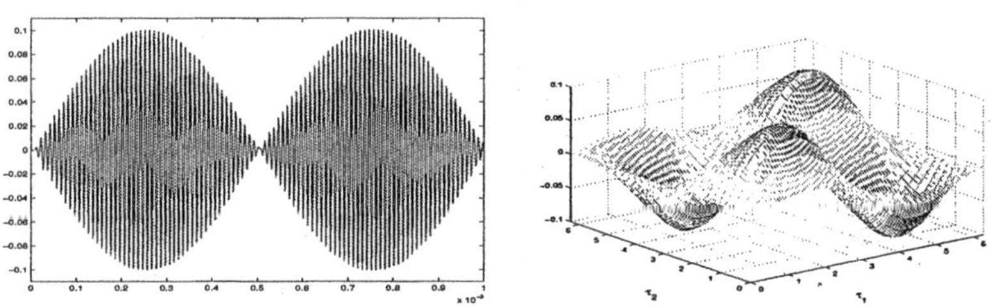

**Fig. 1.** a) Reference solution; b) Solution as a multivariable function

Various embedding methods have been presented in the literature (see [2], [6], [5], [13], [4], [8], [10], [12], [14], [3], [11] and [15] for a survey). They differ in the form of the embedding PDE, the path of evaluation and the considered domain, depending on the objective of the simulation and the assumptions on the special form of the solutions. Here we present a unified approach for the different derivations, such that they can all be considered as special cases of this general embedding method. The form of this general embedding clearly points out the important role of the characteristic lines for this approach.

The relation between the PDE and the initial DAE system can be used to develop new numerical methods to solve the DAE and to give new views on existing methods. Here we suggest a numerical method which stems from a method of characteristics for the PDE, specialized such that it results in a DAE method which works only with the original DAE problem, approximating only every k-th oscillation.

In the following generalized quasiperiodic functions are introduced, the embedding method is described and theoretical results are given. Applications of this embedding method are presented and numerical algorithms are considered.

## 2 Generalized Quasiperiodic Functions

Consider functions $x : \mathbb{R} \to \mathbb{R}^n$ of the form

$$x(t) = \sum_{z \in \mathbb{Z}^m} X_z(t)\, e^{i(\sum_{j=1}^{m} z_j \omega_j(t))t} \tag{2}$$

where $X_z(t) : \mathbb{R} \to \mathbb{R}^n$ and $\omega = (\omega_1, \omega_2, \ldots, \omega_m)$ with $\omega : \mathbb{R} \to \mathbb{R}^m$.

We call them generalized quasiperiodic functions. If the factors $X_z(t)$ and $\omega_j(t)$ do not depend on $t$, they are just the well-known quasiperiodic functions. Throughout this paper these sums are assumed to be finite.

Now let $x$ be a generalized quasiperiodic function. With $\omega$ from the generalized quasiperiodic representation of $x$ above we can build for a given interval $I \subseteq \mathbb{R}$ a path $\gamma$ in $\mathbb{R}^m$ as:

$$\gamma(t) := (\omega_1(t)t, \omega_2(t)t, \ldots, \omega_m(t)t) : I \to \mathbb{R}^m.$$

To each generalized quasiperiodic function $x$ we associate a function $\widehat{x}$ with multidimensional domain of the following form:

$$\widehat{x}(\tau) = \sum_{z \in \mathbb{Z}^m} \widehat{X}_z(\tau)\, e^{i(\sum_{j=1}^{m} z_j \tau_j)},$$

where $\tau = (\tau_1, \tau_2, \ldots, \tau_m) \in \mathbb{R}^m$ and the $\widehat{X}_z(\tau)$ satisfy

$$\widehat{X}_z(\gamma(t)) = X_z(t) \quad \text{for all} \quad z \in \mathbb{Z}^m. \tag{3}$$

Then $\widehat{x}(\gamma(t)) = x(t)$. We denote such a function by: multivariate function associated to the generalized quasiperiodic function. Because different functions $\widehat{X}_z$ could satisfy the property (3) above, an associated multivariate function is in general not unique. It depends on the choice of $\widehat{X}_z$.

Note, that if the $\widehat{X}_z$ and $\omega_j$ are constant the associated multivariate function is $2\pi$-periodic in each coordinate by construction.

## 3 Embedding Method and Theoretical Investigations

For the circuit equation (1) the corresponding embedding system, a system of partial differential equations, is defined by

$$\sum_{j=1}^{m} \alpha_j(\tau)\, \frac{\partial}{\partial \tau_j}\, q(\widehat{x}(\tau)) + f(\widehat{x}(\tau)) + \widehat{b}(\tau) = 0, \qquad \text{(PDE)} \tag{4}$$

where $\widehat{x} : \Omega \to \mathbb{R}^n$ ($\Omega \subseteq \mathbb{R}^m$, $\Omega$ open and connected), $\tau = (\tau_1, \ldots, \tau_m) \in \Omega$, $\alpha_j : \Omega \to \mathbb{R}, \alpha_j(\tau) > 0$, $m \in \mathbb{N}$; in practice often $m \leq 3$. $q$ and $f$ can be taken directly from (1)), but $\widehat{b}$ must be chosen such that if evaluated along the path

$$\gamma : I \to \mathbb{R}^m : t \mapsto \gamma(t) = (\gamma_1(t), \gamma_2(t), \ldots, \gamma_m(t)) \tag{5}$$

it coincides with $b$ ($I \subset \mathbb{R}$ interval with $0 \in I$), i.e. $\widehat{b}(\gamma(t)) = b(t)$. The path $\gamma$ is chosen to be the solution of the initial value problem

$$\dot{\gamma}_j(t) = \alpha_j(\gamma(t)), \quad \text{für alle} \quad j \in \{1, \ldots, m\}, \tag{6}$$

$$\gamma(0) = 0 \in \mathbb{R}^m. \tag{7}$$

Equation (4) is a system of quasilinear partial differential equations of first order. If $Dq$, the Jacobian of q, is invertible, the equation can be decoupled (see [12] for a special case). Supposably this proof can be generalized to problems of index 1.

Then the theorie of characteristics is applicable. It can be shown that the caracteristics do not intersect. The problem of existence of solutions of the PDE can be reduced to the existence of a family of solutions of the DAE with appropriate stimulus. Because the characteristics do not intersect, only a "blow up" can prevent the existence. Conversely the existence of a PDE-solution implies the existence of the DAE-solution.

## 4 Applications

Various embedding methods have been presented in the literature, but they differ in the partial differential equation, the path of evaluation and the considered domain. The approach above allows a general derivation and provides a theoretical framework. In the following the results are summarized.

If we specify the coefficient functions $\alpha_j$, $j = 1, .., m$, the path of evaluation is determined. Together with the specification of the domain of the PDE various circuit simulation problems can be treated. These are e.g. steady states of non-autonomous and mixed autonomous/non-autonomous circuits with quasi periodic stimulus (see [1],[2],[3],[6],[10],[14]). Also transient behavior (see [9],[13],[14]) as well as steady state analysis for autonomous circuits with quasi periodic steady states can be treated. Furthermore one can choose an embedding system to analyze the transient behavior of high Q oscillators (see [2],[4]). In [3] also an embedding system for the study of "injection locking" is presented.

Several numerical methods are considered in the literature as well. Essentially these are "multi-tone harmonic balance" and standard methods to solve the PDE directly.

## 5 Numerical Methods

The following algorithm, which is a modification of the method of characteristics presented in [10], calculates the steady state solution of electric circuits with two superimposed oscillations. It is based on the described embedding method and utilizes the circuit equation directly. The system of partial differential equations provides periodicity conditions on time intervals where the circuit equation is evaluated. At first the frequencies are assumed to be known. Then an extension to the case of unknown frequencies is discussed.

We consider steady state solutions of the form

$$x(t) = \sum_{z \in \mathbb{Z}^2} X_z \; e^{i(z_1\omega_1 + z_2\omega_2)t}, \tag{8}$$

with $\omega_2 \gg \omega_1$ both known. The solution is embedded in the $2\pi$-biperiodic function

$$\widehat{x}(\tau) = \sum_{z \in \mathbb{Z}^m} \widehat{X}_z \; e^{i(z_1\tau_1 + z_2\tau_2)}, \tag{9}$$

which returns $x$ when evaluated on the path $\gamma(t) = (\omega_1\ t, \omega_1\ t)$. The circuit is described by the differential-algebraic system (1) which corresponds to the system of partial differential equations

$$\omega_1 \frac{\partial}{\partial \tau_1} q(\widehat{x}(\tau)) + \omega_2 \frac{\partial}{\partial \tau_2} q(\widehat{x}(\tau)) + f(\widehat{x}(\tau)) + \widehat{b}(\tau) = 0. \tag{10}$$

The PDE is considered on the domain $[0, 2\pi) \times [0, 2\pi)$ with periodic boundary conditions. The evaluation path $\gamma$ is projected on this square, see Fig.2,a. For the algorithm we select every k-th of these projected lines (Fig.2,b). The PDE-solution restricted to the selected lines is exactly the DAE-solution on corresponding time intervals (Fig.2,c). These parts of the DAE-solution are computed by a multiple shooting method. This, together with the PDE's periodicity conditions, yields the non-linear system of equations to be solved (Fig.2,b).

The algorithm was tested on the ring modulator with frequencies $\omega_1 = 2000\pi$ and $\omega_2 = 200000\pi$. Figure 3,a shows the result of the algorithm after one Newton-step. As initial values we chose a 10% perturbation of the steady state solution in 40 points. Again, the calculated oscillations are marked. The oscillations in between are obtained by linear interpolation and therefore the oscillation peaks are piecewise linear which could be improved without

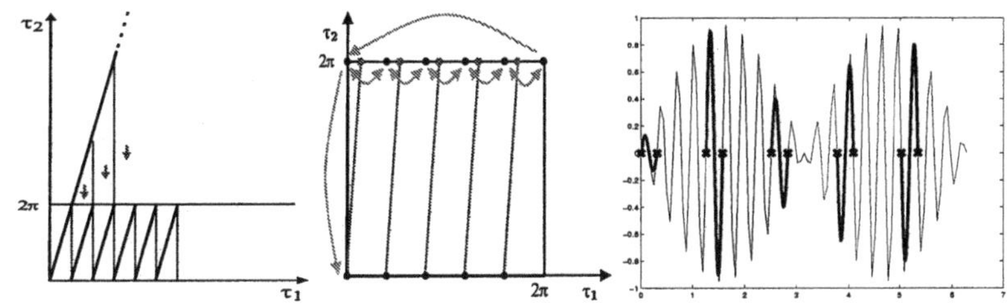

**Fig. 2.** a) Projection of the characteristic $\gamma$; b) Characteristics and periodic conditions; c) Solution: Marked oscillations correspond to the characteristics which are used for calculation

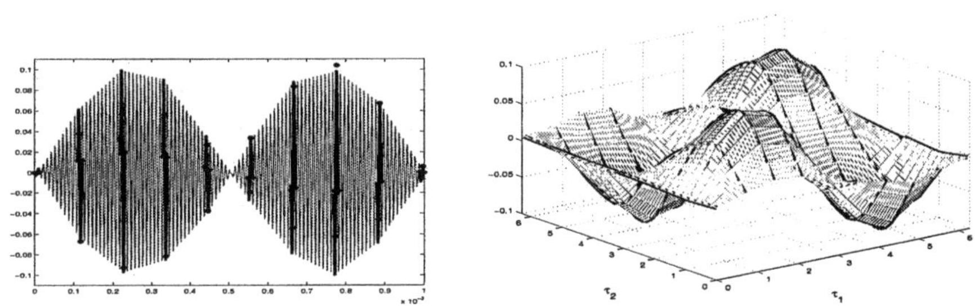

**Fig. 3.** a) Solution by method of characteristics; b) Solution as a multivariate function

significant increase of calculations e.g. by quadratic interpolation. Figure 3,b illustrates the computed result as the corresponding multivariate function.

The method described above can be expanded to the case of unknown frequencies. This problem occurs for autonomous circuits. The frequencies are now treated as additional unknows. Because the characteristics depend on the frequencies, it's necessary to adjust them at certain states and the algorithm must be embedded in an outer iteration loop. As the frequencies add to the number of unknowns, two more equations are needed. The value of one component of the solution can be fixed at a certain time and/or the derivative of some components can be set to zero. The advantage of the first choice is, that the zero-solution is excluded (see [11]). However it cannot be guaranteed that the fixed value can be reached by the solution. Another possibility is to choose a calibration condition of the form

$$\frac{1}{T}\int_0^T x(t)\cos(\frac{2\pi}{T}t)dt = 0 \tag{11}$$

for each parameter. After each Newton-step it has to be checked that the current frequencies are still admissible. Otherwise a next step of the outer loop must be started.

Analogously an algorithm for the transient analysis of oscillators can be developed. Here the principal idea is presented and difficulties are discussed.

The start-up behavior of oscillators is described by the DAE (1) together with the initial condition $x(0) = x_0$. It is assumed that the solution is a fast oscillation which is modulated by a slowly varying function. We use the ansatz

$$x(t) = \sum_{z\in\mathbb{Z}^m} X_z(t)\, e^{i(\sum_{j=1}^{m} z_j\omega_j)t}, \tag{12}$$

where $m = 2$, $z = (0, z_2)$ and known $\omega_2$. The embedding system is

$$\frac{\partial}{\partial\tau_1}q(\widehat{x}) + \omega_2\frac{\partial}{\partial\tau_2}q(\widehat{x}) + f(\widehat{x}) + \widehat{b} = 0 \quad \text{on}[0,2\pi)\times\mathbb{R}_+ . \tag{13}$$

The solution is assumed to be $2\pi$-periodic in $\tau_2$. This yields the boundary condition $\widehat{x}(\tau_1, 0) = \widehat{x}(\tau_1, 2\pi)$. We have the initial condition $\widehat{x}(0,0) = \widehat{x}_0(0) = x_0$ from the DAE. But in addition we need initial values on $\widehat{x}(0, \tau_2) = \widehat{x}_0(\tau_2)$ on $[0, 2\pi)$ to solve the PDE numerically.

With numerical algorithms proceeding in $\tau_1$-direction approximate values on the k-th characteristic can be obtained. (Multiple) shooting methods can correct these extrapolated values. This is then continued in $\tau_1$-direction. This approach is illustrated in Fig.4.

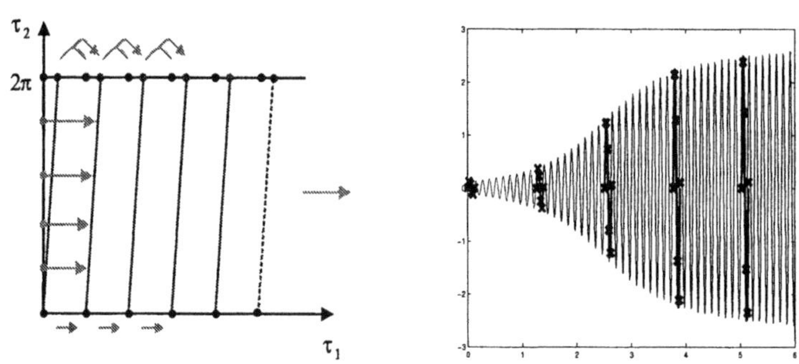


**Fig. 4.** a) Characteristics and periodicity in $\tau_2$-direction; b) Solution: Marked oscillations correspond to the characteristics which are used for calculation

This method has still some problems. First of all, different choices of initial values on the $\tau_2$-axis lead to the same solution of the DAE. It is important for numerical stability and efficiency to choose initial conditions such that the solution is smooth. Extrapolation methods to proceed in $\tau_1$-direction are not efficient. BDF methods for the PDE are highly damping which is unsuitable for solutions with increasing oscillations (see [3]). When correcting with the multiple shooting method, different solutions are possible. Again further calibration conditions have to be used.

## References

1. H. G. Brachtendorf. Simulation des eingeschwungenen Verhaltens elektronischer Schaltungen. Shaker, Aachen, 1994.
2. H. G. Brachtendorf. On the relation of certain classes of ordinary differential algebraic equations with partial differential algebraic equations. Technical Report 1131G0-791114-19TM, Bell-Laboratories, 1997.
3. H. G. Brachtendorf. Theorie und Analyse von autonomen und quasiperiodisch angeregten elektrischen Netzwerken. Habilitationsschrift, Universität Bremen, 2001.
4. H. G. Brachtendorf and R. Laur. Transient simulation of oscillators. Technical Report 1131GO-980410-09TM, Bell-Laboratories, 1998.
5. H. G. Brachtendorf, G. Welsch, and R. Laur. A novel time-frequency method for the simulation of the steady state circuits driven by multitone signals. ProcIEEESCS, June 1997.
6. H. G. Brachtendorf, G. Welsch, R. Laur, and A. Bunse-Gerstner. Numerical steady state analysis of electronic circuits driven by multi-tone signals. Electronic Engineering, 79(2), pp 103–112, April 1996.
7. W. Kampowski, P. Rentrop, and W. Schmidt. Classification and numerical simulation of electric circuit. Surveys on Mathematics for Industry, 2(1), pp 23–65, 1992.
8. O. Narayan and J. Roychowdhury. Multi-time simulation of voltage-controlled oscillators. Proc. Design Automation Conf., pp 629–634,1999
9. E. Ngoya and R. Larchevâque. Envelope transient analysis: A new method for the transient and steady state analysis of microwave communications circuit and systems. In Proc. IEEE MTT-S Int. Microwave Symp., pp 1365–1368, Sanfrancisco, 1996.
10. R. Pulch. Numerische Simulation von PDE-Modellen in der Analyse von RF-Schaltungen, 1999.
11. R. Pulch. PDE techniques for finding quasi-periodic solutions of oscillators. Preprint 09, IWRMM, Universität Karlsruhe, 2001.
12. R. Pulch and M. Günther. A method of characteristics for solving multirate partial differential equations in radio frequency application. Preprint 07, IWRMM, Universität Karlsruhe, 2000.
13. J. Roychowdhury. Efficient methods for simulating highly nonlinear multi-rate circuits. In Proc. IEEE Design Automation Conf., pp 269–274, 1997.
14. J. Roychowdhury. Analyzing circuits with widely separated time scales using numerical pde methods. IEEE Transactions on Circuits and Systems I - Fundamental Theory and Applications, 48, pp 578–594, May 2001.
15. J. Roychowdhury., Multi-time PDEs for dynamical system analysis. In U. van Rienen and M. Guenther and D. Hecht (eds.): Scientific Computing in Electrical Engineering. Lecture Notes in Computational Science and Engineering, Springer, Berlin , pp 3-14, 2001

# Mixed-Hybrid Finite Element Methods for Coupled Problems in Silicon Dioxide Technology

Paola Causin and Riccardo Sacco

MOX - Modeling and Scientific Computing, Dipartimento di Matematica - Politecnico di Milano, via Bonardi, 9 - 20133 Milano - Italy

**Abstract.** In this work we deal with the numerical simulation of thermal oxidation in silicon device technology. This application is a complex coupled phenomen, involving the solution of a diffusion-reaction problem and of a fluid-structure interaction problem. Suitable iterative procedures are devised for handling nonlinearities and strong coupling between the sub-problems to be solved. In particular, we propose a unified dual-mixed hybrid formulation that allows for the simultaneous solution of the compressible/incompressible Navier equations in both solid and fluid domains. The accuracy and the flexibility of the proposed approach are demonstrated on benchmark test problems.

## 1 Introduction and motivation

Thermal oxidation of silicon is one of the several steps involved in the manufacturing of integrated circuits (IC). The silicon dioxide is the product of the following chemical reaction

$$\mathtt{Si} + \mathtt{O}_2 \rightarrow \mathtt{SiO}_2.$$

Silicon dioxide is thermally grown on the silicon wafer bulk to:

- electrically insulate basic devices like transistors and capacitors built on a single wafer
- act as gate oxide in Metal Oxide Semiconductor (MOS) structures or serve as a mask against dopant implantation.

Numerical simulation of the thermal oxidation process is aimed at predicting the oxide shape after oxidation in order to better assess the electrical performance of the device. Moreover, it is of relevant interest to analyze the stress history of the material in order to study its effect on the evolution of the oxidation process and to prevent mechanical failures. Realistic simulations of the process are achieved by taking into account different phenomena arising in a strongly heterogeneous assembling of materials. Fig.1 shows schematically the reduction from a 3D model to a 2D model of the local oxidation structure (LOCOS). The most widely adopted mathematical model of the oxidation process consists in solving two PDE systems, the first

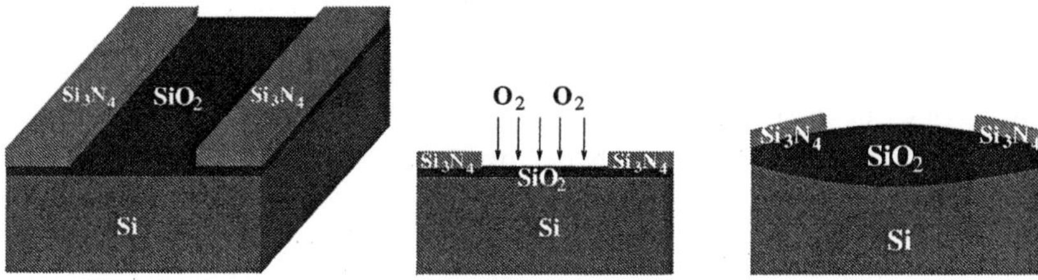


**Fig. 1.** Schematics of the thermal oxidation process in a local oxidation structure (LOCOS): 3D model (left), 2D reduction (center) at the beginning of the oxidation and 2D model (left) after process completion.

being a diffusion-reaction problem for the oxidant and the second a stress analysis in the oxide, nitride and silicon bulk. The two PDE systems are mutually dependent: the diffusion and kinetic reaction coefficients as well as the geometry of the deformed domain depend on the stress distribution; in turn, the chemical reaction forces the oxide-silicon interface to move, driving the mechanical problem. This first level of coupling is handled by using in the diffusion-reaction problem at the new time $t^{n+1}$ the coefficients and the geometry computed from the stress field at time $t^n$. An incremental stress analysis is then performed on the structure subjected to the displacements due to the computed rate of silicon consumption and dioxide expansion (see Fig.2).

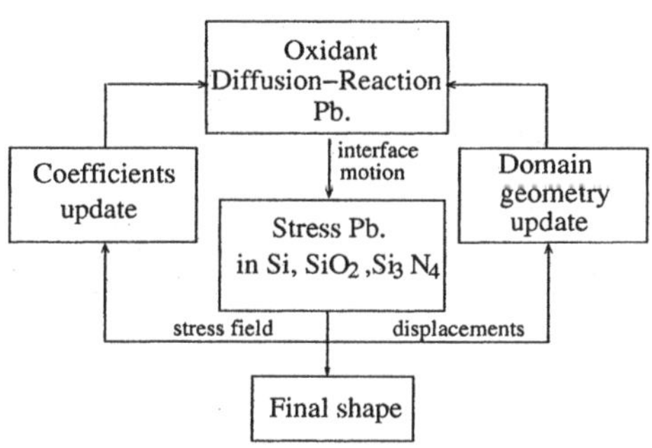


**Fig. 2.** Diagram flux for the full coupled problem.

This incremental stress analysis phase introduces a second nested level of coupling, since it requires solving a set of coupled mechanical problems, each one in a different material: the $Si_3N_4$ mask and the $Si$ bulk are indeed modeled as linear elastic materials, while a non-Newtonian incompressible fluid model with non-linear stress-dependent viscosity is used for the $SiO_2$ [3].

Different strategies may be pursued to handle this second coupled problem. These will be object of discussion in Sect.3.1. The data exchanged between the sub-blocks in both the coupled systems are fluxes and stress dependent quantities, so that the quality of their approximation clearly affects the accuracy of the overall computation. In a standard displacement finite element approach fluxes and stresses are typically post-processed quantities that suffer from a number of limitations. Examples of these limitations are the failure of the post-processed stresses at satisfying self-equilibrium and interelement traction reciprocity, the lack of continuity of the fluxes at the interelement interfaces and the possible onset of locking problems in the incompressible regime. The aim of this work is to investigate and demonstrate the use of alternative finite element formulations specifically tailored to overcome the above mentioned shortcomings. These formulations, known under the comprehensive name of mixed and hybrid finite element techniques and originally developed in the framework of structural analysis, approximate with the same accuracy and physical adherence both the primal fields (*e.g.* the displacements) and the dual fields (*e.g.* stresses and fluxes). A convenient implementation of the proposed methods, based on hybridization and static condensation, yields efficient numerical algorithms with computational effort comparable to standard displacement formulations.

The paper is organized as follows: in Sect.2 we discuss the finite element discretization of the diffusion-reaction problem and of the fluid-mechanical problem. Sect.3 deals with the decoupled algorithm used to iteratively solve the thermal oxidation problem, while Sect.4 demonstrates the performance of the numerical method on several benchmark test-cases. Finally, some concluding remarks are drawn in Sect.5.

## 2 Finite Element Discretization

In this section we discuss the finite element discretization of each subproblem in thermal oxidation.

### 2.1 Notation

In the following, we shall denote by $\Omega$ a bounded open set in $\mathbb{R}^2$ with Lipschitz continuous boundary $\Gamma = \Gamma_D \cup \Gamma_N \cup \Gamma_R$, where $\Gamma_D, \Gamma_N, \Gamma_R$ are the Dirichlet, Neumann and Robin portions of $\Gamma$, respectively, with $\Gamma_R = \emptyset$ in the fluid-mechanical problem. Let $\mathcal{T}_h$ be a regular partition $\mathcal{T}_h$ [2] of $\Omega$ into triangles $K$ such that

$$\overline{\Omega} = \bigcup_{K \in \mathcal{T}_h} \overline{K}.$$

For each element $K \in \mathcal{T}_h$, we denote by $\partial K$ the Lipschitz continuous boundary of $K$, by $\partial K_{int}$ the portion of $\partial K$ such that $\partial K \cap \Gamma = \emptyset$ and by $n_K$ the unit outward normal vector along the boundary $\partial K$. Moreover, if $v$ is any function defined in $\Omega$, we denote by $v^K$ its restriction to the element $K$ and by $v_{\partial K}$ its restriction on the element boundary $\partial K$.

## 2.2 Diffusion-Reaction Problem

The diffusion-reaction problem is solved using the following primal-hybrid finite element nonconforming formulation:

find $C_h \in W_{h,C^*}^{NC}$ such that $\forall v_h \in W_{h,0}^{NC}$ we have

$$\sum_{K \in \mathcal{T}_h} \left( \int_K D \nabla C_h \cdot \nabla v_h \, dx + \int_{\partial K \cap \Gamma_R} D^{-1} k_s \, C_h \, v_h \, ds \right) = \int_K f \, v_h \, dx, \qquad (1)$$

where $W_{h,g}^{NC}$ is the set of affine functions that are:

- mid-point continuous on each edge of the triangulation
- equal to the average of $g$ at the mid-point of each edge of $\Gamma_D$, for any function $g \in L^2(\Gamma_D)$.

In this formulation the primal variable $C_h$ (approximate oxidant concentration) is sought to be a-priori discontinuous and the normal flux of the variable itself arises as a Lagrangian multiplier to enforce interelement continuity. The relaxation of interelement continuity for $C_h$ has the advantage of providing an approximation $p_h$ of the flux $p = -D\nabla C$ that satisfies element-by-element the self-equilibrium condition and that has continuous normal components across interelement edges between neighboring triangles. As a consequence, the normal component of the velocity of the interface is *directly* computed from the normal flux of the concentration as $V_n = -p_h \cdot n / C_h$, where $n$ is the outward unit normal vector on the oxide-silicon interface (see [4] for the physical-mathematical derivation of the above relation). Notice that no post-processing on the computed approximate concentration field $C_h$ is needed, unlike in standard displacement-based finite element methods. Moreover, optimal second order convergence in $L^2(\Omega)$ can be proved for $C_h$ (see [6]).

## 2.3 Fluid-mechanical problem

For the solution of the fluid-mechanical problem, we adopt the novel dual mixed-hybrid finite element formulation introduced and analyzed in [1]. This method provides an accurate stress representation and at the same time handles under a unified formulation both the compressible and incompressible regimes by introducing a pressure function. This avoids resorting to the quasi-incompressible approximation, which is a common approach to deal with incompressible materials in this application field, or using separate computer codes, with a significant saving of software maintenance. The discrete formulation of the problem reads:

find $(\sigma_h, u_h, \lambda_h, p_h, \omega_h) \in (\Sigma_{h,g_N} \times U_h \times \Lambda_{h,g_D} \times Q_h \times W_h)$ such that

$$\begin{cases} \dfrac{1}{2\widehat{\mu}} \displaystyle\int_\Omega \sigma_h : \tau_h \, dx, + \sum_{K \in \mathcal{T}_h} \Big( \int_K u_h \cdot \operatorname{div} \tau_h \, dx - \int_{\partial K_{int}} \lambda_h \cdot (\tau_h \, n) \, ds \Big) \\ + \displaystyle\int_\Omega \frac{\rho_{\widehat{\lambda}}}{2} p_h \operatorname{tr} \tau_h \, dx + \int_\Omega \omega_h \operatorname{as} \tau_h \, dx = \sum_{K \in \mathcal{T}_h} \int_{\partial K \cap \Gamma_D} g_D \cdot (\tau_h \, n) \, ds, \ \forall \tau_h \in \Sigma_{h,0}, \\ \displaystyle\sum_{K \in \mathcal{T}_h} \int_K v_h \cdot \operatorname{div} \sigma_h \, dx = - \int_\Omega f \cdot v_h \qquad \forall v_h \in U_h, \\ \displaystyle\int_\Omega \rho_{\widehat{\lambda}} (\frac{1}{2} \operatorname{tr} \sigma_h + p_h) q_h \, dx = 0 \qquad \forall q_h \in Q_h, \\ \displaystyle\int_\Omega \vartheta_h \operatorname{as} \sigma_h \, dx = 0 \qquad \forall \vartheta_h \in W_h, \\ - \displaystyle\sum_{K \in \mathcal{T}_h} \int_{\partial K_{int}} \mu_h \cdot (\sigma_h n) \, ds = \sum_{K \in \mathcal{T}_h} \int_{\partial K \cap \Gamma_N} g_N \cdot \mu_h \qquad \forall \mu_h \in \Lambda_{h,0}, \end{cases} \tag{2}$$

where $\rho_{\widehat{\lambda}} = \dfrac{\widehat{\lambda}}{\widehat{\mu}(\widehat{\lambda} + \widehat{\mu})}$, $\widehat{\lambda}$ and $\widehat{\mu}$ being the Lamè coefficients of the material. Notice that for $\widehat{\lambda} = +\infty$ system (2) becomes the discrete approximation of the Stokes problem for incompressible fluids. As for the finite element spaces, for $k \geq 0$, we denote by $¶_k(K)$ the space of polynomials in two variables of total degree at most $k$ on the element $K$ and by $R_k(\partial K)$ the space of polynomials of total degree at most $k$ on each edge of $K$. Notice that functions belonging to $R_k(\partial K)$ need not be continuous at the vertices of $\partial K$. Furthermore, we denote by $\mathbb{RT}_0(K)$ the lowest order Raviart-Thomas finite element space [5] on $K$ and by $\mathcal{B}_K = \operatorname{curl}(b_K)$, where $b_K$ is the cubic bubble function on $K$. The finite element spaces in (2) are defined as follows:

$$\begin{aligned} &\Sigma_{h,\xi} = \{\tau^K \in (\mathbb{RT}_0(K) \oplus \mathcal{B}_K)^2, \, \tau^K n = \mathcal{P}\xi \text{ on } \Gamma_N\}, \; V_h = \{v^K \in (¶_0(K))^2\}, \\ &W_h = \{\vartheta \in C^0(\overline{\Omega}) \,|\, \vartheta^K \in ¶_1(K)\}, \qquad\qquad Q_h = \{q^K \in ¶_0(K)\}, \\ &\Lambda_{h,\eta} = \{\lambda^K \in (R_0(\partial K))^2, \, \lambda = \mathcal{P}\eta \text{ on } \Gamma_D\}, \qquad\qquad \forall K \in \mathcal{T}_h, \end{aligned} \tag{3}$$

where $\mathcal{P}$ is the $L^2$ projection over the space of piecewise constant functions and $\xi, \eta$ are given functions in $(L^2(\Gamma_N))^2$ and $(L^2(\Gamma_D))^2$, respectively. Notice that two kinds of Lagrangian multipliers have been introduced in formulation (2). The variable $\omega_h$ is a rotational parameter that avoids requesting the stress tensor to be sought *a priori* in a symmetric function space. The hybrid variable $\lambda_h$ is instead the Lagrangian multiplier that enforces back the continuity of the normal component of the stress tensor across the interelement interfaces. The abstract analysis of the above formulation has been carried out in [1], where in particular a superconvergence result has been shown for $\lambda_h$, as typical of mixed methods with hybridization.

# 3 Iterative algorithm for the coupled problem

The numerical solution of the full coupled problem is achieved by a sequence of successive steps. The computer code implements the following *staggered algorithm*:

> For $n \geq 0$, given at the time level $t^n$ the solution, the geometry and the diffusion and reaction coefficients, at the new time level $t^{n+1}$ we have to:
>
> 1. solve the diffusion-reaction problem and compute the propagation velocity of the $SiO_2$-Si interface;
> 2. perform a mechanical stress-analysis in $SiO_2$, $Si_3N_4$ and Si domains;
> 3. determine the new geometry of the Si and $SiO_2$ and the maximum allowed time step $\Delta t$ according to the present (deformed) configuration;
> 4. set $t^n = t^{n+1}$, $t^{n+1} = t^{n+1} + \Delta t$. Update the stress-dependent $D$ and $k_s$ coefficients;
> 5. update the nodal grid point configuration by relaxing the deformed mesh;
> 6. if $t^{n+1} < t^{max}$ goto 1, else end simulation.

This decoupled procedure has the advantage of splitting the solution of the full problem into several self-contained subproblems of smaller size. Particular attention must be paid to the solution of the stress problem at point 2. of the algorithm, topic that is addressed in the next section.

## 3.1 Fluid-mechanical coupled sub-problem

The stress analysis problem is a fluid-structure interaction problem. In its basic implementation, the Stokes problem in the $SiO_2$ domain is solved first and the computed normal stresses are used to load the elastic problem in the Si and $Si_3N_4$ domains. Eventually, no response from these latter materials is fed back to the $SiO_2$ domain. This strategy, usually referred to as Boundary Loading Method (BLM), is economical and widely adopted in the literature. However, real-life elastic materials do possess memory and as a consequence they tend to "squeeze out" the fluid while relaxing to their initial configuration. Accounting for this behavior by simply loading back the $SiO_2$ domain with the deformations computed from the Si and $Si_3N_4$ domains invariably leads to severe instabilities in the overall numerical procedure, as can be explained by the following argument. Assume that the fluid-mechanical system is modeled by a spring (elastic behavior) placed in series to a damper (fluid behavior). An elementary analysis reveals that solving iteratively the motion of the spring-damper system by a decoupled procedure leads to a conflicting request on the time step $\Delta t$: indeed, in order to ensure stability of the procedure, $\Delta t$ should be small enough when the displacement of the spring is computed from the velocity imposed by the damper, while $\Delta t$ should be large enough when the velocity of the damper is computed from the displacement imposed by the spring. Based on the above considerations, we have adopted the following approach:

1. The interaction between the Si domain and the $SiO_2$-$Si_3N_4$ domains is handled by the standard BLM, since it can be checked that the deformations produced by the Si domain on the $SiO_2$ are negligible.
2. The interaction between the $SiO_2$ and the $Si_3N_4$ domains is handled by a coupled procedure, with an inner iterative map to solve for the nonlinear dependence of the oxide viscosity on the normal stresses. This coupling procedure is implemented exploiting the unified compressible/incompressible formulation discussed in Sect.2.

# 4 Numerical results

As a first test case, we show the results for a simple fluid-elastic structure solved with the unified coupled procedure illustrated in the previous section. The domain $\Omega$ is the unit square, with the upper half behaving like a fluid, while the lower half behaving like an elastic solid (with a very low Young modulus). For a certain time interval a compressive load is applied on the top edge of the fluid domain. Then, the load is released and the elastic solid relaxes recovering its original shape and squeezing out the fluid, as shown in Fig.3 where some phases of the evolution of the phenomenon are displayed. The second benchmark problem that we have considered is the

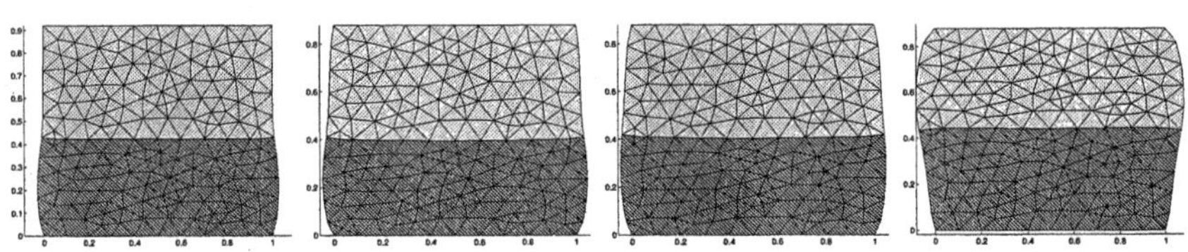

**Fig. 3.** Evolution of the coupled fluid-elastic solid system.

complete simulation of the thermal oxidation process in a LOCOS structure. The computational domain is one half of the domain shown in Fig.1. The geometry and the material properties have been chosen as in [4]. In Fig.4 the deformed configuration and the corresponding pressure field are shown at different time levels. The typical "bird's beak" shape of the final oxide configuration is clearly recognizable. Notice also how the largest stress arise on the junction line between the $SiO_2$ and the $Si_3N_4$ regions and in particular near the lateral edge of the $Si_3N_4$ band.

# 5 Conclusions

Thermal oxidation in silicon device technology is a complex coupled phenomen, involving the solution of a diffusion-reaction problem and of a fluid-structure interaction problem. Special attention has been devoted in the

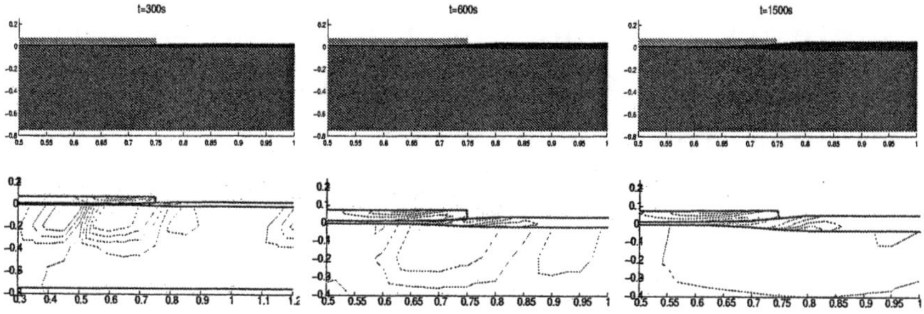


**Fig. 4.** Deformed configuration (top, zoom of the area) and pressure field (bottom) for $t = 300, 600, 1500$s.

present work to devising suitable procedures for handling the two single problems as well as their coupling. In particular, we have proposed a unified dual-mixed hybrid formulation that allows for the simultaneous solution of the compressible/incompressible Navier equations in both solid and fluid domains. Numerical results on benchmark test problems show the accuracy and the flexibility of the approach.

## References

[1] Causin, P., Sacco, R.: A dual-mixed hybrid formulation for simulating mechanical constraints in advanced silicon technology, MOX Technical Report no.5, June 2002, pre-print available at the web page `mox.polimi.it`, submitted for publication.

[2] Ciarlet, P.G.: The Finite Element Method for Elliptic Problems, North Holland, Amsterdam (1978).

[3] Rafferty, C.S.: Stress effects in silicon oxidation - Simulation and experiment, Ph.D. Thesis, Stanford University (1990).

[4] Rao, V.S., Hughes, T.J.R., Garikipati, K.: On modeling thermal oxidation in silicon Part I-II, Int. J. Num. Meth. Engr. **47** (2000) 341-377.

[5] Raviart, P.A., Thomas, J.M.: A mixed finite element method for second order elliptic problems, in Mathematical Aspects of Finite Element Methods (I), Galligani I. and Magenes E. eds., Springer-Verlag Berlin (1977).

[6] Roberts, J., Thomas, J.M.: Mixed and Hybrid Methods, Handbook of Numerical Analysis, Vol.II, Finite Elements Methods (Part 1), P.G. Ciarlet and J.L.Lions Eds., Elsevier Science Publishers (1991).

# Rosenbrock-Methods for Time Step Adaptive Transient Eddy Current Simulations without Nonlinear Iterations

Markus Clemens, Markus Wilke *, and Thomas Weiland

Technische Universität Darmstadt, Dept. Electrical Engineering and Information Technology, Computational Electromagnetics Laboratory (TEMF), Schlossgartenstrasse 8, D-64289 Darmstadt, Germany.

**Abstract.** Nonlinear transient eddy current simulations require the solution of nonlinear differential-algebraic systems of equations of index 1, for which linear-implicit time marching methods of Rosenbrock-type are proposed. These methods avoid the iterative solution of nonlinear systems within each time step due to their built-in Newton procedures. Embedded lower order schemes allow an error-controlled adaptive time step selection to take into account the nonlinear dynamics of the underlying process. Extrapolation methods used for start value generation include a new subspace projection method to improve the numerical performance of the simulations.

## 1 Introduction

Using the Finite Integration Technique (FIT) [16], a discretization method which reformulates the Maxwell's equations in their integral form on a dual grid pair $\{G, \widetilde{G}\}$ into a set of matrix equations, the simulation of transient magnetic fields can be performed with the solution of a nonlinear differential-algebraic system of equations of index 1 (DAE-1)

$$\mathbf{M}_{\kappa}\frac{d}{dt}\widehat{\mathbf{a}}(t) + \widetilde{\mathbf{C}}\mathbf{M}_{\nu}[\widehat{\mathbf{a}}(t)]\mathbf{C}\widehat{\mathbf{a}}(t) = \widehat{\widehat{\mathbf{j}}}_{s}(t), \;\; \widehat{\mathbf{a}}_0 := \widehat{\mathbf{a}}(t_0), \tag{1}$$

where $\widehat{\mathbf{a}}$ denotes the component vector for the line integrals of the modified magnetic vector potential along the edges of the grid $G$, the matrices $\mathbf{C}, \widetilde{\mathbf{C}}$ are discrete curl operators containing the incidence relations of the grid edges, the matrix $\mathbf{M}_{\kappa}$ is the commonly singular matrix of electric conductivities, the matrix $\mathbf{M}_{\nu}$ is the matrix of material reluctivities which are solution dependent in case of ferromagnetic saturation effects, the vector $\widehat{\widehat{\mathbf{j}}}_s$ contains the time dependent excitation source currents. The eddy currents and the magnetic fluxes are given with $\widehat{\widehat{\mathbf{j}}}_e = -\mathbf{M}_{\kappa}\frac{d}{dt}\widehat{\mathbf{a}}$ and $\widehat{\widehat{\mathbf{b}}} = \mathbf{C}\widehat{\mathbf{a}}$, respectively [4]. A computationally slightly more efficient reformulation of (1) introducing the reduced vector potential is given in [5].

* Markus Wilke is supported by the Deutsche Forschungsgemeinschaft (DFG) under grants WE1239/13-1 and WE1239/17.

Since the material coefficients in the reluctivity matrix commonly have measurement errors in the range of several percent, low accuracy requirements can be applied for the solution of (1). In [3], [15] and [7] the usage of a third order singly diagonal implicit Runge-Kutta (SDiRK3(2)) method with an embedded second order method was proposed for an error-controlled variable step length time integration of system (1). In this method, however, an additional error control scheme is required for the solution of the nonlinear algebraic systems of equations of each internal stage. In this paper we propose to alternatively use linear-implicit Rosenbrock methods [12], [13] for the solution of system (1), which allow to avoid the application of such additional linearization schemes.

## 2 Linear-Implicit Rosenbrock Time Integration Methods

Rosenbrock methods belong to the family of one-step linear-implicit time integration methods that differ from SDiRK-methods in such a way that they already incorporate the Jacobi-matrix $\frac{1}{\gamma \Delta t}\mathbf{M}_\kappa + \widetilde{\mathbf{C}}\mathbf{M}_{\nu_d}[\widehat{\mathbf{a}}^n]\mathbf{C}$ of the Newton-Raphson-method in the $s$-stage scheme. The Jacobi matrix contains the material matrix $\mathbf{M}_{\nu_d}[\widehat{\mathbf{a}}^n]$ of the differential reluctivities $\nu_d = \frac{\partial H}{\partial B}$, i.e., the inverse of the local derivative in the B-H-material curve evaluated for the magnetic fluxes $\widehat{\widehat{\mathbf{b}}}^n = \mathbf{C}\widehat{\mathbf{a}}^n$ at time $t^n$. The application of a Rosenbrock method to (1) yields the linear algebraic systems of equations

$$\left[\frac{1}{\gamma \Delta t}\mathbf{M}_\kappa + \widetilde{\mathbf{C}}\mathbf{M}_{\nu_d}[\widehat{\mathbf{a}}^n]\mathbf{C}\right]\widehat{\mathbf{v}}_{ni} = -\widetilde{\mathbf{C}}\mathbf{M}_\nu[\widehat{\mathbf{v}}_i]\mathbf{C}\widehat{\mathbf{v}}_i + \widehat{\widehat{\mathbf{j}}}_s(t^i) - \mathbf{M}_\kappa \cdot \sum_{j=1}^{i-1}\frac{c_{ij}}{\Delta t}\widehat{\mathbf{v}}_{nj} + \Delta t\gamma_i \frac{d}{dt}\widehat{\widehat{\mathbf{j}}}(t^n), \qquad (2)$$

which have to be solved for $i = 1,\ldots,s$ with $t^i = t^n + \alpha_i \Delta t$, $\widehat{\mathbf{v}}_i = \widehat{\mathbf{a}}^n + \sum_{j=1}^{i-1} a_{ij}\widehat{\mathbf{v}}_{nj}$ and the new time solution results from $\widehat{\mathbf{a}}^{n+1} = \widehat{\mathbf{a}}^n + \sum_{i=1}^{s} m_i \cdot \widehat{\mathbf{v}}_{ni}$. The coefficients $a_{ij}$, $c_{ij}$, $\alpha_i$ ,$\gamma_i$, $\gamma$ and $m_i$ determine the specific Rosenbrock method [13].

An embedded lower order solution is delivered by a second coefficient set $\widehat{m}_i$, which provides an error criterion by comparison to the higher order solution. For the error-controlled time integration of (1) only one accuracy threshold has to be specified for the nonlinear transient time marching process. Standard implicit DAE-1 time integration schemes such as the $\vartheta-$methods, the backward differentiation schemes or embedded SDiRK methods require the separate application of a nonlinear iteration scheme such as the (modified) Newton-Raphson method or Quasi-Newton methods. These linearization methods, however, need additional control parameters for the nonlinear iteration accuracy and the corresponding adaption of the linear solvers, the maximum numbers of iterations, a relaxation parameter optimization. For the convergence of the Newton-Raphson scheme in magnetodynamic simulations problems were reported in [9] in some practical applications.

In practical simulations with linear-implicit methods the evaluation of the excitation current vector time derivative, which is required in the applied linear-implicit schemes, can be a problem since the current functions may be non-differentiable e.g. in case of ramped piecewise linear current excitations or if the shape is provided as a list of discrete measurements. In practical calculations [7], however, it was found to be sufficient to provide an approximation of the one-sided derivative $\frac{d}{dt}\widehat{\widehat{\mathbf{j}}}(t^n) := \frac{1}{\delta t}(\widehat{\widehat{\mathbf{j}}}(t^n) - \widehat{\widehat{\mathbf{j}}}(t^n - \delta t))$. Linear-implicit schemes exist for which the evaluation of the time derivative vector of the righthand side is not required at the cost of additional internal stages [12].

## 3 Adaptive Time-Stepping

Adaptive time stepping can be implemented elegantly with embedded Implicit Runge-Kutta (IRK) methods, which deliver for each time step both a solution of a given order and an embedded solution of lower order. Their difference $\widehat{\mathbf{y}} = \widehat{\mathbf{a}}^{(p)} - \widehat{\mathbf{a}}^{(\widehat{p})}$ with the orders $p$ and $\widehat{p}$ of the Runge-Kutta method can be used to produce an error criterion [12] to yield an estimation for the truncation error of the time integration.

### 3.1 Error Norms and Error Tolerances

The norm of the error vector $\widehat{\mathbf{y}}$ can be used to predict the required time step length to achieve an a priori user defined accuracy. A suitable error vector norm is selected in [7] to be $\|\widehat{\mathbf{y}}\|_{err} = \max_i |\widehat{y}_i/(|\widehat{a}_i| + \widetilde{a}_i)|$, where $\widetilde{a}_i$ is an absolute tolerance for the component $\widehat{\mathbf{a}}_i$, which is typically chosen to correspond to one global absolute error parameter $\widetilde{a}$ [7]. The introduction of this parameter $\widetilde{a}$ is necessary to avoid numerical underflow as well as an unnecessarily fine time discretization, if the magnetic flux component vectors at that time are small in value compared to their maxima during the whole transient process [4].

The influence of the absolute tolerance parameter on the adaptive scheme is large [1, p.131], [6], because the systems resulting from (1) are commonly not scaled to receive algebraic solutions with a 2-norm in the interval $[0, 1]$ – a typical silent assumption in the mathematical literature. The parameter $\widetilde{a}$ can either be chosen by an *a priori* magnetostatic field simulation for the maximum current $\widehat{\widehat{\mathbf{j}}}_{s,\max}(t)$ during the time process or in a more conservative, adaptive way by monitoring the maximum norm $\widetilde{a} := \vartheta \cdot \max_{t\in[t_0,t_i]}\{\|\widehat{\mathbf{a}}(t)\|_\infty\}$ with $\vartheta \in [10^{-2}, 10^{-1}]$ of the previous solution vectors [7].

Though closely resembling each other, numerical experience shows that the specific choice of the error norms and the absolute tolerance used in the problems will severely effect the efficiency of simulations (see e.g. [4], [7]). Coupling the relative time step error tolerance $rtol$ and the entries of the Butcher scheme of the SDiRK method with $tol_{\text{NR}} = 0.5||(b^T - \widehat{b}^T)A^{-1}||^{-1} rtol$ according to [14], the nonlinear Newton-Raphson iterations are terminated once $||\delta\widehat{\mathbf{a}}_{i+1}|| \leq tol_{\text{NR}}||\widehat{\mathbf{a}}_i||$ holds. The preconditioned conjugate gradient (PCG) iterations for the linearized systems are stopped by $||\widehat{\mathbf{r}}_n|| \leq tol_{\text{PCG}}||\widehat{\mathbf{r}}_0||$, involving the residual vectors $\widehat{\mathbf{r}}_i$. Here the tolerance $tol_{\text{PCG}}$ is adaptively coupled to $tol_{\text{NR}}$ following [10]. In linear–implicit Rosenbrock methods only the iterations for the linear algebraic systems are to be terminated accordingly.

### 3.2 Time Step Selection

Based on the error norm $||\widehat{\mathbf{y}}||_{\text{err}}$ as an estimation of the local truncation error, the prediction of a new time step, either for the next simulation step, or for a repetition of the last step in case of insufficient accuracy, is performed by a step size controller. Hereby a solution is rejected, if a situation with $rtol < \nu||\widehat{\mathbf{y}}||_{\text{err}}$ occurs, where $\nu$ is an acceleration factor typically set as 1.2. A controller scheme involving a maximum and minimum error and a linear step size variation has been proposed in [15].

More advanced schemes can be received using control-theory models [11]. Understanding the simulation as a control loop leads to a re-interpretation of a standard time step predictor originating from mathematical literature (e.g. [12]) as I-controller scheme [11]. Numerical tests in [6] with more sophisticated PI- or hybrid PI/I-controller schemes [11], however, show no significant advantage over the I-controller scheme for common simulation situations involving (1).

## 4 Extrapolation Techniques

In [8] simple extrapolation techniques, i.e., schemes to provide starting values for the iterative linear system solver, mainly based on Taylor series expansion as well as more sophisticated schemes for the SDiRK scheme [2] have been compared. To attain a problem independent robustness of the different strategies, hybrid schemes were proposed where the evaluation of the minimal residual norm decides about the choice of the specific start vector for the iterative solution process at the new time step.

A new and even more refined hybrid scheme uses a subspace projection strategy: Given two extrapolated start vectors $\widehat{\mathbf{a}}_{0,1}^{(n+1)}$ and $\widehat{\mathbf{a}}_{0,2}^{(n+1)}$ resulting from the non-hybrid extrapolation techniques proposed in [8]. An orthonormalization of these vectors yields the vectors $\widehat{\mathbf{v}}_1$ and $\widehat{\mathbf{v}}_2$ which set up the matrix $\mathbf{V} = \{\widehat{\mathbf{v}}_1, \widehat{\mathbf{v}}_2\}$ onto which the algebraic linear system $\mathbf{A}^{(n+1)}\widehat{\mathbf{a}}^{(n+1)} = \widehat{\mathbf{r}}^{(n+1)}$ of the time stepping process for time $t^{(n+1)}$ (as e.g. (2)) is projected

$$\mathbf{V}^T\mathbf{A}^{(n+1)}\mathbf{V}\mathbf{z} = \mathbf{V}^T\widehat{\mathbf{r}}^{(n+1)}. \tag{3}$$

Exact solution of the $2 \times 2$-system (3) for the coefficient vector $\mathbf{z}$ allows to define the extrapolated vector $\widehat{\mathbf{a}}_0^{(n+1)} := \mathbf{Vz}$. In case of singularity of the system (3) the projection can be reduced to a scalar equation $\widehat{\mathbf{v}}_1^T \mathbf{A}^{(n+1)} \widehat{\mathbf{v}}_1 z = \widehat{\mathbf{v}}_1^T \widehat{\mathbf{r}}^{(n+1)}$ to yield $\widehat{\mathbf{a}}_0^{(n+1)} := z\widehat{\mathbf{v}}_1$. For the positive definite linear systems this procedure provides linear combinations of two (or eventually more) different extrapolated solution vectors which are optimal with respect to the spanned subspace $V_E := \text{span}\{\widehat{\mathbf{v}}_1, \widehat{\mathbf{v}}_2\}$ and the linear system to be solved, since

$$\mathbf{z} = \arg \min_{\mathbf{z}:\mathbf{Vz} \in V_E} \left( \frac{1}{2} \mathbf{z}^T \mathbf{V}^T \mathbf{A}^{(n+1)} \mathbf{Vz} - \mathbf{z}^T \mathbf{V}^T \widehat{\mathbf{r}}^{(n+1)} \right) \tag{4}$$

holds. This optimality result can be extended to the consistently singular algebraic linear systems resulting from the non-gauged magnetodynamic formulation (1).

## 5 Numerical Results

Here, we mainly investigate the behavior of a linear-implicit Rosenbrock-Wanner method RODAS3(2). This 4-stage method is of third order and has an embedded scheme of second order [13]. Both the time integrator of third order and the embedded scheme are stiffly accurate and L-stable and thus applicable to the magnetodynamic DAEs of index 1.

Comparison of this scheme will be performed with a singly diagonal implicit Runge-Kutta scheme (SDiRK3(2)) presented in [3], which was successfully adapted to time step adaptive magneto-quasistatic simulations with (1) in [7] and [15]. Similar to the RODAS3(2) scheme, this 4-stage method is also of order 3 with an embedded solution of second order and also stiffly accurate and L-stable for both orders. Whereas in [15] a Successive Approximation technique is used for the evaluation of the 4 nonlinear intermediate stage solutions per time step in the SDiRK3(2), here we use a Newton-Raphson scheme augmented with a scalar under-relaxation process to ensure stability.

Both the linear-implicit method and the standard SDiRK scheme are applied to a test configuration consisting of an eddy current iron plate with a hole (11.750 degrees of freedom) with different current excitation forms using the B-H-curve of the TEAM 25 benchmark problem. The results in Fig. 1 show that both the linear-implicit RODAS3(2) method and SDiRK3(2) scheme robustly follow the prescribed relative error tolerances $rtol = 10^{-1} - 10^{-4}$, where minor transgressions are due to safety factors in the norms [7]. For low accuracy requirements ($rtol \geq 10^{-3}$) the RODAS3(2) scheme exhibits smaller simulation times on a Pentium III 933 MHz PC than the SDiRK3(2) scheme, which appears to be faster for the smallest accuracy $rtol = 10^{-4}$. Note, that even the ramped piece-wise linear current excitation function in Fig. 1(b) could be integrated without problems with both schemes.

The results of a time step adaptive transient simulation of the TEAM 21b problem, a 50 Hz current driven nonlinear magnetic field benchmark

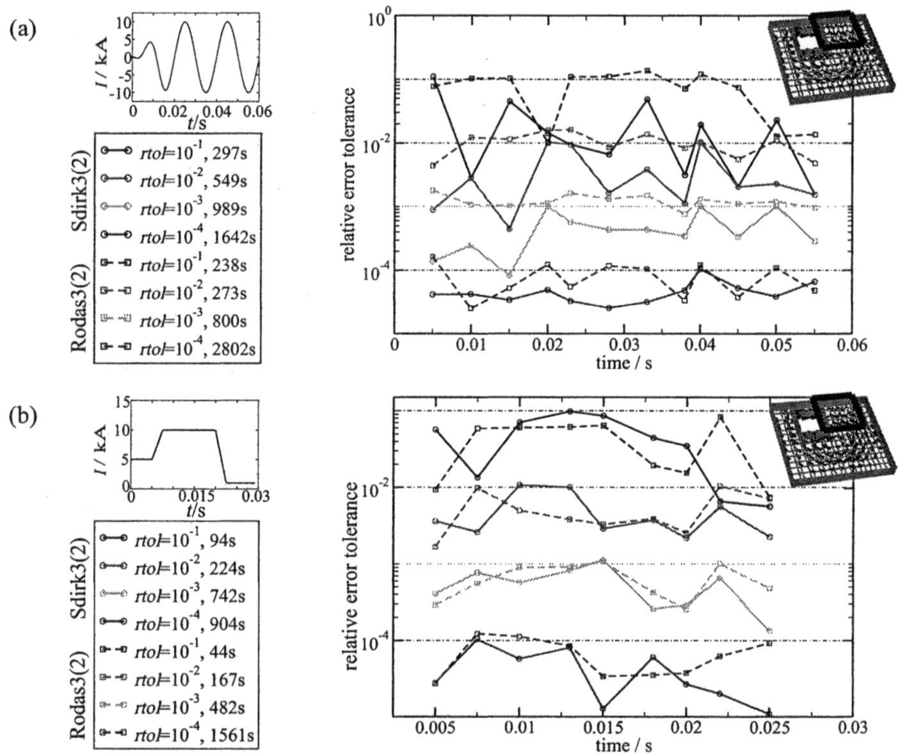


**Fig. 1.** Comparison of different rel. error thresholds *rtol* for RODAS3(2) and SDIRK3(2) method with a nonlinear eddy current test example (Plate with hole; simulation times on 933 MHz PC): (a) Sinusoidal current excitation function. (b) Ramped excitation curve.

problem, are depicted in Fig. 2 for both 4-stage schemes. For a FIT model with 475.000 degrees of freedom [9] the number of required matrix vector multiplications (MxV) and simulation times on a Pentium III 933 MHz PC are given in Table 1 for different relative accuracy values $rtol = 10^{-1} - 10^{-2}$.

The results in Table 1 indicate, that for lower accuracy again the linear-implicit RODAS3(2) appears to be more efficient than the SDiRK3(2) scheme, which benefits for higher accuracy from the local quadratic convergence of its Newton-Raphson process used for the interior stages.

Finally, the results of numerical experiments with extrapolated start values are shown in Table 2 for the plate with a hole problem (linear material) and a sinusoidal ramped excitation (see Fig. 1), the TEAM 11 hollow conductive sphere in an abruptly started magnetic field and again the nonlinear time-harmonic TEAM 21b benchmark problem. Here the total number of required matrix-vector-multiplications (MxV) corresponding to the compu-

**Table 1.** TEAM 21b Problem: Comparison of the linear-implicit RODAS3(2) and the SDIRK3(2) scheme. (Number of rejected steps in brackets.)

| | RODAS | SDiRK | RODAS | SDiRK | RODAS | SDiRK | RODAS | SDiRK |
|---|---|---|---|---|---|---|---|---|
| *rtol* = | $1.0 \cdot 10^{-1}$ | | $5.0 \cdot 10^{-2}$ | | $2.5 \cdot 10^{-2}$ | | $1.0 \cdot 10^{-2}$ | |
| MxV | 31910 | 56869 | 39658 | 65727 | 87277 | 90196 | 427599 | 205270 |
| Cpu-time | 3581 s | 5769 s | 3992 s | 5961 s | 8434 s | 8172 s | 41240 s | 20620 s |
| Steps | 12(3) | 11(6) | 16(5) | 15(7) | 30(20) | 22 (12) | 256(25) | 97(16) |
| NR-cycles | - | 141 | - | 185 | - | 289 | - | 894 |

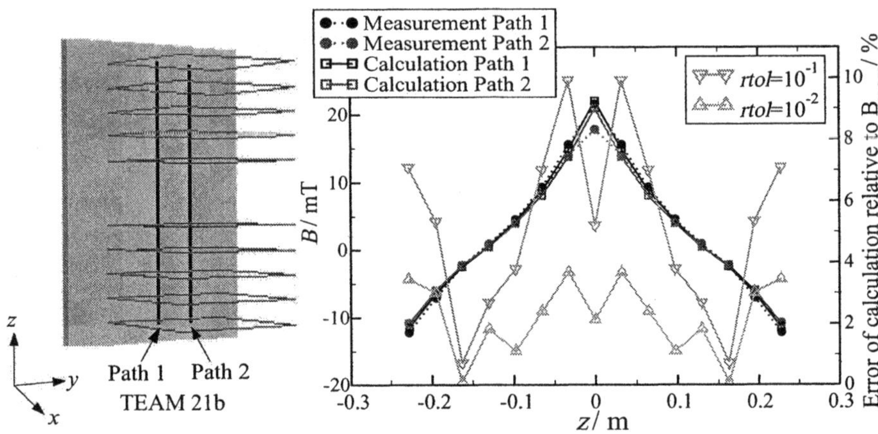


**Fig. 2.** Calculated and measured results of magnetic flux densities for TEAM 21b problem for $rtol = 10^{-2}$, for which both RODAS3(2) and SDiRK3(2) yield identical results. Included are also the relative error distributions along measurement path 1 for the simulations with $rtol = 10^{-1}$ and $rtol = 10^{-2}$.

tational costs are compared for several extrapolation strategies described in [8] and in (3). Time integration schemes used here are conventional methods as a fixed time stepping Euler backward differentiation (BDF1) scheme and the time step adaptive SDiRK3(2) method compared to the linear-implicit RODAS3(2) scheme. Using the sophisticated extrapolation schemes leads to a clear advantage in terms of computational costs.

## 6 Conclusion

Linear-implicit Rosenbrock-type time integration methods were proposed for nonlinear transient magnetic field simulations using the Finite Integration Technique as spatial discretization method. These methods, featuring a built-in Newton procedure, allow for an error-controlled adaptive time stepping

**Table 2.** Comparison of extrapolation strategies using fixed and adaptive time stepping schemes: Total number of matrix vector multiplications.

| Meth. Extrapolator | EC plate | TEAM 11 | TEAM 21b |
|---|---|---|---|
| BDF1, fixed time steps: $tol_{\mathrm{PCG}} =$ | $1.0 \cdot 10^{-5}$ | $1.0 \cdot 10^{-5}$ | $1.0 \cdot 10^{-5}$ |
| 1) $\widehat{\mathbf{a}}_0^{(n+1)} = 0$ | 3951 | 16085 | 136736 |
| 2) $\widehat{\mathbf{a}}_0^{(n+1)} = \widehat{\mathbf{a}}^{(n)}$ | 3424 | 6663 | 133657 |
| 3) $\widehat{\mathbf{a}}_0^{(n+1)} = \widehat{\mathbf{a}}^{(n)} + \Delta t \frac{d}{dt} \widehat{\mathbf{a}}^{(n)}$ | 2554 | 16730 | 129022 |
| 4) 2) or 3) by $\min \lVert \mathbf{M} \widehat{\mathbf{a}}_0^{(n+1)} - \widehat{\mathbf{r}}^{(n+1)} \rVert_2$ | 2714 | 6703 | 129167 |
| 5) 2) and 3) by eqn. (3) | 923 | 2889 | 118516 |
| SDiRK3(2), adaptive: $rtol =$ | $10^{-3}$ | $2.5 \cdot 10^{-4}$ | $2.5 \cdot 10^{-2}$ |
| 6)=1) | 4824 | 17155 | 99552 |
| 7)=3) | 4473 | 10925 | 69946 |
| 8) Continuous extension [2] | 3877 | 7466 | 58799 |
| 9) Stage extrapolation [2] | 4145 | 10787 | 66660 |
| 10) 8) or 9) by $\min \lVert \mathbf{M} \widehat{\mathbf{a}}_0^{(n+1)} - \widehat{\mathbf{r}}^{(n+1)} \rVert_2$ | 3777 | 6688 | 61478 |
| RODAS3(2), adaptive: $rtol =$ | $10^{-3}$ | $2.5 \cdot 10^{-4}$ | $2.5 \cdot 10^{-2}$ |
| 11)≈1) | 5123 | 8135 | 67883 |

of nonlinear magnetodynamic problems. A comparison to standard SDiRK schemes of the same order shows advantages of the Rosenbrock-schemes in terms of numerical efficiency for the low accuracy required in magnetodynamic field simulations. For the generation of start values for the iterative solution of the algebraic systems an optimal subspace projection extrapolation strategy was introduced and tested with good results.

## References


1. Brenan, K. E., Campbell, S. L., Petzold, L. R.: Numerical Solution of Initial-Value Problems In Differential-Algebraic Equations volume 14. SIAM, Philadelphia (1989) Classics in applied mathematics edition
2. Cameron, F.: Low-Order Runge-Kutta Methods for Differential Algebraic Equations. Ph.D. thesis Tampere Univ. of Technology (1999)
3. Cameron, F., Piché, R., Forsman, K.: Variable step size time integration methods for transient eddy current problems. IEEE Trans. Magn. **34** (1998) 3319–3323
4. Clemens, M., Drobny, S., Weiland, T., Wilke, M.: Numerical algorithms for the calculation of magneto-quasistatic fields using the Finite Integration Technique. In van Rienen, U., Günther, M., Hecht, D., (eds.) Scientific Computing in Electrical Engineering 63–70 Springer Verlag, LNCSE series, Heidelberg (2001)
5. Clemens, M., Weiland, T.: Reduced vector potential formulation for FI²TD schemes. IEE Proc. Sci. Meas. Techn. **149** (2002) 232–236

6. Clemens, M., Wilke, M., Weiland, T.: Advanced $FI^2TD$ algorithms for transient 3d eddy current problems. Compel **20** (2001) 365–379
7. ______ 3D transient eddy current simulations using $FI^2TD$ with variable time step size selection schemes. IEEE Trans. Magn. **38** (2002) 605–608
8. ______ Extrapolation strategies in numerical schemes for transient magnetic field simulations. In Conf. Proc. of the CEFC'02, Perugia, Italy (2002) 331. Accepted for publication in IEEE Trans. Magn. (2003)
9. Drobny, S., Weiland, T.: Numerical calculation of nonlinear transient field problems with the Newton-Raphson method. IEEE Trans. Magn. **36** (2000) 809–812
10. Fu, W. N., Ho, S. L., Li, H. L., Wong, H. C.: An effective method to reduce the computing time of nonlinear time-stepping finite-element magnetic field computation. IEEE Trans. Magn. **38** (2002) 441–444
11. Gustafsson, K.: Control-theoretic techniques for stepsize selection in implicit Runge-Kutta methods. ACM Trans. Math. Softw. **20** (1994) 496–517
12. Hairer, E., Wanner, G.: Solving Ordinary Differential Equations II, Stiff and Differential-Algebraic Problems. Springer, Wien, New York (1996)
13. Lang, J.: Adaptive Multilevel Solution of Nonlinear Parabolic PDE Systems: Theory, Algorithm and Application. Springer-Verlag, Berlin, Heidelberg, New York (2001)
14. Nørsett, S. P., Thomsen, P. G.: Local Error Control in SDIRK-Methods BIT **26** (1986) 100-113
15. Wang, H., Taylor, S., Simkin, J., Biddlecomb, C., Trowbridge, B.: An adaptive-step time integration method applied to transient magnetic field problems. IEEE Trans. Magn. **37** (2001) 3478–3481
16. Weiland, T.: Time domain electromagnetic field computation with finite difference methods. Int. J. Num. Mod.: ENDF **9** (1996) 259–319

# A Finite Element Model for Rutherford Cable in Superconductive Magnets

Herbert De Gersem and Thomas Weiland

Technische Universität Darmstadt, Dept. Electrical Engineering and Information Technology, Computational Electromagnetics Laboratory, Schloßgartenstraße 8, D-64289 Darmstadt, Germany

**Abstract.** The finite element model of a superconductive dipole magnet is equipped with a specialised conductor model which accounts for the inter-strand currents caused by the ramping of the magnet without explicitly meshing the individual strands.

## 1 Introduction

Superconductive magnets are planned for the new accelerator facilities in the "Gesellschaft für Schwerionenforschung" (GSI) in Darmstadt, Germany. The ramping of the magnets with 4 T/s will cause eddy currents in the coils of the magnet which diminish the quality of the aperture field and generate additional losses. Three different mechanisms are distinguished: persistent currents in the superconductive filaments, the coupling between filaments and the coupling between the strands of the Rutherford cable [6]. Persistent currents are simulated in [5]. This paper focuses on the simulation of inter-strand eddy currents within finite element (FE) magnet models.

## 2 Finite element magnet model

Eddy current phenomena are described by the partial differential equation (PDE)

$$\nabla \times (\nu \nabla \times \mathbf{A}) + \sigma \frac{\partial \mathbf{A}}{\partial t} = \mathbf{J}_\mathrm{s} \tag{1}$$

in terms of the magnetic vector potential $\mathbf{A}$, the source current density $\mathbf{J}_\mathrm{s}$, the reluctivity $\nu$ and the conductivity $\sigma$. Here, the magnetic field is computed in a 2D cross-section $\Omega_\mathrm{fe}$ of the superconductive magnet (Fig. 1). The Cartesian coordinate system parallel to $\Omega_\mathrm{fe}$ is denoted by $(x, y)$. For convenience, some formulae are expressed in terms of the equivalent polar coordinate system $(r, \vartheta)$. The currents are perpendicular to the cross-section. Because of the longitudinal magnet geometry, the magnetic flux can be assumed to lie in the considered plane. The 2D formulation is

$$-\frac{\partial}{\partial x}\left(\nu \frac{\partial A_z}{\partial x}\right) - \frac{\partial}{\partial y}\left(\nu \frac{\partial A_z}{\partial y}\right) + \sigma \frac{\partial A_z}{\partial t} = J_{\mathrm{s},z} \tag{2}$$

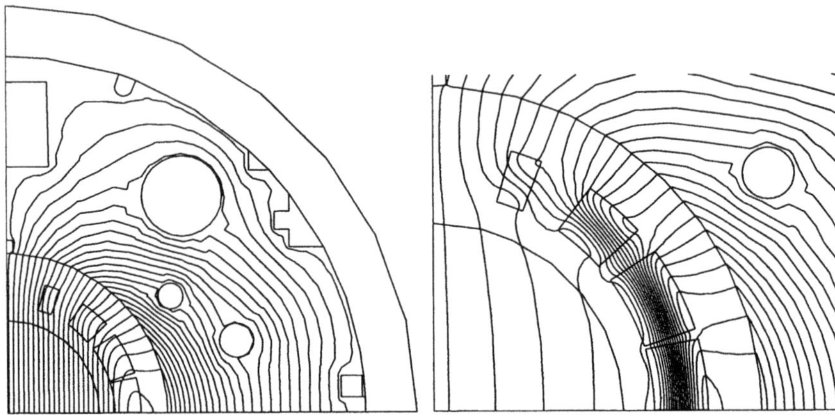

**Fig. 1.** (a) Magnetic flux lines for the $\cos\phi$-type dipole superconductive magnet and (b) magnetisation due to cross-over eddy currents in the coils.

where $A_z$ and $J_{s,z}$ are the $z$-components of $\mathbf{A}$ and $\mathbf{J}_s$ respectively.

The discretisation of (2) by e.g. linear triangular FE shape functions $N_i(x,y)$ yields the system of equations

$$Ku + g(u) = f \tag{3}$$

where $u$ contains the FE degrees of freedom for $A_z$,

$$K_{ij} = \int_{\Omega_{\mathrm{fe}}} \left( \nu \frac{\partial N_i}{\partial x} \frac{\partial N_j}{\partial x} + \nu \frac{\partial N_i}{\partial y} \frac{\partial N_j}{\partial y} \right) \mathrm{d}\Omega \,, \tag{4}$$

$$f_i = \int_{\Omega_{\mathrm{fe}}} J_{s,z} N_i \,\mathrm{d}\Omega \,. \tag{5}$$

Because the strands of the Rutherford cable are not considered in full detail in the FE model, discretising the eddy current term in (2) is not straightforward. Different eddy current mechanisms give rise to several eddy current contributions of the form $g(u)$ as described below.

## 3 Rutherford cable

The $\cos\phi$-type dipole magnet (Fig. 1) has coils featuring windings of Rutherford-type cable (Fig. 2). The cable originally has a rectangular cross-section, it consists of two twisted layers of strands and is keystoned in order to provide a better fit to the magnet geometry. Each strand consists of a copper wire with several embedded superconductive NbTi-filaments. All strands are connected in series and in normal operation, they only carry current in the filaments. If the filaments are not saturated, the voltage drop along the cable is zero in static regime. The copper matrix takes over the current if the

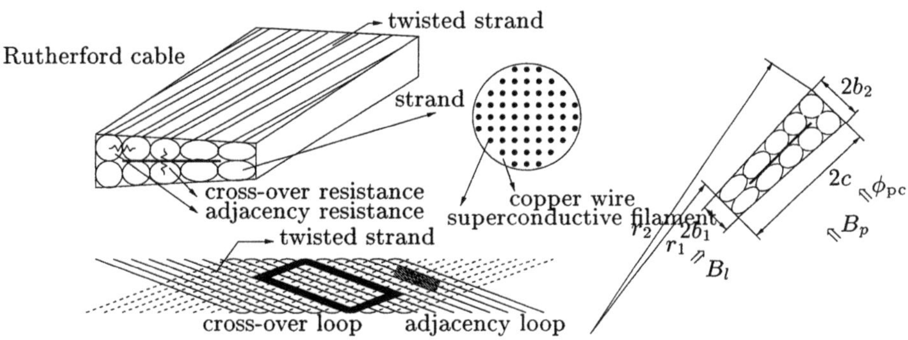


**Fig. 2.** Geometry of a Rutherford cable.

filaments get saturated. The strands are insulated from each other. To ensure a sufficient robustness against the quenching of the magnet, however, a current redistribution between the strands has to be possible up to a certain extent. Therefore, the insulation between the strands is required to be partially conductive. During the ramping of the magnet, the time-varying field induces eddy currents crossing the insulation and forming loops over several strands. The eddy currents cause a deterioration of the quality of the magnetic field in the magnet aperture. This phenomenon has to be simulated accurately in order to predict magnet operation during ramping at an early stage in the magnet design. Two different kinds of current paths connecting several strands are distinguished: rectangular paths formed by neighbouring strands carrying so-called *adjacency eddy currents* and diamond-shaped paths formed by strands of different layers which carry so-called *cross-over eddy currents*. The corresponding inter-strand resistances are called *adjacency* and *cross-over resistance* (Fig. 2). The Rutherford cable itself features a better insulation. Therefore, currents do not migrate between the windings of the coil.

It is not possible to model all geometric details of the Rutherford cable within the overall FE magnet model. The dimensions of the single strands and especially those of the single filaments are very small compared to the dimensions of the overall magnet, which prohibits the use of a detailed FE mesh. It is not recommended to consider the individual windings as well. More efficient FE models are obtained if the discretisation goes beyond the geometrical barriers of single filaments, strands and windings. Problem-tailored modelling techniques are developed here, dealing with the particular geometric properties of the Rutherford cable as applied in $\cos\phi$-type superconductive magnets.

## 4 Finite element model for Rutherford cable

### 4.1 Keystoning

The keystoning of the Rutherford cable will cause the density of superconductive filaments to be larger at the inner side of the coil compared to the outer side. As a consequence, the applied current density $J_{s,z}(r,\vartheta)$ depends on the spatial coordinate $r$ in the considered cross-section:

$$J_{s,z}(r,\vartheta) = \frac{N_{\text{turns}} I_{\text{app}}}{(r_2 - r_1)^2} \left( \frac{r_2 - r}{2b_1} + \frac{r - r_1}{2b_2} \right) \tag{6}$$

where $r_1$ and $r_2$ are the inner and outer radii of the coil, $2b_1$ and $2b_2$ are the thicknesses of the Rutherford cable at $r_1$ and $r_2$ respectively, $N_{\text{turns}}$ is the number of strands per winding and $I_{\text{app}}$ is the applied current (Fig. 2). Keystoning is taken into account in the FE model by accurately integrating (5) with the current density defined by (6).

### 4.2 Adjacency eddy currents

Analytical formulae dealing with adjacency eddy current effects are reported in e.g. [4,6]. There, a distinction is made between adjacency eddy currents $J_{\text{pa},z}$ closing in neighbouring strands due to the time-varying magnetic field $B_{\text{p}}$ perpendicular to the long side of the Rutherford cable, and eddy currents $J_{\text{la},z}$ closing in neighbouring strands due to the time-varying parallel magnetic field $B_{\text{l}}$. In analytical models, the additional magnetisation due to eddy currents is characterised by the time constants $\tau_{\text{pa}}$ and $\tau_{\text{la}}$. The eddy currents $J_{\text{pa},z}$ and $J_{\text{la},z}$ flow in cross-sections of the Rutherford cable with different orientation and shape. Their treatment within the FE cable model developed here, is nevertheless the same. Specialised conductor models have already been developed for foil windings and multi-conductor windings in [1] and [2], respectively. In this paper, a similar modelling technique is developed for the particular Rutherford cable layout. Another possible approach is the embedding of a detailed cable model as a macro-element into the magnet model as is done for machine windings in [3].

Consider a *current redistribution zone* $\Omega_q$ consisting of a single layer of all strands either along the short side or along the long side of the cable. Due to the finite resistance of the insulation between the individual strands, the current can redistribute within $\Omega_q$. Further migration between the cable windings is prevented by the cable insulation which has a substantially higher resistivity. The redistribution of the current in the direction perpendicular to the considered layer is not considered since this redistribution is treated by a separate current distribution zones with orientation perpendicular to the considered current redistribution zone $\Omega_q$. This splitting allows to consider anisotropic adjacency resistances as e.g. in the case where the Rutherford cable has a resistive core between the two longitudinal layers

(Fig. 2). The conductivity $\sigma_{\mathrm{pa}}$ experienced by the adjacency eddy currents can be computed analytically and depends on the true adjacency resistance of the cable, which is commonly obtained by measurement, the dimensions of the strands and the twist pitch of the Rutherford cable. The electric field $E_z$ in the $z$-direction is unknown. The eddy current density in layer $\Omega_q$ due to the ramping of the magnetic field is

$$J_{\mathrm{pa},z}(r,\vartheta) = \sigma_{\mathrm{pa}} E_z - \sigma_{\mathrm{pa}} \frac{\partial A_z}{\partial t} \,. \tag{7}$$

The netto current

$$I_{z,q} = \int_{\Omega_q} J_{\mathrm{pa},z}(r,\vartheta)\mathrm{d}\Omega \tag{8}$$

through the current redistribution zone $\Omega_q$ is zero.

Since the Rutherford cable is comparatively long in the $z$-direction and since the current can close through the perfectly conductive filaments of the twisted strands, the electric field can be assumed to be constant over $\Omega_q$. For each current redistribution zone $\Omega_q$, a constant shape function $M_q(x,y)$ is defined to have the value 1 in $\Omega_q$ and 0 in $\Omega_{\mathrm{fe}} \setminus \Omega_q$. The electric field can be expressed by

$$E_z(x,y) = \sum_q E_{z,q} M_q(x,y) \,. \tag{9}$$

The adjacency current density vector due to the perpendicular time-varying magnetic field, $g_{\mathrm{pa}}$, is obtained by weighing (7) by the FE shape functions $N_i(x,y)$ associated with the magnetic vector potential:

$$g_{\mathrm{pa}} = M_{\mathrm{pa}} \frac{\partial u}{\partial t} - Z_{\mathrm{pa}} e_{\mathrm{pa}} \tag{10}$$

where

$$M_{\mathrm{pa},ij} = \int_{\Omega_{\mathrm{fe}}} \sigma_{\mathrm{pa}} N_i(x,y) N_j(x,y)\mathrm{d}\Omega \,, \tag{11}$$

$$Z_{\mathrm{pa},iq} = \int_{\Omega_{\mathrm{fe}}} \sigma_{\mathrm{pa}} N_i(x,y) M_q(x,y)\mathrm{d}\Omega \tag{12}$$

$$\tag{13}$$

and $e_{\mathrm{pa}}$ is the vector with all degrees of freedom $E_{z,q}$. The additional constraint forcing the netto current (8) to zero is weighted by the FE shape functions $M_p(x,y)$ associated with the electric field unknowns yielding

$$-Z_{\mathrm{pa}}^T \frac{\partial u}{\partial t} + G_{\mathrm{pa}} e_{\mathrm{pa}} = 0 \tag{14}$$

where

$$G_{\mathrm{pa},pq} = \int_{\Omega_{\mathrm{fe}}} \sigma_{\mathrm{pa}} M_p(x,y) M_q(x,y)\mathrm{d}\Omega \,. \tag{15}$$

The FE conductor model representing adjacency eddy currents in neighbouring strands due to the time-varying magnetic field, is inserted in the global FE model:

$$\begin{bmatrix} M_{\mathrm{pa}} & 0 \\ -Z_{\mathrm{pa}}^T & 0 \end{bmatrix} \frac{\partial}{\partial t} \begin{bmatrix} u \\ e_{\mathrm{pa}} \end{bmatrix} + \begin{bmatrix} K & -Z_{\mathrm{pa}} \\ 0 & G_{\mathrm{pa}} \end{bmatrix} \begin{bmatrix} u \\ e_{\mathrm{pa}} \end{bmatrix} = \begin{bmatrix} f \\ 0 \end{bmatrix} \tag{16}$$

which, after discretisation in time and scaling the constraint equations by an appropriate factor, results in a symmetric, positive definite system of equations.

The correct modelling of eddy currents closing through adjacent strands depends on the choice of the set of current redistribution zones $\Omega_q$ which are represented in the coupled FE model by the constant shape functions $M_q(x, y)$. Two types of current redistribution zones are selected: zones corresponding to layers of strands along the long side of the Rutherford cable and those corresponding to layers of strands along the short side of the cable. The set of shape functions $M_q(x, y)$ constitutes a discretisation of the unknown electric field. For large coils, this approach may still introduce a too large number of additional unknowns. The formulation (16), however, allows for a further coarsening of the FE model. The numerical current redistribution zones $\Omega_q$ can consist of several physical current redistribution zones. It is shown in [2] that the FE model already provides reliable results for numerical electric field discretisations which are substantially coarser than the true geometry of the cable strands. It is also possible, and sometimes even explicitly recommended, to select FE shape functions other than piecewise constants for $M_q(x, y)$. A set of wavelets containing the frequencies according to the periodical distribution of strands, may yield extremely efficient FE models.

### 4.3 Cross-over eddy currents

Due to the typical twisting of the strands in a Rutherford cable, diamond-shaped current loops arise if the current migrates between both layers and follows paths with different orientations (Fig. 2). Only current paths with two cross-over points are considered. Due to twisting, the cross-over currents close without experiencing additional resistance at the short sides of the Rutherford cable. Since such a cross-over loop necessarily extends to the borders of the Rutherford cable, the cross-over eddy current phenomenon can not be considered as a local eddy current effect.

Cross-over eddy currents are induced by a time-varying, perpendicular magnetic field. The magnetic flux through a cross-section at the azimuthal coordinate $\vartheta$ of the Rutherford cable with an $r$-$z$-plane is

$$\phi_{\mathrm{p}}(\vartheta) = -\ell_z \left( A_z \left( r_2, \vartheta \right) - A_z \left( r_1, \vartheta \right) \right) \tag{17}$$

where $\ell_z$ is the length of the magnet in the $z$-direction. The magnetising flux $\phi_{\mathrm{pc}}(\vartheta)$ induced by the cross-over eddy currents is

$$\phi_{\mathrm{pc}}(\vartheta) = -\tau_{\mathrm{pc}} \frac{\partial \phi_{\mathrm{p}}(\vartheta)}{\partial t} \tag{18}$$

where $\tau_{\mathrm{pc}}$ is the time constant for cross-over magnetisation which depends on the cross-over resistance and the geometry of the Rutherford cable [4]. The magnetising flux is represented in the FE model by current sheets at $r = r_1$ and $r = r_2$. The corresponding current densities are

$$J_z(r,\vartheta) = -\nu \frac{\phi_{\mathrm{pc}}}{\ell_z\,(r_2 - r_1)} \frac{r_1 + r_2}{2} \left( \frac{\delta_{r_1}(r)}{\gamma r_1} - \frac{\delta_{r_2}(r)}{\gamma r_2} \right) \tag{19}$$

where $\delta_{r_w}(r)$ denotes the Kronecker delta function located at $r_w$ and $\gamma$ represents the thickness of the current sheets. Weighing (19) by the FE shape functions $N_i(x,y)$ results in an additional term representing the cross-over eddy current effect:

$$g_{pc} = M_{\mathrm{pc}} \frac{\partial u}{\partial t} \tag{20}$$

where, with $\mathcal{N}_k(\vartheta) = (N_k\,(r_2,\vartheta) - N_k\,(r_1,\vartheta))$,

$$M_{\mathrm{pc},ij} = \int_0^{2\pi} \nu \tau_{\mathrm{pc}} \frac{r_1 + r_2}{2\,(r_2 - r_1)} \mathcal{N}_i(\vartheta)\mathcal{N}_j(\vartheta)\mathrm{d}\vartheta \,. \tag{21}$$

The integrations in (21) are performed over the inner and outer boundaries of the coil. The load term (20) only affects FE nodes at these boundaries. This feature reflects the non-local character of cross-over eddy currents. The coupled FE model accounting for eddy currents in Rutherford cable both in adjacent strands and due to cross-over coupling is

$$\begin{bmatrix} M_{\mathrm{pa}} + M_{\mathrm{pc}} & 0 \\ -Z_{\mathrm{pa}}^T & 0 \end{bmatrix} \frac{\partial}{\partial t} \begin{bmatrix} u \\ e_{\mathrm{pa}} \end{bmatrix} + \begin{bmatrix} K & -Z_{\mathrm{pa}} \\ 0 & G_{\mathrm{pa}} \end{bmatrix} \begin{bmatrix} u \\ e_{\mathrm{pa}} \end{bmatrix} = \begin{bmatrix} f \\ 0 \end{bmatrix} . \tag{22}$$

This system is discretised in time by Galerkin-type linear time steps. The systems of equations are solved by the Conjugate Gradient method with a Symmetric Successive Over-Relaxation preconditioner.

## 5 Application

The 2D transient FE model of a quarter of a superconductive dipole magnet is equipped with the specialised Rutherford cable model. The magnetic dipole field is increased from 0.2 T to 2 T with a ramp rate of 4 T/s. In the simulations carried out here, it is assumed that the superconductive filaments are not saturated and hence perfectly conductive. The ramping of the magnetic field causes eddy currents in the Rutherford cable and hence, disturbances to the magnetic field in the aperture. The adjacency and cross-over eddy current densities are shown at the time instant when the magnetic dipole field is 2 T. Due to the high perpendicular magnetic field at $\vartheta = 0$, significant eddy currents appear there. In contrast to the adjacency eddy current density, the cross-over eddy current density reflects a symmetry along the center line of the coil (Fig. 3). The magnetisation due to cross-over coupling is shown in

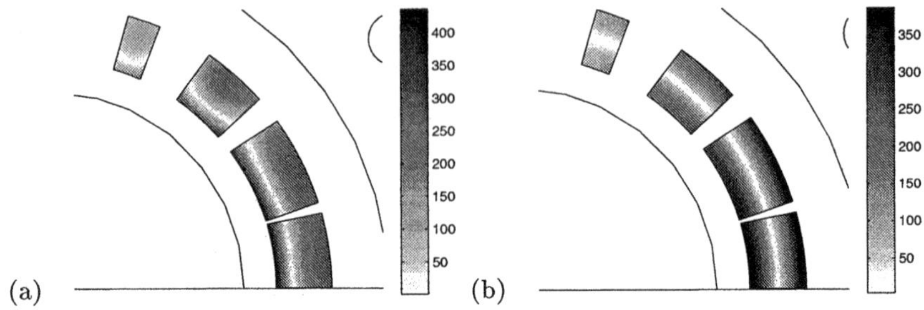


**Fig. 3.** Absolute values of (a) the adjacency eddy current density due to the time-varying azimuthal field and (b) the cross-over eddy current density in the Rutherford cable.

Fig. 1b. Thanks to the neighbouring iron yoke, only a small fraction of the cross-over magnetising flux enters the aperture of the magnet. The magnetisation due to eddy currents causes a small deterioration of the field quality of the magnetic dipole field in the aperture. The specialised cable model avoids the construction of detailed meshes in the winding area and hence yields small FE models for which transient simulation following a ramped excitation cycle becomes feasible.

## 6 Conclusions

The specialised FE cable model accounts for the adjacency eddy currents in neighbouring strands and the cross-over eddy currents between the twisted layers of Rutherford cable without explicitly meshing the individual strands of the cable. This results in considerably smaller FE models for superconductive magnets and enables an accurate, transient analysis of the magnetic field during the ramping of the magnets.

## Acknowledgments

Herbert De Gersem is working in the cooperation project "DA-WE1 (TEMF/-GSI)" with the "Gesellschaft für Schwerionenforschung (GSI)", Darmstadt.

## References

1. H. De Gersem and K. Hameyer. A finite element model for foil winding simulation. *IEEE Transactions on Magnetics*, 37(5):3427–3432, September 2001.
2. H. De Gersem and K. Hameyer. A multi-conductor model for finite element eddy current simulation. *IEEE Transactions on Magnetics*, 38(2):533–536, March 2002.

3. Á. Szűcs and A. Arkkio. Consideration of eddy currents in multi-conductor windings using the finite element method and the elimination of inner nodes. *IEEE Transactions on Magnetics*, 35(3):1147–1150, May 1999.
4. A.P. Verweij and H.H.J. ten Kate. Coupling currents in Rutherford cables under time varying conditions. *IEEE Transactions on Applied Superconductivity*, 3(1):146–149, March 1993.
5. C. Völlinger, M. Aleksa, and S. Rüssenschuck. Calculation of persistent currents in superconductive magnets. *COMPEL*, 20(2):562–569, 2001.
6. M.N. Wilson. Superconducting magnets for accelerators: a review. *IEEE Transactions on Applied Superconductivity*, 7(2):727–732, June 1997.

# Best of Both Worlds: Combining EM Accuracy with the Speed of Analytical Models

Tom Dhaene[1 2]

[1] University of Antwerp, Middelheimlaan 1, 2020 Antwerp, Belgium; tom.dhaene@ua.ac.be
[2] Agilent Technologies, EEsof EDA, Lammerstraat 20, 9000 Ghent, Belgium; tom_dhaene@agilent.com

**Abstract.** New modeling technology is developed that allows engineers to define the frequency range, layout parameters, material properties and desired accuracy for automatic generation of simulation models of general passive electrical structures. It combines electromagnetic (EM) accuracy of parameterized passive models with the simulation speed of analytical models. The adaptive algorithm does not require any a priori knowledge of the dynamics of the system to select an appropriate sample distribution and an appropriate model complexity. With this technology, designers no longer must put up with legacy modeling techniques or invest resources in examining new ones.

## 1 Introduction

Component and circuit models are a cornerstone of EDA (Electronic Design Automation) technology. With wireless and wireline designs constantly increasing in complexity and operating at higher frequencies, design engineers push the limits of their EDA tool's passive analytical models. Often, these passive models are used outside their operational range, causing the EDA tool to return inaccurate simulation results. The inconsistencies of legacy modeling techniques from the 1970s and 1980s hinder the accuracy of these models when applied to different processes and frequencies. Exceeding a model's frequency limit causes errors due to the model's failure to account for higher-order propagation modes. Limitations of the equivalent circuit model, such as frequency independent inductive or capacitive elements, also lead to simulation errors. Since most EDA tools do not proactively report such errors, they propagate through the design flow and may not be discovered until a prototype fails to perform as expected. To avoid errors and inconsistencies, full-wave EM simulation is required to fully characterize the structure and produce an accurate S-parameter model of the discontinuity that is then used by the circuit simulator.

Developing new models is not a trivial task! To model a single parameter over a range of values, several sample points are required. Since the model can be a function of many layout parameters (line width, length, metal thickness, dielectric constant, substrate thickness, loss tangent, etc.) there is an exponential growth in the number of samples as the number of layout parameters

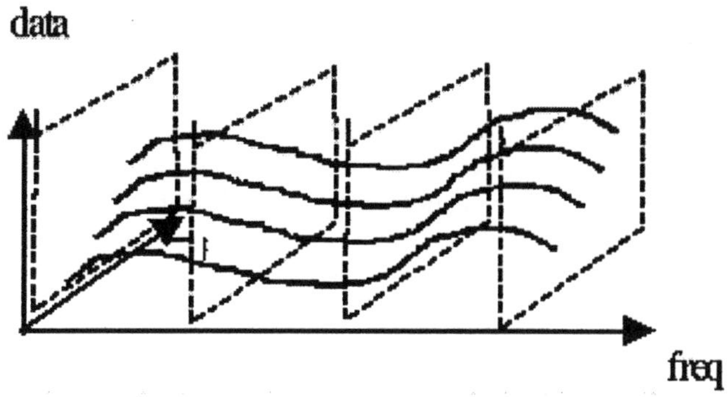


**Fig. 1.** *Step One:* AFS rational models over the desired frequency range, derived from full-wave EM simulation.

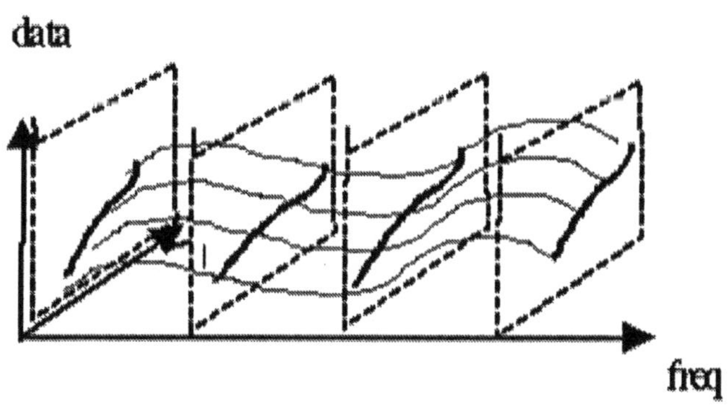


**Fig. 2.** *Step Two:* Multinomial models are created at discrete frequencies.

increases. Also, developing a new model usually requires a highly skilled person working for an extended period-several weeks or even months-to build, test and produce the desired analytical model. If the requirement is for a complete library of models, the total effort is multiplied by the number of models sought. This task needs to be weighed against measurement-based or EM-based modeling on a case-by-case basis.

Some common approaches to modeling issues have limiting factors [1,2]. Methods using pre-calculations of equivalent circuits, using a variety of look-up tables, fitting equations and interpolation techniques can have a limited number of samples and have insufficient interpolation methods. One clear example where the dependability of these techniques comes into question is with high-Q resonant circuits such as those used in narrow band filters. Using discrete data grids and interpolation techniques with such circuits might cause the generated model to suffer from either "oversampling" or "undersampling." With oversampling, too many data samples are collected and model generation is inefficient; on the other hand, with undersampling, too few data samples are collected and the model is not completely defined.

As an alternative to building classic analytical models, engineers can utilize a full-wave EM modeling tool to fully characterize a given passive component. This method permits accurate characterization of the actual passive structure to be used, accounting for higher-order mode propagation, dispersion and other parasitic effects. However, the calculation time required for full-wave EM simulation of a given component makes real-time circuit tuning impossible. A new efficient adaptive sampling and modeling technique addresses this model accuracy dilemma. The *Multidimensional Adaptive Parameter Sampling* algorithm (MAPS) selects a limited set of data samples in consecutive iterations, and interpolates all S-parameter data using rational and multinomial fitting models. This algorithm allows important details to be modeled by automatically sampling the response of the structure more densely where the S-parameters are changing more rapidly. The goal is minimizing the total number of samples needed, while maximizing the information provided by each new sample. The new modeling technique combines the speed and flexibility of analytical models, and the accuracy and generality of full-wave EM simulation in one compact parameterized passive model [3,4].

## 2 Adaptive modeling and sampling technique

The MAPS technique builds a global fitting model of the chosen parameters, handling frequency and geometrical dependencies separately. Multidimensional polynomial (or multinomial) fitting techniques are used to model the geometrical dependencies, while rational fitting techniques [5] are used to handle frequency dependencies. The modeling process does not require any *a priori* knowledge of the circuit under study. Different adaptive algorithms are combined to efficiently generate a parameterized fitting model that meets the

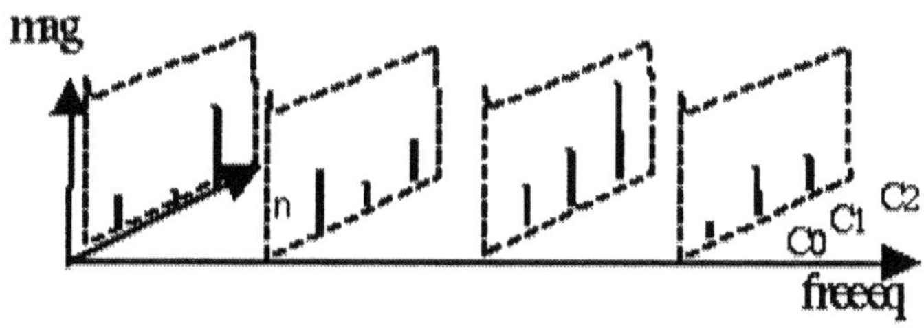


**Fig. 3.** *Step Three:* Creation of the coefficients of orthogonal multinomials at discrete frequencies.

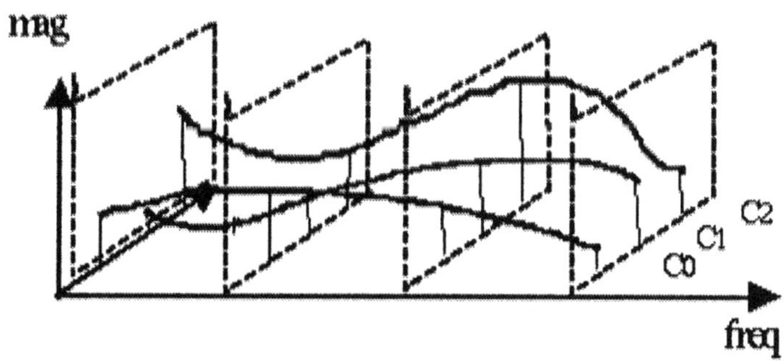


**Fig. 4.** *Step Four:* Calculation of coefficients of orthogonal multinomials over the entire frequency range.

predefined accuracy. This includes the adaptive selection of an optimal number of data samples along the frequency axis and in the geometrical parameter space, and adaptive selection of the optimal order of the multinomial-fitting model.
The number of data points is selected to avoid oversampling and undersampling. The process of selecting data points and building models in an adaptive way is called *reflective exploration* [6]. Reflective exploration is useful when the process that provides the data is very costly, which is the case for full-wave EM simulators. Reflective exploration requires *reflective functions* that are used to select new data points. For example, the difference between two fitting models can be used as a reflective function. Also, some physical rules, such as a passivity-check, can be used as a reflective function. The modeling process starts with an initial set of data points. New data points are selected near the maximum of the reflective function until the desired accuracy is reached.
The model complexity is automatically adapted to avoid overmodeling (overshoot or ringing) and undermodeling, and the model covers the whole parameter and frequency space and can easily be used for optimization purposes.
The MAPS modeling technique follows four steps to adaptively build a model.

- *Step 1:* The frequency response of the circuit is calculated at a number of discrete sample points (using the Agilent Momentum full-wave EM simulator [7]). The Adaptive Frequency Sampling (AFS) algorithm [5] selects a set of frequencies and builds a rational model for the S-parameters over the desired frequency range (Figure 1).
- *Step 2:* A multinomial is fitted to the S-parameter data at multiple discrete frequencies (Figure 2).
- *Step 3:* This model is written as a weighted sum of orthonormal multinomials. The multinomials only depend on the layout parameters. The weighting coefficients preceding the orthonormal multinomials in the sum are only frequency dependent (Figure 3).
- *Step 4:* Using the AFS models built in step one, the coefficients can be calculated over the whole frequency range (Figure 4). These coefficients, together with the orthonormal multinomials, are stored in a database for use during extraction afterwards.

## 3 Example

The automated modeling technique was used to generate analytical circuit models for *all* sub-parts (*transmission line, open end, slot coupler, step in width, corner-fed patch*) of a *slot-coupled microstrip-fed patch antenna* structure (figure 5). This modeling step is a one-time, up-front time investment. A double-sided duroid substrate was used (thickness = 31 mil & 15 mil, $\varepsilon_r$

**Fig. 5.** Slot-coupled microstrip-fed patch antenna structure

= 2.33, tg $\delta$ = 0.0012).
First, parameterized circuit models were built for *all* substructures of the circuit. For example, the corner-fed patch (figure 6) circuit model was built over the following parameter range (tabel 1):

| variable | min | max |
|---|---|---|
| $L_{\text{patch}}$ | 320 mil | 400 mil |
| $W_{\text{feed}}$ | 5 mil | 30 mil |
| $f$ | 5 GHz | 15 GHz |

**Tab.1** Parameter ranges of corner-fed patch.

The automated modeling tool (ADS Model Composer) selected 25 data points (= discrete layouts) in an adaptive way, and grouped all S-parameter data all in one global, compact, analytical model. ADS Momentum was used as planar EM simulator [7]. The desired accuracy level was set to 55 dB. In figure 7, the reflection coefficient $S_{11}$ of the corner-fed patch is shown as a function of frequency and width.
Then, the parameterized circuit models were used to simulate the overall antenna structure (figure 5). Figure 8 shows $S_{11}$ simulated with Momentum, and with the new analytical circuit models for all sub-components (divide and conquer approach). Both results correspond very well. However, the simulations based on the circuit models easily allow optimization and tuning, and took only a fraction of the time of the fullwave simulation (2 s versus 96 min on a 450 MHz Pentium II).

## 4 Conclusions

An advanced modeling technique was presented for building parameterized models for general passive microwave and RF structures. The models are based on fullwave EM simulations, and have a userdefined accuracy. Once generated, the analytical models can be grouped in a library, and incorporated in an EDA tool where they can be used for simulation, design and optimization purposes. A patch antenna example was given to illustrate the technique. The results based on the parameterized models correspond very

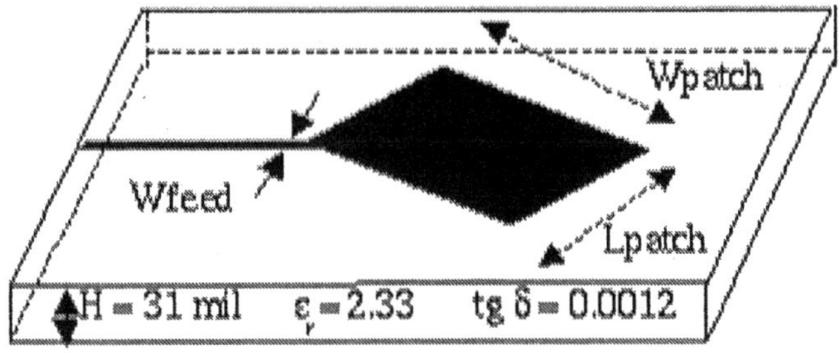


**Fig. 6.** Layout of *corner-fed patch*

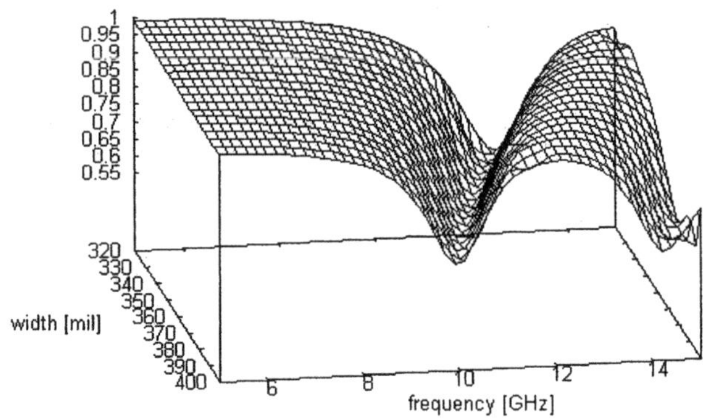


**Fig. 7.** Reflection coefficient $S_{11}$ of corner-fed patch ($W_{feed}$ = 8 mil)

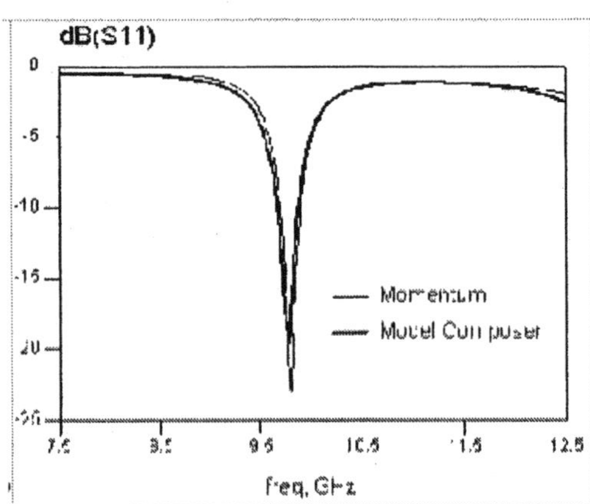


**Fig. 8.** Reflection coefficient $S_{11}$ of slot-coupled microstrip-fed patch antenna

well with the global fullwave simulations. However, the time required for a simulation using the compact analytical circuit models was only a fraction of the time required for a global fullwave simulation.

## References


1. Chaki S., Aono S., Andoh N., Sasaki Y., Tanino N., Ishihara O.: Experimental Study on Spiral Inductors, Proceedings of the IEEE Symposium on Microwave Theory and Techniques, (1995) pp. 753–756.
2. Liang JF., Zaki K. A.: CAD of Microwave Junctions by Polynomial Curve Fitting, Proceedings of the IEEE Symposium on Microwave Theory and Techniques, (1993) pp. 451–454.
3. De Geest J., Dhaene T., Fache N., De Zutter D.: Adaptive CAD-Model Building Algorithm for General Planar Microwave Structures, IEEE Transactions on Microwave Theory and Techniques, vol. **47**, no. **9**, (1999) pp. 1801–1809.
4. Dhaene T., De Geest J., De Zutter D.: EM-based Multidimensional Parameterized Modeling of General Passive Planar Components, Proceedings of the IEEE Symposium on Microwave Theory and Techniques, Vol. **3**, (2001) pp. 1745–1748.
5. Dhaene, T., Ureel, J., Fache, N., De Zutter, D.: Adaptive Frequency Sampling Algorithm for Fast and Accurate S-parameter Modeling of General Planar Structures, Proceedings of the IEEE Symposium on Microwave Theory and Techniques (1995)
6. Beyer, U., and Smieja, F.: Data Exploration with Reflective Adaptive Models, Computational Statistics and Data Analysis, vol. **22**, pp. 193–211, 1996.
7. Momentum software, Agilent EEsof Comms EDA, Agilent Technologies, Santa Rosa, CA.

# Simulation of Magnetic Circuits Including Hysteresis Nonlinearity

Sinan Güngör and Saffet Altay

Istanbul Technical University, Dept. of Electrical Eng., 80626 Istanbul - Turkey

**Abstract.** Magnetic circuits can be represented with a topological dual circuit. In the dual circuit, flux paths are modelled by hysteretic permeances instead of reluctances. Hysteresis effect is taken into account by using the Jiles-Atherton (JA) approach. In addition, iron losses due to eddy current are also included to the model. Comparison of simulated results with the experimental results from a core type transformer demonstrates the capability of the proposed method.

## 1 Introduction

Magnetic circuits can be represented with a topological dual circuit [1]. By choosing the time derivative of flux ($d\phi/dt$) as potential quantity and the mmf drop ($v$) as flow quantity, the flux paths are modeled by permeances. The comparison of the reluctance and permeance networks is summarized in Table 1. The mesh equations of the reluctance network are written according to Ampère's law. The total currents enclosed by a mesh, $\sum i$ is considered as the mmf source and the mmf drop across a flux path is defined as $v = Hl$, where $H$ is the magnetic field strength, $l$ is the length of the flux path. Flux paths store magnetic energy and also have losses. Magnetic energy storage property is not directly modelled with reluctance network. In [2], a permeance network for switched reluctance machines is given without considering the magnetic hysteresis. In this work, the method is expanded to include the hysteresis nonlinearity into the model. Iron losses due to eddy current are also separately modelled. Network equations of permeance network are a set of differential equations which are to be integrated to determine the fluxes of network element.

**Table 1.** Comparison of reluctance and permeance networks ($v = Hl$: mmf drop, $\phi$: flux, $i$: current)

| | Reluctance network | Permeance network |
|---|---|---|
| Potential quantity | $v$ | $\frac{d\phi}{dt}$ |
| Flow quantity | $\phi$ | $v$ |
| Mesh equation | $\sum v = \sum i$ | $\sum \frac{d\phi}{dt} = 0$ |
| Node equation | $\sum \phi = 0$ | $\sum v = \sum i$ |

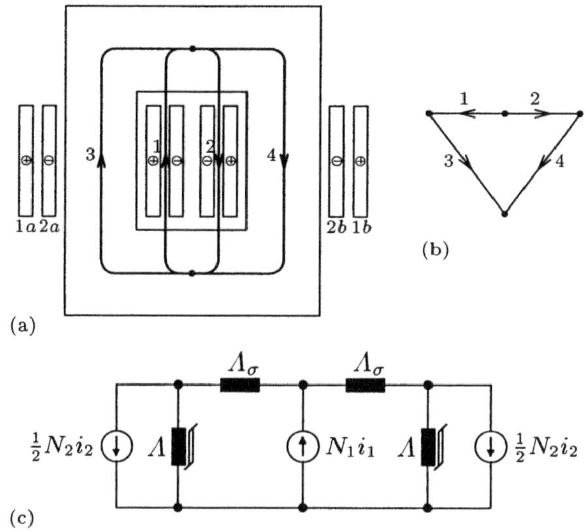


**Fig. 1.** Topological construction of permeance network for a core type transformer. (a) Transformer and original graph of the assumed flux path. (b) Dual graph. (c) Permeance network.

## 2 Network Topology and Dual Networks

A network is characterized by its topological graph. By means of a tree chosen on the directed graph of network, fundamental loops and fundamental cut-sets can be constructed. By an appropriate numbering of the graph branches, the fundamental loop matrix **B** and the fundamental cut-set matrix **C** can be written in the following form:

$$\mathbf{B} = [\,\mathbf{1}_l \mid \mathbf{H}\,] \qquad \mathbf{C} = [\,-\mathbf{H}^T \mid \mathbf{1}_t\,] \tag{1}$$

where $t$ is the number of tree element, and $l$ is the number of link elements.

A network graph is defined as a dual graph to the another if its node matrix **N** is identical to the mesh matrix **M** of the concerning network and vice versa.

$$\mathbf{N_{dual}} = \mathbf{M_{orig.}} \quad \mathbf{M_{dual}} = \mathbf{N_{orig.}} \tag{2}$$

## 3 Magnetic Circuit Modeling

By means of a network tree on the dual graph of the flux paths (Fig.1), the fundamental mesh and cut-set equations can be written as the follow:

$$[\,\mathbf{1}_l \mid \mathbf{H}\,]\ \frac{d}{dt}[\phi_1, \ldots, \phi_l, \phi_{l+1}, \ldots, \phi_{l+t}]^T = \mathbf{0} \tag{3}$$

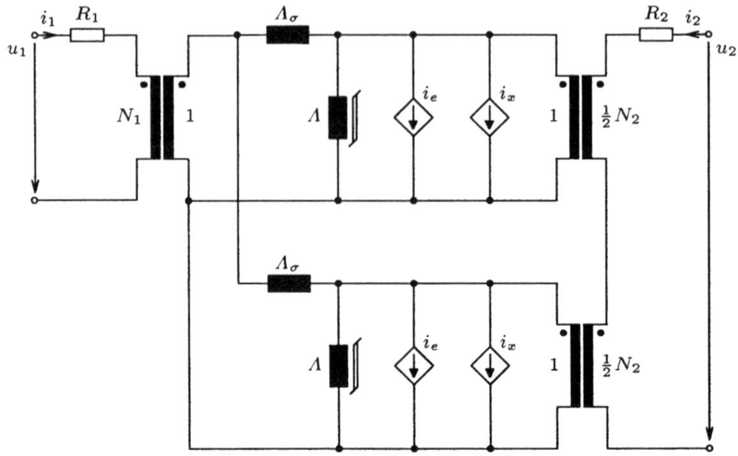


**Fig. 2.** Coupling to Electrical Network. Voltage dependent current sources represent dynamic iron losses.

$$[\,-\mathbf{H^T} \mid \mathbf{1}_t\,]\ [v_1, \ldots, v_l, v_{l+1}, \ldots, v_{l+t}]^T = \mathbf{Ei} \tag{4}$$

where $\phi_i$, $v_i$ are fluxes and mmf drops, $i$ is the current vector of windings and $\mathbf{E}$ is the excitation matrix. These can be furthermore reduced to the following equation

$$[\,\mathbf{\Lambda_{d(l)}} + \mathbf{H\Lambda_{d(t)}H^T}\,]\frac{\mathrm{d}}{\mathrm{dt}}[v_1, \ldots, v_l]^T = -\mathbf{H\Lambda_{d(t)}E}\frac{\mathrm{di}}{\mathrm{dt}} \tag{5}$$

where

$$\mathbf{\Lambda_{d(l)}} = \mathbf{diag}(\frac{\partial\phi_1}{\partial v_1}, \frac{\partial\phi_2}{\partial v_2}, \ldots, \frac{\partial\phi_l}{\partial v_{l)}})$$

and

$$\mathbf{\Lambda_{d(t)}} = \mathbf{diag}(\frac{\partial\phi_{l+1}}{\partial v_{l+1}}, \frac{\partial\phi_{l+2}}{\partial v_{l+2}}, \ldots, \frac{\partial\phi_{l+t}}{\partial v_{l+t}})$$

are the matrices of the differential permeances.

The integration of (5) yields the mmf drops of the link elements. Then the mmf drops of the tree elements are calculated from following equation:

$$[v_{l+1}, \ldots, v_{l+t}]^T = \mathbf{H^T}[v_1, \ldots, v_l]^T + \mathbf{Ei} \tag{6}$$

By considering $B = \mu_0(H + M)$, the differential permeances in (5) can be approximated as

$$\Lambda_d = k_\Lambda(1 + \frac{dM}{dH}) \tag{7}$$

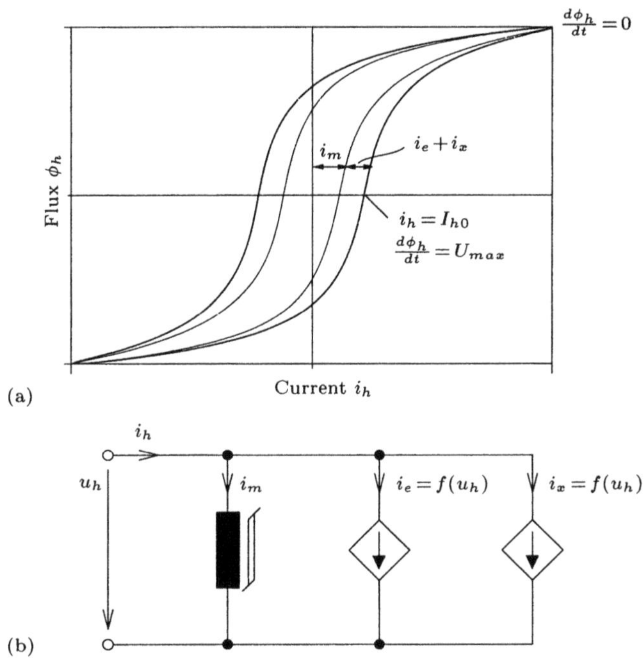


**Fig. 3.** Modelling of iron losses. (a) Hysteresis and dynamic iron losses for sinusoidal flux variation. (b) Equivalent circuit of the hysteresis and dynamic losses.

where the derivative $dM/dH$ is calculated by using the Jiles-Atherton hysteresis model [3]. The increment of the fluxes is calculated:

$$\Delta\phi_i = \Lambda_{d\,i}\,\Delta v_i \tag{8}$$

For $k$th integration step is written:

$$\phi_i^k = \phi_i^{k-1} + \Delta\phi_i \tag{9}$$

### 3.1 Coupling to Electrical Network

The electrical equations of the windings are

$$\mathbf{u} = \mathbf{Ri} + \frac{d}{dt}\,[\psi_1, \ldots, \psi_n]^T \tag{10}$$

where $\mathbf{R}$ is the diagonal matrix of the winding resistances, $n$ is the number of windings. The flux linkages of the windings $\psi_i$ are written in terms of fluxes $\phi_i$:

$$[\psi_1, \ldots, \psi_n]^T = \mathbf{k}_\psi\,[\phi_1, \ldots, \phi_{l+t}]^T \tag{11}$$

As shown in Fig. 2 the permeance network can be coupled to the electrical network via ideal transformers. Equation 5 and 10 are to be solved together.

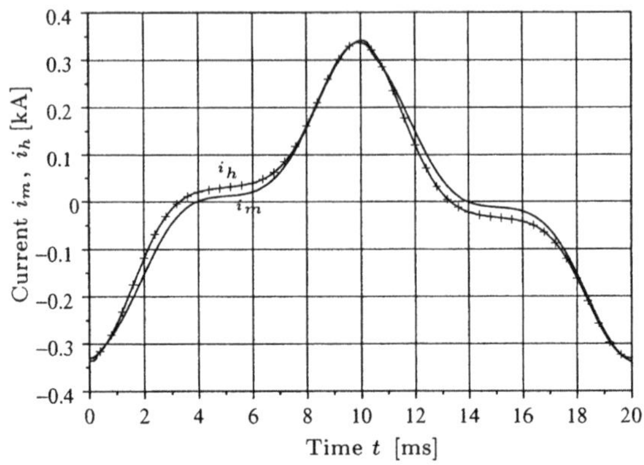


**Fig. 4.** Total current $i_h$ obtained from measurement results and the magnetization current $i_m$ calculated using JA model.

Because of mmf drop has dimension of current, afterwards it will be represented with the same symbol as the current $i$.

### 3.2 Iron Losses

Iron losses which dissipate as heat in magnetic circuit material are split in magnetization hysteresis losses, associated with the pure magnetic hysteresis loop, and the dynamic iron losses due to local eddy currents. Hysteresis and dynamic iron losses all contribute to the width of the hysteresis loop (Fig. 3). The area of the magnetization hysteresis loop $\phi_h$ - $i_m$ is proportional to the magnetization hysteresis loss:

$$P_m = k_m \oint i_m d\phi_h \tag{12}$$

The hysteresis loss $P_m$ is modelled with a hysteretic permeance. For the mathematical representation of the hysteresis loop, the JA approach is used. The dynamic iron losses themselves are distinguished between classical eddy current losses $P_e$ and excess losses $P_x$. Classical eddy current losses are proportional to the square of the rate of change in flux density,

$$P_e = \frac{1}{T} \int_0^T k_e \left( \frac{d\phi}{dt} \right)^2 dt \tag{13}$$

however excess losses are expressed as follows [4]

$$P_x = \frac{1}{T} \int_0^T k_x \left( \left| \frac{d\phi}{dt} \right| \right)^{3/2} dt \ . \tag{14}$$

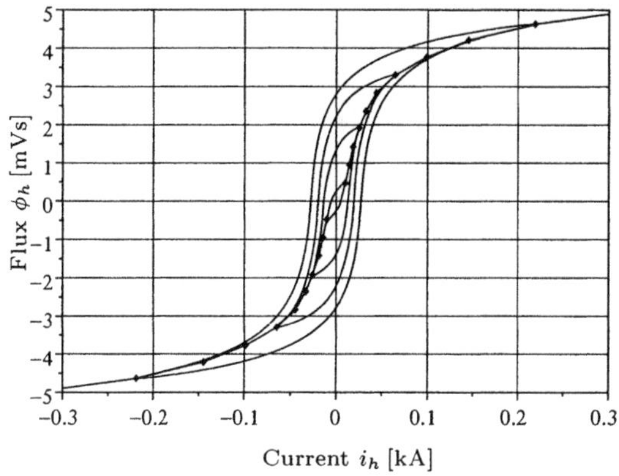


**Fig. 5.** Measured hysteresis loops and the commutation curve (trajectory of loop tips) at $f = 50\,Hz$.

The parameter $k_e$ is calculated as

$$k_e = \frac{ld^2}{12\rho} \tag{15}$$

where $l$ is the length of the flux path, $d$ is the thickness of lamination, and $\rho$ is the resistivity of the material. The excess losses depend on specific material parameters which are not supplied by manufacturers [5], therefore $k_x$ is determined from the experimental loss data. Eddy current and excess losses are modelled with a voltage dependent current source connected in parallel to the permeance of the corresponding flux path. Then the current due to the excess losses is written:

$$i_x = i_h - i_m - i_e \tag{16}$$

Fig. 4 shows the current $i_h$ obtained from measurement results and the magnetization current $i_m$ calculated using JA model.

## 4 Magnetization Hysteresis Modeling

Experimental results (see Sect. 5) show that the commutation curves at different measurement frequencies are coincident. On the commutation curve the derivative $d\phi/dt$ and the mmf drop for dynamic iron losses are zero. Therefore, the upper increasing part of the magnetization hysteresis loop is on the commutation curve (Fig. 6). For the separation of the magnetization hysteresis current $i_m$, the magnetization hysteresis loops $\phi_h$ - $i_m$ are modelled by JA approach. In JA hysteresis model, the relationship between magnetization $M$ and magnetic field $H$ is described as follow:

$$\frac{dM}{dH} = \frac{1}{(1+c)} \frac{M_{an} - M}{k\delta - \alpha(M_{an} - M)} + \frac{c}{(1+c)} \frac{dM_{an}}{dH} \tag{17}$$

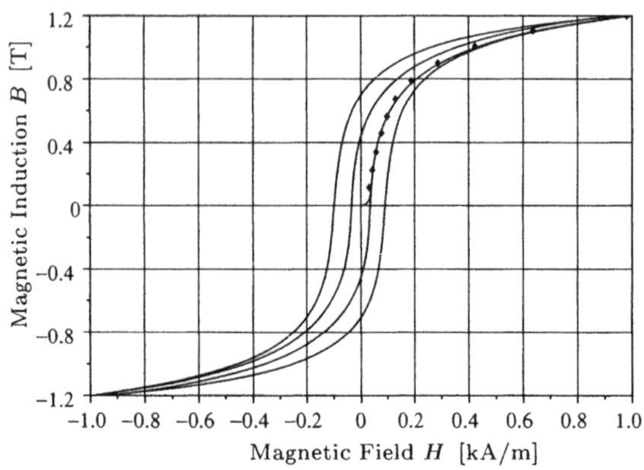


**Fig. 6.** Hysteresis modeling with JA approach: The pure hysteresis loop modelled by JA approach (inner loop), the measured hysteresis loop with dynamic iron losses and the commutation curve, $f = 50\,Hz$.

where

$$M_{an} = M_s \left[ \coth\left( \frac{H + \alpha M}{a} \right) - \frac{a}{H + \alpha M} \right] \tag{18}$$

is the Langevin anhysteretic curve. The symbol $\delta$ indicates a directional parameter equal to $+1$ or $-1$ for the ascending or the descending branch of hysteresis loop.

The parameters of JA model are so determined that the loop tips are on the commutation curve and this parameters for the magnetic core material of the transformer used in experiments are identified as follow [4]: $M_s = 1.225 \times 10^6\,A/m$, $a = 425\,A/m$, $k = 36\,A/m$, $\alpha = 1.0 \times 10^{-3}$, $c = 0.2$.

## 5 Experimental Results and Simulation

A core type transformer is supplied from a synchronous generator with variable voltage and frequency. The flux level is kept constant at different values to limit the current at low frequencies ($U/f = constant$). The current is measured on the primary winding which is supplied and the voltage on the secondary which is open. The measured voltage and current are sampled by means of a two channel digital oscilloscope with 20 MHz bandwidth and 8 bit resolution. The $\phi_h$ - $i_h$ loops of a core type transformer are calculated from measured voltage and currents on the terminals of the transformer (Fig. 7). Simulations are carried out by solving (5) and (10). The simulation results are compared with the measured results (Fig. 7).

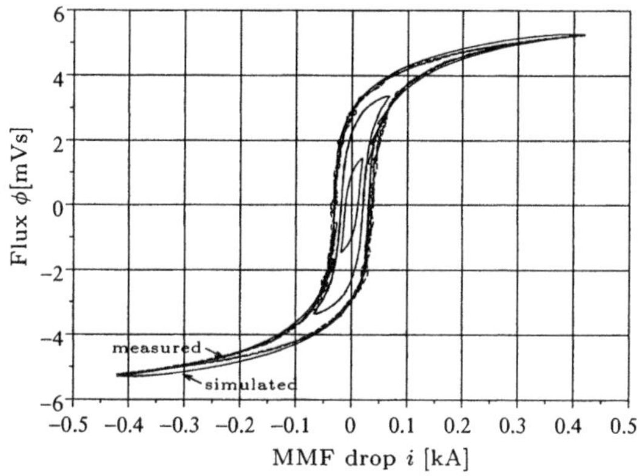


**Fig. 7.** Hysteresis loops in the transformer core (Simulation result is only given for outermost loop), $f = 50\,Hz$.

# 6 Conclusions

The hysteresis effect in magnetic circuit of a core type transformer is modelled by using a dual magnetic equivalent circuit. The proposed equivalent circuit which consists of permeances allows the modelling the magnetization hysteresis loss and dynamic iron losses separately. For the modelling of the magnetization hysteresis, the JA approach is used. The simulation results show that the proposed method can be satisfactorily used to take into account the hysteresis effect and iron loses in magnetic circuits. The inaccuracy in simulation results which occurs at low flux density levels under $0.2\,T$ can be reduced by using a modified JA model.

# References


1. G. Slemon: *Magnetoelectric Devices, Transducers, Transformers, and Machines*, John Wiley and Sons, Inc., New York, 1966.
2. S. Güngör and M. Stiebler: "A permeance network for magnetic circuit modeling of switched reluctance machines", *Record of the 13th Compumag Conference*,Vol. 3, pp.64-65, 2000.
3. D. C. Jiles and D. L. Atherton: "Theory of Ferromagnetic Hysteresis", *Journal of Magnetism and Magnetic Materials*,Vol. 61, pp.48-60, 1986.
4. D. Lederer, H. Igarahsi, A. Kost and T. Honma: "On the parameter identification and application of the Jiles-Atherton hysteresis model for numerical modeling of measured characteristics", *IEEE Transactions on Magnetics*,Vol. 35, pp.1111-1114, 1999.
5. G. Bertotti: "General properties of power losses in soft ferromagnetic materials", *IEEE Transactions on Magnetics*, Vol. 24, pp.621-630, 1988.

# Eigenmode Computation of Microwave and Laser Structures Including PML

Georg Hebermehl[1], Friedrich-Karl Hübner[1], Rainer Schlundt[1], Thorsten Tischler[2], Horst Zscheile[2], and Wolfgang Heinrich[2]

[1] Weierstrass Institute for Applied Analysis and Stochastics, Mohrenstr. 39, D-10117 Berlin, Germany
[2] Ferdinand-Braun-Institut für Höchstfrequenztechnik, Albert-Einstein-Str. 11, D-12489 Berlin, Germany

**Abstract.** The field distribution at the ports of the transmission line structure is computed by applying Maxwell's equations to the structure. Assuming longitudinal homogeneity an eigenvalue problem can be derived, whose solutions correspond to the propagation constants of the modes. The nonsymmetric sparse system matrix is complex in the presence of losses and Perfectly Matched Layer. The propagation constants are found solving a sequence of eigenvalue problems of modified matrices with the aid of the invert mode of the Arnoldi method. Using coarse and fine grids, and a new parallel sparse linear solver, the method, first developed for microwave structures, can be applied also to high dimensional problems of optoelectronics.

## 1 Introduction

The fields of applications are mobile communications, radio links, automobile radar systems, optical communications and material processing. The commercial applications of microwave circuits cover the frequency range between 1 GHz and about 100 GHz, special applications in radioastronomy use even higher frequencies up to 1 THz. For optoelectronic devices frequencies about several hundred THz are common.

Basic elements of the structures are their transmission lines, whose propagation behavior has to be determined accurately. The propagation behavior of the transmission lines can be calculated by applying Maxwellian equations to the infinitely long homogeneous transmission line structure and solving an eigenvalue problem [1].

Only a few modes of smallest attenuation are able to propagate and have to be taken into consideration. Using a conformal mapping between the plane of propagation constants and the plane of eigenvalues the task is to compute all eigenmodes in a region, bounded by two parabolas. The region is covered by a number of overlapping circles. The eigenmodes in these circles are found solving a sequence of eigenvalue problems of modified matrices [2] with the aid of the invert mode of the Arnoldi iteration using shifts.

For numerical treatment, the computational domain has to be truncated by electric or magnetic walls or by a so-called absorbing boundary condition simulating open space. A very efficient formulation for the latter case is the Perfectly Matched Layer (PML) [3]. Introducing the complex, anisotropic material PML leads to an increased computational time.

Due to the fact, that only small fractions of a microwave circuit can be simulated, the pressure to larger problem sizes is evident. Especially, the application of the method for optoelectronic devices requires new strategies to reduce the numerical effort and storage requirement. The computation of large cross sections combined with an extension of specific material layers in the sub $\mu$m-range yields high dimensional problems. Additionally, due to the high wavenumber in semiconductor lasers the region containing potential propagating modes grows substantially. That means, a significant higher number of eigenvalue problems have to be solved. To reduce the execution times, in a first step the problem is solved using a coarse grid in order to find approximately the locations of the interesting propagation constants. The accurate modes are calculated in a second step for an essentially reduced region using a fine grid. In addition, the method is optimized reducing the storage requirement and the computing times applying a new linear sparse solver, that can be used serial or parallel.

## 2 Boundary Value Problem

We start from a three-dimensional structure. The structure under investigation can be described as an interconnection of infinitely long transmission lines. The junction, the so-called discontinuity, may have an arbitrary structure. The transmission lines have to be longitudinal homogeneous. Ports are defined on the transmission lines. A three-dimensional boundary value problem can be formulated using the integral form of Maxwell's equations in the frequency domain in order to compute the electromagnetic field:

$$\oint_{\partial\Omega} \boldsymbol{H} \cdot d\boldsymbol{s} = \int_{\Omega} \jmath\omega[\epsilon]\boldsymbol{E} \; d\boldsymbol{\Omega}, \qquad \oint_{\cup\Omega} ([\epsilon]\boldsymbol{E}) \cdot d\boldsymbol{\Omega} = 0, \tag{1}$$

$$\oint_{\partial\Omega} \boldsymbol{E} \cdot d\boldsymbol{s} = -\int_{\Omega} \jmath\omega[\mu]\boldsymbol{H} \cdot d\boldsymbol{\Omega}, \qquad \oint_{\cup\Omega} ([\mu]\boldsymbol{H}) \cdot d\boldsymbol{\Omega} = 0, \tag{2}$$

$$\boldsymbol{D} = [\epsilon]\boldsymbol{E}, \;\; \boldsymbol{B} = [\mu]\boldsymbol{H}, \;\; [\epsilon] = \operatorname{diag}(\epsilon_x, \epsilon_y, \epsilon_z), \;\; [\mu] = \operatorname{diag}(\mu_x, \mu_y, \mu_z). \tag{3}$$

In the left-hand sides of formulae (1) and (2) $\Omega$ is an open surface surrounded by a closed contour $\partial\Omega$, while in the right-hand sides of (1) and (2) $\cup\Omega$ is a closed surface with an interior volume. The direction of the element $d\boldsymbol{s}$ of the contour $\partial\Omega$ is such that when a right-handed screw is turned in that direction, it will advance in the direction of the vector element $d\boldsymbol{\Omega}$.

The transverse electric mode fields at the ports are the solutions of an eigenvalue problem for the transmission lines. All other parts of the surface of the computation domain are assumed to be an electric or a magnetic wall. The PML's are filled with an artificial material with complex anisotropic material properties. Therefore, the quantities are diagonal complex tensors.

# 3 Maxwellian Grid Equations

The Maxwellian equations are discretized using staggered nonequidistant rectangular grids. Using the Finite Integration Technique (FIT) [4], [5], [6] with the lowest order integration formulae

$$\oint_{\partial\Omega} \boldsymbol{f} \cdot d\boldsymbol{s} \approx \sum(\pm f_i s_i), \quad \int_{\Omega} \boldsymbol{f} \cdot d\boldsymbol{\Omega} \approx f\Omega \tag{4}$$

equations (1), (2) are transformed into a set of Maxwellian grid equations

$$A^T D_{s/\mu} \boldsymbol{b} = \jmath\omega\epsilon_0\mu_0 D_{A_\epsilon} \boldsymbol{e}, \qquad B D_{A_\epsilon} \boldsymbol{e} = 0, \tag{5}$$

$$A D_s \boldsymbol{e} = -\jmath\omega D_{A_\mu} \boldsymbol{b}, \qquad \widetilde{B} D_{A_\mu} \boldsymbol{b} = 0. \tag{6}$$

The vectors $\boldsymbol{e}$ and $\boldsymbol{b}$ contain the components of the electric field intensity and the magnetic flux density of the elementary cells, respectively. The diagonal matrices $D_{s/\mu}$, $D_{A_\epsilon}$, $D_s$, and $D_{A_\mu}$ contain the information on cell dimension and material. $A$, $B$, and $\widetilde{B}$ are sparse.

Eliminating the components of the magnetic flux density from the two equations of the left-hand side of (5), (6) we get the system of linear algebraic equations

$$(A^T D_{s/\mu} D_{A_\mu}^{-1} A D_s - k_0^2 D_{A_\epsilon}) \boldsymbol{e} = 0, \quad k_0 = \omega\sqrt{\epsilon_0\mu_0}, \tag{7}$$

which have to be solved using the boundary conditions. $k_0$ is the wavenumber in vacuum.

# 4 Eigenvalue Problem

The field distribution at the ports is computed assuming longitudinal homogeneity for the transmission line structure. Thus, any field can be expanded into a sum of so-called modal fields which vary exponentially in the longitudinal direction

$$\boldsymbol{E}(x,y,z) = \underline{\boldsymbol{E}}(x,y) e^{\mp\jmath k_z z}. \tag{8}$$

A substitution of ansatz (8) into the system of linear algebraic equations (7) and the elimination of the longitudinal electric field intensity components by means of the electric-field divergence equation $B D_{A_\epsilon} \boldsymbol{e} = 0$ (see (5)) gives an eigenvalue problem

$$C\underline{\mathrm{e}} = \gamma\underline{\mathrm{e}}, \quad \gamma = -4\sin^2(hk_z). \tag{9}$$

$\underline{\mathrm{e}}$ consists of components of the discretized eigenfunctions $\underline{\boldsymbol{E}}$. $2h$ is the length of an elementary cell in $z$-direction. The sparse matrix $C$ is in general nonsymmetric complex. The order of $C$ is $n = 2n_x n_y - n_b$. $n_x n_y$ is the number of elementary cells at the port. The size $n_b$ depends on the number of cells with perfectly conducting material. The relation between the propagation constants $k_z$ and the eigenvalues $\gamma$ is nonlinear, and can be expressed as

$$k_z = \frac{\jmath}{2h} \ln \left( \frac{\gamma}{2} + 1 + \sqrt{\frac{\gamma}{2}\left(\frac{\gamma}{2} + 2\right)} \right) = \beta - \jmath\alpha. \tag{10}$$

We are interested only in a few modes with the smallest attenuation. These are the modes with the smallest magnitude of imaginary part, but possibly with large real part of their propagation constant. The computation of all eigenvalues in order to find a few propagation constants must be avoided for the high-dimensional problem. For numerical treatment we have to limit the search for propagation constants by a maximum value $k_f$ of their real part. This $k_f$ value depends on the highest permittivity $[\epsilon]$ and permeability $[\mu]$ values of the waveguide, though regions with metallic or PML filling are ignored, see [2]. Using the limited $k_f$ and a preset maximum value $\alpha_m$ of the imaginary part of the propagation constants the region containing the interesting constants is defined as a rectangle $\widehat{F}$ bounded by the lines

$$\beta = \pm k_f \quad \text{and} \quad \alpha = \alpha_m. \tag{11}$$

In an additional step all computed modes that are related to the PML boundary are neglected, using the power part criterion given with [8]. We can use the approximation $\sin(x) \approx x$ in (9) if we choose $h$ to be small enough, which is necessary anyway to get small discretization errors:

$$\gamma = -4\sin^2(hk_z) \approx -4(hk_z)^2 = u + \jmath v. \tag{12}$$

With aid of the approximation (12) we get a conformal mapping between the plane of eigenvalues ($\gamma$-plane) and the plane of propagation constants ($k_z$-plane, see (10)):

$$u = -4h^2(\beta^2 - \alpha^2), \quad v = 8h^2\alpha\beta. \tag{13}$$

Using this mapping the rectangle $\widehat{F}$ of the $k_z$-plane is transformed into a region $F$ of the $\gamma$-plane bounded by the two parabolas

$$v = \pm 4hk_f\sqrt{u + 4h^2{k_f}^2} \quad \text{and} \quad v = \pm 4h\alpha_m\sqrt{-u + 4h^2{\alpha_m}^2}. \tag{14}$$

That means, we have to find all eigenvalues of the region bounded by the parabolas.

## 5 Computation of Eigenmodes

We need an algorithm that computes just a few selected eigenvalues and eigenvectors of a complex sparse matrix. A state-of-the-art algorithm for such problems is the Arnoldi method [9], [10]. In general the Arnoldi method converges for our problem only using the invert mode and looking for eigenvalues of largest magnitude. Thus, a simple way to find the eigenvalues located in the region $F$ would be to look for all eigenvalues of smallest magnitude, which are located in a circle centered on the origin and covering the region $F$. Caused by the high wavenumber $k_f$, the number of eigenvalues located in this circle is in general too large for a feasible computation using an iterative method. We can solve this problem covering the region $F$ with $s \geq 1$ circles $C_i,\ i = 1(1)s$, centered on the u-axis and calculating the eigenvalues located in these circles. That is done in the following way. $s$ points

$$\widehat{P}_i(\beta_i, \alpha_m),\ i = 1(1)s,\ \beta_1 = \frac{k_f}{s} \geq \overline{\beta},\ \beta_s = k_f, \quad \text{with} \quad \overline{\beta} = \sqrt{3}\alpha_m \quad (15)$$

are defined on the interval $[0, k_f]$ of the line $\alpha = \alpha_m$. The distance between the points need not be equidistant and is controlled as shown below. Even the meaning of the distance $\overline{\beta}$ is discussed below. The points $\widehat{P}_i$ are transformed into the points $P_i$ of the $\gamma$-plane. They are located on the parabola ((14), right formula). The $s$ circles $C_i$ of the $\gamma$-plane

$$(u - m_i)^2 + v^2 = {r_i}^2,\ r_i = \sqrt{(\Re(P_i) - m_i)^2 + (\Im(P_i))^2}, \quad i = 1(1)s, (16)$$

with

$$m_1 = 0,\ m_i = \frac{(\Re(P_i))^2 - (\Re(P_{i-1}))^2 + (\Im(P_i))^2 - (\Im(P_{i-1}))^2}{2(\Re(P_i) - \Re(P_{i-1}))} \quad (17)$$

are centered on the u-axis, covering the region bounded by the parabolas.

In order to find all eigenvalues, located in the circle $C_i$, $l$ points $Q_j$ are defined on the periphery of $C_i$. The matrix $C$ is extended by the diagonal matrix $Q$. The diagonal elements of $Q$ are the $l$ complex elements $Q_j$:

$$\overline{C} = \begin{pmatrix} Q & \\ & C \end{pmatrix}, \quad Q = diag(Q_1, ..., Q_l). \quad (18)$$

The $s$ eigenvalue problems

$$(\overline{C} - m_i I)\underline{e} = (\gamma - m_i)\underline{e}, \quad i = 1(1)s, \quad (19)$$

are solved with the aid of the implicitly restarted Arnoldi method using the invert mode. The eigenvalue problems can be solved separably. The number $m$ of eigenvalues to be computed for this circle must be $l$ on the first call to the Arnoldi procedure. The main idea is to raise $m$ by $l$ for so long until at least one value $Q_j$ was found. But, since $m \ll \overline{n}$ ($\overline{n}$ order of matrix $\overline{C}$) for a feasible computation, one has to restrict the number $m$ of required eigenvalues by $m_{max}$. If $m$ exceeds $m_{max}$ in case of $i \geq 2$, we insert a point $\widehat{P}_{i-\frac{1}{2}}$ between $\widehat{P}_{i-1}$ and $\widehat{P}_i$ and restart with $m = l$. The same procedure is used if a given number $\nu_{max}$ of iterations in the Arnoldi method is exceeded. If the condition

$$\Delta\widehat{P} = \Re(\widehat{P}_{i-\frac{1}{2}}) - \Re(\widehat{P}_{i-1}) \geq \frac{\alpha_m}{2} \quad (20)$$

cannot be fulfilled, we restart with new parameters $m_{max}$, $\nu_{max}$ and possibly $\alpha_m$. If all eigenvalues $Q_j$ are found in case of $m > l$, we look for the eigenvalue $\gamma_{max}$ of largest magnitude. If $\sqrt{|\gamma_{max}|} > r_i$, a new circle $\widetilde{C}_i$ of radius $\sqrt{|\gamma_{max}|}$ with the same center as $C_i$ is defined. The left intersection point of this circle with the parabola ((14), right formula) is used as new point $P_i$, and $\Delta\widehat{P} = \Re(\widehat{P}_i) - \Re(\widehat{P}_{i-1})$ as distance for the next step. $m$ is reduced by the number of eigenvalues with $\sqrt{|\gamma|} > r_i$ for the next circle.

Separating the new values on each eigenvalue problem $i$, we are sure to have found all eigenvalues which are located in the corresponding circles $C_i$. Applying the mapping (13) the circles $C_i$ (see (16)) are transformed into Cassinian curves $\widehat{C}_i$

$$(\beta^2 + \alpha^2)^2 - \frac{m_i}{2h^2}(\beta^2 - \alpha^2) = \frac{{r_i}^2}{16h^4} - \frac{{m_i}^2}{16h^4}, \tag{21}$$

which cover the rectangle $\widehat{F}$ containing all desired propagation constants. Propagation constants outside of $\widehat{F}$ and PML-Modes are eliminated. The Cassinian curves $\widehat{C}_i$, $i = 2(1)s$, consist of two separated ovals, if $r_i < m_i$. Using $\overline{\beta}$ as minimum distance between the origin and $\widehat{P}_1$ (see (15)) other shapes of Cassinian curves (e.g. waisted ovals), which would lead to higher execution times, are avoided.

## 6 Optoelectronic Devices

The maximum cell size of discretization should be less than $\lambda/10$, where $\lambda$ denotes the wavelength in the material with the highest $\Re(\epsilon)$. Essentially finer grids have to be used for regions of the circuit with highly variable electric fields. That means, the problems become high dimensional, and only small fractions of a circuit can be simulated. Especially the application of the method to optoelectronic devices requires new strategies. The dimension of the eigenvalue problem to be solved increases essentially in this case caused by the short wavelength. In addition, due to the high wavenumber in optoelectronic devices the length of the rectangle $\widehat{F}$ containing potential propagation constants grows substantially. That means, we have to calculate a significant higher number of eigenvalue problems. Due to electric and magnetic walls behind the PML undesired modes are generated inside the computation domain. The non physical modes can be eliminated by examining the eigenfunctions. Anyway, the number of eigenmodes to be calculated increases caused by the shifted modes. Due to the significant difference between the magnitude of the real and imaginary part of the propagation constant a high computational accuracy has to be required. To overcome these problems two strategies have been realized.

(1) To reduce the execution times, in a first step the problem is solved using a coarse grid with lower accuracy requirements in order to find approximately the locations of the interesting propagation constants. Finally the modes are

calculated in a second step for an essentially reduced region using a fine grid, that fulfills higher accuracy requirements.
(2) Because in general the Arnoldi method does not converge using the regular mode for our eigenvalue problem the invert mode with shifting (see (19)) is applied. A time and memory consuming system of linear algebraic equations has to be solved on each iteration step in this case. The storage requirement and the computing times could be reduced substantially, applying the new linear sparse solver PARDISO [11], [12], rather than the formerly used UMFPACK [13]. The fill in is reduced approximately by a factor of 4.75. Moreover, the dynamic memory allocation of PARDISO allows to diminish the memory requirements. The computing times for the numerical factorization and forward and backward solve are reduced on the average by a factor of 15 and 4 for our problem, respectively. The algorithm is split into three phases: symbolic factorization, numerical factorization, and forward and backward solve. The symbolic factorization can be used for all modified matrices of our problem. The numerical factorization has to be repeated for every new shift. The typical ratio of factorization time to solution time on a single CPU can be used to define $\nu_{max}$ in the subinterval control process (see section 5). This ratio amounts on the average 20. That means, the costs using $\nu_{max} = 60$ Arnoldi iterations for the computation of $m$ eigenmodes in a circle $C_i$ defined by the points $P_{i-1}, P_i$ are comparable with the costs, defined by the costs for two circles defined by the points $P_{i-1}, P_{i-\frac{1}{2}}$ and $P_{i-\frac{1}{2}}, P_i$ using $\nu_{max} = 20$ iterations. On the other hand the time is lost, interrupting the computation of $m$ eigenmodes after $\nu_{max} = 60$ iterations and starting a new iteration process for two reduced circles. Thus, we use a greater $\nu_{max}$. Moreover, due to the significant difference between the length and the height of the rectangular region $\widehat{F}$ in the $k_z$-plane we have to solve a large number $s$ of eigenvalue problems (see section 7). In order to diminish this number we use Cassinian curves with relatively large diameters. That means, a number of non desired eigenvalues outside of the area $\widehat{F}$ has to be calculated. In general the computation of a large number $m$ of eigenvalues in one circle needs more iterations than a small number.

The $s$ eigenvalue problems (19) could be solved independently and in parallel using $s$ processors and single CPU mode of PARDISO. But that means, the amount of memory increases nearly by a factor of $s$, and the maximum problem size which could be managed is reduced. Contrary to that the parallel CPU mode of PARDISO provides an additional possibility to reduce the computing times for high dimensional problems on shared memory multiprocessors without essential additional memory requirements.

## 7 Laser Application

As an example we have calculated the guided mode of an optoelectronic device. A so called self aligned stripe (SAS) laser is investigated, see Figure 1.

This laser structure contains an additional, so called antiguided layer (marked with $\epsilon_x = 12 - i * 0.1$ in Figure 1) outside the emitting stripe (marked with $\epsilon_x = 11.3 + i * 0.05$ in Figure 1). This high power laser diode excites only the fundamental mode, the active region is useful for wavelengths shorter than 800 nm. The frequency is fixed to $299.7925 * 10^{12}$ Hz. In our eigenmode computation of the laser structure a graded mesh of 283 times 345 elementary cells, including 10-cell PML regions, is used as a fine grid. The maximum cell size amounts $\lambda/12 = 25$nm, where $\lambda$ denotes the wavelength in the material with the highest $\Re(\epsilon)$. The minimum cell size is 1 nm. Maximum cell size is scaled down exponentially in the vertical direction near the 100 nm zones and in the horizontal direction near the material cut 118 and 119 (see Figure 1). The dimension of the eigenvalue problem is 192 423. The eigenvalues and eigenvectors have been solved with the relative accuracy $tol = 10^{-10}$, and with $m_{max} = 16$, $l = 5$ (see section 5). 84 Cassinian curves have been used to cover the long small region of the complex plane ($\alpha_m = 2500\,\mathrm{m}^{-1}$, $k_f = 21\,765\,592\,\mathrm{m}^{-1}$, see (11)) containing potential guided modes. A maximum number $\nu_{max} = 200$ of Arnoldi iteration has been used. The total computational time amounts approximately 3h and 23 minutes using a Compaq Professional Workstation with processor XP1000 alpha 667 MHz.

**Fig. 1.** Laser (amplifier)

One guided mode according to the lasers fundamental mode, was found. The computed complex propagation constant is given by $k_z = 20\,817\,578 + j\,1\,488$.

A graded mesh of 121 times 127 elementary cells is used as a coarse grid. The maximum cell size amounts 80nm, and the minimum cell size 4nm. The dimension of the eigenvalue problem is 29 625. The total computational time amounts approximately 19 minutes using the relative accuracy $tol = 10^{-7}$. The circle that contains the guided mode is known after this step. The time to find the accurate value $k_z$ using the fine grid amounts only 142 s. Thus, the computational time is reduced by a factor of 1/9 for the given structure.

## References

1. Christ, A., Hartnagel, H. L.: Three-Dimensional Finite-Difference Method for the Analysis of Microwave-Device Embedding. IEEE Transactions on Microwave Theory and Techniques **35** (1987) 688–696
2. Hebermehl, G., Hübner, F.-K., Schlundt, R., Tischler, T., Zscheile, H., Heinrich, W.: Numerical Simulation of Lossy Microwave Transmission Lines Including PML. in: Scientific Computing in Electrical Engineering (ed. U. van Rienen, M. Günther, D. Hecht), Lecture Notes in Computational Science and Engineering, Springer Verlag, (2001) 267–275
3. Sacks, Z. S., Kingsland, D. M., Lee, R., Lee, J.-F.: A Perfectly Matched Anisotropic Absorber for Use as an Absorbing Boundary Condition. IEEE Transactions on Antennas and Propagation **43** (1995) 1460–1463
4. Weiland, T.: A Discretization Method for the Solution of Maxwell's Equations for Six-Component Fields. Electronics and Communication (AEÜ) **31** (1977) 116–120
5. Beilenhoff, K., Heinrich, W., Hartnagel, H. L.: Improved Finite-Difference Formulation in Frequency Domain for Three-Dimensional Scattering Problems. IEEE Transactions on Microwave Theory and Techniques **40** (1992) 540-546
6. Hebermehl, G., Schlundt, R., Zscheile, H., Heinrich W.: Improved Numerical Methods for the Simulation of Microwave Circuits, Surveys on Mathematics for Industry, **9** (1999) 117–129
7. Hebermehl, G., Hübner, F.-K., Schlundt, R., Tischler, T., Zscheile, H., Heinrich, W.: On the Computation of Eigen Modes for Lossy Microwave Transmission Lines Including Perfectly Matched Layer Boundary Conditions. The International Journal for Computation and Mathematics in Electrical and Electronic Engineering **20** (2001) 948–964
8. T. Tischler and W. Heinrich: The Perfectly Matched Layer As Lateral Boundary in Finite-Difference Transmission-Line Analsysis. IEEE Transactions on Microwave Theory and Techniques **48** (2000) 2249-2253
9. Lehoucq, R. B.: Analysis and Implementation of an Implicitly Restarted Arnoldi Iteration. Rice University, Department of Computational and Applied Mathematics, Technical Report **13** (1995) 1-135
10. Sorensen, D. C.: Implicit Application of Polynomial Filters in a k-Step Arnoldi Method. SIAM J. Matr. Anal. Apps. **13** (1992) 357–385

11. Schenk, O., Gärtner, K., Fichtner, W.: Efficient Sparse LU Factorization with Left-Right Looking Strategy on Shared Memory Multiprocessors. BIT 40 (2000) 158-176
12. Schenk, O., Gärtner, K.: Two-Level Dynamic Scheduling in PARDISO: Improved Scalability on Shared Memory Multiprocessing Systems. Parallel Computing 28 (2002) 187–197
13. Davis, T. A., Duff, I. S.: A Combined Unifrontal/Multifrontal Method for Unsymmetric Sparse Matrices. University of Florida, Technical Report **16** (1997) 1–18

# Interconnect Optimization by Semidefinite Programming

Tino Heijmen and Jaap van Staalduinen

Philips Research Laboratories, Prof. Holstlaan 4, 5656 AA, Eindhoven, The Netherlands

**Abstract.** A method to optimize delay and power dissipation in on-chip interconnect is reported. Propagation delay can be represented by the dominant time constant of the corresponding RC circuit or as a $p$–$q\%$ delay [1]. The optimization problem is formulated as a sequence of semi-definite programming problems. The method is applied to interconnect with inclusion of the fringing capacitance and capacitive coupling between wires. Shapes of single wires and models of real-life bus designs are optimized. It is shown that the optimal wire shape depends on the chosen delay metric and that it can be described accurately with a linear model. The differences between wire sizing and wire tapering are discussed. The importance of capacitive coupling in the optimization of multi-wire buses is demonstrated. Future extensions of the approach are discussed.

## 1 Introduction

As process technology scales into deep submicron dimensions, interconnect delay increasingly dominates over gate delay. This is because the trend of process technology scaling has led to the increase of the resistance per unit length of the interconnect, while the capacitance per unit length remains approximately constant [2]. The gate delay, on the other hand, decreases. In addition, whereas the bottom capacitance reduces, the capacitive coupling between neighboring wires is becoming increasingly important, because the wire spacings decrease [3]. Figure 1 depicts the capacitances in interconnect [1].

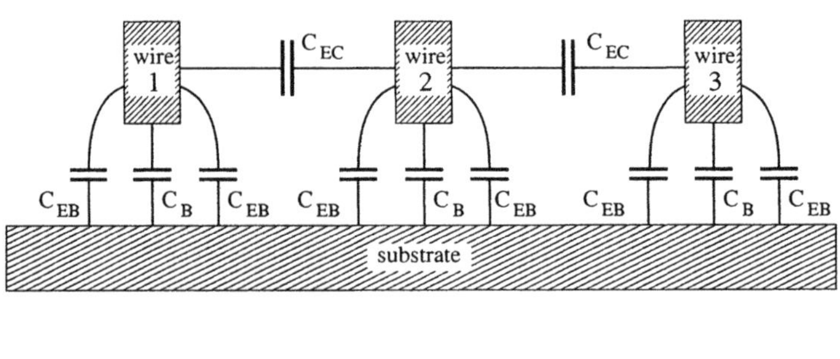


**Fig. 1.** Interconnect capacitances. The symbols $C_{\mathrm{B}}$, $C_{\mathrm{EB}}$, and $C_{\mathrm{EC}}$ denote the bottom, fringing, and coupling capacitances, respectively [1].

Therefore, deep submicron design shows a growing interest in optimization techniques that size wires and/or insert and size buffers [3,4]. Many of these methods apply the Elmore delay model to determine the signal propagation delay [1,4]. Unfortunately, Elmore delay ignores the resistive shielding of downstream capacitance and is therefore unacceptably inaccurate in some cases. Furthermore, the application of Elmore delay is restricted to resistor-capacitor (RC) trees, which means that non-grounded capacitances cannot be handled. This restriction inhibits the inclusion of coupling capacitances between neighboring wires [5]. Capacitive coupling, however, has to be incorporated if cross-talk effects on signal delays are to be taken into account.

An alternative delay metric for RC circuits was proposed by Vandenberghe et al. [5], who used the dominant time constant ($T_{\text{dom}}$) of the circuit as a measure of signal delay. Circuit sizing problems can then be solved with a semidefinite programming (SDP) method. Their approach was shown to be applicable to general nontree topologies, such as meshes of resistors and buses with coupling capacitances between the wires.

However, problems in which delay is minimized were not considered in [5], because these cannot be directly expressed as semidefinite programming problems. Instead, examples were studied in which $T_{\text{dom}}$ is constrained. Second, only the largest time constant was taken into account, the effect of the other time constants was neglected. Furthermore, the applied wire model did not include fringing capacitance, which is a non-negligible component in deep-submicron technologies, see Fig. 1. Also, the used values for the model parameters were not related to real-life process technologies.

In the current work, the approach of Vandenberghe et al. [5] is extended to include problems in which the propagation delay is minimized. The extended method is capable of optimizing $T_{\text{dom}}$ or other delay metrics in which all time constants are incorporated. An interconnect model is used in which fringing capacitances are included and the applied parameter values correspond to an existing 0.18 $\mu$m process technology. The approach is used to study optimal wire shapes of single wires and the optimization of multi-wire bus models. Future extensions of the method are treated.

## 2 Delay Optimization

The approach considers general RC circuits composed of two-terminal resistors and capacitors, and independent voltage sources. If the branch capacitors and conductances in the circuits are nonnegative, the capacitance matrix (**C**) and the conductance matrix (**G**) describing the circuit are positive semidefinite. Further, if the branch capacitors and conductances are affine functions of some design parameters $x \in \mathbb{R}^m$, the matrices **C** and **G** are also:

$$\mathbf{C}(x) = \mathbf{C}_0 + x_1\mathbf{C}_1 + \cdots + x_m\mathbf{C}_m \ , \tag{1}$$

$$\mathbf{G}(x) = \mathbf{G}_0 + x_1\mathbf{G}_1 + \cdots + x_m\mathbf{G}_m \ . \tag{2}$$

Typically, the parameters $\boldsymbol{x}$ correspond to widths of wire segments and transistors and the spacings between wire segments.

The dominant time constant of the circuit can be expressed as [5],

$$T_{\text{dom}}(\boldsymbol{x}) = \min\{T \mid T\,\mathbf{G}(\boldsymbol{x}) - \mathbf{C}(\boldsymbol{x}) \geq 0\} \ . \tag{3}$$

The inequality means that the left-hand side is a positive semidefinite matrix. The work in [5] considered problems that can be expressed as SDP problems, such as the minimization of area, dynamic power dissipation, or bus width, with a linear matrix inequality (LMI) constraining the dominant time constant. In the present work problems have been studied that can be cast as generalized eigenvalue minimization problems (GEVPs), such as the minimization of the dominant time constant, subject to variable bounds,

$$\begin{aligned} &\text{minimize} && T \\ &\text{subject to} && T\,\mathbf{G}(\boldsymbol{x}) - \mathbf{C}(\boldsymbol{x}) \geq 0 \\ & && x_i^{\min} \leq x_i \leq x_i^{\max} \ , \end{aligned} \tag{4}$$

with variables $T$ and $\boldsymbol{x}$. Additional constraints on the wire widths and spacings can be expressed as LMIs also. This is the case, for instance, when a set of parallel wires is optimized [5].

The most convenient way to solve a GEVP is to solve a sequence of SDPs [6]. If $T$ is fixed, the problem of (4) becomes a feasibility problem with variables $\boldsymbol{x}$, which can solved by solving the SDP,

$$\begin{aligned} &\text{minimize} && w \\ &\text{subject to} && T\,\mathbf{G}(\boldsymbol{x}) - \mathbf{C}(\boldsymbol{x}) + w\mathbb{1} \geq 0 \\ & && x_i^{\min} \leq x_i \leq x_i^{\max} \ . \end{aligned} \tag{5}$$

The constraints in (4) are fulfilled if and only if the solution $w^*$ of (5) is non-positive. The GEVP of (4) can therefore be solved by finding the value of $T$ for which $w^*$ is equal to zero.

The approach of (5) is general, i.e., delay metrics other than the dominant time constant can be minimized as well. A possible alternative is the $p$–$q$% propagation delay, $D(T)$, defined as the difference between the times at which the response to a step input signal has completed $p$% and $q$% of its transition, with $q > p$ [1]. The problem is then cast as,

$$\begin{aligned} &\text{minimize} && D(T) \\ &\text{subject to} && w^*(T) \leq 0 \ , \end{aligned} \tag{6}$$

where $w^*(T)$ is the solution of (5) for given $T$ and $D(T)$ is calculated by including all time constants of the circuit. In the case of a rising transition, the $p$–$q$% propagation delay can be expressed as,

$$\begin{aligned} D(T) = \ &\min\left\{ T_q \mid |v_{\text{out}}(t;T)| \geq \frac{q}{100} V_{\text{DD}}, \text{ for } t \geq T_q \right\} \\ &- \max\left\{ T_p \mid |v_{\text{out}}(t;T)| \leq \frac{p}{100} V_{\text{DD}}, \text{ for } t \leq T_p \right\} \ , \end{aligned} \tag{7}$$

where $v_{\text{out}}(t;T)$ denotes the voltage at the receiver node, which can be computed from $\mathbf{C}(\boldsymbol{x})$ and $\mathbf{G}(\boldsymbol{x})$ [5], and $V_{\text{DD}}$ is the supply voltage. In the following sections, $p=0$ and $q=80$ will be used.

In the current work, the SeDuMi toolbox [7] has been applied to solve SDPs such as (5). Additionally, a nonlinear constrained optimization algorithm that was developed recently in our department [8,9] has been used to solve (6).

## 3 Interconnect Modeling

Each wire segment $i$ is modeled as a $\pi$-type RC circuit (a resistance $r_i$ between two capacitors with value $c_i/2$ to ground). The resistance and capacitance parameters $r_i$ and $c_i$ are functions of the design parameters. If $l_i$ and $w_i$ denote the length and the width of the segment, respectively, then, in the case of a single wire, the values of $r_i$ and $c_i$ are given by,

$$r_i = R_{\text{s}}\frac{l_i}{w_i}, \tag{8}$$

$$c_i = (c_{\text{b}}w_i + c_{\text{f}})l_i \ , \tag{9}$$

where $R_{\text{s}}$, $c_{\text{b}}$, and $c_{\text{f}}$ denote the sheet resistance, the unit area bottom capacitance, and the unit length fringing capacitance, respectively. The fringing capacitance has a component that depends on the wire spacing. The most accurate results are obtained if the spacing-dependent term of the fringing capacitance is added to the coupling capacitance between the wires, which also depends on spacing, and the sum of the two is modeled.

Figure 2 illustrates that the length of the wire segments has only a relatively weak effect on the accuracy of the model. $T_{\text{dom}}$ is plotted as a function of the number of segments for a 10 mm long wire. The curve converges rapidly

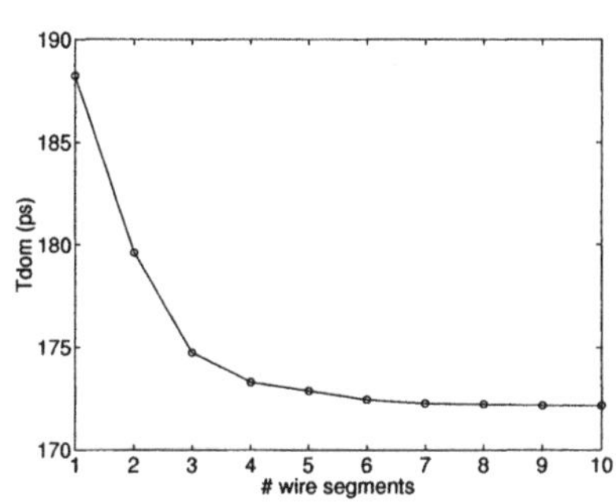


**Fig. 2.** Effect of the segment length on the accuracy of the computed delay for a 10 mm long wire.

towards an almost constant value. In the present work, a typical segment length of 1 mm was used, resulting in an accuracy within 0.1%.

It is assumed that the signal wires are driven by a driver with a strength of $R_\mathrm{d} = 100\ \Omega$. The capacitive load connected to the receiver end of the wires is neglected, which is legitimate for long wires.

# 4 Optimization of a Single Wire

Figure 3 shows wire shapes optimized with respect to $T_\mathrm{dom}$ and to $D(T)$ (0–80% delay). Minimizing $D(T)$ results in a more blunt wire shape, in which the capacitive load at the receiver end is higher, while the resistance of the downstream path is lower. The values of $D(T)$ and $T_\mathrm{dom}$ are 2.1% lower and 1.9% higher, respectively, compared to the case where $T_\mathrm{dom}$ is minimized. The dynamic power dissipated by a signal propagation, expressed as $P_\mathrm{dyn} = \mathbf{1}^T\mathbf{C}(\boldsymbol{x})\mathbf{1}$, is 3.4% higher, because a larger capacitance has to be charged. Significant differences between the optimization results for the two delay metrics are obtained only for very long wires with a high upper bound to the wire width.

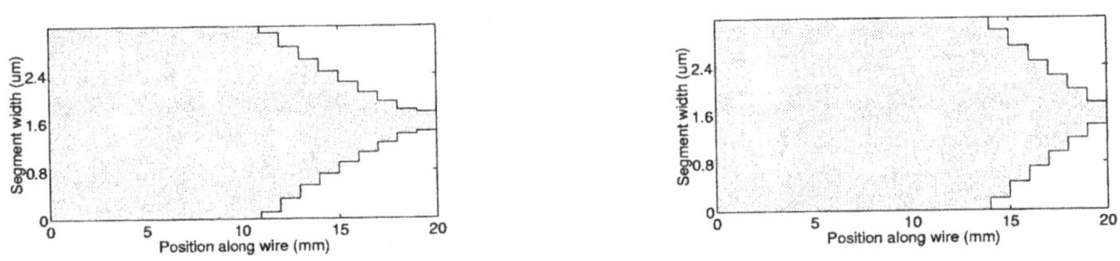


(a) (b)

**Fig. 3.** Optimal wire shapes resulting from the minimization of (a) $T_\mathrm{dom}$ and (b) $D(T)$. Signals are propagated from left to right.

Theoretical studies [10,11] have resulted in expressions for optimal wire shapes with respect to (Elmore) delay. Our method provides a tool to optimize wire shapes and to compare to results with analytic models. We investigated a linear model for the wire width,

$$
\begin{aligned}
w(z) &= \min\left(w_\mathrm{max}, \max\left(w_\mathrm{min}, w_0 + b_\mathrm{w} z\right)\right) \ , \\
w_0 &= a_0 + a_1 L_\mathrm{w} + \frac{a_2}{R_\mathrm{d}} + a_3 w_\mathrm{max} \ , \\
b_\mathrm{w} &= b_0 + \frac{b_1}{L_\mathrm{w}} + \frac{b_2}{R_\mathrm{d}} + \frac{b_3}{L_\mathrm{w} R_\mathrm{d}} + \frac{b_4 w_\mathrm{max}}{L_\mathrm{w}} + \frac{b_5 w_\mathrm{max}}{R_\mathrm{d}} \ ,
\end{aligned}
\tag{10}
$$

where $z$ denotes the position along the wire, $L_\mathrm{w}$ is the wire length, $w_\mathrm{max}$ is the upper bound to the width, and $R_\mathrm{d}$ denotes the driver strength, respectively.

The parameters $a_i$ and $b_i$ were determined by least-squares fits to the optimal wire shapes for various combinations of $L_w$, $w_{max}$, and $R_d$ values. It was found that the delay and dynamical power dissipation computed from the linear model (10) agreed within 1% with results for the optimized wire shapes, for practical ranges of $L_w$, $w_{max}$, and $R_d$. This shows that the complicated expressions of [10,11] are not necessary to accurately model wire shapes.

Figure 4 shows that wire tapering, in which the width of each wire segment is varied independently, offers only a relatively small advantage over wire sizing, in which all segments have equal width. The performance of tapered and sized wires differs significantly only if the upper bound on the wire width, $w_{max}$, is loose. This observation is in agreement with other studies [2] comparing wire sizing and wire tapering.

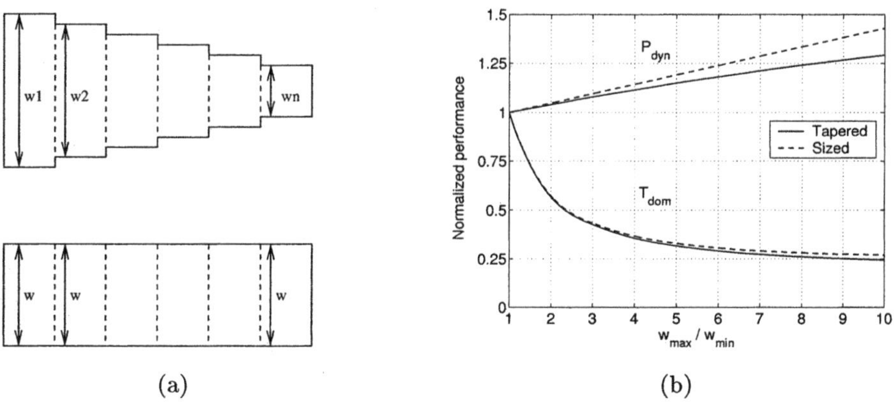


**Fig. 4.** (a) Wire tapering (upper drawing) compared with wire sizing (lower drawing). (b) Performance as a function of the upper bound ($w_{max}$) on the width of the wire segments.

## 5 Optimization of Bus Models

Our method can also be applied to multi-wire models that are used for real-life bus designs. Figure 5 shows the results of the optimization of $D(T)$ for a bus consisting of three parallel wires, where a signal is propagated through the middle wire. In Fig. 5(a), the signal is "boosted" by letting the neighboring wires carry the same signal [12]. Cross-talk between the wires then speeds up the signal propagation. In Fig. 5(b), it is assumed that the neighboring wires switch in the opposite direction. In this case, the capacitive coupling between the wires is effectively doubled, due to the Miller effect. While the optimal width and shape of the middle wire is approximately the same, the wire spacing is increased at the cost of narrower neighboring wires. For the bus model used here, situation (b) results in a 42% increase in propagation delay compared to the boosted situation (a).

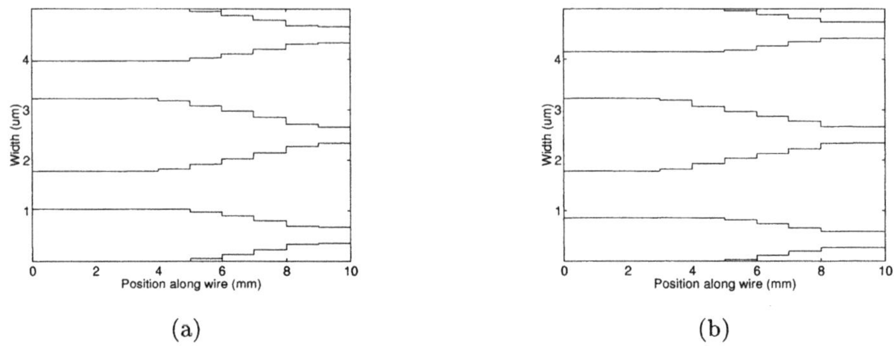


**Fig. 5.** Optimal wire shapes for a three-wire bus in the case that (a) the central wire is boosted by the neighbors and (b) the neighbors switch in the opposite direction.

## 6 Future Extensions

The impact of inserting repeaters in long wires can be increased by optimizing the repeater sizes. Because the delay of a repeater depends on the capacitive load of the wire connected to its output, interconnect including repeaters can be optimized by resizing the repeater after each iteration in the sequence of SDPs that is used to solve (4) or (6). Also the optimization of shielding wires can be used to further decrease signal delay.

As in [5], in the present work it is assumed that the signal wire is stimulated by a step input signal. However, responses to input signals with different shapes can be studied as well. Also, the approach can be further extended by including optimization with respect to the input delays of the signals in a multi-wire bus [13].

## 7 Conclusions

In the present work, we have demonstrated that the semidefinite programming (SDP) approach to interconnect sizing, as proposed by Vandenberghe et al. [5], can be successfully extended to include the minimization of the dominant time constant or more general delay metrics. Using a wire model including fringing capacitance and realistic parameter values, the method has been applied to study optimal wire sizes and the optimization of a multi-wire bus model. The minimization of either the dominant time constant $T_{\mathrm{dom}}$ or the more general 0–80% delay provides different results, in particular for very long and wide wires. A simple linear expression for the optimal wire shape provides results that agree within 1%. Complicated wire-shape expressions reported elsewhere [10,11] are therefore of limited value in practical situations. Wire tapering results in shorter delay times than wire sizing, but the effect is only significant if the upper bound on the wire width is loose. The optimization of a three-wire bus model shows the important effect of capacitive coupling on propagation delay and the optimal wire sizes and spacing. In summary, the presented approach offers a general and flexible framework for accurate studies on the sizing of various types of interconnect.

## Acknowledgement

The authors are thankful to Peter Meijer and Steven van Dijk for critically reading the manuscript and to Manish Garg for helpful discussions.

## References

1. Weste, N., Eshraghian, K.: Principles of VLSI design: a systems perspective (2nd ed.). Reading: Addison–Wesley (1993).
2. Alpert, C., Devgan, A., Fishburn, J., Quay, S.: Interconnect synthesis without wire tapering. IEEE Trans. Computer-Aided Design **20** (2001) 90–104 ; ibid., **21** (2002) 497 (Erratum)
3. Cong, J., Pan, Z.: Wire width planning for interconnect performance optimization. IEEE Trans. Computer-Aided Design **21** (2002) 319–329
4. Cong, J., He, L., Koh, C.-K., Madden, P.: Performance optimization of VLSI interconnect layout. Integr. VLSI J. **21** (1996) 1–94
5. Vandenberghe, L., Boyd, S., El Gamal, A.: Optimizing dominant time constant in RC circuits. IEEE Trans. Computer-Aided Design **17** (1998) 110–124
6. Vandenberghe, L.: private communication.
7. Sturm, J.: Using SeDuMi 1.02, a MATLAB toolbox for optimization over symmetric cones. Optimization Methods and Software **11–12** (1999) 625–653
8. Heijmen, T., Kevenaar, T., Pranger, H.-J.: Adapt: an interactive tool for analog synthesis. Submitted for publication
9. Lin, C., Heijmen, T., ter Maten, J., Kole, M.: Adapt: design assistance for iterative analogue synthesis. Proc. Analog 2002 (appeared in GMM-Fachbericht **38**) (2002) 195–200
10. Fishburn, J.: Shaping a VLSI wire to minimize Elmore delay. Proc. ED&TC (1997) 244–251
11. Chen, C.-P., Wong, D.: Optimal wire-sizing function with fringing capacitance consideration. Proc. DAC (1997) 604–607
12. Van Dijk, S., Hély, D.: Reduction of interconnect delay by exploiting cross-talk. Proc. ESSCIRC (2001) 316–319
13. Hirose, K., Yasuura, H.: A bus delay reduction technique considering crosstalk. Proc. DATE (2000) 141–145

# Electromagnetic Force Densities in a Continuous Medium

François Henrotte and Kay Hameyer

Katholieke Universiteit Leuven, Dep. ESAT, Div. ELECTA,
Kasteelpark Arenberg 10, B-3001 Leuven, Belgium.

**Abstract.** The paper introduces a systematic procedure to derive the expression of the Maxwell stress tensor associated with a given expression of the electromagnetic energy density.

## 1 Introduction

There exist numerous formulae for the computation of electromagnetic (EM) forces. They can be sorted into two distinct families, depending on whether they are based on the definition of a *Maxwell stress tensor* or on the application of the *virtual work principle.* A look into the literature shows that expressions of *Maxwell stress tensors* (See e.g. [1]) are generally obtained by algebraic and differential operations, starting from the Maxwell equations and assuming an *a priori* knowledge of the expression of the force density in the medium under consideration. The *virtual work principle*, on the other hand, relies more clearly in theory on the required energy concepts [2–4], but the formulae proposed in [5–7] for its implementation are all obtained by a roundabout way involving the jacobian matrix of a mapping at the finite element level. In both cases, the underlying thermodynamic concepts are buried into an overwhelming algebra.

## 2 The Euclidean case

The first step towards a thermodynamic analysis of an electromechanical system is to define the magnetic and mechanical state variables in such a way that they are independent of each other. Whereas it sounds obvious to anybody that one can freely modify the magnetic field [1] in a system without deforming it, by increasing the imposed currents for instance, it is much less clear to imagine how the system can be deformed without modifying the magnetic field. One feels indeed that any deformation of the system will affect the magnetic field.

Let first $M$ be the *material manifold*, i.e. a continuous set of points each representing a material particle of a given electromechanical system. Let $C_1(M)$ be the set of all regular curves in $M$ and $C_2(M)$ be the set of all

[1] Considered in this section as a vector field, *not* as a differential form.

regular surfaces. Let $E$ be the Euclidean space $\mathbb{R}^3$. Following [8], the magnetic state variable of the electromechanical system is defined by the *magnetic flux map*

$$\phi : C_2(M) \mapsto \mathbb{R} \tag{1}$$

that associates a real number, the *magnetic flux*, to any surface in $M$. Similarly, the kinematics of the system is defined by the *placement map*

$$p : M \mapsto \Omega \subset E, \tag{2}$$

which associates to any point of $M$ its position in $E$.

Although the magnetic flux map (1) determines completely the fluxes in the system, it does not give the local value of the *induction field.* The latter is a secondary quantity that requires the definition of an *interpolation operator* noted $\boldsymbol{b}(\phi, p)$, as it may depend on both $p$ and $\phi$. This is the reason why the interpolated induction field is not suitable as a primary variable in a thermodynamic representation. The properties of this vector-valued interpolation operator are not trivial. It may involve, for instance, the selection of a set of particular facets for the representation of the field, and an accuracy and convergence analysis. The interpolation with Whitney facet elements in a mesh is an example of such an interpolation tool. Another example is given below. A similar interpolation operator, denoted by $\boldsymbol{x}(p)$, is associated with the placement map.

Since the maps $\phi$ and $p$ are independent of each other, they are suitable variables for the definition of the *energy functional* of the system $\Omega$. One has

$$\Psi(\phi, p) = \int_{\Omega(p)} \rho^{\Psi}(\boldsymbol{b}(\phi, p), \boldsymbol{x}(p), p) \tag{3}$$

where the *energy density* $\rho^{\Psi}$ depends on the *interpolated* vector fields $\boldsymbol{b}$ and $\boldsymbol{x}$.

If the problem is more easily posed in terms of the magnetic field $\boldsymbol{h}$, than in terms of $\boldsymbol{b}$, the available thermodynamic state function is the *coenergy functional*

$$\Phi(I, p) = \int_{\Omega(p)} \rho^{\Phi}(\boldsymbol{h}(I, p), \boldsymbol{x}(p), p) \tag{4}$$

with

$$I : C_1(M) \mapsto \mathbb{R} \tag{5}$$

the *magnetomotive force map*, which associates a real number, the *magnetomotive force*, to any curve in $M$, and $\boldsymbol{h}(I, p)$ the interpolation operator for the magnetic field.

The definition of forces follows now from the variation of those energy functionals, $\delta\Psi(\phi, p)|_{\delta\phi=0}$ or $\delta\Phi(I, p)|_{\delta I=0}$, and the factorization under the form of a mechanical work monomial (e.g. $\int_\Omega \rho^{\boldsymbol{f}} \cdot \delta\boldsymbol{x}$, $\int_\Omega \sigma : \varepsilon$, $\boldsymbol{f} \cdot \delta\boldsymbol{x}, \ldots$). If *total* energy functionals are used, the computed forces will be *total forces*, i.e. of mechanical *and* electromagnetic nature together. In order to define specifically *electromagnetic forces*, we have to use *electromagnetic* energy functionals instead of *total* energy functionals. The definition of such restricted functionals is however not an obvious matter in a magnetostrictive material for instance. But, as this is not the topic of this paper, it will be assumed in the following that all energy functionals are electromagnetic ones.

### 2.1 Application at the local level

Let us consider as *material manifold* $M$ a unit cube with coordinates $\{\alpha, \beta, \gamma\}$, Fig. 1. Let $O \equiv (0,0,0)$, $A \equiv (1,0,0)$, $B \equiv (0,1,0)$ and $C \equiv (0,0,1)$ be four particular points of $M$. The interpolated field $\boldsymbol{x}(p)$ is defined by the affine combination of the placement of those four points

$$\boldsymbol{x} = (1 - \alpha - \beta - \gamma)\, p(O) + \alpha\, p(A) + \beta\, p(B) + \gamma\, p(C). \tag{6}$$

This determines in $E$ a parallelepiped region $p(M) = \Omega$ of volume

$$V = (\boldsymbol{r} \times \boldsymbol{s}) \cdot \boldsymbol{t} = (\boldsymbol{s} \times \boldsymbol{t}) \cdot \boldsymbol{r} = (\boldsymbol{t} \times \boldsymbol{r}) \cdot \boldsymbol{s}, \tag{7}$$

where $\boldsymbol{r} = p(A) - p(O)$, $\boldsymbol{s} = p(B) - p(O)$ and $\boldsymbol{t} = p(C) - p(O)$ are three linearly independent vectors of $E$. One can check that they verify

$$(\boldsymbol{r} \times \boldsymbol{s})\, \boldsymbol{t} + (\boldsymbol{s} \times \boldsymbol{t})\, \boldsymbol{r} + (\boldsymbol{t} \times \boldsymbol{r})\, \boldsymbol{s} = V\mathbb{I} \tag{8}$$

where $\mathbb{I}$ is the *identity matrix*. Note the use of the *dyadic (undotted) product.*

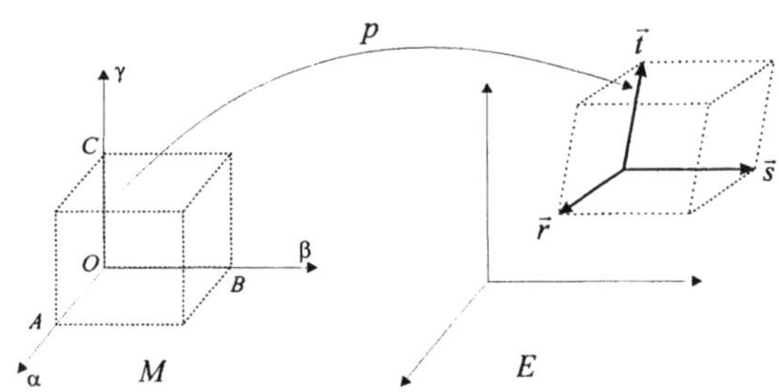


**Fig. 1.** Theoretical setup for the electromechanical coupling in a continuous medium.

Taking the gradient of (6), one gets $\boldsymbol{\nabla x} \equiv \mathbb{I} = \boldsymbol{\nabla}\alpha\, \boldsymbol{r} + \boldsymbol{\nabla}\beta\, \boldsymbol{s} + \boldsymbol{\nabla}\gamma\, \boldsymbol{t}$, and after identification with (8),

$$V\boldsymbol{\nabla}\alpha = \boldsymbol{s} \times \boldsymbol{t}, \quad V\boldsymbol{\nabla}\beta = \boldsymbol{t} \times \boldsymbol{r}, \quad V\boldsymbol{\nabla}\gamma = \boldsymbol{r} \times \boldsymbol{s}. \tag{9}$$

The parallelepiped $\Omega$ is deformed by perturbing the placement of the points $A$, $B$ and $C$. The displacement field in $\Omega$ and its gradient are

$$\boldsymbol{u} \equiv \delta\boldsymbol{x} = \delta p(O) + \alpha\delta\boldsymbol{r} + \beta\delta\boldsymbol{s} + \gamma\delta\boldsymbol{t}, \quad \boldsymbol{\nabla u} = \boldsymbol{\nabla}\alpha\, \delta\boldsymbol{r} + \boldsymbol{\nabla}\beta\, \delta\boldsymbol{s} + \boldsymbol{\nabla}\gamma\, \delta\boldsymbol{t}. \tag{10}$$

## 2.2 Formulation in $b$

Let $\phi_\alpha$, $\phi_\beta$ and $\phi_\gamma$ be the fluxes obtained by applying the magnetic flux map (1) to the particular parallelogram facets $OBC$, $OCA$ and $OAB$ of $M$. If the parallelepiped is small enough, the (uniform) induction field in $\Omega$ is by definition the vector $\boldsymbol{b}$ that verifies the left-hand equation in (11). The right-hand equation is the expression of the interpolated induction field $\boldsymbol{b}(\phi, p)$, which is obtained by inversion of the former equation.

$$\begin{bmatrix} \phi_\alpha \\ \phi_\beta \\ \phi_\gamma \end{bmatrix} = \begin{bmatrix} \boldsymbol{s} \times \boldsymbol{t} \\ \boldsymbol{t} \times \boldsymbol{r} \\ \boldsymbol{r} \times \boldsymbol{s} \end{bmatrix} \cdot \boldsymbol{b} \quad \Leftrightarrow \quad \boldsymbol{b} = \frac{\boldsymbol{r}}{V}\phi_\alpha + \frac{\boldsymbol{s}}{V}\phi_\beta + \frac{\boldsymbol{t}}{V}\phi_\gamma. \tag{11}$$

The parallelepiped $\Omega$ is now deformed by perturbing the placement of the point $C$, i.e. by perturbing the vector $\boldsymbol{t}$, leaving $\boldsymbol{r}$ and $\boldsymbol{s}$ unchanged. The variation of $\boldsymbol{b}$ with fluxes held constant is

$$\delta \boldsymbol{b}|_{\delta\phi=0} = -\frac{\delta V}{V^2}(\boldsymbol{r}\phi_\alpha + \boldsymbol{s}\phi_\beta + \boldsymbol{t}\phi_\gamma) + \frac{\delta \boldsymbol{t}}{V}\phi_\gamma. \tag{12}$$

Once the variation is done, it is allowed to substitute back for $\boldsymbol{b}$. Using (11) and (7), on finds

$$\delta \boldsymbol{b}|_{\delta\phi=0} = \frac{1}{V}\left\{-\boldsymbol{b}\ (\boldsymbol{r} \times \boldsymbol{s}) \cdot \delta \boldsymbol{t} + \boldsymbol{b} \cdot (\boldsymbol{r} \times \boldsymbol{s})\ \delta \boldsymbol{t}\right\}. \tag{13}$$

In case of a non-magnetic material, the energy function (3) is

$$\Psi = V \varrho^\Psi(\boldsymbol{b}) \quad , \quad \varrho^\Psi(\boldsymbol{b}) = \frac{|\boldsymbol{b}|^2}{2\mu_0} \tag{14}$$

and its variation with fluxes held constant is

$$\delta\Psi|_{\delta\phi=0} = \frac{V}{\mu_0}\boldsymbol{b}\ \ \delta \boldsymbol{b}|_{\delta\phi=0} + \delta V \frac{|\boldsymbol{b}|^2}{2\mu_0} - V\nabla \boldsymbol{u} : \sigma_M(\boldsymbol{b}) \tag{15}$$

with, by using (9), (10) and the property $(\boldsymbol{a} \cdot \boldsymbol{c})(\boldsymbol{b} \cdot \boldsymbol{d}) = (\boldsymbol{a}\ \boldsymbol{b}) : (\boldsymbol{c}\ \boldsymbol{d})$,

$$\nabla \boldsymbol{u} = \frac{1}{V}(\boldsymbol{r} \times \boldsymbol{s})\ \delta \boldsymbol{t} \quad , \quad \sigma_M(\boldsymbol{b}) = \frac{1}{\mu_0}\left(\boldsymbol{b}\ \boldsymbol{b} - \frac{|\boldsymbol{b}|^2}{2}\mathbb{I}\right), \tag{16}$$

which is the expression of the *Maxwell stress tensor* [1] associated with the expression (14) of the energy density.

## 2.3 Formulation in $h$

Let $I_r$, $I_s$ and $I_t$ be the circulations obtained by applying the magnetomotive force map (5) to the edges $OA$, $OB$ and $OC$ of $M$. If the parallelepiped $\Omega$ is small enough, the (uniform) magnetic field in $\Omega$ is by definition the vector that verifies the left-hand equation in (17). The right-hand equation is the interpolated magnetic field $\boldsymbol{h}(I,p)$, which is obtained by inversion of the former equation.

$$\begin{bmatrix} I_r \\ I_s \\ I_t \end{bmatrix} = \begin{bmatrix} \boldsymbol{r} \\ \boldsymbol{s} \\ \boldsymbol{t} \end{bmatrix} \cdot \boldsymbol{h} \quad \Leftrightarrow \quad \boldsymbol{h} = \frac{\boldsymbol{s} \times \boldsymbol{t}}{V} I_r + \frac{\boldsymbol{t} \times \boldsymbol{r}}{V} I_s + \frac{\boldsymbol{r} \times \boldsymbol{s}}{V} I_t. \tag{17}$$

The variation of $\boldsymbol{h}$ with currents held constant is

$$\delta \boldsymbol{h}|_{\delta I=0} = -\frac{\delta V}{V}\boldsymbol{h} + \frac{1}{V}\left\{\boldsymbol{s} \times \delta \boldsymbol{t}\; I_r + \delta \boldsymbol{t} \times \boldsymbol{r}\; I_s\right\} = -\frac{1}{V}(\boldsymbol{r} \times \boldsymbol{s})\; \boldsymbol{h} \cdot \delta \boldsymbol{t} \tag{18}$$

by (17) and using the variation of (8).

Considering again a non-magnetic material, the coenergy function (4) is

$$\Phi = V\varrho^{\Phi}(\boldsymbol{h}) \quad , \quad \varrho^{\Phi}(\boldsymbol{h}) = \mu_0 \frac{|\boldsymbol{h}|^2}{2}, \tag{19}$$

and its variation with currents held constant gives

$$\delta\Phi|_{\delta I} = V\mu_0 \boldsymbol{h} \cdot \delta \boldsymbol{h}|_{\delta I=0} + \delta V \mu_0 \frac{|\boldsymbol{h}|^2}{2} = -V \nabla \boldsymbol{u} : \sigma_M(\boldsymbol{h}) \tag{20}$$

with

$$\nabla \boldsymbol{u} = \frac{1}{V}(\boldsymbol{r} \times \boldsymbol{s})\; \delta \boldsymbol{t} \quad , \quad \sigma_M(\boldsymbol{h}) = \mu_0 \left(\boldsymbol{h}\; \boldsymbol{h} - \frac{|\boldsymbol{h}|^2}{2}\mathbb{I}\right). \tag{21}$$

The classical expression of the Maxwell stress tensor in terms of $\boldsymbol{h}$ is found [1], although the intermediary steps were somewhat different.

## 3 Magnetic materials

In case of a *permanent magnet material*, the constitutive law and the energy density are

$$\boldsymbol{b} = \mu_0(\boldsymbol{h} + \boldsymbol{m}) \quad , \quad \rho^{\Psi}(\boldsymbol{b}, \boldsymbol{m}) = \frac{|\boldsymbol{b}|^2}{2\mu_0} - \boldsymbol{b} \cdot \boldsymbol{m} \tag{22}$$

where the magnetisation vector $\boldsymbol{m}$ does not depend on the magnetic field but may in general be a function of the deformation tensor $\varepsilon$. In order to determine the Maxwell stress tensor associated with this particular expression of the energy density, we need to choose the geometrical nature of $\boldsymbol{m}$.

If one decides to have the magnetisation, denoted by $\boldsymbol{m}^{(1)}$, represented by a map $C_1(M) \mapsto \mathbb{R}$, like the magnetomotive force map (5), one finds by applying the procedure described in the previous section that the term $\boldsymbol{b} \cdot \boldsymbol{m}^{(1)}$ will have no contribution to the Maxwell stress tensor, except the one due to the explicit dependence of $\boldsymbol{m}$ with $\varepsilon$ [2]. Under this assumption, the Maxwell stress tensor of a permanent magnet material is

$$\sigma_M(\boldsymbol{b}, \boldsymbol{m}^{(1)}) = \frac{1}{\mu_0}\left(\boldsymbol{b}\,\boldsymbol{b} - \frac{|\boldsymbol{b}|^2}{2}\mathbb{I}\right) - \boldsymbol{b}\cdot\partial_\varepsilon \boldsymbol{m}^{(1)} \tag{23}$$

since $\delta\varepsilon = [\boldsymbol{\nabla u}]_{SYM}$.

If one decides on the contrary to have the magnetisation, denoted by $\boldsymbol{m}^{(2)}$, represented by a map $C_2(M) \mapsto \mathbb{R}$, like the magnetic flux map (1), one finds then

$$\sigma_M(\boldsymbol{b}, \boldsymbol{m}^{(2)}) = \frac{\boldsymbol{b}\,\boldsymbol{b}}{\mu_0} - \boldsymbol{b}\,\boldsymbol{m}^{(2)} - \boldsymbol{m}^{(2)}\,\boldsymbol{b} - \left(\frac{|\boldsymbol{b}|^2}{2\mu_0} - \boldsymbol{b}\cdot\boldsymbol{m}^{(2)}\right)\mathbb{I} - \boldsymbol{b}\cdot\partial_\varepsilon \boldsymbol{m}^{(2)}, \tag{24}$$

which is different from the former.

This shows that the distinction between 1–forms and 2–forms, which is irrelevant for the expression of the *magnetic* constitutive law (22), becomes essential when the *magneto-mechanical* coupling is considered. There is however no mathematical reason to favour one of these expressions. The first one might be better, for instance, when the magnetisation is actually due to the presence of microscopic magnetic dipoles, whereas the second one might better fit a magnetisation due to microscopic flux carriers (such as Abrikosov vortices in HTc Type II superconductors).

In case of an *isotropic reversible magnetic material*, the magnetisation writes $\boldsymbol{m} = \chi(|\boldsymbol{h}|, \varepsilon)\ \boldsymbol{h}$ with $\chi$ the *magnetic susceptibility*. This gives a good representation of polycrystalline saturable materials, like iron and non-laminated steel. The magnetisation being a function of $\boldsymbol{h}$, it is more natural in this context to work with the coenergy:

$$\Phi = V\varrho^{\Phi}(\boldsymbol{h}) \quad , \quad \varrho^{\Phi}(\boldsymbol{h}) = \int_0^{|\boldsymbol{h}|} \mu_0\,(1 + \chi(x, \varepsilon))\, x\; dx. \tag{25}$$

The associated Maxwell stress tensor is

$$\sigma_M(\boldsymbol{h}) = \mu_0(1 + \chi(|\boldsymbol{h}|, \varepsilon))\ \boldsymbol{h}\ \boldsymbol{h} - \rho^{\Phi}\mathbb{I} - \int_0^{|\boldsymbol{h}|} \mu_0\ \partial_\varepsilon\chi(x, \varepsilon)\ x\ dx. \tag{26}$$

## 4 The Riemannian case

It is worth showing how easier the definition of EM forces is in differential geometry terms. Let us consider the case of a non-magnetic material, for which the coenergy density is the following function

$$\rho^{\Phi}(h) = \frac{\mu_0}{2} h \wedge \star h = \frac{\mu_0}{2}(h|h)\pi = \frac{\mu_0}{2}\sqrt{g}\ h_i g^{ij} h_j\ \pi, \tag{27}$$

[2] This is due to the fact that the exterior product of a 1–form and a 2– form does not involve the metric and is therefore insensitive to deformation.

of the 1−form $h$, the *Hodge operator* $\star$, the *volume form* $\pi \equiv dx \wedge dy \wedge dz$ and the *metric* $g$, with components $g_{ij}$. Let $\widehat{g}$ be the determinant of the *metric matrix*, and the $g^{ij}$'s the components of its inverse.

The variation of the coenergy, with the field $h$ held constant, writes simply

$$\pounds_{\boldsymbol{v}}\rho^{\Phi}\big|_{\pounds_{\boldsymbol{v}}h=0} \tag{28}$$

with $\boldsymbol{v} \equiv \partial_t p(t)$ the *velocity field.* Thanks to the Lie derivative [9], there is no need here to define the magnetomotive map and the interpolated magnetic field as distinct concepts.

Applying the rules of the Lie derivative and using the condition

$$\pounds_{\boldsymbol{v}}h = 0 \quad \Rightarrow \quad v^k \frac{\partial h_i}{\partial x^k} = -h_k \frac{\partial v^k}{\partial x_i} \tag{29}$$

to eliminate the terms like $v^k \frac{\partial h_i}{\partial x^k}$, one ends up with

$$\pounds_{\boldsymbol{v}}\rho^{\Phi}\big|_{\pounds_{\boldsymbol{v}}h=0} = \left\{ \frac{v^k}{2\widehat{g}} \frac{\partial \widehat{g}}{\partial x^k} + \frac{\partial v^k}{\partial x^k} \right\} \rho^{\Phi} + \mu_0 \sqrt{\widehat{g}}\, h_i (\partial_t \varepsilon^{\sharp})^{ij} h_j \; \pi \tag{30}$$

where the tensor $\partial_t \varepsilon^{\sharp}$ is related with the time derivative of the *strain tensor* $\partial_t \varepsilon$ by

$$\partial_t \varepsilon^{\sharp} = \frac{1}{2} \pounds_{\boldsymbol{v}} g^{-1} \quad , \quad (\partial_t \varepsilon^{\sharp})^{ij} = -g^{ip} (\partial_t \varepsilon)_{pq} g^{qj} \quad , \quad \partial_t \varepsilon = \frac{1}{2} \pounds_{\boldsymbol{v}} g. \tag{31}$$

Equation (30) factorizes as

$$\pounds_{\boldsymbol{v}}\rho^{\Phi}\big|_{\pounds_{\boldsymbol{v}}h=0} = -\left\{ \rho^{\boldsymbol{f}} \cdot \boldsymbol{v} + \sigma_M : \partial_t \varepsilon \right\} \pi, \tag{32}$$

which defines the *force density* $\rho^{\boldsymbol{f}}$ and the *Maxwell stress tensor* $\sigma_M$ associated with the expression (27) of the coenergy:

$$\rho^{\boldsymbol{f}}_k = -\frac{1}{2\widehat{g}} \frac{\partial \widehat{g}}{\partial x^k} \varrho^{\Phi} \quad , \quad \sigma_M^{mn} = \mu_0 \sqrt{\widehat{g}}\, h_i g^{im} g^{nj} h_j - \varrho^{\Phi} \delta^{mn} \tag{33}$$

with $\varrho^{\Phi} = \rho^{\Phi}_{123}$.

Similar developments can be done for a $b$−formulation. One has

$$\rho^{\Psi}(b) = \frac{b \wedge \star b}{2\mu_0} = \frac{(b\,|\,b)}{2\mu_0} \pi = -\frac{\sqrt{\widehat{g}}}{2\mu_0}\, b_{pq} g^{pi} g^{qj} b_{ij} \; \pi \tag{34}$$

and

$$\pounds_{\boldsymbol{v}}\rho^{\Psi}\big|_{\pounds_{\boldsymbol{v}}b=0} = -\left\{ \frac{v^p}{2\widehat{g}} \frac{\partial \widehat{g}}{\partial x^p} + \frac{\partial v^p}{\partial x^p} \right\} \rho^{\Psi} - \frac{\sqrt{\widehat{g}}}{\mu_0}\, b_{pq} (\partial_t \varepsilon^{\sharp})^{pi} g^{qj} b_{ij} \; \pi. \tag{35}$$

Equation (35) factorizes as

$$\pounds_{\boldsymbol{v}}\rho^{\Psi}\big|_{\pounds_{\boldsymbol{v}}b=0} = \left\{ \rho^{\boldsymbol{f}} \cdot \boldsymbol{v} + \sigma_M : \partial_t \varepsilon \right\} \pi, \tag{36}$$

which defines the *force density* $\rho^{\boldsymbol{f}}$ and the *Maxwell stress tensor* $\sigma_M$ associated with the expression (34) of the energy

$$\rho^{\boldsymbol{f}}_k = -\frac{1}{2\widehat{g}} \frac{\partial \widehat{g}}{\partial x^k} \varrho^{\Psi} \quad , \quad \sigma_M^{mn} = -\frac{\sqrt{\widehat{g}}}{\mu_0}\, b_{ij} b_{pq} g^{pm} g^{ni} g^{qj} - \varrho^{\Psi} \delta^{mn}. \tag{37}$$

**Comments** The expressions (33) and (37) of the Maxwell stress tensor of a non-magnetic material in a general Riemannian metric reduce to 21 and 16 in case of an *Euclidean metric*, $g_{ij} = \delta_{ij}$. On the other hand, the force density $\rho^f$ exists only where the metric field is not uniform. But, as we have seen, the computation of a Maxwell stress tensor by this approach requires quite an involved algebra.

## 5 Conclusion

The thermodynamic definition of EM forces in a continuous medium has been recalled. In a Riemannian metric, it can be applied directly thanks to the Lie derivative. However, in order to facilitate the calculation of the Maxwell stress tensor of more complex materials, a formalism expressed in classical terms, but yet coordinate-free, has been proposed, which is based on the same principles but applies only in the case of an Euclidean metric. In this case, the Maxwell stress tensor has been found to be the fundamental expression of the local electromechanical coupling in a continuous medium. It can be readily used as an applied stress in structural computations. It can also serve to compute electromagnetic forces: *force densities* are obtained by taking the divergence of it, *nodal forces* by multiplying it with the gradient of a nodal shape function [7] and *resultant forces* by integrating it over an enclosing surface. According to the approach proposed in this paper, the Maxwell stress tensor of a particular medium is not *inferred* from a postulated expression of the force density, which is usually the case, but on the contrary systematically *derived* from the expression of the electromagnetic energy or coenergy density of the medium under consideration. The proposed procedure has been applied for permanent magnet materials and isotropic saturable magnetic materials.

## References

1. Woodson, H. and Melcher, J.: Electromechanical dynamics I, II and III. John Wiley, 1968.
2. Bossavit A.: Edge-element computation of the force field in deformable bodies. IEEE Transactions on Magnetics **28(2)** (1992) 1263–1266.
3. Bossavit A.: Forces in magnetosatics and their computation. J. Appl. Phys. **67** **(9)**, 1 May 1990.
4. Bossavit A.: On local computation of the electromagnetic force field in deformable bodies. International Journal of Applied Electromagnetics in Materials **2** (1992) 333-343. Elsevier.
5. Coulomb, J.L.: A methodology for the determination of global electromechanical quantities from a finite element analysis and its application to the evaluation of magnetic forces, torques and stiffness. IEEE Transactions on Magnetics **19(6)** (1983) 2514–2519.
6. Ren Z., Razek A.: Local force computation in deformable bodies using edge elements. IEEE Transactions on Magnetics **28(2)** (1992) 1212–1215.

7. Kameari A.: Local force calculation in 3D FEM with edge elements. International Journal of Applied Electromagnetics in Materials, **3** (1993) 231–240.
8. Bossavit A.: 'Generalized Finite Differences' in Computational Electromagnetics. Progress in Electromagnetics Research, PIER 32 (2001), F.L. Texeira ed., EMW (Cambridge, Ma), 45-64.
9. Choquet-Bruhat Y., DeWitt-Morette C., Dillard-Bleick M.: Analysis, Manifolds and Physics, Elsevier Sciences, 1982.

# Reduced Order Modelling of RLC-networks Using an SVD-Laguerre Based Method

Pieter Heres[1] and Wil Schilders[1,2]

[1] Eindhoven University of Technology, Department of Mathematics and Computing Science, Scientific Computing Group
[2] Philips Research Laboratory, Eindhoven, The Netherlands

**Abstract.** With interconnect increasingly contributing to the electrical behaviour of integrated circuits, both by higher frequencies and smaller dimensions, it becomes increasingly important to incorporate its behaviour into simulations of ICs. This can be done rather elegantly by summarizing interconnect behaviour into a compact or reduced order model which is then co-simulated with the circuit. A similar approach can be used in the case of more conventional printed circuit boards. The SVD-Laguerre algorithm proposed by Knockaert and De Zutter [4] can be used for this purpose. In this paper, we describe an efficient implementation of the algorithm for multiple inputs, and show how the mathematical reduced order models can be translated into realizable circuit elements.

## 1 Introduction

To increase their performance, the characteristic dimensions of ICs and printed circuit boards (PCBs) are decreased and will decrease even further in the future. Higher speed makes the effect of higher frequency modes on the interconnect more important. Therefore, the analysis of signal propagation on the interconnect system is important. However, this requires the solution of Maxwell's equations which is rather demanding from the point of view of computation times. In addition, accurate modelling leads to large systems which can hardly be used in conventional circuit simulations.
To be able to work with models for interconnect structures, a technique known as reduced order modelling is employed. This class of mathematical techniques is able to reduce the sizes of models while preserving their essential features. Classical techniques in this area are the asymptotic waveform evaluation (AWE) method and the Pade-via-Lanczos (PVL) method. The latter is an efficient and robust implementation of the former. Recently, a new reduction method was proposed by Knockaert and De Zutter [4]. We will take a closer look at this method and will show how this method can be used to make realizable circuits.

The paper is built up as follows. In section 2, we briefly show how the discretized Maxwell equations lead to an RLC model for the interconnect system. Then, in section 3, the concept of transfer function is introduced, relating the area of reduced order modelling to concepts used in systems and control engineering where frequent use is made of state space models. Section

4 discusses several points which are of interest when using reduced order modelling techniques in the context of ICs and PCBs. In section 5 the SVD-Laguerre method is explained, whereas in section 6 the efficient treatment of multiple inputs is presented. The translation of the mathematical results to a realizable circuit is discussed in section 7. In the last section, some numerical results are given.

## 2 Discretization procedure

The modelling of interconnect systems has gradually developed over the years. For DC situations interconnect can be modelled as a short, but as losses and inductances are becoming more important increased use is made of representations using RLCG circuits. It is sufficient to restrict ourselves to the case in which the components R, L, G, and C are frequency independent.
Both ICs and PCBs can be modelled by (large!) RLC-circuits. These models can be obtained via a discretization of the Maxwell equations. As an illustration of this, we will very briefly review how this is done in [7].
To calculate the electromagnetic fields in an electronic system, the Maxwell equations must be solved:

$$
\begin{aligned}
\nabla \times \mathbf{E} &= -\tfrac{\partial \mathbf{B}}{\partial t} & \mathbf{J} &= \sigma \mathbf{E} \\
\nabla \times \mathbf{H} &= \mathbf{J} + \tfrac{\partial \mathbf{D}}{\partial t} & \mathbf{B} &= \mu \mathbf{H} \\
\nabla \cdot \mathbf{B} &= 0 & \mathbf{D} &= \epsilon \mathbf{E} \\
\nabla \cdot \mathbf{D} &= \rho & &
\end{aligned}
$$

After introducing a magnetic vector potential $\mathbf{A}$ and an electric scalar potential $\varphi$ the system can be rewritten as follows:

$$
\begin{aligned}
(\Delta + k^2)\mathbf{A} &= -\mu \mathbf{J}, \\
\nabla \cdot (\epsilon \nabla \varphi) + \epsilon k^2 \varphi &= -\rho, \\
\mathbf{J} = \sigma \mathbf{E} &= \sigma(-\nabla \varphi + i\omega \mathbf{A}), \\
\nabla \cdot \mathbf{J} - i\omega\rho &= 0
\end{aligned} \tag{1}
$$

with suitable boundary conditions. This system is discretized using a boundary integral method [7] making frequent use of Green's functions. The final discrete system can then be written into a form which is familiar to IC and PCB designers:

$$
\begin{aligned}
(\mathbf{R} - i\omega \mathbf{L})I - \mathbf{P}V &= 0, \\
-\mathbf{P}^T I + i\omega \mathbf{M} Q &= 0 \\
\mathbf{M}^T V - \mathbf{D} Q &= 0
\end{aligned} \tag{2}
$$

Here, the elements of the vector $V$ are the potentials of the elements. The vector $I$ consists of the current through the edges. $Q$ contains the weights of the surface charge density, therefore its elements are the charges of the elements of the circuit.

## 3 Transfer Functions, Approximation and Reduced Order Modelling

It is convenient to rewrite RLC models in terms of a state space formulation. Such formulations are of the form

$$\begin{aligned} \mathbf{C}\dot{\mathbf{x}} &= \mathbf{G}\mathbf{x} + \mathbf{B_i}\mathbf{u} \\ \mathbf{y} &= \mathbf{B_o}^T\mathbf{x}, \end{aligned} \tag{3}$$

where $\mathbf{B_i}$ and $\mathbf{B_o}$ are the matrices selecting the input and output, respectively. This reformulation can be done both for MNA formulations of circuits, and for the discretized system of Maxwell equations as derived in the previous section.

An efficient and commonly used way to solve the state space system is via the Laplace transform. Within this methodology, the so-called transfer function is introduced. It is the function $\mathbf{H}(s)$ giving the direct relation between input and output, in the frequency (denoted by $s$) domain. It is obtained by eliminating the state space vector $\mathbf{x}$. The $s$-parameter can be considered as the complex frequency $i\omega$. For (3) we have:

$$\mathbf{H}(s) = \mathbf{B_o}^T(\mathbf{G} + s\mathbf{C})^{-1}\mathbf{B_i}, \tag{4}$$

such that $\mathbf{y} = \mathbf{H}(s)\mathbf{u}$. A model which approximates the original model, can be called accurate if the transfer function of the original model is approximated well by the transfer function of the approximating model.

As can be understood from the procedure summarized in section 2, the models obtained for interconnect systems on ICs or PCBs consist of (very) large systems of equations. This is not very convenient, and a coupling of these large systems with circuit equations is almost out of the question. With Reduced Order Modelling the original model is replaced by a model which is smaller, but has (approximately) the same properties. There is a danger, however, that some essential properties are lost during the mathematical procedure. Ideally, these properties should be preserved.

In our search for a smaller circuit, describing approximately the same behaviour, an important issue is the preservation of stability and passivity. An RLC-circuit is passive, because it has no active components. Passivity is stronger than stability. A stable circuit, can become unstable when non-linear components are attached to its terminals. In contrast, a passive circuit remains stable under all conditions.

The behaviour of a circuit and the transfer function are uniquely determined by the poles and their associated residue. Poles can be calculated by:

$$\frac{-1}{\sigma(\mathbf{G}^{-1}\mathbf{C})},$$

with $\sigma(\mathbf{G}^{-1}\mathbf{C})$ the eigenvalues of $-\mathbf{G}^{-1}\mathbf{C}$. Because the poles are determining the behaviour of the system, also the poles can be approximated. This is why methods from the area of eigenvalue approximations are often well-suited for these problems also. Examples of this are the Krylov subspace methods, PVL and PRIMA. In this paper we consider a new Krylov subspace method which is very similar to the others, but with some very attractive properties: SVD-Laguerre [4].

## 4 Some Theory Behind the Laguerre Method

The idea behind the Laguerre method is that the transfer function can be expanded in terms of Laguerre functions. We consider the Laguerre functions in the $s$-domain:

$$\Phi_n^\alpha(s) = \frac{\sqrt{2\alpha}}{s+\alpha}\left(\frac{s-\alpha}{s+\alpha}\right)^n \quad \text{for } n = 0, 1, 2, ... \tag{5}$$

These functions form a uniform bounded orthonormal basis in the frequency domain for $s = i\omega$, with $\omega \in (0, \infty)$, for the space $\mathcal{H}_2$. The transfer function can be expanded in terms of these functions:

$$\mathbf{H}(s) = \frac{2\alpha}{s+\alpha}\mathbf{L}^T \sum_{n=0}^{\infty}\Big((\mathbf{G}+\alpha\mathbf{C})^{-1}(\mathbf{G}-\alpha\mathbf{C})\Big)^n(\mathbf{G}+\alpha\mathbf{C})^{-1}\mathbf{B}\left(\frac{s-\alpha}{s+\alpha}\right)^n \tag{6}$$

Due to a lack of space we are forced to reference to [3], where the derivation of this expression can be found. The matrices used in this Laguerre expansion can be used to build up Krylov subspaces. An $n$-dimensional Krylov subspace is defined by:

$$\mathcal{K}_n(\mathbf{b}, \mathbf{A}) = [\mathbf{b}, \mathbf{A}\mathbf{b}, \dots, \mathbf{A}^{n-1}\mathbf{b}] \tag{7}$$

The main part of the Laguerre method consists of building a Krylov subspace, with the vector $(\mathbf{G} + \alpha\mathbf{C})^{-1}\mathbf{B_i}$ and matrix $(\mathbf{G} + \alpha\mathbf{C})^{-1}(\mathbf{G} - \alpha\mathbf{C})$.

The columns of the Krylov subspace are made orthogonal. In the original article of the SVD-Laguerre method this orthogonalisation is done after all columns of the Krylov-space are created. This is done with a Singular Value Decomposition (SVD). But in repetitive multiplication with a matrix the vectors tend to one dominant direction. To avoid numerical artefacts, we propose to perform orthogonalisation this during the generation of the columns. The system matrices are projected onto this Krylov subspace, spanned by $\mathbf{V}$:

$$\widetilde{\mathbf{G}} = \mathbf{V}^T\mathbf{G}\mathbf{V} \quad \widetilde{\mathbf{C}} = \mathbf{V}^T\mathbf{C}\mathbf{V} \quad \widetilde{\mathbf{B_i}} = \mathbf{V}^T\mathbf{B_i} \quad \widetilde{\mathbf{B_o}} = \mathbf{V}^T\mathbf{B_o}$$

For a single input column $\mathbf{B_i}$, the algorithm can be summarized as follows:

Solve $(\mathbf{G} + \alpha\mathbf{C})\mathbf{v}_1 = \mathbf{B_i}$
$\mathbf{v}_1 = \frac{\mathbf{v}_1}{\|\mathbf{v}_1\|}$
for j=1,...,k-1
  Solve $(\mathbf{G} + \alpha\mathbf{C})\mathbf{t} = (\mathbf{G} - \alpha\mathbf{C})\mathbf{v}_j$

for $i = 1, ..., j$
  $h_{i,j} = \mathbf{v}_i^H \mathbf{t}$
  $\mathbf{t} = \mathbf{t} - h_{i,j}\mathbf{v}_i$
end
$h_{j+1,j} = ||\mathbf{t}||$
$\mathbf{v}_{j+1} = \frac{\mathbf{t}}{h_{j+1,j}}$
end
$\widetilde{\mathbf{G}} = \mathbf{V}^T\mathbf{G}\mathbf{V} \quad \widetilde{\mathbf{C}} = \mathbf{V}^T\mathbf{C}\mathbf{V} \quad \widetilde{\mathbf{B_i}} = \mathbf{V}^T\mathbf{B_i} \quad \widetilde{\mathbf{B_o}} = \mathbf{V}^T\mathbf{B_o}$

Although, solving the matrix equation in this equation is quite expensive, it has to be done for one choice of $\alpha$, so we can for instance invest in an LU-decomposition, to solve the system efficiently. Further note, that the following holds, during the algorithm: $(\mathbf{G} - \alpha\mathbf{C})^{-1}(\mathbf{G} + \alpha\mathbf{C})\mathbf{V}_{k-1} = \mathbf{V}_k\mathcal{H}$ where $\mathcal{H}$ is a Hessenberg matrix. The small matrix $\mathcal{H}^T\mathcal{H}$ can be used to approximate the singular values of the matrix $(\mathbf{G} - \alpha\mathbf{C})^{-1}(\mathbf{G} + \alpha\mathbf{C})$ and can therefor be used in a stopping criterium.

## 5 The Laguerre Algorithm for Multiple Input

If an RLC-model is considered with more than one input, the matrix $\mathbf{B_i}$ obviously has more than one column. All of these columns describe one specific input. The approximate model should then also allow more inputs and give an accurate approximation for all of these. In fact, the transfer function has become a transfer matrix, and we should have accurate approximations for all entries of this matrix.

Multiple inputs implies that the Krylov subspaces are also larger. For example, if two inputs are considered, the Krylov subspaces have a dimension which is twice as large compared with the subspaces generated for one input:

$$\mathcal{K}_n(\mathbf{B_i}, \mathbf{A}) = [\underbrace{\mathbf{B_i}}_{2}, \underbrace{\mathbf{AB_i}}_{2}, \ldots, \underbrace{\mathbf{A}^{n-1}\mathbf{B_i}}_{2}] \tag{8}$$

Projecting onto these larger subspaces leads to system matrices which are correspondingly larger. Hence, the reduction obtained is less, and we have to be very careful with the number of columns generated. To find an appropriate space which contains the information needed for several inputs, we propose the following algorithm:

For every input column $\mathbf{B}_m$
  Calculate $\mathbf{t} = (\mathbf{G} + \alpha\mathbf{C})^{-1}\mathbf{B}_m$
  Make this vector orthogonal to the already existing columns
  Do $k$ iterations of the Laguerre algorithm
    every columns is put orthogonal to every other
  end
end
Project onto the Krylov-space

Because every column is treated independently and every generated column is made orthogonal to the others, we expect the Krylov subspace to contain less redundant information.

## 6 A Disadvantage of Reduction Methods

The Kirchhoff's current laws and the branch equations, describing the RLC-model, can be derived directly from the state space formulation. Unfortunately, after reduction this is not possible anymore. For instance, if a state space vector $\mathbf{x}$ is used, in general it will consist of branch currents and node voltages. After projecting this vector onto a Krylov subspace, the rows have lost their physical meaning. The input and output terminals are kept, but the others may disappear. All well-know reduction methods (AWE [1], PVL [2], PRIMA [5]) suffer from this problem. This includes the Laguerre method presented in the previous section(s).

The problem mainly consists of the fact that we cannot make use of a circuit simulator in a direct way. Furthermore, starting from a model consisting of resistances, inductances and capacitors, it is desirable to have a reduced system which also consists of realizable or even passive components. This can not be done directly. However, there is a way to solve this problem via the Laguerre method, and we shall present this now.

Reconsider the Laguerre expansion given before:

$$\mathbf{H}(s) = \frac{2\alpha}{s+\alpha} \sum_{n=0}^{\infty} \mathbf{L}^T \Big( (\mathbf{G}+\alpha\mathbf{C})^{-1} (\mathbf{G}-\alpha\mathbf{C}) \Big)^n (\mathbf{G}+\alpha\mathbf{C})^{-1} \mathbf{B} \left( \frac{s-\alpha}{s+\alpha} \right)^n \tag{9}$$

The advantage of this formulation is that the $s$ parameter is not part of the inversion process for large matrices anymore. The parts of this equation which depend on $s$ can be represented by small filters. These filters are shown in Figure 1.

The voltages implied by these filters have to be multiplied by a factor $\mathbf{L}^T((\mathbf{G}+\alpha\mathbf{C})^{-1}(\mathbf{G}-\alpha\mathbf{C}))^n(\mathbf{G}+\alpha\mathbf{C})^{-1}\mathbf{B}$ for every $n$ and then added, in order to obtain the weighted summation. The circuit shown in Figure 2(a) stops at $n$ terms. But this series does converge very slow, so all elements, or at least many elements in the sum must be taken into account. This can be done by implementing a loop of filters, as shown in Figure 2(b). This realizable circuit can be implemented in a circuit simulator. We used `Pstar` which is the Philips proprietary circuit simulation programme.

## 7 Experimental Results

We applied the proposed algorithm to a PEEC model, and to some PCB examples which were modelled as RLC-circuits. The PEEC method gives an approximation for the behaviour of interconnect, the method was developed

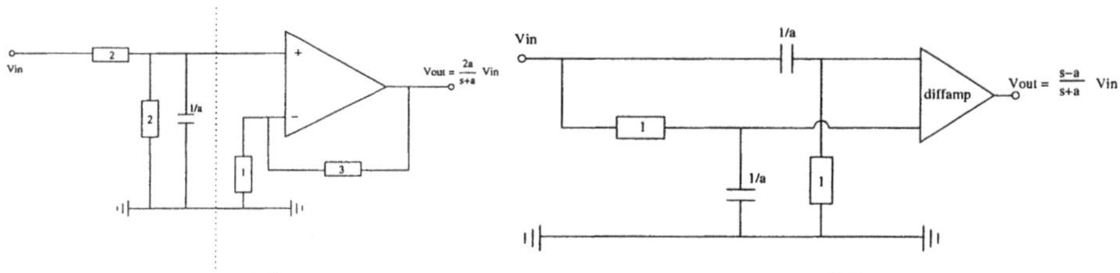


(a) (b)

**Fig. 1.** (a) The filter representing $\frac{2\alpha}{s+\alpha}$, (b) The filter representing $\frac{s-\alpha}{s+\alpha}$

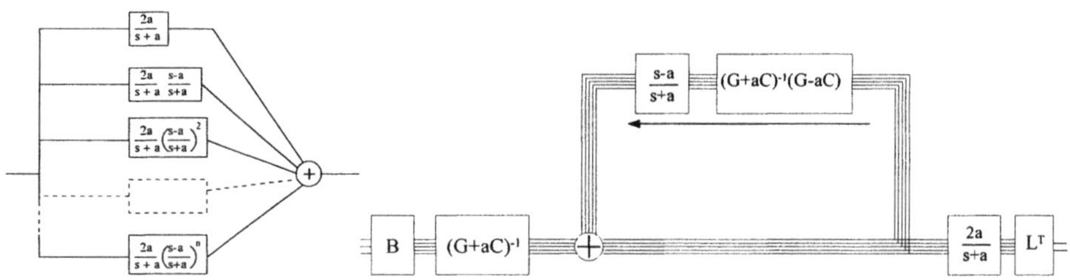


(a) (b)

**Fig. 2.** (a) The filter circuit, (b) The loop circuit

by Ruehli [6]. The PEEC model we used is a nice example, because the graph of the transfer function is rather intricate and hard to approximate. The proposed Laguerre algorithm can approximate this example very well. In Figure 3(a) the approximation of the Laguerre algorithm with $\alpha = 5\ 10^{10}$ and $q = 92$ is shown. For the given frequency range no difference can be observed. Of course we have to be careful with this result. An approximation in the frequency domain does not guarantee a good approximation in the time domain. Transient analysis should be applied to be sure, that the result is accurate.

The other example is not chosen for its complexity, but to show that it is possible to combine our filter realization with non-linear components. We consider a PCB board (see Fig. 3(b)) which, after discretization, can be described by system matrices of $460 \times 460$ entries. This representation can be reduced with the proposed method. We used a reduced order model consisting of system matrices of size $60 \times 60$, in order to get approximation up to 1 GHz. The loop filter representation of these kind of models were implemented in Pstar and combined with each other and other component. The preliminary results are fine. Sometimes, (as yet unexplained) strange behaviour is observed for lower frequencies. Further, some practical issues have to be solved.

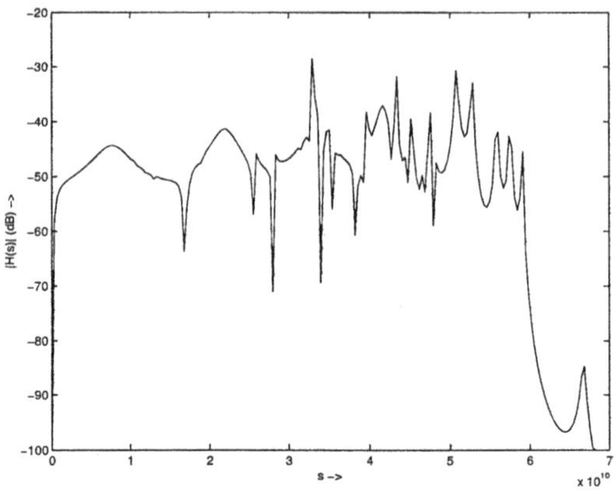


(a)

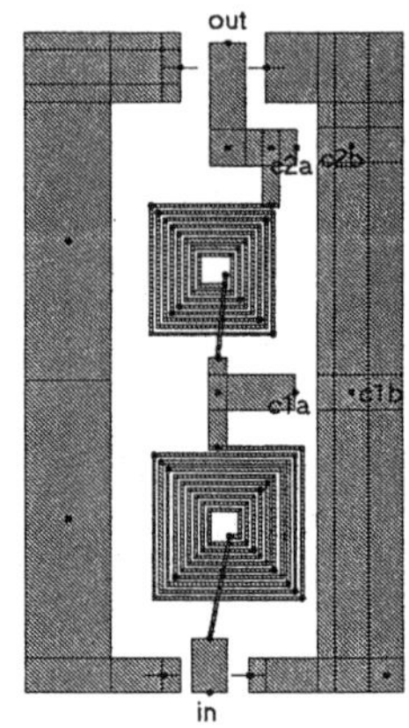


(b)

**Fig. 3.** (a) The transfer function of the PEEC model: the original and the Laguerre approximation, for $\alpha = 5\ 10^{10}$ and $q = 92$(b) The PCB used in the second example

# 8 Conclusion

We have shown a modified implementation of the SVD-Laguerre algorithm. We are now able to deal with multiple input in an efficient way and we orthogonalize during the proces. The algorithm is stable and passive and leads to an accurate solution.

# References


1. Eli Chiprout and Michel S. Nakhla, Asymptotic Waveform Evaluation and Moment Matching for Interconnect Analysis, Kluwer Academic Publishers, 1994.
2. Peter Feldmann and Roland W. Freund, "Efficient Linear Circuit Analysis by Padé Approximation via the Lanczos Process". *IEEE Trans. Computer-Aided Design*, vol 14, 1993, pp. 137-158
3. Pieter Heres, "Reduced Order Modelling of RLC-networks using an SVD-Laguerre based method." *Master's thesis, Eindhoven Univ. of Tech., 2001*
4. Luc Knockaert and Daniel De Zutter, "Passive Reduced Order Multiport Modeling: The Padé-Arnoldi-SVD Connection", *Int. J. Electronics and Communications*, nr. 53, 1999, pp. 254-260.
5. A. Odabasioglu and M. Celik, "PRIMA: Passive Reduced-Order Interconnect Macromodeling Algorithm". *IEEE. Trans Computer-Aided Design*, vol.17, 1999, pp. 645-654
6. A.E. Ruehli, "Equivalent Circuit Models for three dimensional multiconductor systems", *IEEE Trans. Microwave Theory*, Mar 1974.
7. A.J.H Wachters and R.A.J.M. de Best, "Evaluation Methods for Interaction Integrals, used in the Quasi-static Boundary Element Method for Electromagnetic Simulation of PCB's", *Philips Research Report 153/95*, 1995.

# Uniform Treatment of Numerical Time-Integrations of the Maxwell Equations

Róbert Horváth

Technische Universiteit Eindhoven, Den Dolech 2, 5600 MB Eindhoven, the Netherlands

**Abstract.** In this paper we show how the existing numerical time-integration methods of the Maxwell equations can be handled in the framework of the theory of operator splitting methods. We consider the classical Yee-method, the Namiki-Zheng-Chen-Zhang alternating direction implicit method (NZCZ) and the Kole-Figge-De Raedt method (KFR). The unconditional stability of the NZCZ-method has been proven only by means of extensive use of computer algebraic tools. We give a pure mathematical proof. We compare the methods from the point of view of accuracy and computational speed.

## 1 Introduction

The mathematical model of electromagnetic problems can be written in the form of the so-called Maxwell equations (in source-free case)

$$-\nabla \times \mathbf{H} + \varepsilon \partial_t \mathbf{E} = \mathbf{0}, \quad \nabla \times \mathbf{E} + \mu \partial_t \mathbf{H} = \mathbf{0}, \tag{1}$$

$$\nabla(\varepsilon \mathbf{E}) = 0, \quad \nabla(\mu \mathbf{H}) = 0, \tag{2}$$

where $\mathbf{E}$ and $\mathbf{H}$ are the electric and magnetic field strengths, respectively, $\varepsilon$ is the electric permittivity and $\mu$ is the magnetic permeability. We have to compute the fields $\mathbf{E}$ and $\mathbf{H}$ with some given boundary and initial conditions and material parameters. Because $\mathbf{E}$ and $\mathbf{H}$ are supposed to satisfy (2) at $t = 0$, we have to solve only system (1).

In real-life problems the exact solution of system (1) is very complicated or even impossible, this is why numerical methods are generally applied. The most frequently used one is the so-called Yee-method, which was introduced for the Maxwell equations in 1966 ([13]). This is a finite difference method, which applies a staggered spatial discretization and a so-called leap-frog scheme in time-variable. The Yee-method is computationally equivalent with the so-called Finite Integration Technique (FIT, see e.g. [11,12]), when this technique is applied to Cartesian grids and combined with the leap-frog time-integration scheme. The Yee-method suffers from a very strict stability condition, namely, it is stable if and only if the condition

$$\Delta t < \frac{1}{c\sqrt{(1/\Delta x)^2 + (1/\Delta y)^2 + (1/\Delta z)^2}} \tag{3}$$

is fulfilled, where $c = 1/\sqrt{\varepsilon\mu}$ is the maximal speed of light in the computational domain, $\Delta x$, $\Delta y$ and $\Delta z$ are the grid-sizes and $\Delta t$ is the time-step (see e.g. [9]).

Because of the very large system integrity of electronic devices and the applied very wide frequency range, a new claim has been established to the simulation methods. They have to be able to model, among others, the coupling effects between interconnect systems and electronic circuits (such as cross-talk), and the effects inside the wires (e.g. skin-effect), etc. To simulate the skin-effect we must choose very small grid-size. This implies, according to the upper bound (3), that we have to choose very small time-step too, which increases the computational time dramatically. A lot of effort has been invested during the last decade to bridge the stability problem of the Yee-method. The main goal was the construction of methods, where $\Delta t$ can be chosen based only on accuracy conditions instead of stability reason. The first paper was [15] that showed an unconditionally stable method (called NZCZ-method) with a detailed proof of the stability. The method was also considered for more general problems in [8]. It is also given an unconditionally stable numerical scheme in [5] (KFR-method). In [4] we showed that the above methods have common features and they all are based on matrix exponential approximation. In this paper we investigate the applicability of the operator splitting theory for the numerical solution of the Maxwell equations. This allows us the comparison and to find the way to the construction of new simulation methods in the future. The stability of the NZCZ-method was proven with the help of significant use of computer algebra (namely software MAPLE V). We give the proof of unconditional stability for nonhomogeneous media with pure mathematical tools.

The operator splitting method is very efficient in solving initial and boundary value problems for differential equations. The basic idea of the method is the splitting of the original problem into sub-problems according to the physical processes involved. Then these sub-problems, which can be handled more easily, are solved sequentially using some appropriate methods. The solution of the original problem is approximated by the solutions of the sub-problems.

We demonstrate the method on the system of ordinary differential equations (ODEs)

$$\Psi'(t) = \mathbf{A}\Psi(t), \; t \in (0,T], \;\; \Psi(0) \text{ is given,} \tag{4}$$

where $\mathbf{A} \in \mathbb{R}^{m\times m}$ and $\Psi : \mathbb{R} \to \mathbb{R}^m$. It is known that the solution of (4) can be written in the form $\Psi(t) = \exp(t\mathbf{A})\Psi(0)$, where the exponential of the matrix $t\mathbf{A}$ is defined by the series of the exponential function. This shows that to compute the exact solution of (4) we have to determine the exponential $\exp(t\mathbf{A})$ exactly. This is generally very difficult because of the infinite series of the exponential function. Splitting the matrix into the form $\mathbf{A} = \mathbf{A}_1 + \mathbf{A}_2$

and choosing a time-step $\tau$ we can define the sequence of systems of ODEs (so-called sequential or S-splitting)

$$\left(\Psi_k^{(1)}\right)'(t) = \mathbf{A}_1\Psi_k^{(1)}(t),\ t \in (t_{k-1}, t_k],\ \Psi_k^{(1)}(t_{k-1}) = \Psi_{k-1}^{(2)}(t_{k-1}), \tag{5}$$

$$\left(\Psi_k^{(2)}\right)'(t) = \mathbf{A}_2\Psi_k^{(2)}(t),\ t \in (t_{k-1}, t_k],\ \Psi_k^{(2)}(t_{k-1}) = \Psi_k^{(1)}(t_k), \tag{6}$$

where $k = 1, 2, \ldots, n$ $(T = n\tau)$, $t_{k-1} = (k-1)\tau$ and $\Psi_0^{(2)}(0) = \Psi(0)$. Solving these systems, the solution $\Psi$ of (4) can be approximated at time-level $t_k$ by $\Psi_k^{(2)}(t_k)$. The error of the splitting methods is defined as the difference between the exact and the approximated solution at time-level $t_1 = \tau$. For the S-splitting method we have

$$\begin{aligned} Err_{Spl.}(\tau) &= (\exp(\tau(\mathbf{A}_1 + \mathbf{A}_2)) - \exp(\tau\mathbf{A}_2)\exp(\tau\mathbf{A}_1))\Psi(0) = \\ &= \frac{\tau^2}{2}[\mathbf{A}_1, \mathbf{A}_2]\Psi(0) + \mathcal{O}(\tau^3). \end{aligned} \tag{7}$$

It is clear from the above expression that the splitting error is zero for all initial vector $\Psi(0)$ if and only if the matrices $\mathbf{A}_1$ and $\mathbf{A}_2$ commute, that is their commutator, defined by $[\mathbf{A}_1, \mathbf{A}_2] = \mathbf{A}_1\mathbf{A}_2 - \mathbf{A}_2\mathbf{A}_1$, is zero. This is the ideal case, but the splitting error generally occurs. It can be seen from (7) that there is a strong connection between equations (5), (6) and the exponential approximation $\exp(\tau(\mathbf{A}_1 + \mathbf{A}_2)) \approx \exp(\tau\mathbf{A}_2)\exp(\tau\mathbf{A}_1)$. In similar manner, second order splitting procedure can be defined applying the approximation

$$\exp(\tau(\mathbf{A}_1 + \mathbf{A}_2)) \approx S_\tau := \exp((\tau/2)\mathbf{A}_2)\exp(\tau\mathbf{A}_1)\exp((\tau/2)\mathbf{A}_2) \tag{8}$$

(so called Strang-splitting). Fourth order splitting can be obtained with the approximation (see [14])

$$\exp(\tau(\mathbf{A}_1 + \mathbf{A}_2)) \approx S_{\vartheta\tau}S_{(1-2\vartheta)\tau}S_{\vartheta\tau},\quad \vartheta = (2 - \sqrt[3]{2})^{-1}. \tag{9}$$

The splitting method can be extended applying more than two matrices in the splitting, for partial differential equations or for non-linear problems (see e.g. [3,6,7,10]).

## 2 Splitting of the Semi-Discretized Maxwell Equations

In this section the Yee-, NZCZ- and KFR-methods are analyzed and compared in the framework of the operator splitting theory. The comparison is done in the computational speed keeping the accuracy acceptable.

With the usual staggered semi-discretization of the Maxwell equations, dividing the computational domain into $N$ so-called Yee-cells, we arrive at a system of first order linear ordinary differential equations in the form (4). The matrix $\mathbf{A} \in \mathbb{R}^{6N\times 6N}$ is a sparse (at most four elements per row) skew-symmetric matrix and the function $\Psi : \mathbb{R} \to \mathbb{R}^{6N}$, $t \mapsto \Psi(t)$ gives the field components (multiplied by the square roots of the material parameters) at the point of time $t$ (see e.g. [4,5]). In the solution of (4), an additional difficulty is the very large size of $\mathbf{A}$. Let us investigate how the operator splitting method can be applied here in order to split the original system into easier (or exactly) solvable ones.

## 2.1 Splitting in the Yee-Scheme

We split the matrix $\mathbf{A}$ into the form $\mathbf{A} = \mathbf{A}_{1Y} + \mathbf{A}_{2Y}$, where the splitting is done according to the magnetic and electric fields, respectively. $\mathbf{A}_{1Y}$ is composed from the matrix $\mathbf{A}$ changing the rows belonging to the electric field variables to zero rows, and $\mathbf{A}_{2Y}$ can be derived, in similar manner, zeroing the rows belonging to the magnetic field variables (see [4]). From the operator splitting point of view the Yee-method is based on this splitting. The exponential of the matrices $\mathbf{A}_{1Y}$ and $\mathbf{A}_{2Y}$ can be computed easily.

**Lemma 1.** *We have the relations* $\exp(\mathbf{A}_{1Y}) = \mathbf{I} + \mathbf{A}_{1Y}$ *and* $\exp(\mathbf{A}_{2Y}) = \mathbf{I} + \mathbf{A}_{2Y}$, *where* $\mathbf{I}$ *is the identity matrix.*

*Proof.* It is enough to notice that, because of their structure, the powers of the matrices are equal to the zero matrix. Thus their exponential can be computed with the first two terms of the series of the exponential function. □

Applying the S-splitting method, the sub-systems (5) and (6) can be solved exactly because of Lemma 1. Thus the approximate solution can be determined by the iteration scheme

$$\Psi^{k+1} = \exp(\tau\mathbf{A}_{2Y})\exp(\tau\mathbf{A}_{1Y})\Psi^k = (\mathbf{I} + \tau\mathbf{A}_{2Y})(\mathbf{I} + \tau\mathbf{A}_{1Y})\Psi^k, \tag{10}$$

where $\Psi^0 = \Psi(0)$ is given and $\Psi^k$ stands for the approximation of $\Psi$ at time-level $k\tau$. The time-integration error comes purely from the operator splitting error. The method is reasonably fast in the calculation of one iteration step, because the matrix exponential is approximated in explicit manner in (10). The drawback of the method is its conditional stability, $\tau$ has to be very small. So the Yee-method is not efficient in calculations on a relatively long time-interval, which can not be improved by the application of the Strang or the fourth order splitting either.

## 2.2 Splitting in the KFR-Scheme

In the KFR-method the matrix $\mathbf{A}$ is split into the sum of skew-symmetric matrices, for which additionally the matrix exponential can be computed exactly using the identity

$$\exp\left(\begin{bmatrix} 0 & \alpha \\ -\alpha & 0 \end{bmatrix}\right) = \begin{bmatrix} \cos\alpha & \sin\alpha \\ -\sin\alpha & \cos\alpha \end{bmatrix}, \tag{11}$$

where $\alpha$ is an arbitrary constant. In order to apply the above equality, we have to split $\mathbf{A}$ into two, four and six parts, in one, two and three dimension (see [5]), respectively. Thus we have the splitting $\mathbf{A} = \mathbf{A}_{1K} + \cdots + \mathbf{A}_{pK}$ ($p = 2, 4$ or $6$). Similar to the method discussed in the previous section we can apply the S-, the Strang- or the fourth order splitting for system (4). So we obtain systems of ordinary differential equations with the coefficient matrices $\mathbf{A}_{1K}, \ldots, \mathbf{A}_{pK}$. Because the exponentials of these matrices can be computed with sinus and cosinus functions, we need not solve the sub-systems numerically. This was the case in the Yee-method too. Thus, for example, for the S-splitting we have the iteration

$$\Psi^{k+1} = \exp(\tau \mathbf{A}_{pK}) \cdot \cdots \cdot \exp(\tau \mathbf{A}_{1K}) \Psi^k. \tag{12}$$

Considering the fact that the exponential of a skew-symmetric matrix always has unit 2-norm (namely it is orthogonal), the KFR-method is unconditionally stable by construction. Because of the exact computation of the exponentials, the time-integration error comes only from the operator splitting.

## 2.3 Splitting in the NZCZ-Scheme

The first terms in system (1) can be written in the form of the difference of two spatial derivatives. According to these two derivatives we can split the matrix $\mathbf{A}$ into the form $\mathbf{A} = \mathbf{A}_{1N} + \mathbf{A}_{2N}$, where $\mathbf{A}_{1N}$ and $\mathbf{A}_{2N}$ are skew-symmetric matrices (see [4,15]). This splitting is applied in the NZCZ-method. Applying the Strang-splitting with the matrices $\mathbf{A}_{1N}$, $\mathbf{A}_{2N}$ we have to solve three systems of ordinary differential equations (see expression (8)) with the coefficient matrices $(\tau/2)\mathbf{A}_{2N}$, $\tau \mathbf{A}_{1N}$ and $(\tau/2)\mathbf{A}_{2N}$, respectively. These systems cannot be solved exactly so we must apply numerical methods. To do this we split the second system employing $(\tau/2)\mathbf{A}_{1N}$ two times. Solving the systems by the explicit, implicit, explicit and implicit Euler-methods, respectively, we obtain the iteration $\Psi^{k+1} = U(\Delta t \mathbf{A})\Psi^k$, where

$$U(\Delta t \mathbf{A}) = \tag{13}$$

$$= (\mathbf{I} - (\Delta t/2)\mathbf{A}_{2N})^{-1} \cdot (\mathbf{I} + (\Delta t/2)\mathbf{A}_{1N}) \cdot (\mathbf{I} - (\Delta t/2)\mathbf{A}_{1N})^{-1} \cdot (\mathbf{I} + (\Delta t/2)\mathbf{A}_{2N}).$$

(Here $\Delta t$ is the time-step of the numerical methods. This is set to be $\Delta t = \tau$.) In practice, the above iteration can be simplified to the solution of two systems of linear equations with symmetric tridiagonal matrices in each iteration step. In the following, a pure mathematical proof of the stability of the NZCZ-method (with non-homogeneous material parameters) will be given. We remark that a similar proof appeared in [2] after the submission of this paper.

**Theorem 2.** *Let* $h = \min\{\Delta x, \Delta y, \Delta z\}$ *and let* $q = c\Delta t/h$ *be an arbitrary fixed positive number. Using staggered spatial discretization and the NZCZ time-integration method, the numerical solution of the Maxwell equations is unconditionally stable in 2-norm.*

*Proof.* The unconditional stability means that for all arbitrary $q = c\Delta t/h$ the relation

$$\| \Psi^k \|_2 \leq K \| \Psi^0 \|_2 \tag{14}$$

is true for all $k \in \mathbb{N}$ with a constant $K$ (independent of $k$). Since

$$\| (\mathbf{I} + s\mathbf{C}) \cdot (\mathbf{I} - s\mathbf{C})^{-1} \|_2 = 1 \tag{15}$$

is valid for any skew-symmetric matrix $\mathbf{C}$ and scalar $s$ (see [4]), (13) implies the relation

$$\| \Psi^k \|_2^2 = \| (U(\Delta t\mathbf{A}))^k \Psi^0 \|_2^2 \leq \tag{16}$$

$$\leq \| (\mathbf{I} - (\Delta t/2)\mathbf{A}_{2N})^{-1} \|_2^2 \cdot \| (\mathbf{I} + (\Delta t/2)\mathbf{A}_{2N})) \|_2^2 \cdot \| \Psi^0 \|_2^2 .$$

Because of the skew-symmetricity of $\mathbf{A}_{2N}$, its eigenvalues can be written in the form $\pm \mathrm{i}\lambda_l$ ($l = 1, \ldots, 3N$, $\lambda_l \geq 0$, $\mathrm{i} = \sqrt{-1}$). Applying this, we have the estimations

$$\| (\mathbf{I} - (\Delta t/2)\mathbf{A}_{2N})^{-1} \|_2^2 = \varrho((\mathbf{I} + (\Delta t/2)\mathbf{A}_{2N})^{-1}(\mathbf{I} - (\Delta t/2)\mathbf{A}_{2N})^{-1}) =$$

$$= \varrho((\mathbf{I} - (\Delta t/2)^2\mathbf{A}_{2N}^2)^{-1}) = \frac{1}{\min_l\{|1 - (\Delta t/2)^2(\pm \mathrm{i}\lambda_l)^2|\}} =$$

$$= \frac{1}{1 + (\Delta t/2)^2\lambda_{min}^2} \leq 1, \tag{17}$$

$$\| \mathbf{I} + (\Delta t/2)\mathbf{A}_{2N} \|_2^2 = \varrho((\mathbf{I} - (\Delta t/2)\mathbf{A}_{2N})(\mathbf{I} + (\Delta t/2)\mathbf{A}_{2N})) =$$

$$= \varrho(\mathbf{I} - (\Delta t/2)^2\mathbf{A}_{2N}^2) = 1 + (\Delta t/2)^2\lambda_{max}^2 \leq 1 + \left(\frac{c\Delta t}{h}\right)^2 = 1 + q^2, \tag{18}$$

where

$$\lambda_{max} = \max\{\lambda_1, \ldots, \lambda_{3N}\}, \quad \lambda_{min} = \min\{\lambda_1, \ldots, \lambda_{3N}\} \tag{19}$$

and $\varrho(.)$ denotes the spectral radius. Furthermore, the Gerschgoren-theorem and the $1/(\sqrt{\varepsilon_{.,.,.}\mu_{.,.,.}}\Delta.)$ form of the elements of $\mathbf{A}_{2N}$ are applied to get an upper bound for $\lambda_{max}$. In the end we get that $\| \Psi^k \|_2^2 \leq (1 + q^2) \| \Psi^0 \|_2^2$, that is the choice $K = \sqrt{1 + q^2}$ is satisfactory. □

*Remark 3.* For 3D problems the constant $q$ must be chosen according to the inequality $q < 1/\sqrt{3}$ (here $h = \Delta x = \Delta y = \Delta z$) in the classical Yee-method. In the NZCZ-method the parameter $q$ can be set arbitrarily, which shows the unconditional stability of the method.

## 3 Comparison of the methods

The NZCZ- and KFR-methods are unconditionally stable methods, which allows us arbitrary $\Delta t$ choice. Of course, enlarging $\Delta t$, the splitting error and the error of the explicit and implicit Euler methods will increase. The required accuracy results in a practical upper bound for $\Delta t$. We compare the methods from the point of view of the computational speed, keeping the accuracy of the methods acceptable.

For the sake of brevity we consider the 1D Maxwell equations with the exact solution

$$E_z(t,x) = \sin(\pi x)\sin(\pi t), \quad H_y(t,x) = -\cos(\pi x)\cos(\pi t). \tag{20}$$

The numerical results with the Yee-, NZCZ- and KFR-methods are listed in Table 1 ($x \in [0,1]$, $\Delta x = 1/500$). We denote the KFR-method with fourth order splitting by KFR4 and the sign " $-$ " means that the error of the method is not acceptable (larger than 0.1). The error is given in $l_2$-norm (at time level 0.8) and the CPU-time in seconds, respectively. We can notice that

| $\Delta t$ | 0.8 | 0.08 | 0.008 | 0.004 | 0.002 | 0.001 | 0.0005 | 0.00005 | 0.000005 |
|---|---|---|---|---|---|---|---|---|---|
| YEE | - | - | - | - | $2.0510 \times 10^{-9}$ | 2.2865 | 2.3454 | 2.3648 | 2.3648 |
| $\times 10^{-6}$ | | | | | 0.36 | 0.70 | 1.42 | 14.14 | 148.48 |
| NZCZ | - | 7463.5 | 78.034 | 21.285 | 7.0949 | 3.5475 | 2.6606 | 2.3680 | 2.3652 |
| $\times 10^{-6}$ | | 0.04 | 0.43 | 0.88 | 1.83 | 3.67 | 7.34 | 72.15 | 735.51 |
| KFR4 | - | - | - | - | 54022 | 3792.2 | 244.40 | 2.3893 | 2.3415 |
| $\times 10^{-6}$ | | | | | 0.62 | 1.22 | 2.49 | 25.31 | 247.87 |

**Table 1.** *Computational results with the exact solutions in (20).*

the KFR-method with fourth order splitting behaves much poorer than the Yee-method, although, the Yee-method applies only the S-splitting. While we get a very accurate solution with the Yee-method in 0.36 seconds (see [9] regarding the magic time-step), the KFR4-method obtains an unusable solution in the same computational time. What is the reason of this? We can compare easily the Yee-method and the KFR1-method (KFR with S-splitting) applying them for the above 1D problem. The two methods have similar structure. They split the original system (4) into two sub-systems which are solved exactly. Let us calculate the norm of the leading term in expression (7). We get the results

$$\| \frac{\tau^2}{2}[\mathbf{A}_{1Y}, \mathbf{A}_{2Y}]\Psi^0 \|_2 = 1.25 \times 10^{-3}, \quad \| \frac{\tau^2}{2}[\mathbf{A}_{1K}, \mathbf{A}_{2K}]\Psi^0 \|_2 = 0.73. \tag{21}$$

This difference (almost three order) makes the KFR-method relatively inaccurate. The difference can be explained by the presence of the initial vector in the error expression and with the rather unnatural splitting in the KFR-method. Although, the KFR-method is unconditionally stable, to keep the accuracy acceptable, very small step-size must be chosen. This step-size must be much smaller than the maximal step-size of the Yee-method. Thus from the computational fastness point of view the KFR-method is not better than the Yee-method.

The NZCZ-method is slower with a factor about five than the Yee-method. This drawback can be compensated by the increase of the time-step. So, in the long run, the NZCZ-method solves the equations faster than the Yee-method ([1,4,15]). According to Table 1, the NZCZ-method is the computationally most efficient one.

## References


1. C.C.-P. Chen, Tae-Woo Lee, N. Murugesan, S.C. Hagness, *Generalized FDTD-ADI: An Unconditionally Stable Full-Wave Maxwell's Equations Solver for VLSI Interconnect Modeling Computer Aided Design*, ICCAD-2000. IEEE/ACM International Conference, (2000), pp. 156 -163.
2. M. Darms, R. Schuhmann, H. Spachmann, T. Weiland, *Dispersion and asymmetry effects of ADI-FDTD*, IEEE-MWCL 12, 12 (2002), pp. 491-493.
3. Á. Havasi, J. Bartholy, I. Faragó, *Splitting method and its application in air pollution modelling*, Időjárás, Hungarian Meteorological Service, 2001.
4. R. Horváth, I. Faragó, W.H.A. Schilders, *Investigation of Numerical Time-Integrations of the Maxwell Equations Using the Staggered Grid Spatial Discretization*, Technische Universiteit Eindhoven, RANA-report 02-15, 2002.
5. J. S. Kole, M. T. Figge, H. De Raedt, *Unconditionally Stable Algorithms to Solve the Time-Dependent Maxwell Equations*, Phys. Rev. E 64, 066705, 2001.
6. D. Lanser, J.G. Verwer, *Analysis of Operator Splitting for Advection-Diffusion-Reaction Problems in Air Pollution Modelling*, J. Comput. Appl. Math. 111, No. 1-2 (1999), pp. 201-216.
7. G.I. Marchuk, *Methods of Splitting*, Nauka, Moscow, 1988.
8. T. Namiki, *A New FDTD Algorithm Based on Alternating-Direction Implicit Method*, IEEE Transactions on Microwave Theory and Techniques, Vol. 47, No. 10, Oct. 1999, pp. 2003-2007.
9. A. Taflove, *Computational Electrodynamics: The Finite-Difference Time-Domain Method*, 2 ed., Artech House, Boston, MA, 2000.
10. V. S. Varadarajan, *Lie Groups, Lie Algebras and Their Representations*, Prentice-Hall Inc., Englewood Cliffs, New Jersey 1974.
11. T. Weiland, *A Discretization Method for the Solution of Maxwell's Equations for Six-Component Fields*, Electronics and Communication (AEÜ), Vol. 31 (1977), pp. 116.
12. T. Weiland, *Time Domain Electromagnetic Field Computation with Finite Difference Methods*, International Journal of Numerical Modelling, Vol. 9 (1996), pp. 295-319.
13. K. S. Yee, *Numerical Solution of Initial Boundary Value Problems Involving Maxwell's Equations in Isotropic Media*, IEEE Transactions on Antennas and Propagation, vol. 14, no. 3, 302-307, March 1966.

14. H. Yoshida, *Construction of Higher Order Symplectic Integrators*, Phys. Lett. A, 150 (1990), pp. 262-268.
15. F. Zheng, Z. Chen, J. Zhang, *Toward the Development of a Three-Dimensional Unconditionally Stable Finite-Difference Time-Domain Method*, IEEE Trans. Microwave Theory and Techniques, Vol. 48, No. 9, September 2000, pp. 1550-1558.

# Simulating Multi-tone Free-Running Oscillators with Optimal Sweep Following

S.H.M.J. Houben

Eindhoven University of Technology

**Abstract.** A new method for the simulation of circuits with widely-varying time scales is given. The method makes a splitting of the behaviour of the circuit in a fast-varying and a slowly-varying component. The method is attractive because it can handle frequency modulated (FM) circuits, unlike existing methods. Numerical results are given.

## 1 Introduction

With ever increasing operating frequencies of electric circuits, widely varying time scales become increasingly common. The fast frequency $f_f$ can be up to $10^3$ times the slow frequency $f_s$. A typical example would be a circuit which generates a high-frequency carrier wave, which is modulated with a low-frequency data signal. In such a situation, the period and/or amplitude changes of the high-frequency signal changes slowly compared to the high-frequency oscillation itself.Traditional simulation techniques have great difficulty with such widely separate time scales, since they require an amount of computation time $O(f_f/f_s)$. In order to simulate a complete waveform of the low-frequency signal, such techniques have to simulate many waveforms of the high-frequency signals.

Several approaches have been suggested to speed up the simulation of such a circuit with widely varying time scales. They can be split into two groups.

1. Methods that are based on partitioning a circuit into a fast and a slow part, such as those described in [1]. These methods can be very effective when a good partitioning can be made, but this is not always possible.
2. Methods that simulate the circuit as a whole, but attempt to split the low-frequency behaviour from the high-frequency behaviour. Examples of such methods are envelope-following techniques [4,5], and the Multi-rate Partial Differential Equation (MPDE) technique [2,8].

In this paper, a new method of type 2 is discussed. The method is named Optimal Sweep Following New in this method is that it handles autonomous circuits in which Frequency Modulation (FM) takes place. MPDE as described in [8] and Envelope Following only handle Amplitude Modulated (AM) circuits. Modifications to MPDE to handle FM modulation have been proposed; see [7].

The Sweep Following method splits the behaviour of the system in a slowly-varying and a fast-varying component. The splitting is optimal in its class.

The essential new feature of the method is that the frequency of the fast-varying signal does not have to be provided a priori; rather, it is determined by the method itself. As a result, the method can detect and handle changes in the period of the fast-varying signal. This makes it suitable for simulating Frequency Modulated circuits. Other methods require the period $f_f$ to be provided a priori, and cannot deal with the situation where $f_f$ changes over time. The fact that our method computes $f_f$ locally is also an advantage in that this information is often desired by the circuit designer.

This paper is built up as follows. In section 2, we explain the notion of a sweep and show how a sweep can efficiently be represented. In section 3, we outline the Sweep Following method. Finally, in section 4 we present a test problem and show that Sweep Following solves it. The test problem is an FM-modulated problem; it cannot be handled by MPDE or Envelope Following.

## 2 Efficient representation of solutions with widely-varying time-scales

Consider the following formulation of a Differential-Algebraic Equation (DAE).

$$\frac{d}{dt}\mathbf{q}(\mathbf{x}) + \mathbf{j}(\mathbf{x}) = \mathbf{s}(t). \tag{1}$$

The following definition is of importance throughout this paper.

**Definition 1.** A *sweep* of a set $\mathcal{C}_0 \subset \mathbb{R}^N$ by (1) is the set $\{\xi(\mathbf{x}_0, t) \,|\, \mathbf{x}_0 \in \mathcal{C}_0, t \geq 0\}$, where $\xi(\mathbf{x}_0, t)$ is defined as the solution of (1) with initial condition $\mathbf{x}(0) - \mathbf{x}_0$ at timc $t$.

The solution $\mathbf{x}$ of (1) may change at two very different time scales, as is the case in Fig. 1. Thus we have a fast oscillation, which itself slowly changes due to the low-frequency signal. This gives rise to "telephone cord" solutions as shown in Fig. 2. We call these "telephone cord" solutions because they are tightly wound in a spiral, just like the cord of a telephone. The key idea of the Sweep Following method is not to simulate the "telephone cord" itself, but rather the following two entities.

1. The "tube" that is spanned by the cord.
2. The speed of the rotation of the cord around the tube.

It turns out that it is often cheaper to represent tube and speed rather than to directly represent the solution ("telephone cord") itself.

We now give a rigorous definition of the "tube" that represents the solution 2. Suppose we have a set $\mathcal{C}_0$ of initial values to (1), parametrised by a

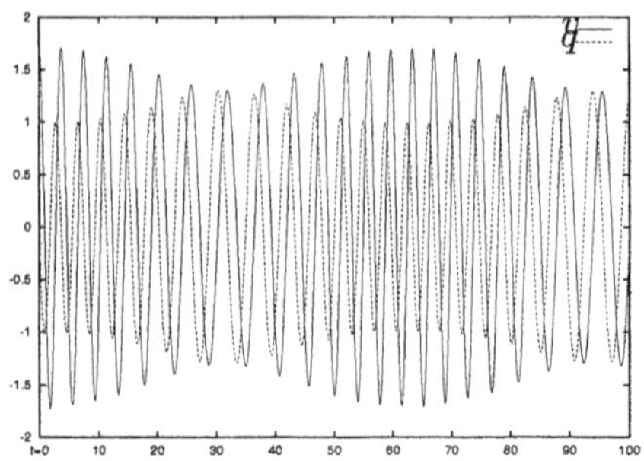

**Fig. 1.** Transient solution of a modulated LC-oscillator

parameter $s \in [0, 2\pi]$, i.e. $\mathcal{C}_0 = \{\widehat{\mathbf{x}}_0(s) \mid 0 \leq s \leq 2\pi\}$. Moreover, we require that the parametrisation $\widehat{\mathbf{x}}_0$ is continuous, differentiable and periodic with period $2\pi$. In addition, the right-hand side $\mathbf{s}$ is also a continuous, differentiable function of $s$ and $t$.

For a given $s \in [0, 2\pi]$, consider the solution $\widehat{\mathbf{x}}(s,t)$ to (1) with initial condition $\widehat{\mathbf{x}}_0(s)$, i.e. the solution of the problem

$$\frac{d}{dt}\mathbf{q}(\widehat{\mathbf{x}}(s,t)) + \mathbf{j}(\widehat{\mathbf{x}}(s,t)) = \mathbf{s}(s,t), \tag{2a}$$

$$\widehat{\mathbf{x}}(s,0) = \widehat{\mathbf{x}}_0(s). \tag{2b}$$

The function $\widehat{\mathbf{x}}$ depends on two variables $s$ and $t$ and parametrises a surface consisting of all solutions of (1) that have initial values in $\mathcal{C}_0$. This surface is the "tube" discussed earlier. Observe that this tube is exactly the sweep of $\mathcal{C}_0$ by (1). Since our method follows this sweep, rather than a single solution, it is called "Sweep Following". The parametrisation $\widehat{\mathbf{x}}$ of the sweep induces a coordinate system $(s,t)$ on the sweep. However, this coordinate system may be very skew. This implies that information is not efficiently represented in this coordinate system. Therefore, we investigate coordinate transformations of the form

$$u = s + \alpha(t), \quad v = t, \tag{3}$$

for some differentiable function $\alpha$. The parametrisation in $(u,v)$ coordinates is called $\widehat{\mathbf{x}}_\alpha$. With this definition we have

$$\widehat{\mathbf{x}}_\alpha(u,v) = \widehat{\mathbf{x}}_\alpha(s+\alpha(t),t) := \widehat{\mathbf{x}}(s,t), \tag{4}$$

$$\text{and} \quad \mathbf{s}_\alpha(u,v) = \mathbf{s}_\alpha(s+\alpha(t),t) := \mathbf{s}(s,t). \tag{5}$$

From (4) and (2), we can derive

$$\frac{d}{dt}\mathbf{q}(\widehat{\mathbf{x}}_\alpha(u,v)) + \mathbf{j}(\widehat{\mathbf{x}}_\alpha(u,v)) = \mathbf{s}_\alpha(u,v), \tag{6a}$$

$$\widehat{\mathbf{x}}_\alpha(u,0) = \widehat{\mathbf{x}}_0(s). \tag{6b}$$

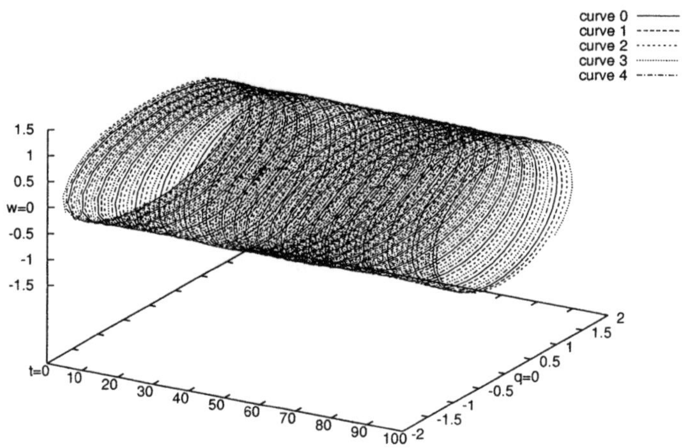


**Fig. 2.** Solutions of the modulated LC-oscillator in $(v, q, t)$-space for various initial conditions.

By using (3), we can rewrite (6) into

$$\alpha'(v)\frac{\partial}{\partial u}\mathbf{q}(\widehat{\mathbf{x}}_\alpha(u,v)) + \frac{\partial}{\partial v}\mathbf{q}(\widehat{\mathbf{x}}_\alpha(u,v)) + \mathbf{j}(\widehat{\mathbf{x}}_\alpha(u,v)) = \mathbf{s}_\alpha(u,v), \tag{7a}$$

$$\widehat{\mathbf{x}}_\alpha(u,0) = \widehat{\mathbf{x}}_0(u - \alpha(0)). \tag{7b}$$

The following observations are important to make.

- The resulting formulation is a partial differential equation (PDE) in $u$ and $v$. Obviously, (7) reduces to an ordinary DAE along lines of constant $s$; that is, lines of the form $u - \alpha(v) =$ constant. Lines of this form are the characteristics of (7). See also [6].
- The formulation is not unique: it depends on a certain choice for $\alpha$.
- From the way that (7) has been derived, it follows that it is well posed if and only if (2) is well-posed.

The characteristics are all of the form

$$u = \alpha(v) + c, \quad \text{for some } c \in \mathbb{R}. \tag{8}$$

This means that characteristics can never intersect each other, and therefore no shocks in the solution can occur.

## 3 Outline of the method

The basic idea of the method is now as follows. We start at some initial sweep $\mathcal{C}_0$. We select a number of points $\widehat{\mathbf{x}}_\alpha(u_i, 0)$ on $\mathcal{C}_0$, and use (7) as an evolution equation in $v$ to find $\widehat{\mathbf{x}}_\alpha(u_i, v_j)$ at later points $\{v_j\}$. Finally, we use the relation (4) to find the solution of the original problem (1).

At first, it may appear that solving (7) is much more expensive than solving (1). However, for a specific class of problems and for a suitable choice of $\alpha$, the solution of (7) changes much slower than the solution of (1). In this case, solving (7) allows much larger step sizes to be taken than if we solve (1). This is the motivating idea behind the method.

This raises of course the question how $\alpha$ should be selected. We want to choose $\alpha$ in such a way that the step-size $\Delta v$ of the time stepping in the $v$ direction can be taken as large as possible, without compromising accuracy. In essence, we want the factor $\frac{\partial}{\partial v}\mathbf{q}(\widehat{\mathbf{x}}_\alpha(u,v))$ from (7) to be small; the smaller this factor is, the larger step-sizes $\Delta v$ can be taken. Using (7) we find

$$\int_0^{2\pi} \|\frac{\partial}{\partial v}\mathbf{q}(\widehat{\mathbf{x}}_\alpha(u,v))\|^2\, du = \int_0^{2\pi} \|\mathbf{s}_\alpha(u,v) - \mathbf{j}(\widehat{\mathbf{x}}_\alpha(u,v)) - \alpha'(v)\frac{\partial}{\partial u}\mathbf{q}(\widehat{\mathbf{x}}_\alpha(u,v))\|^2\, du. \tag{9}$$

We want to look for an $\alpha'(v)$ for which (9) becomes minimal. Basic calculus shows that this happens when

$$\alpha'(v) = \frac{\int_0^{2\pi} \left(\mathbf{s}_\alpha(u,v) - \mathbf{j}(\widehat{\mathbf{x}}_\alpha(u,v)), \frac{\partial}{\partial u}\mathbf{q}(\widehat{\mathbf{x}}_\alpha(u,v))\right) du}{\int_0^{2\pi} \|\frac{\partial}{\partial u}\mathbf{q}(\widehat{\mathbf{x}}_\alpha(u,v))\|^2\, du}. \tag{10}$$

From (7a), we find that

$$\mathbf{s}_\alpha(u,v) - \mathbf{j}(\widehat{\mathbf{x}}_\alpha(u,v)) = \alpha'(v)\frac{\partial}{\partial u}\mathbf{q}(\widehat{\mathbf{x}}_\alpha(u,v)) + \frac{\partial}{\partial v}\mathbf{q}(\widehat{\mathbf{x}}_\alpha(u,v)). \tag{11}$$

With that, we find that

$$\begin{aligned}\alpha'(v) &= \frac{\int_0^{2\pi} \left(\alpha'(v)\frac{\partial}{\partial u}\mathbf{q}(\widehat{\mathbf{x}}_\alpha(u,v)) + \frac{\partial}{\partial v}\mathbf{q}(\widehat{\mathbf{x}}_\alpha(u,v)), \frac{\partial}{\partial u}\mathbf{q}(\widehat{\mathbf{x}}_\alpha(u,v))\right) du}{\int_0^{2\pi} \|\frac{\partial}{\partial u}\mathbf{q}(\widehat{\mathbf{x}}_\alpha(u,v))\|^2\, du} \\ &= \alpha'(v) + \frac{\int_0^{2\pi} \left(\frac{\partial}{\partial v}\mathbf{q}(\widehat{\mathbf{x}}_\alpha(u,v)), \frac{\partial}{\partial u}\mathbf{q}(\widehat{\mathbf{x}}_\alpha(u,v))\right) du}{\int_0^{2\pi} \|\frac{\partial}{\partial u}\mathbf{q}(\widehat{\mathbf{x}}_\alpha(u,v))\|^2\, du}.\end{aligned} \tag{12}$$

So we see that for this choice of $\alpha'$, we obtain

$$\int_0^{2\pi} \left(\frac{\partial}{\partial v}\mathbf{q}(\widehat{\mathbf{x}}_\alpha(u,v)), \frac{\partial}{\partial u}\mathbf{q}(\widehat{\mathbf{x}}_\alpha(u,v))\right) du = 0. \tag{13}$$

I.e. there is an orthogonality between $\frac{\partial}{\partial v}\mathbf{q}$ and $\frac{\partial}{\partial u}\mathbf{q}$. Equation (10) gives us a differential equation for $\alpha$. The system formed by (7) and (10) is a complete evolution equation for $\widehat{\mathbf{x}}_\alpha$ and $\alpha$.

In order to compute a solution of the system (7), (10), we have to discretise the system. We treat $u$ as a space parameter and $v$ as the time parameter. Then we use the method of lines, i.e. we first discretise in the space direction $u$. The resulting semi-discrete system can then be solved in the $v$ direction by a choice of ODE methods.

For the discretisation in the $u$ direction, a number of points $\{u_1, \dots, u_N\}$ is selected. The operator $\frac{\partial}{\partial u}$ can be discretised by a simple symmetric difference scheme. The semi-discretised system now becomes

$$\frac{\partial}{\partial v}\mathbf{q}(\widehat{\mathbf{x}}_\alpha(u_i, v)) = \mathbf{s}_\alpha(u_i, v) - \mathbf{j}(\widehat{\mathbf{x}}_\alpha(u_i, v)) - \alpha'(v)\mathbf{d}_i(v) \text{ for } i = 1, \dots, N, \tag{14a}$$

$$\alpha'(v) = \frac{\sum_{i=1}^N (\mathbf{d}_i(v), \mathbf{s}_\alpha(u_i, v) - \mathbf{j}(\widehat{\mathbf{x}}_\alpha(u_i, v)))}{\sum_{i=1}^N ||\mathbf{d}_i(v)||^2}. \tag{14b}$$

where the $\mathbf{d}_i$ are defined as

$$\mathbf{d}_i(v) := \frac{\mathbf{q}(\widehat{\mathbf{x}}_\alpha(u_{i+1}, v)) - \mathbf{q}(\widehat{\mathbf{x}}_\alpha(u_{i-1}, v))}{u_{i+1} - u_{i-1}}. \tag{15}$$

Unfortunately, the computation of $\alpha'(v)$ is somewhat expensive, and the computation of $d\alpha'(v)/dv$ is even more expensive. Recall that this choice of $\alpha(v)$ was only taken because this leads to a coordinate system that is optimal in some sense. However, if this particular choice is too expensive, it makes sense to take a slightly sub-optimal but much cheaper alternative. This alternative is to keep $\alpha'$ constant between two successive time steps $v_n$ and $v_{n+1}$, as follows.

$$\alpha'(v) = \frac{\sum_{i=1}^N (\mathbf{d}_i(v_n), \mathbf{s}_\alpha(u_i, v_n) - \mathbf{j}(\widehat{\mathbf{x}}_\alpha(u_i, v_n)))}{\sum_{i=1}^N ||\mathbf{d}_i(v_n)||^2}, \text{for } v \in [v_n, v_{n+1}). \tag{16}$$

## 4 Test problem

A simple test problem has been constructed, which captures the essence of the problems we want to solve. We take a simple modulated LC oscillator. The capacitor is controlled by some external source.

$$C\dot{w} = \Theta q, \tag{17a}$$

$$L\dot{q} = -w, \tag{17b}$$

$$\Theta = \Theta_{\text{avg}} + \Theta_{\text{amp}} \cos \omega_\Theta t. \tag{17c}$$

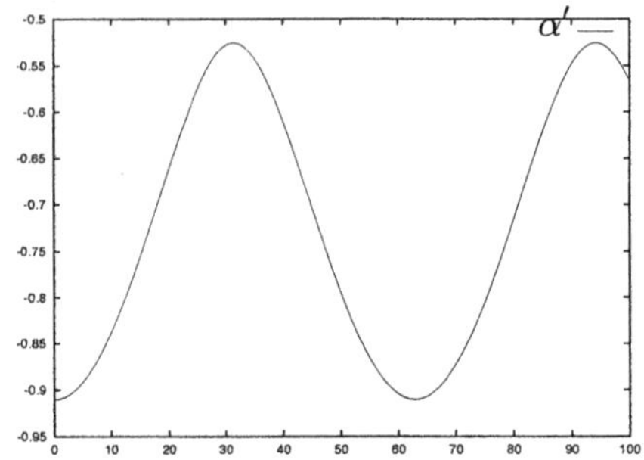


**Fig. 3.** The $\alpha'$ computed for (17)

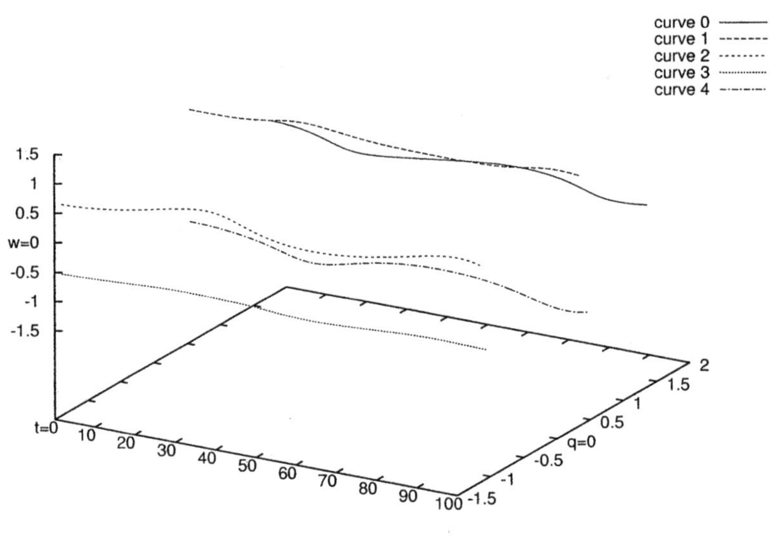


**Fig. 4.** Solution of (17) with sweep following.

The choice of parameters in the simulations is $L = C = 1$, $\Theta_{\text{avg}} = 2$, $\Theta_{\text{amp}} = 1$ and $\omega_\Theta = \frac{1}{10}$. A transient simulation of (17) is shown in Fig. 1. We find that if we take $\Theta$ to be constant $\Theta_{\text{avg}} + \Theta_{\text{amp}}$ in problem (17), and initial values $w_0 = 1$ and $q_0 = 0$, we obtain a solution where $(w, q)$ describe an ellipse around $\mathbf{0}$ with radii 1 and $\sqrt{3}$. This ellipse will be taken as our initial set $\mathcal{C}_0$. If we take a few points on this ellipse, and compute the solution for each of these initial values, we obtain Fig. 2. This figure gives a good idea of the sweep that is tracked. Note that the computed solutions all vary very rapidly.

We now apply the Sweep Following idea, i.e. we compute $\alpha'$. The computed curve is shown in Fig. 3. This $\alpha'$ is then used to compute a solution $\widehat{\mathbf{x}}_\alpha$, which is shown in Fig. 4. Note the smooth, flat behaviour of $\widehat{\mathbf{x}}_\alpha$ as a function of $v$. Since the computed curves vary much slower in Fig. 4 than in Fig. 2, we see that it is now possible to take much larger step-sizes in the $v$ direction. This in turn leads to a considerable gain in efficiency.

So far, the method has not yet been tested on realistic circuits. Therefore, two questions may arise.

1. Has the system (7), (10) always a solution?
2. Will there be a speed-up compared to traditional transient analysis?

In [3], we show that the answer to the first question is affirmative, provided that the underlying DAE (1) has solutions for all initial values in $\mathcal{C}$. The second questions remains open.

## References

1. A. Bartel. Multirate ROW methods of mixed type for circuit simulation. In *Scientific Computing in Electrical Engineering*. Springer, 2000.
2. H.G. Brachtendorf, G. Welsch, R. Laur, and A. Bunse-Gerstner. Numerical steady state analysis of electronic circuits driven by multi-tone signals. *Electronic Engineering*, 79(2):103–112, 1996.
3. S.H.M.J. Houben. *Circuits in motion – the Simulation of Electrical Oscillators*. Ph.D. Thesis, Eindhoven University of Technology, 2003.
4. K. Kundert, J. White, and A. Sangiovanni-Vincentelli. An envelope-following method for efficient transient simulation of switching power and filter circuits. In *Proc. IEEE Int. Conf. Computer-Aided Design*, pages 446–449, 1988.
5. Kartikeya Mayaram, David C. Lee, Shahriar Moinian, David A. Rich, and Jaijeet Roychowdhury. Computer-aided circuit analysis tools for RFIC simulation: Algorithms, features and limitations. *IEEE Transactions on Circuit and Systems – II: Analog and Digital Signal Processing*, 47(4):274–286, 2000.
6. R. Pulch and M. Günther. A method of characteristics for solving multirate partial differential equations in radio frequency applications. *Universität Karlsruhe, Institut für Wissenschaftliches Rechnen und Mathematische Modellbildung, preprint Nr. 00/07*, 2000.
7. Roland Pulch. Numerical techniques for solving multirate partial differential algebraic equations. *Presented at Scientific Computing in Electrical Engineering SCEE*, 2002.
8. Jaijeet Roychowdhury. Analysing circuits with widely-separated time scales using numerical PDE methods. *IEEE Transactions in Circuit and Systems – I: Fundamental Theory and Applications*, 48(5):578–594, 2001.

# Fast Extraction of Static Electric Parameters with Accuracy Control

Daniel Ioan, Marius-Cristian Radulescu, Gabriela Ciuprina

"Politehnica" University of Bucharest, Numerical Methods Laboratory Spl. Independentei 313, 77206 Bucharest, ROMANIA, E-mail: lmn@lmn.pub.ro

**Abstract.** The paper presents an efficient numerical method to extract accurate R, L, C parameters of passive on-chip structures. This method is based on two main original ideas. First, the accuracy is controlled by using two complementary approaches based on scalar and vector potential, which provide lower and upper bounds for the extracted parameter. The convergence is accelerated by using the Richardson extrapolation of the average value of the two complementary bounds. Second, the field equations are solved by multigrid finite element method with local adaptive mesh subgriding. The refining process is stopped as soon as the desired accuracy is reached.

## 1 Introduction

An important step in the integrated circuit design verification is the extraction of the equivalent circuit to be simulated with SPICE, starting from the layout description. The present design tools are based on very simple electromagnetic field models (usually, uniform field is supposed). The importance of parasitic effects in high frequency integrated circuit urges the demand of accurate modeling of field effects.

Extraction of R, L, C parameters of on-chip components, p.u.l. values in the case of interconnect transmission lines, is apparently a simple static field problem, based on the Laplace equation satisfied by the scalar potential. However, in real life circuits, millions of different geometric problems should be solved to design an integrated circuit. The main difficulty is related to the computational speed and then to the solution accuracy. These two aspects are closely related, the decrease of the acceptable error requiring a larger CPU time to solve the problem. Therefore, the reliable control of the numerical solution accuracy is essential to be carried out in order to extract the desired parameters with minimum CPU time, and to avoid computation resources wasting.

## 2 Complementary formulation of static field problems

To simplify the presentation only the case of resistance extraction for plane homogeneous conducting layers is considered. The problem can be formulated in terms of scalar potential $V = V(x, y)$, where $\mathbf{E} = -gradV$, and $V$ is the solution of Laplace equation: $\Delta V = 0$, with Dirichlet boundary conditions: $V = 0$ on $S_{D_1}$, $V = V_0$ on $S_{D_2}$, and Neumann boundary conditions: $\frac{\partial V}{\partial n} = 0$ on $S_{N_1}$, where $S_{D_1}$, $S_{D_2}$ are the two terminals having potentials $V = 0$ and $V = V_0$ respectively and $S_{N_1} = \partial D - (S_{D_1} \cup S_{D_2})$.

The problem can also be formulated in terms of the vector potential $\mathbf{T} = \mathbf{k}T(x,y)$, where $\mathbf{J} = curl\mathbf{T}$, and $\mathbf{T}$ is solution of Laplace equation $\Delta T = 0$, with Dirichlet boundary conditions: $T = 0$ on $S_{D_3}$, $T = T_0$ on $S_{D_4}$, and Neumann boundary conditions: $\frac{\partial T}{\partial n} = 0$ on $S_{N_2}$. The parameter $T_0 = I/g$, where $I$ is the total current and $g$ is the thickness of the conductive layer.

The energy functional is in both cases equal to the power losses:

$$W_V = \sigma \int_D (grad\, V)^2 \mathrm{d}v; \tag{1}$$

$$W_T = \rho \int_D (curl\, \mathbf{T})^2 \mathrm{d}v = \rho \int_D (grad\, T)^2 \mathrm{d}v, \tag{2}$$

where $\sigma = 1/\rho$ is the electric conductivity.

According to the variational formulation, the value of the energy functional is minimal for the real field, being bigger for any other field distribution [1]. If $W_V$ and $W_T$ represent the power losses associated to the numerical solution of scalar potential and vector potential problems, the real power loss $W$ satisfies: $W \leq W_T,\ \ W \leq W_V$. It follows that:

$$R_V = \frac{V_0^2}{W_V} \leq \frac{V_0^2}{W} = R_e, \tag{3}$$

$$R_T = \frac{W_T}{I^2} \leq \frac{W}{I^2} = R_e, \tag{4}$$

so the complementary bounds are obtained: $R_V \leq R_e \leq R_T$, where $R_e$ is the exact value of the resistance and $R_V$, $R_T$ are the values of the resistance extracted from scalar potential and vector potential, respectively.

# 3 The convergence acceleration

## 3.1 Averaging the bounds

The two complementary bounds provide a reliable method to control the accuracy of the numerical result, using the relative error:

$$\varepsilon_r = \frac{R_T - R_V}{R_a}. \tag{5}$$

After computing the two bounds, the returned result is $R_a \pm \varepsilon_r$, where $R_a = (R_V + R_T)/2$ is the average value of the two bounds.

The two potentials will be computed using a numerical method for Laplace equations, such as: Finite Difference Method (FDM), Finite Element Method (FEM) or Finite Integration Method (FIT), all based on a rectangular mesh, which discretize the computational domain $D$. The larger is the number of nodes $n$ of the that mesh, the lower is the upper bound and the higher is the lower bound, then the exact value is localized more and more accurately.

Therefore, we purpose the following approach that control the accuracy, uses a coarse mesh to start and recursively refine until the relative error becomes lower than imposed value. This technique is illustrated in fig. 1, where $h$ is the mesh norm.

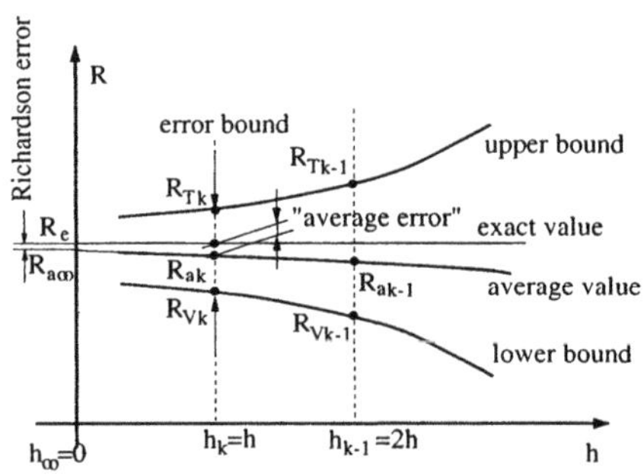


**Fig. 1.** Richardson extrapolation

The first technique we propose to accelerate the convergence of the iterative process is based on the use of the average value $R_a$ instead of any of the bounds $R_T$ or $R_V$.

To evaluate the efficiency of this technique was extracted the resistance of a L shaped plate using FEM with rectangular first order elements. At each step $k$ the number of nodes (DOFs) $n_k$ increases about four times $n_k \approx 4n_{k-1}$. The exact resistance $R_e$ of a such plate can be computed using conformal transforms [2]. The relative errors of the extracted resistance $R_V$, $R_T$ or their average $R_a$ is:

$$\varepsilon_V = \frac{R_V - R_e}{R_e}, \quad \varepsilon_T = \frac{R_T - R_e}{R_e}, \quad \varepsilon_a = \frac{R_a - R_e}{R_e}, \tag{6}$$

are represented versus DOFs number in fig. 2. The acceleration effect of the averaging is obvious.

If the relation between $\varepsilon$ and $n$ in double logarithmic scale (as in fig. 2) is approximated by a line:

$$\lg\varepsilon = \lg\varepsilon_0 - R\lg n, \tag{7}$$

the following relation between error and DOFs number is obtained:

$$\varepsilon = \varepsilon_0 n^{-R}, \tag{8}$$

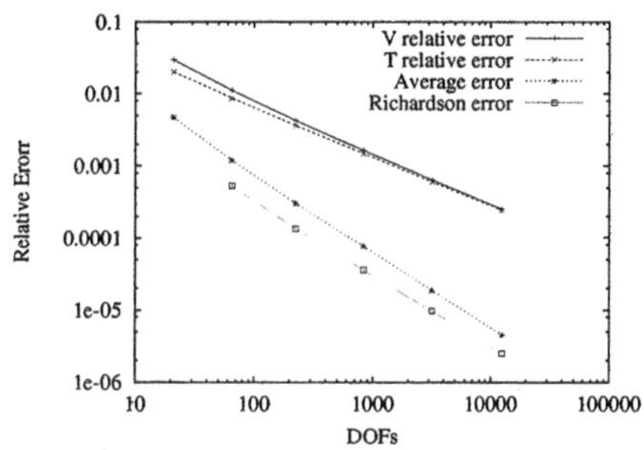


**Fig. 2.** Relative errors versus number of DOFs.

where $\varepsilon_0$ is the "initial error" and $R$ is the convergence rate. The values of relative errors and their convergence rates for several methods are given in table 1.

### 3.2 The Richardson extrapolation

The second method used to accelerate the convergence is based on Richardson extrapolation technique. Based on (8), suppose the following approximation for average resistance at iteration $k$: $R_a^{(k)} = R_e(1 + \varepsilon_0 n_k^{-R})$, and at iteration $k+1$: $R_a^{(k+1)} = R_e(1 + \varepsilon_0 4^{-R} n_k^{-R})$, because $n_{k+1} = 4n_k$. Eliminating $n_k$, the following Richardson extrapolation for exact resistance $R_e$ is obtained:

$$R_\infty = \frac{4^R R_a^{k+1} - R_a^k}{4^R - 1}. \tag{9}$$

In the FEM case, the relative error of the Richardson extrapolation $\varepsilon_\infty = (R_\infty - R_e)/R_e$ is approximatively 2 times lower than average error (table 1).

## 4 Performance evaluation of iterative methods

The method chosen for solving the linear system of equations obtained from discretization method has a very important influence on the algorithm global performances. The linear system obtained from FEM, FDM or FIT is in general of high dimensions ($10^4$ - $10^6$ unknowns), but the matrix is symmetric, positive defined and sparse (with maximum 9 elements per line).

The reference method for solving such systems is the conjugate gradient (CG) method [3]. The main idea of this method is that the of the solution linear system $\mathbf{A}x = \mathbf{b}$, minimizes the functional $f(\mathbf{u}) = \frac{1}{2}\mathbf{u}^T\mathbf{A}u - \mathbf{u}^T\mathbf{b}$. The method is a realization of orthogonal projection on Krylov subspace. The advantage of this method, is that on each iteration $k$ the main computational effort consists of the multiplication between a vector $d_k$ and the matrix $\mathbf{A}$. CG is a semi-iterative method, because on an ideal computer, the solution is obtained after a number of iteration $n = size(\mathbf{A})$,

**Table 1.** Relative error of extracted results, convergence rates and initial errors

| | V Scalar potential $\varepsilon_V$ | | | T Vector potential $\varepsilon_T$ | | |
|---|---|---|---|---|---|---|
| n (DOF) | FIT | FDM | FEM | FIT | FDM | FEM |
| 65 | -0.0136122 | -0.0076849 | -0.0111396 | 0.0041446 | 0.0026230 | 0.0087394 |
| 833 | -0.0023611 | -0.0009849 | -0.0016455 | 0.0009287 | 0.0006715 | 0.0014923 |
| 12545 | -0.0003948 | -0.0001446 | -0.0002533 | 0.0001602 | 0.0001235 | 0.0002443 |
| $R$ | 0.6594671 | **0.7074254** | 0.6899686 | 0.6479846 | **0.6672742** | 0.6243531 |
| $\varepsilon_0$ | -0.0308983 | **-0.0226166** | -0.0295839 | **0.0046511** | 0.018581 | 0.0201327 |
| | Average $\varepsilon_a$ | | | Richardson $\varepsilon_\infty$ | | |
| 65 | -0.0047338 | -0.0025310 | -0.0012001 | -0.0006038 | 0.0013325 | 0.0005354 |
| 833 | -0.0007162 | -0.0001567 | -0.0000766 | -0.0001790 | 0.0000762 | 0.0000358 |
| 12545 | -0.0001173 | -0.0000106 | -0.0000045 | -0.0000329 | 0.0000034 | 0.0000025 |
| $R$ | 0.6671075 | 0.9931547 | **1.04516** | 0.6245892 | **1.146584** | 0.9814212 |
| $\varepsilon_0$ | -0.0131236 | -0.0103792 | - **0.0047256** | -0.0006038 | 0.0013325 | **0.0005354** |

**Table 2.** CPU Time and number of iterations ($R_V/R_T$)

| | No. Iteration | | | CPU Time [s] | | |
|---|---|---|---|---|---|---|
| DOFs | *GS* | *DPGC* | *ICCG* | *GS* | *DPGC* | *ICCG* |
| 65 | 373/53 | 29/19 | 12/8 | 0.00 | 0.01 | 0.00 |
| 833 | 5420/779 | 110/76 | 36/25 | 0.9 | 0.06 | 0.35 |
| 12545 | 68473/10169 | 419/292 | 129/90 | 274.86 | 4.9 | 84.71 |

if $\mathbf{A} = \mathbf{A}^T$ and $\mathbf{A} > 0$. In practice, less iterations (10 - 15) are needed, but this number of iterations depends on the conditioning number of matrix $\mathbf{A}$.

To improve the convergence rate of CG preconditioning techniques can be applied. Among them, the best known are Diagonal Preconditioning (DPCG) and Incomplete Cholesky preconditioning (ICCG).

Unfortunately, complicated preconditioning schemes, such as ICCG, are not efficient in all cases. Even if the iteration number decreases, the total CPU time could increase, due to the computational effort of *Cholesky* factorization, as it can be seen in table 2. Therefore, it is much more useful, to measure the efficiency of an iterative method not in number of iterations, but in CPU time (preferable expressed in *MFlops*, and not in seconds, since the result can be dependent of the computer used).

Fig. 3 shows the variation of the CPU time with respect to the refinement levels of the mesh, which can be approximated by: $\lg(T) = \lg(T_0) + R_T \lg(n)$, or equivalent:

$$T = T_0 n^{R_T} = O(n^{R_T}), \tag{10}$$

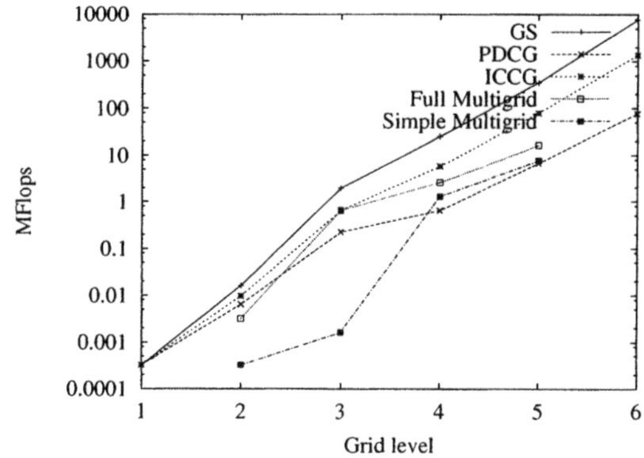


**Fig. 3.** Times for Iterative Methods

relation that gives the complexity order $R_T$ of the studied iterative methods.

From the user's point of view, the relation between the computational time $T$ and the relative error obtained, is more important than the relations between $\varepsilon$ and $n$ (8) or $T$ and $n$ (10).

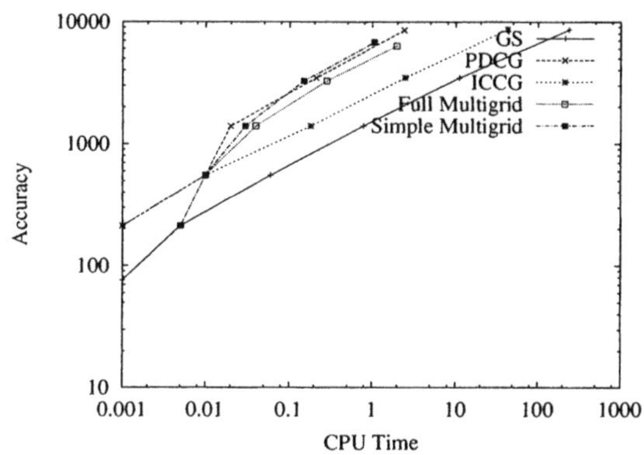


**Fig. 4.** Performance of iterative methods

Eliminating $n$ between (8) and (10) is obtained following relation (represented in fig. 4):

$$\alpha = \alpha_0 T^P, \tag{11}$$

where $\alpha = 1/\varepsilon$ is define as the "solution accuracy", $P = R_T/R$ is the "performance rate" and $\alpha_0 = 1/(\varepsilon_0 T_0^P)$ is "initial accuracy".

**Table 3.** Performance parameters

| Method | $\varepsilon_0$ | R | $T_0$ | $R_T$ | $\alpha_0$ | P |
|---|---|---|---|---|---|---|
| FIT + GS | 0.0131236 | 0.666 | 6.360E-09 | 2.2518 | 20214 | 0.29 |
| FIT + DPCG | 0.0131252 | 0.660 | 1.176E-07 | 1.7835 | **27915** | **0.37** |
| FIT + ICCG | 0.0131234 | 0.6711 | 1.238E-07 | 2.0826 | 12807 | 0.32 |
| FEM + FVM | 0.0047334 | 0.8063 | 3.927E-10 | 2.3663 | 1986 | 0.34 |
| FEM + SMG | 0.0047334 | 0.8839 | 2.104E-10 | 2.3663 | 4655 | 0.37 |

The performance of a method is characterized by the values of $\alpha_0$ and $P$. When comparing two iterative methods from the performance point of view, the most efficient will be the one with greater values for both parameters. If this is not the case, say method A has $\alpha_0$ greater than method B but $P$ less than method B, then method A is appropriate for carrying out fast but approximative computations while method B is appropriate for high accuracy computations.

In table 3 is presented the performance parameters values for different solving techniques: $GS$ = Gauss Siedel, $DPCG$ = Diagonal Preconditioning Conjugate Gradient, $ICCG$ =Incomplete Cholesky Conjugate Gradient, $FVM$ = Full V Cycle Multigrid and $SMG$ = Simple Multigrid [4].

## 5 Adaptive Multigrid Method

The algorithms described above were based on a hierarchy of uniform grids having increasing number of nodes. Their purpose was to allow a more exact representation of the field in critical areas, where their spatial distribution is highly nonuniform. However, a fine uniform grid over the whole domain is useless, since there are subdomains where the field is smooth and for its representation a coarser grid could be enough.

In order to improve even more the performance of the extraction algorithms, we propose the use of multigrid method in an original form. The process starts from an initial coarse grid, recursive refined only in critical zones, where the field has high nonuniformities. To find the subdomains where the subgridding should be used, the error indicator of the linear interpolation along x and y directions, is used:

$$\varepsilon_{ij} = \frac{|\varepsilon_x| + |\varepsilon_y|}{2} = \frac{|V_{i-1,j} - 2V_{i,j} + V_{i+1,j}| + |V_{i,j-1} - 2V_{i,j} + V_{i,j-1}|}{16}. \tag{12}$$

The refining will be carried out where this error indicator has relatively high values.

Fig. 5 and 6 shows the variation of this error in the case of the L shape domain, for two different grids. It can be noticed that for the fine grid the error has a sharp maximum near the inner corner and much lower values on the rest of the domain. This conduces to a subgridding domain of small size placed in the neighborhood of the inner corner.

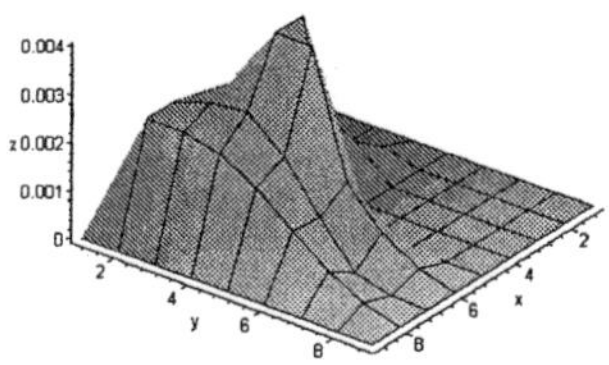

Fig. 5. The error in 65 DOFs

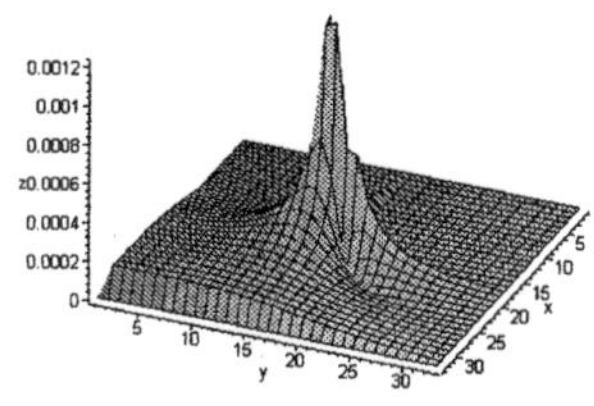

Fig. 6. The error in 833 DOFs

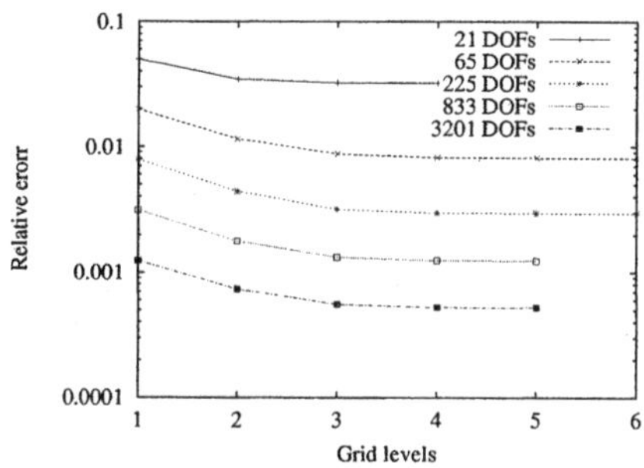


Fig. 7. Adaptive multigrid errors

The variation of the error with respect to the adaptive refining level for initial grids with different DOFs is shown in Fig. 7.

By applying the proposed adaptive local subgridding, the performance parameters of the multigrid FEM becomes $\alpha_0 = 27158$, $P = 0.17$.

# 6 Conclusions

Complementarity is a powerful principle, which allows the extraction of parameters of passive on-chip components with accuracy control. Thus, the waste of computational resources is avoided and fast extraction algorithms are obtained.

The proposed techniques based on averaging the complementary bounds followed by Richardson extrapolation proved to be extremely efficient in conjunction with the finite element method.

In the case of the studied test structure, the acceleration yielded a decrease of the error by two order of magnitudes (100 times). To obtain the same error without using acceleration, grids with 100 times bigger number of nodes should be used, leading to an important increase in the CPU time.

An even more decrease of the computing time can be obtained by using appropriate solvers for the linear sparse system of equations. In this respect, the best results were obtained with the simple-descent multigrid method, where the smoothing is carried out by means of the conjugate gradient method with diagonal preconditioning. The adaptive multigrid method, based on local subgridding is a very promising one, with many possible developments.

## References

[1] Dr. A. Alain Bossavit: Numerical Modelling in Electromagnetics. Ac. Press, 1997, Bucharest, May 19-23.
[2] Sergey Polstyanko and Jin-Fa Lee: Adaptive Finite Element Electrostatic Solver IEEE Transaction on Magnetics, 2001, vol 37, no. 5, pp. 3120 - 3124.
[3] Yousef Saad: Iterative Methods for Sparse Linear Systems 1996.
[4] William L. Briggs and Van Emden Henson and Steve F. McCormick: A Multigrid Tutorial, 2000.

# Optimization of Inductive Coupling between Qbit Rings

Christoph Kerner[(a)], Wim Magnus[(a)], Wim Schoenmaker[(a)], and Chris Van Haesendonck[(b)]

[(a)]IMEC, Kapeldreef 75, B-3001 Heverlee, Belgium
[(b)]Katholieke Universiteit Leuven, Celestijnenlaan 200 D, Heverlee, Belgium

**Abstract.** We are investigating inductive coupling optimization schemes and quantization effects for microscopic metal rings as a possible basis for a quantum bit (qbit). Faraday induction is proposed to provide electromagnetic coupling between the rings, therefore acting as an information carrier. Quantizing this information will produce distinguishable ring states that can be denoted by $|0\rangle$ and $|1\rangle$, representing the logic states of the qbit. We have set up simulation case studies with the aim of reducing signal loss between the rings. Further, different quantization mechanisms are investigated analytically. A combination of the two concepts can in theory be used to design qbits, consisting of metal rings with I/O facilities.

## 1 Introduction

Quantum computing [1,2] has recently become an active field of research with increasing interest, both in the theoretical and practical aspects. Quantum mechanics offers the phenomenon of superposition of states, enabling parallel computing. Unlike in classical computing, where a single bit is always in either one of the two logic states $|0\rangle$ and $|1\rangle$, linked to physical, measurable states, in quantum mechanics it is possible to bring a bit in a state that is an arbitrary linear combination of these basic states. Therefore, information space itself is larger as it extends to the full Hilbert space spanned by the basic states. During computation, a two-state system with orthogonal basis states $|0\rangle$ and $|1\rangle$ can then generally be found in a global state $|q\rangle$ as any combination of $|0\rangle$ and $|1\rangle$,

$$|q\rangle = \alpha\,|0\rangle + \beta\,|1\rangle\ , \quad |\alpha|^2 + |\beta|^2 = 1\ , \tag{1}$$

where the complex numbers $\alpha$ and $\beta$ are representing the probability amplitudes of finding the system in states $|0\rangle$ and $|1\rangle$ respectively, while the probabilities $|\alpha|^2$ and $|\beta|^2$ add up to one.
The power of quantum computing lies in the fact that the different qbits can be coherently connected to each other during computation until the result is read out. This coherence is provided by an entangled state of superposition which the qbit register can take. For a register of $n$ qbits, a global state $|Q\rangle$ can be described by

linearly combining the $2^n$ qbits basis vectors,

$$|Q\rangle = \sum_{i_1=0}^{1} \sum_{i_2=0}^{1} \dots \sum_{i_n=0}^{1} \alpha(i_1, i_2, \dots i_n) \; |i_1\rangle \otimes |i_1\rangle \otimes \dots \otimes |i_n\rangle \; . \tag{2}$$

In general, any qbit has to satisfy the following requirements [3]

- Good description of the devices and circuits that represent the qbit.
- High isolation from the environment to insure that coherence times are longer than computation times.
- A universal, reproducible initial state.
- The possibility of accessing the qbit to perform a controlled sequence of unitary transformations in order to realize quantum algorithms.
- The ability to link up measurements with quantum probabilities.

What's the gain with respect to classical computing? One should focus on problems for which classical computers would either not find a solution or require extremely long computation times and then find appropriate quantum algorithms for solving them. By implementing the parallel computing phenomenon provided by quantum mechanics, great computational speed-ups can be gained. Shor's famous quantum algorithm [4] for factoring large numbers has recently been realized experimentally using *NMR* techniques [5].
Factoring an arbitrarily large integer $n$ can only be done by using a quantum parallelism.

$$f(x)_{n,a} = a^x mod\ n \; . \tag{3}$$

The Shor algorithm tries to find the period $r$ of the function $f(x)$, where $n$ is the number to be factored and $a < n$ being a coprime to $n$, i.e. it has no common factors to n. Number theory allows us to rewrite the algorithm as $(a^{r/2} - 1)(a^{r/2} + 1) = 0 \; mod \; n$ provided that $r$ is even. Any $r$ that kills the product, thus producing an integer multiple of $n$ for an arbitrary value of $a$ is a solution and can be used to extract factors of $n$;

$$\gcd(a^{r/2} - 1, n) \; \times \; \gcd(a^{r/2} + 1, n) = n \; , \tag{4}$$

where "gcd" denotes the greatest common divisor.
Calculating this function for a large number of $a$'s would take an exponentially long time on a classical computer but can be done in one step (polynomial time) on a quantum computer, by calculating all values in superposition.

## 2 Microscopic metal rings as qbit basis

The hydrogen atom has been studied intensively over the past century and can be described very well quantum mechanically, thus seemingly providing a qbit application. However, the smallness of the atomic scale and the solid-state nature of computer hardware prohibit us from doing so. If we desire a stable qbit register inside a computer, including the necessary in- and output facilities, we will have to construct *artificial atoms*. One interesting candidate for such a device, are microscopic metal rings, communicating via electromagnetic induction. In- and output ports can exist in the form of ring-like structures connected to the outside world, accessible for applying or reading out alternating current signals. A ring, carrying such a current produces a time-varying magnetic field which can induce a current signal in another ring. Thus, a three-ring arrangement, consisting of an input ring, a free ring and an output ring could be used in principle for transferring a signal as well as using it for a logic operation.

The free ring or *computing element* will be our operating system. Classically, we can distinguish between a ring containing an induced current or not where the amplitude of the induced signal changes linearly with respect to the driving signal. At this stage, it is important to set the classical basis for information transfer. The magnetic field lines produced by the input ring will spread out in space and only a limited amount ($\simeq 20\%$) will go through the free ring. Thus, ways of guiding or focusing the magnetic flux have to be found. We have investigated three approaches (see section 3):

- sandwiching the ring system between superconducting plates, in order to explore the Meissner effect to bundle the flux,
- introducing ferromagnetic cores to guide the flux similar to a transformer,
- putting the free ring in a higher plane where the flux coupling is stronger compared to the radial direction.

After optimizing the classical behavior, different quantization mechanisms have to be investigated. There are two ways of achieving information quantization, one, by using nano-sized rings with discrete energy-eigenstates, the other, by using superconducting rings carrying persistent currents.

# 3 Simulation results

Electromagnetic induction is governed by Faraday's law : $V_{\mathrm{emf}} = -\mathrm{d}\Phi/\mathrm{d}t$. Induced currents depend on input frequency and system impedance. Thus, high flux coupling is necessary to ensure the desired physical effects.

## 3.1 Superconducting plates

The magnetic effect of sandwiching an input ring and a free ring between two superconducting plates was studied first, neglecting the influence of the free ring. After reaching the superconducting state, all magnetic field lines will be expelled from the plates due to the Meissner effect. Consequently, the magnetic field lines get squeezed and are confined to the region bounded by the plates. A resulting boost in the induced signal of the free ring is expected. The rings can be described as small tori of inner radius $a$, outer radius $b$ and height $c$ and are separated by a minimal distance $R$. The circulating current density inside such a torus reads in cylindrical coordinates $(\rho, \phi, z)$

$$\mathbf{J}(\rho,\phi,z) = \begin{cases} J(\rho,z)\mathbf{e}_\phi & \text{for } a \le \rho \le b\,, \quad |z| \le \frac{c}{2}\,, \\ 0 & \text{elsewhere.} \end{cases} \tag{5}$$

While the current density is confined to the torus area, the magnetic field lines are squeezed within a region between the two plates, separated by a distance $d$ (larger than the penetration depth of the superconductor) due to the Meissner effect, which is reflected in the boundary condition

$$B_z(\rho, \pm\frac{d}{2}) = 0 \tag{6}$$

The latter suggests a convenient Fourier expansion for $B_z(\rho, z)$ :

$$B_z(\rho, z) = \sum_{n=1}^{\infty} B_n(\rho) \sin k_n(z + \frac{d}{2}) \ , \ k_n = \frac{n\pi}{d} \ , \ n = 1, 2, 3, \ldots \tag{7}$$

and similarly for $J((\rho, z)$. Substituting the Fourier series into the Maxwell equations $\nabla \times \mathbf{B} = \mu_0 \mathbf{J}$ and $\nabla \cdot \mathbf{B} = 0$ , we obtain a second order differential equation for each Fourier component,

$$\frac{1}{\rho}\frac{\mathrm{d}}{\mathrm{d}\rho}(\rho\frac{\mathrm{d}B_n(\rho)}{\mathrm{d}\rho}) - k_n^2 B_n(\rho) = -\mu_0 \frac{1}{\rho}\frac{\mathrm{d}}{\mathrm{d}\rho}\Big(\rho J_n(\rho)\Big) \ . \tag{8}$$

The second ring is located in the same plane ($z = 0$) as the input ring. Taking the current density equal to a constant value $J_0$ within the input ring, we may solve Eq. (8) in terms of the modified Bessel functions $I_0$ and $K_0$ and obtain an expression for the magnetic field component $B_z(\rho, 0)$ for studying the effect of the input ring on the second ring :

$$B_z(\rho, 0) = -\frac{4\mu_0 J_0 d}{\pi^2} \sum_{n(\mathrm{odd})}^{\infty} \frac{1}{n^2} \sin(\frac{k_n c}{2}) K_0(k_n \rho) \int_{k_n a}^{k_n b} t I_1 dt \ . \tag{9}$$

The magnetic flux induced in the area $S$ of the second ring due to the current flowing through the first ring, is then given by :

$$\begin{aligned} \Phi_{12} &= \int_S \mathbf{B} \cdot \mathbf{dS} = \int_S B_z(\rho, 0)\, \mathrm{d}^2 r \\ &= -\frac{8\mu_0 J_0 \mathrm{d}^3}{\pi^4} \sum_{n(\mathrm{odd})}^{\infty} \frac{1}{n^4} \sin\frac{k_n c}{2} \int_0^{\phi_0} \mathrm{d}\phi \int_{k_n \rho_1(\phi)}^{k_n \rho_2(\phi)} x K_0(x) \mathrm{d}x \int_{k_n a}^{k_n b} I_1(t) \mathrm{d}t. \end{aligned} \tag{10}$$

with $\phi_0 = \sin^{-1}(b/R)$, $\rho_{1,2}(\phi) = R\cos\phi \mp \sqrt{b^2 - R^2 \sin^2\phi}$. Numerical simulations based on Eq. (10) were carried out using a Fortran 90 program. The following parameters were selected : $a = 1.5$ µm, $b = 2.5$ µm, $c = 0.2$ µm, R = 0.1 µm, and a static input current of I = 1 mA.
There seems to exist a balanced relation between the ring separation and the plate separation. Close-by rings between close plates do not necessarily experience a

signal boost. For each fixed value of either separation, there exists an optimal value for the other. Figures 1(a) to (c) show the magnetic effect (field and flux) on the second ring due to the first ring for a fixed ring separation $R$ and increasing distance $d$ between the plates. The slope first increases until a maximum is reached, after which it decreases asymptotically to the free field value (no plates). Physically, this is an interesting observation as it suggests that although the field lines are indeed squeezed as intended, their focusing point shifts. For closely spaced plates, the rings have to be far away from each other to get influenced at all.
Further, the boosted up signal in the second ring is only 20% higher in the optimal case ($\Phi_{\text{opt}} \simeq 5 \cdot 10^{-16}$ Wb) than the free field signal and coupling itself is about 24%. In a three-ring system, the total coupling (I/O) is $\sim$ 5%.

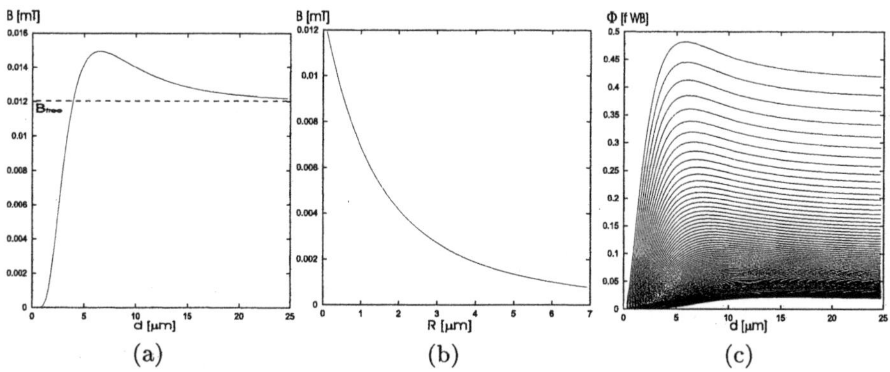


**Fig. 1.** Magnetic behavior: (a) B-field vs. $d$ for $R = 0.1$ μm (b) Free field (c) Flux vs. $d$ for different $R$

### 3.2 Ferromagnetic cores

The effect of using ferromagnetic core structures was studied next. In the three-ring system, the free ring shares one core with each of the other two rings without galvanic contact between the cores and/or the rings to avoid electronic coupling. The cores use the magnetic field produced by the input ring to switch their permanent magnetic moments, provided the input frequency is below $\sim$ 0.5 GHz, above which the switching cannot keep up anymore. Another critical quantity is the saturation field of the ferromagnetic material which should not be exceeded in order to enable controlled field reversal. The presence of the ferromagnetic cores confining the magnetic field lines to the core region produces a transformer-like action which is exploited to achieve efficient electromagnetic coupling between subsequent rings. Static 2D simulations were carried out for a symmetrical system, using the parameters of 3.1 [6]. The cores are separated from each other and the ring edges by 0.1 μm, lie 0.5 μm above the rings and are 0.5 μm thick. A flux coupling of 99.5% seems possible between the first two rings with a core error of the order of 0.1 %. The total coupling in the system is about 49% and proofs logical as half of the flux in the second ring can be transferred to the output from geometrical considerations. In the simulations, a general core material (any soft permalloy) with a medium permeability $\mu_r = 1000$ was used (some permalloys have $\mu_r \simeq 100000$). Magnetic fields of around 350 mT were reached inside the cores (figure 2b), providing a flux of around $2 \times 10^{-13}$ Wb threading the central ring, with an improvement factor of the order of 100-1000 compared to the free-field case (figure 2a).

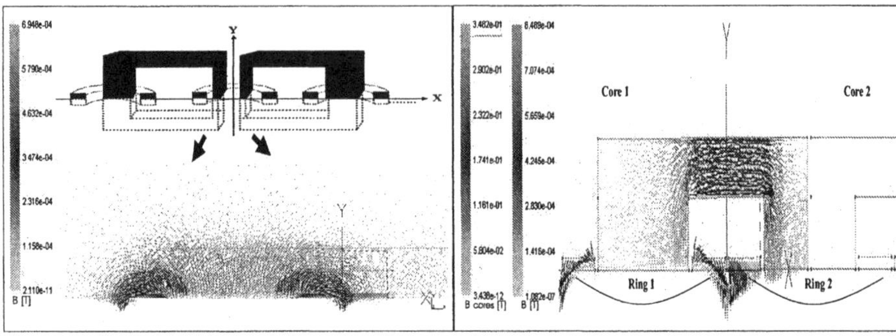


(a) (b)

**Fig. 2.** Magnetic flux lines : without (a) and with cores (b)

### 3.3 Planar arrangements

A seemingly obvious way of improving the flux coupling could be using a different planar arrangement. The in- and output rings are kept in the ground plane, while the free ring is placed in a higher plane such that half of its area lies above half of the in- and the output ring respectively. The magnetic field lines of the input ring

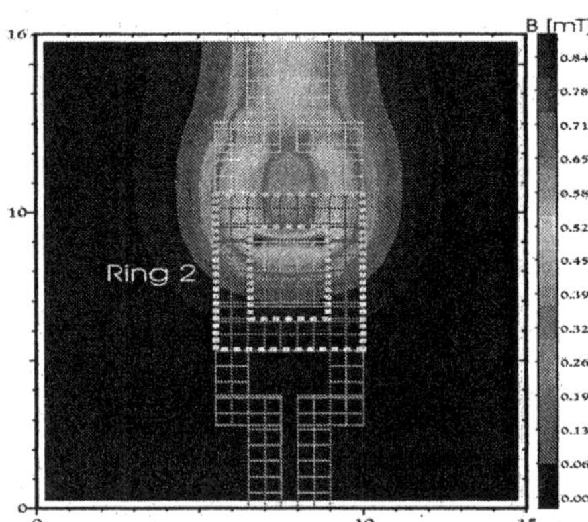


**Fig. 3.** Planar effect

are very dense in a region perpendicular to the ring plane. Direct coupling between the in- and output rings is low due to a rapid decrease of magnetic interaction in the radial direction ($\propto \rho^{-2}$, free field coupling $\simeq$ 20%). Using the same input parameters as above, and putting the free ring 1 nm above the other two rings, static simulations were carried out [7]. A flux coupling of about 70 % is achieved between the input ring and the free ring and a total system coupling of about 49 %, with a corresponding flux of about $7 \times 10^{-15}$ Wb through the free ring (see Fig. 3). However, this system requires good calibration for error control, providing difficulties during experimental realizations.

## 4 Quantization effects

This section states a brief analytical summary of the two different quantization concepts. The final equations will be used for future simulations [8].

### 4.1 Discrete states for a nano-ring

Electrons in a nano-ring of dimensions $a$ = 15 nm, $b$ = 50 nm and $c$ = 5 nm are described by discrete eigenstates $|\Psi_{m,n,p}\rangle$ and eigenenergies $E_{m,n,p}$ . Starting from the Schrödinger equation and a Hamiltonian that includes a vector potential term, the current density in the ring can be derived. For the geometries chosen in this paper, the total current is described in good approximation by

$$I = \int \mathbf{J}\cdot d\mathbf{S} = 2\sum_{m,n,p} \frac{I_{m,n,p}}{1+\exp(\frac{E_{m,n,p}-\mu}{kT})} ,$$
$$I_{m,n,p} = -\frac{e\hbar}{2M_e\pi\rho_0^2}(m+\frac{\Phi_B}{\Phi_0}) ; \tag{11}$$

$m, n, p$ are the quantum numbers labeling the eigenstates, $\Phi_B$ is the driving flux and $\Phi_0 = h/e$ denotes the Dirac flux quantum while $S$ now denotes a cross section of the ring. The chemical potential $\mu$ is determined by the total number of electrons residing in the ring.

### 4.2 Persistent current in a superconducting ring

A superconducting ring, independently of its size traps magnetic field lines inside its central hole for as long as the temperature stays below the transition point. This results in a persistent current flowing around the edges of the ring. Starting from the Ginzburg-Landau theory, the current density can be derived by minimizing the Gibbs free energy with respect to the order parameter $\Psi(\rho,\phi,z) = R(\rho,z)e^{i\vartheta(\phi)}$ and the vector potential $\mathbf{A}$,

$$\mathbf{J}_s = \frac{ie\hbar}{2M_e}(\Psi^*\nabla\Psi - \Psi\nabla\Psi^*) - \frac{2e^2}{M_e}\mathbf{A} = -\frac{eR^2}{M_e}(\hbar\nabla\vartheta + 2e\mathbf{A}) \tag{12}$$

where $-2e$ and $2M_e$ are the charge and the mass of a Cooper pair. In the bulk part of the ring, both the magnetic field and the current density are known to vanish. Hence, putting $\vartheta(\phi) = m\phi$ for integer $m$, we have

$$J_s(\text{bulk}) = -\frac{e\hbar R^2}{M_e \rho}(m + \frac{\Phi_B}{\Phi_0}) = 0 \tag{13}$$

reflecting flux quantization with respect to the London flux quantum $\Phi_0 = h/2e$ while the supercurrent $\mathbf{J_s}$ manifests itself as a surface current.

## 5 Conclusions and acknowledgments

Our studies indicate that ring structures offer interesting physical properties and can be used to transfer information. Quantization of information can be achieved by using a superconducting ring material below its transition temperature. This seems feasible from a production point of view as there are no requirements on the dimensions. The difficult part lies in optimizing the flux coupling between the rings and the best solution is to use ferromagnetic cores like in a macroscopic transformer; 99.5% of the input flux reaches the computing element with low system error. Combining these mechanisms results in a system that can theoretically exploit the necessary effects needed for a qbit. The advantage of such a qbit over other qbit layouts is that it is in solid-state, easily scalable and reproducible, thus making it suitable for chip design and silicon processing technology.
We are very much indebted to F. Henrotte from ESAT, University of Leuven.

## References

1. Preskil, J.: Lecture notes for physics 229: Quantum information and computation (1998); http://www.theory.caltech.edu/people/preskill/ph229/notes/book.ps
2. Barenco, A.: Quantum physics and computers. Contemporary Physics **37** (1996) 375; http://xxx.lanl.gov/pdf/quant-ph/9612014
3. DiVincenzo, D. P.: The physical implementation of quantum computation. Fortschritte der Physik **48** (2000) 771–83; arXiv:quant-ph/0002077
4. Shor P.: Algorithms for quantum computation: Discrete logarithms and factoring, *Proceedings of the 35th Annual Symposium on Foundations of Computer Science*, Santa Fe, NM, November 20-22 edited by IEEE Computer Society Press. (1994) 124–134
5. Vandersypen L.: Experimental realization of Shor's quantum factoring algorithm using nuclear magnetic resonance (NMR) techniques. Nature **414** (2001) 883–887
6. Geuzaine C., Henrotte F.: *Gmsh-GetDp software.* http://www.geuz.org
7. Meuris P., Schoenmaker W., Magnus W.: IEEE Transactions on CAD **20** (2001) 753–762
8. Kerner C., Magnus W., Schoenmaker W.: *Internal Report* (2000)

# Calculation of the Self-inductance of a Rectangular Magnetizer Coil

Meinolf Klocke

University of Dortmund, Institute of Electrical Machines, Drives and Power Electronics (held by Prof. Dr.-Ing. Dr.-Ing. S. Kulig), Emil-Figge-Str. 70, 44221 Dortmund, Germany. E-mail: klocke@mal.e-technik.uni-dortmund.de

**Abstract.** The self-inductance of the operating coil of a magnetizing device is calculated using different methods. The winding of the coil under investigation basically consists of copper sheets with rectangular concentric inner and outer contours. These plates form the turns of a Bitter coil. They are stacked together with an electric insulation between them and connected in series to form a helix-like winding. Analytical formulae for cylindrical coils can only be applied as a coarse approximation due to the rectangular cross-section and because of exact geometric measures of current paths such as for arrangements of filamentary wires not being available. More reliable results are obtained, if first self- and mutual inductances of all turns are determined according to Neumann's formula and the resulting self-inductance is determined afterwards with respect to the connection for all turns in series. A 3D-FEM analysis is carried out in order to verify the method described above and to judge the influence of eddy-current phenomena, i.e. the skin-effect, which might become important in the usual transient operation mode.

## 1 Introduction

In industrial magnetizers a battery of capacitors is discharged via an operating coil, in the interior of which unmagnetic workpieces are placed. The no-load self-inductance of this coil plays a certain role among the data to be known for dimensioning purposes.

Figure 1 shows a scheme of the operating Bitter coil under investigation including relevant measures. Ferromagnetic materials are not contained within the arrangement under no-load condition.

## 2 Coarse Analytical Estimations

The coarsest estimation for the self-inductance of the coil in Fig. 1 is based on assuming a homogeneous magnetic field in the interior and using an average cross-section. Thus $L$ becomes:

$$L = N^2 \mu_0 \frac{\overline{a} \cdot \overline{b}}{l_z}, \quad \overline{a} = \frac{1}{2}\left(a_{\text{ext}} + a_{\text{in}}\right), \quad \overline{b} = \frac{1}{2}\left(b_{\text{ext}} + b_{\text{in}}\right) \quad . \tag{1}$$

Inserting the given data into (1) one obtains $L = 6.76\,\text{mH}$, which must be considered an upper boundary only. Since $l_z \gg a$ and $l_z \gg b$ do not hold, the assumed field uniformity is not given, flux linkage contributions decrease towards the axial ends of the coil and thus the real value for $L$ is less.

By taking into account the decrease of the flux density along the positive and negative z-axis a more accurate result can be calculated. For solenoids the following formula can be applied, see e.g. [1]:

$$L = N^2 \mu_0 \frac{\pi R^2}{l_z} [\sqrt{1 + (\frac{R}{l_z})^2} - \frac{R}{l_z}] \ . \tag{2}$$

In order to use (2) for the given geometry the equivalent radius $R$ of a solenoid to be comparable to the rectangular coil should be chosen with respect to equality of average cross-sectional areas, $\pi R^2 = \overline{a} \cdot \overline{b}$ . With (2) $L$ then becomes $4.99\,\text{mH}$.

## 3 Semi-Analytical Approach

In the semi-analytical approach Neumann's formula is evaluated for the mutual inductance between filamentary rectangular wire loops. For two parallel conductors with different lengths $l_1$ and $l_2$ symmetrically placed at a distance $v$ the integral can be solved analytically giving the contribution $\Delta M$ of this pair of conductors to the resulting mutual inductance $M$.

By contrast to [2], where a program for the calculation of mutual inductances between arbitrary polygonal current loops is described including expressions for all possible positions of straight conductors, (3) suffices in the present situation. The program developed and used in [3] for the determination of end zone leakage inductances of induction motors also contains a subroutine for the general treatment of arbitrary polygonal loops.

By (3) the magnitude of the contribution $\Delta M$ to $M$ is calculated. Its actual sign for the summation depends on the current orientation. Pairs of conductors with perpendicular orientation do not contribute to $M$ anyway. For calculating $M$ between two turns the turn considered to be the exciting current loop is imagined to be concentrated in the rectangular middle line, whereas the flux receiving turn

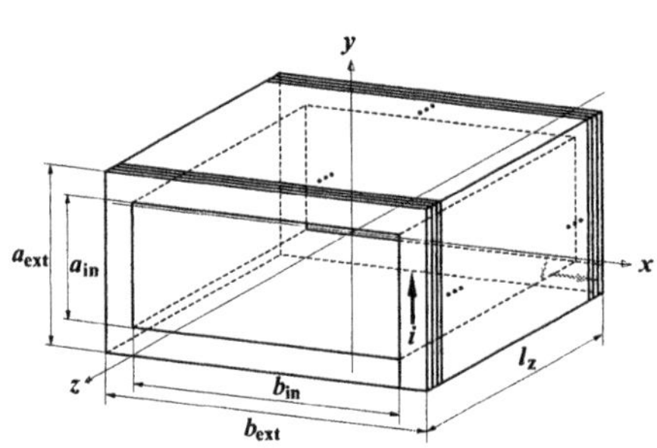


**Fig. 1.** Scheme of the magnetizer coil investigated here. Measures: Outer height $a_{ext} = 204\,\text{mm}$, inner height $a_{in} = 140\,\text{mm}$, outer width $b_{ext} = 384\,\text{mm}$, inner width $b_{in} = 320\,\text{mm}$, axial length $l_z = 450\,\text{mm}$. Number of turns: $N = 200$.

is represented by its inner and outer contour.

$$
\begin{aligned}
\Delta\, M &= \frac{\mu_0}{4\pi} \int\limits_{C_1}\int\limits_{C_2} \frac{\mathbf{dl_1}\cdot\mathbf{dl_2}}{r_{1,2}} = \frac{\mu_0}{4\pi} \int\limits_{-l_1/2}^{l_1/2}\int\limits_{-l_2/2}^{l_2/2} \frac{dy_1\cdot dy_2}{\sqrt{v^2+(y_1-y_2)^2}} = \\
&= \frac{\mu_0}{2\pi}\left[\frac{l_1+l_2}{2}\operatorname{arsinh}(\frac{l_1+l_2}{2v}) - \sqrt{(\frac{l_1+l_2}{2})^2+v^2} + \right. \\
&\left. -\frac{l_1-l_2}{2}\operatorname{arsinh}(\frac{l_1-l_2}{2v}) + \sqrt{(\frac{l_1-l_2}{2})^2+v^2}\right] = \Delta M\,(l_1,l_2,v)\ . \qquad (3)
\end{aligned}
$$

The turns are numbered consecutively, and the mutual inductance between two of them, denoted $M_{\mathrm{t},k}$ here, only depends on the index difference $k$. Referring to the expression given in (3) the mutual inductances between the excited rectangular current loop and the flux receiving contours, $M_{\mathrm{in},k}$ and $M_{\mathrm{ext},k}$, can be determined as shown in (4). The expressions occurring as arguments for (3) are illustrated in Fig. 2.

$$
\begin{aligned}
M_{{\text{ext} \atop \text{in}}\},k} &= 2\cdot\sum_{i=0}^{1}\sum_{j=0}^{1}(-1)^j\,\Delta M\,(x_i\pm d,\ \ x_i\,,\ \ v_{i,j})\ \ \text{with}\ x_0=\overline{a},\ \ x_1=\overline{b}, \\
v_{i,j} &= \sqrt{(\frac{1-(-1)^j}{2}\cdot x_{1-i}\pm\frac{d}{2})^2+(\frac{l_\mathrm{z}}{N-1}\cdot k)^2}\ . \qquad (4)
\end{aligned}
$$

The values for $M_{\mathrm{ext},k}$ and $M_{\mathrm{in},k}$ are averaged, which is considered an adequate approximation for the mutual inductance $M_{\mathrm{t},k}$ between the two turns under consideration.

The analytical calculation of mutual inductances between filamentary rectangular loops is also described in [4], where besides the axial distance also a lateral displacement $\Delta$ of the loops is taken into account. This just equals zero in the present arrangement. But unlike here the width $b$ in [4] is assumed to be the same for the two loops.

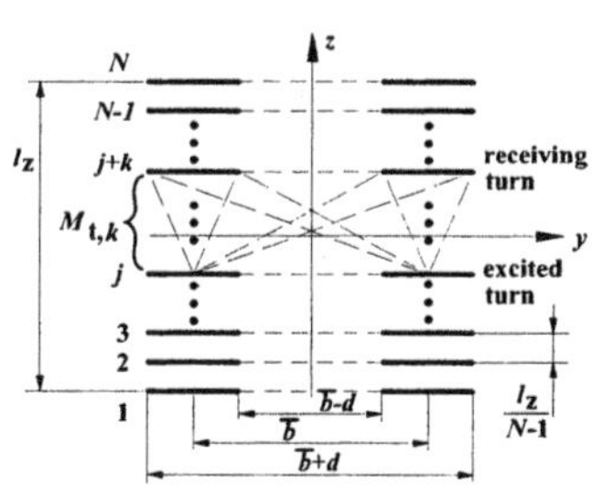


**Fig. 2.** Logitudinal section of the coil investigated. Illustration of turn numbering and relevant distances and measures for the calculation of turn to turn mutual inductances.

By formally including an index difference $k = 0$ in (4) also the self-inductance $L_t$ of a single turn can be determined, where the inner inductance is taken into account as an additional contribution. The summation of all inductances results in:

$$L_{\mathrm{res}} = \underbrace{N\left(M_{\mathrm{t},0} + \frac{\mu_0}{2\pi}(\overline{a}+\overline{b})\right)}_{L_\mathrm{t}} + 2 \cdot \sum_{k=1}^{N-1} (N-k)\, M_{\mathrm{t},k} \ . \tag{5}$$

A suitable algorithm for (3), (4), and (5) can be easily programmed even on a pocket calculator. With the data given in Fig. 1 a value of $L_{\mathrm{res}} = 4.52\,\mathrm{mH}$ is obtained.

The method outlined above can be improved by discretising the conducting plates, i.e. the flux receiving turn as well as the current excited one, by a number of concentrically arranged partial rectangular filamentary loops, e.g. $N_\mathrm{p} = 200$, and then calculating the flux-linkage and mutual inductance $M_{\mathrm{t},k}$ resp., between them. The formula for $M_{\mathrm{t},k}$ in this case must be modified to become the double sum expression in (6), which is still based on (3).

$$\begin{aligned}
M_{\mathrm{t},k} &= \frac{2}{{N_\mathrm{p}}^2} \cdot \sum_{i=1}^{N_\mathrm{p}} \sum_{j=1}^{N_\mathrm{p}} [\Delta M\,(a_i, a_j, v_{\mathrm{b},i,j,+}) - \Delta M\,(a_i, a_j, v_{\mathrm{b},i,j,-}) \\
&\qquad + \Delta M\,(b_i, b_j, v_{\mathrm{a},i,j,+}) - \Delta M\,(b_i, b_j, v_{\mathrm{a},i,j,-})]\,, \\
v_{\mathrm{x},i,j,\pm} &= \sqrt{(\frac{x_i \mp x_j}{2})^2 + (\frac{l_\mathrm{z}}{N-1}\cdot k)^2}, \quad (v_{\mathrm{a},i,j,+} = v_{\mathrm{b},i,j,+})\,, \\
\text{where} \quad & x_n = \overline{x} - d + (2n-1)\cdot \frac{d}{N_\mathrm{p}} \\
\text{with} \quad & x \ \text{standing for} \ a \ \text{or} \ b \ \text{and} \ n \ \text{for} \ i \ \text{or} \ j \ .
\end{aligned} \tag{6}$$

The self-inductance of a single turn cannot be determined by admitting $i = j$ during the summation for $k = 0$. Instead, this particular case is treated in a manner similar to the considerations for $M_{\mathrm{t},0}$ above. The imaginary inner and outer contours of a partial conductor are considered to be located at a distance of $d/(2N_\mathrm{p})$ inwards and outwards from that partial conductor. Its self-inductance is therefore determined by evaluating (4) for $k = 0$ with $a_i$ and $b_i$ from (6) substituting $\overline{a}$ and $\overline{b}$ in (4) and $d/N_\mathrm{p}$ instead of $d$. For $i \neq j$ the sum remains the same as in (6). Because of this calculation of single turn self-inductances and a relatively fine discretisation additional inner inductances of partial conductors appear to be negligible here.

The result for $L_{\mathrm{res}}$ then becomes 4.80 mH. In this approach current paths are considered to be rectangular even in the corners of the plates. Since a uniform current distribution on the turn cross-sections is assumed, skin-effect influences occurring in ac- and transient operation are not taken into account.

## 4 3D-FEM-Computation

The FEM-program EMAS [7–9] is used for determining the 3D-field distribution. AC-computations for three different frequencies are carried out, $10^{-3}$ Hz for dc-operation, 100 Hz and 2.5 kHz for higher frequencies. The calculated supply voltage for a prescribed exciting current $I_{\text{eff}}$ of 1 kA is a direct measure for the impedance of the coil.

Due to the symmetry of the arrangement the FEM-model may be restricted to the first octant of the rectangular coordinate system. The vectors of current density are perpendicular to the $xz$- and the $yz$-planes The magnetic field is tangential to them and the vector potential on them is therefore constrained to have a normal component only.

Discretising each single turn would probably result in an excessive number of nodes and thus huge computational effort. Therefore the windings are modelled as a massive conducting block as shown in Fig. 3. An effective conductivity is obtained by averaging insulating and conducting cross-sections of the real arrangement. Even for ac-considerations this appears to be justified by the one-dimensional skin-effect in rectangular slot conductors. There, the slot width $w_S$ and the conductor width $w_C$ are taken into account in the reduced conductor height [5] $\xi = \beta h$, which occurs in Field's formulae, by the factor $\beta$ also mentioned in [6]:

$$\beta = \sqrt{\frac{w_C}{w_S}\pi f \gamma \mu} \tag{7}$$

In (7) the quotient $w_C/w_S$ can be considered a weighting factor for the conductivity $\gamma$ of the conductive bar: In a slot filled over its full width with a material of this reduced conductivity the same reciprocal depth of penetration $\beta$ would occur. In the present arrangement the situation at least for winding plates in the centre part of the coil should be comparable due to the mainly axial field transversally oriented to the winding plates and thus similar to the field inside a slot conductor.

With an exemplary thickness of $t = 2\,\text{mm}$ and $N = 200$ turns of copper plates with $\gamma_{Cu} = 56 \cdot 10^6\,\text{S/m}$ on an axial lenght of $l_z = 450\,\text{mm}$ the conductivity prescribed in the FEM-model becomes $\gamma_{\text{mod}} = \gamma_{\text{Cu}} \cdot t/(l_z/N) = 49.78 \cdot 10^6\,\text{S/m}$.

In order to prescribe current paths regarding the winding structure, i.e. lying parallel to the $xy$-plane, the conductivity has to be anisotropic with only the tensorial $xx$- and $yy$-components being different from 0. Since field exciting current densities only have $x$- and $y$-components, the vectorpotential may also be restricted to these components, $A_z = 0$.

The winding characteristic has to be modelled by combining two different constraint techniques. On the one hand for each winding plate in the general ac-case the current-density distribution may be nonuniform, but the cross-section of the turn is equipotential. Therefore degrees-of-freedom No. 4, i.e. time-integrated electric potentials $\Psi$, of each layer of nodes perpendicular to the $z$-axis on the input cross-section are equalized with the unknown independent time-integrated electric potential $\Psi$ of an external collective node assigned to this layer. On the other hand the current is the same in every winding plate, but their contributions to the total voltage may differ depending mainly on their flux-linkage contributions and partially in the ac-case on different ohmic voltage drops. Therefore another external collective node is inserted into the model with the constraint that its time-integrated electric potential $\Psi_{\text{tot}}$ is the sum of the contributions of all winding plates. However, as mentioned above single turns do not occur in the FEM-model. But with respect to the constant axial winding density the potentials of the layer collective nodes can be used in the constraint equation taking also into account a desirable nonuniform discretisation in axial direction. If there are $n+1$ layers of nodes on the input cross-section in subsequent distances $\Delta z_1 \ldots \Delta z_n$ with collective potentials $\Psi_1 \ldots \Psi_{n+1}$, the constraint equation for the total collective potential $\Psi_{\text{tot}}$ can be written as follows:

$$\Psi_{\text{tot}} = \frac{N}{l_z} \cdot \left(\frac{\Delta z_1}{2}\Psi_1 + \sum_{k=1}^{n-1} \frac{1}{2}\left(\Delta z_k + \Delta z_{k+1}\right)\Psi_{k+1} + \frac{\Delta z_n}{2}\Psi_{n+1}\right) \tag{8}$$

Note that this approach is not applicable for a complete electrodynamic analysis including transient electric field calculation, which would be desirable for determining inter-turn electric fields and judging insulation behaviour. Since electric fields are not of interest in this example except inside the conducting region, $\Psi = 0$ is set for all nodes outside. The vector potentials of all external collective nodes are set to zero. The surrounding infinite space is modelled by a spherically shaped layer of open-boundary elements with a Radius of $R = 350\,\text{mm}$ in the given example.

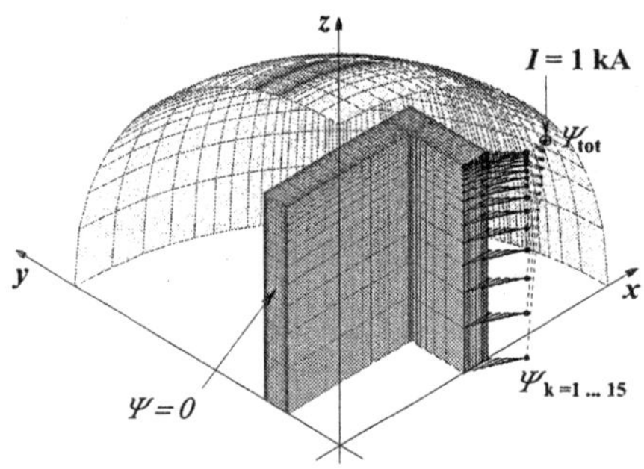


**Fig. 3.** Finite element grid of winding block and outer surface, discretisation of surrounding air not included. External collective nodes of input cross-section layers and for total voltage. Constraints for electric scalar potentials of winding block. Total number of nodes 33130, total number of elements 32940.

For the quasi dc-case with $10^{-3}$ Hz an inductance $L$ of 4.76 mH is calculated. As expected this value is diminished by the skin-effect at higher frequencies, which makes the current concentrate towards the inner surface of the coil and thus reduces its flux-linkage: $L = 4.34\,\text{mH}$ at 100 Hz and 4.11 mH at 2.5 kHz. An increase of the ohmic resistance is also observed.

The current concentration can approximately be taken into account in the semi-analytical calculation by modifying the geometric measures with respect to the depth of penetration $\beta^{-1}$ from (7). Replacing $d$ by $\delta = \beta^{-1}$ and inserting $\widetilde{a} = a_{\text{in}} + \delta(f)$ instead of $\overline{a}$ as well as $\widetilde{b} = b_{\text{in}} + \delta(f)$ for $\overline{b}$ in (6) one obtains $L = 4.34\,\text{mH}$ at $f = 100\,\text{Hz}$ and 4.16 mH at 2.5 kHz. These values deviate less than 2 % from the FEM results. The procedure appears to be justified by the low values of $\delta$ for the selected frequencies, 7.13 mm and 1.43 mm, which are much less than the winding block thickness $d$.

The current density distribution for $f = 100\,\text{Hz}$ as shown in Fig. 4 clarifies the quasi one-dimensional skin-effect close to the centre cross-section of the coil, but also indicates a planar distribution of current-density varying in axial direction near the coil end, where the magnetic field weakens and spreads out.

As an illustration of the field distribution a vector plot of the flux density **B** for $f = 100\,\mathrm{Hz}$ in the $xz$-plane is shown in Fig. 5. From the axis to the inner surface of the coil a slight increase of the magnitude is observed. Inside the winding block $B$ decreases to zero and finally changes its orientation within a distance of a few millimeters. This also indicates the skin-effect inside the winding. Significant deviations from the main axial direction of the inner coil field only occur in the end zone of the coil and inside the winding block.

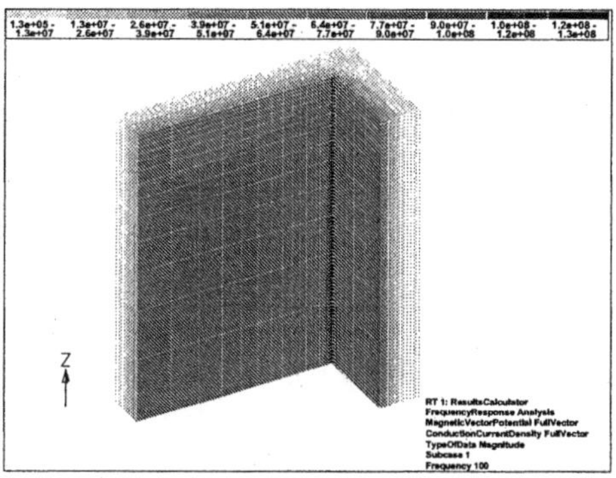

**Fig. 4.** Distribution of effective magnitudes of the current density in the winding block of the FEM-model at a supply frequency of $f = 100\,\mathrm{Hz}$. Exciting current $I_{\mathrm{eff}} = 1\,\mathrm{kA}$. Average current density $J_{\mathrm{av}} = 13.89\,\mathrm{A/mm^2}$. Axial current density $N \cdot i/l_z$ not depending on $z$, similar to an equivalent current sheet of constant 444 A/mm.

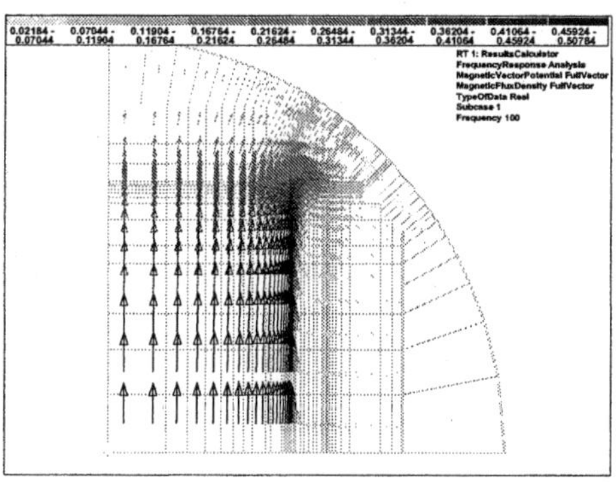

**Fig. 5.** Vector plot of magnetic flux density (real part) in the $xz$-plane of the FEM-model at supply frequency of $f = 100\,\mathrm{Hz}$. Exciting current $I_{\mathrm{eff}} = 1\,\mathrm{kA}$. (By contrast to 2D-problems fieldlines even in symmetry planes cannot be obtained for real 3D-arrangements by just keeping the normal component of **A** constant. Instead, $(\nabla \times \mathbf{A}) \times d\mathbf{l} = 0$ would have to be solved directly.)

# 5 Conclusion

Rectangularly shaped Bitter coils used in magnetizers require adapted methods for the calculation of the no-load inductance. By simple analytical formulae only estimations of unreliable accuracy are obtained.

An inductance calculation based on Neumann's formula is easily programmed, but in order to achieve a reliable value for $L$ later confirmed by FEM the winding plates have to be represented by a sufficiently high number of partial conductors.

Concerning the 3D-FEM calculation of coil arrangements a helpful modeling technique for special kinds of winding regions is presented. The equipotential constraint for turn cross-sections is combined with a summation of the voltage contributions over all turns and the prescription of an unisoptropic conductivity with respect to possible current paths.

## 6 Acknowledgement

The author would like to thank MAGSYS Magnet Systeme GmbH and in particular Mr. Dipl.-Ing. S. Wältring for providing coil data and information about magnetizing devices.

## References

1. Smythe, W. R.: Static and Dynamic Electricity. McGraw-Hill Book Company, 1968.
2. Hannakam, L., Nolle, E.: Programm zur Bestimmung der Gegeninduktivitäten räumlich polygonartiger Leiterschleifen. Archiv für Elektrotechnik 64 (1981), pp. 21-25.
3. Dornau, U.: Berechnung und Messung der Stirnstreuung von Asynchronmaschinen mit Käfigläufer, PhD thesis, University of Dortmund. Institute of Theoretical Electrical Engineering and Electrical Machines, 1990.
4. Oberretl, K. , Kratki, N.: Ausgleichsvorgänge und Schwingungen beim elektrodynamischen Magnetkissen-System. Archiv für Elektrotechnik 57 (1975), pp. 59-64.
5. Richter, R.: Elektrische Maschinen, Band 1, Birkhäuser Verlag Basel und Stuttgart, 1967.
6. Küpfmüller, K.: Einführung in die theoretische Elektrotechnik. 13., verbesserte Auflage, Springer-Verlag, 1990.
7. Brauer, J. R.; Brown, B. S.: EMAS, User's Manual - Version 4, Ansoft Corporation, 1997.
8. Brauer, J. R.: EMAS, Application Manual - Version 4, Ansoft Corporation, 1997.
9. Brauer, J. R.; MacNeal, B. E.: Finite element modeling of multiturn windings with attached electric circuits. IEEE Trans. Magn., vol. 29, March 1993.

# The Synthesis of an EXOR Function by Using Modulo Functions Implemented by SET Circuits

Roelof H. Klunder and Jaap Hoekstra

Delft University of Technology,
Faculty of Information Technology and Systems,
Department of Electrical Engineering,
P.O. Box 5031, 2600 GA Delft, The Netherlands
e-mail: J.Hoekstra@ITS.TUDelft.NL
http://nanocom.et.tudelft.nl/circuits/index.htm

**Abstract.** In this paper we will discuss the EXOR function synthesized with a modulo function. As an implementation of the modulo function we will use the single-electron tunneling (SET) electron-box as a basic structure. A SET electron-box consists of one SET junction in series with a normal capacitor. This basic electron-box structure is extended with a number (equal to the number of inputs) of normal capacitors.

Simulations are carried out with the commonly used simulator in SET-electronics: SIMON [1], indeed showing the expected outcome.

## 1 Introduction

An electron can tunnel through an insulator if the distance between the conductors is small enough. A metal insulator metal structure is called a single-electron tunneling (SET) junction. Due to this layout a SET junction can be modelled as a capacitor, when no tunnel events occur.

When only a number of junctions and capacitors are connected to one node, this node becomes a floating node, which is called an island. On such a floating node electrons can be stored or erased by tunneling through one of the connected SET junctions, creating an independent island charge. Due to the storage of discrete independent charge a periodic behaviour arises (discussed in more details in Sect. 2).

Because the relevant device dimensions are in the nanometer range, SET junctions belong to the class of nano-electronic components. Generally, the use of nanoelectronics is motivated by several arguments [2,3] such as: basic devices can be very small, nanoelectronics has the potential to operate with very low supply power, and quantum properties that appear at nanometer scale in principle represent an increase in signal-processing power. Periodicity is one example which can increase the signal-processing power of a structure.

Only few people use the periodic behaviour of SET junctions for design purposes. Examples of the use of this periodicity are

- Weight factors of neural nets [4]
- Fault tolerance by using periodicity as redundancy [5–7]

Modulo functions are another example of the use of periodic behaviour. Many articles have been written about applications of modulo functions, Chinese Remainder Theorem (CRT), and Residue Number Systems (RNS) [8]. In those articles the usability of the proposed circuits and architectures depends on the implementation of the necessary modulo functions. The advantages of those applications in speed and chip area compared to a standard circuit solution, depend on the way the modulo functions are implemented.

With the help of a modulo function it is easy to create an EXOR function. For a binary EXOR function with $k$ inputs we can write

$$out = \left[\sum_{i=1}^{k} in_i\right] \pmod 2 \tag{1}$$

In Sect. 2 we explain the origin of the periodicity of SET circuits. In Sect. 3 the basic structure which is used in this paper, the electron-box, is described as well as its behaviour as a modulo function. With the help of the electron-box structure an EXOR function is synthesized (Sect. 4) and simulated (Sect. 5). We end this paper with the conclusions (Sect. 6).

## 2 Origin of Periodic Behaviour

A floating node (a so called island) can contain independent island charges $q_{node}$. As we have shown in [9] these independent island charge can be modelled by a current source, with value $i = q_{node}\delta(t)$ (see Fig. 1a). To keep the node a floating node the input voltage $u_{in}$ should be capacitively coupled (shown in Fig. 1a). The voltage source can be replaced by a current source using the source transformation principle, as shown in Fig. 1b.

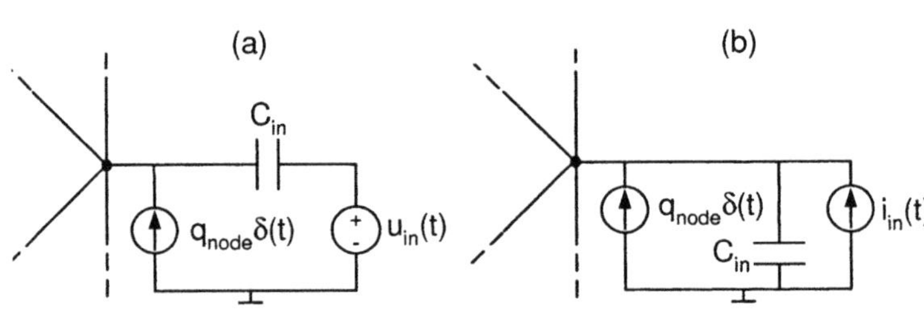


**Fig. 1.** a) An island of which the independent charge $q_{node}$ is modelled by a current source. To the same island the input voltage source $u_{in}$ is capacitively coupled. b) The voltage source $u_{in}$ is changed into a current source $i_{in}$ using source transformation (in the Laplace domain), $i_{in}(t) = \mathcal{L}^{-1}\{sC_{in}U_{in}(s)\}$.

We assume here that the independent charge on the node, $q_{node}$, only consists of an integer number of electrons $n$. This is a correct assumption when a SET junction is connected to the island under investigation. Through a SET junction only an integer number of electrons can tunnel. Let $n$ be the number of independent electrons, and $e$ the electron charge, then we can write for the independent island charge $q_{node} = ne$.

The sum of the two current sources drawn in Fig. 1b can become equally large for different island charges, when the discrete independent island charge is decreased by one electron *and* the input voltage $u_{in}$ is changed from situation $z$ to situation $z+1$. This can be written in Laplace as:

$$U_{in}^{z}(s)sC_{in} + ne = U_{in}^{z+1}(s)sC_{in} + (n-1)e \tag{2}$$

To calculate the necessary change in input voltage we can rewrite this equation into:

$$\Delta u_{in} \equiv u_{in}^{z+1} - u_{in}^{z} = \frac{e}{C_{in}} \tag{3}$$

Hence the periodicity of the transfer function of a structure which contains an input voltage source capacitively coupled to an island is equal to

$$P = \frac{e}{C_{in}} \tag{4}$$

So the periodicity (expressed in voltages) is always determined by the capacitance the input voltage is coupled with.

## 3 The Electron-Box

One of the basic SET structures is the electron-box. The electron-box consists of a SET junction with a capacitance $C_j$ and one 'normal' (non tunneling) capacitor $C_c$. Between these two components an island is created.

For the output voltage $u_{out}$ of the electron-box of Fig. 2a the following relation holds:

$$u_{out} = \frac{u_{in}C_c + q_i}{C_\Sigma} \quad \text{with} \quad C_\Sigma = C_c + C_j \tag{5}$$

The charge $q_i$ is the total independent charge on the island, which is equal to $ne$ (see Fig. 2c).

The critical voltage $u_j^{cr}$, which is the voltage at least necessary to let an electron tunnel [10], is equal to

$$u_j^{cr} = \frac{e}{2C_\Sigma} \tag{6}$$

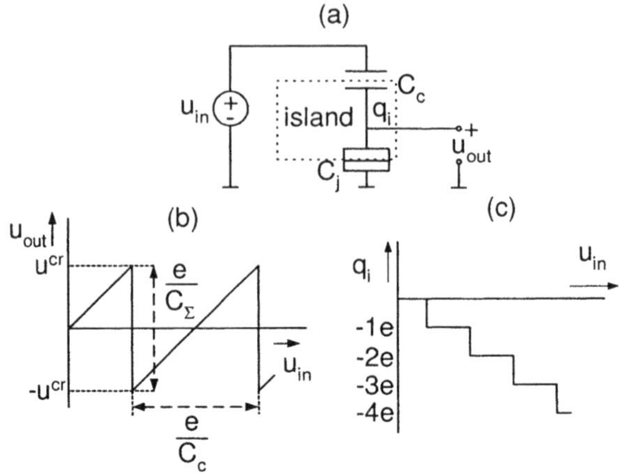


**Fig. 2.** a) The basic electron-box structure, b) its periodic voltage transfer function, which can be used to implement a 'mod' function, and c) the independent island charge, which can be used to implement a 'div' function.

In [11] we have shown that we can use one tunnel transition in the transfer function to implement negative logic, like the NAND, NOR, and the inverter. In this way we use only a small fraction of the complete transfer function. In stead of looking at one transition, we can use the periodic behaviour of the transfer function to implement more complex digital logic based on modulo functions, like the EXOR.

When the input consists of discrete possible amplitudes, the continuous transfer function of Fig. 2b changes into a discrete function, which as a consequence of the periodic behaviour can be transferred into a modulo function (see Fig. 3).

The applied discrete input values create several small regions within the total transfer function. In those regions the input signal can be expected. We will call those regions operating windows. The width of the operating windows in case of digital circuitry gives the allowed error in amplitude of the applied signals.

# 4 The EXOR Function

Up to now most of the (digital) circuits made with SET junctions, are a transformation from the standard CMOS logic, in which many SET junctions are needed. For example Jeong et al. [12] proposed an EXOR function with 16 SET junctions, which is a direct transformation from a CMOS EXOR function. There are many limitations of this design approach, for example many SET junctions are needed, which are still hard to make.

We propose an EXOR which is based on the periodicity obtained from the transfer function of a SET electron-box structure. The EXOR function implemented with the electron-box has no gain, which can simply be solved by buffering the EXOR before coupling it to other structures.

To implement an EXOR function, the operating windows should be placed in such a way that the output voltage is low when the sum of the inputs is even and the output should be high when the sum is odd. In Fig. 3c the transfer function of a multi input electron-box is shown, the number indicates the sum of the inputs that are high.

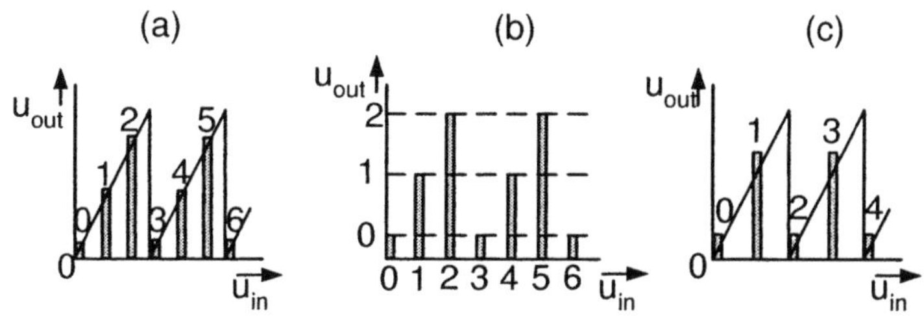


**Fig. 3.** a) Only some input voltages are possible (grey areas). b) The discrete transfer functions showing a modulo 3 implementation. c) The transfer function of a multi input electron-box with the chosen operating windows to create an EXOR function.

The distance between two operating windows is constant, and should be equal to the amplitude of the input signal. The electron box should be biased such that it reaches operating window '0' without signals at the input.

We wish to implement an EXOR function with $k$ inputs (see Fig. 4a for $k = 2$ inputs). The SET structure of Fig. 4a will have a transfer function as shown in Fig. 4b. For the output voltage we can write

$$u_{out} = \frac{C_j u_x + C_c u_c + ne + \sum_{i=1}^{k} u_{in,i} C_{in,i}}{C_\Sigma} \tag{7}$$

with $C_\Sigma = C_j + C_c + \sum_{i=1}^{k} C_{in,i}$. Let $u_0$ be the output voltage when all the input voltages are zero, so (5) can be rewritten into

$$u_0 = \frac{C_j}{C_\Sigma} u_x + \frac{C_c}{C_\Sigma} u_c + \frac{ne}{C_\Sigma} \tag{8}$$

And for the critical output voltage we can write (see (6))

$$u_{out}^{cr} = u_j^{cr} + u_x \tag{9}$$

The voltage drop due to a tunnel event is equal to $\frac{e}{C_\Sigma}$. With the help of these equations we can determine all the values of the components.

We assume that we have an input signal with an amplitude of $1mV$. The EXOR function is modulo 2, so we need two input signals with a high amplitude to go to the next period (see Fig. 3c). Using input signals with an amplitude of $1mV$ the period should be equal to $2mV$. For simplicity we assume that the desired EXOR

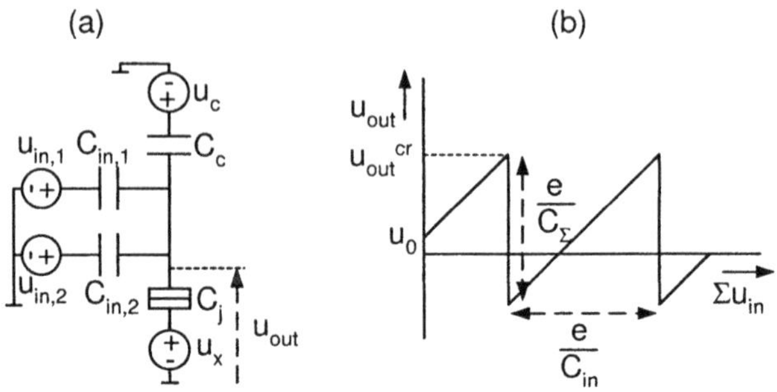


**Fig. 4.** a) A two input electron-box with two extra bias sources $u_x$ and $u_c$ to shift the transfer function. b) The transfer function of the two input electron-box with bias sources.

has two inputs, of which the input capacitances are equal ($C_{in,1} = C_{in,2} = C_{in}$). The period is equal to the derived (4), with the value $\frac{e}{C_{in}}$, which makes it possible to derive the value of the input capacitors: $C_{in} = 80aF$.

If we want to use only positive signals, the whole transfer function of Fig. 4b should be lifted, so

$$u_{out}^{cr} = \frac{e}{C_\Sigma} \tag{10}$$

Combining (10) with (9) we find for the voltage source $u_x$:

$$u_x = \frac{e}{2C_\Sigma} \tag{11}$$

Because a starting point $u_0 = 0$ is desired, the voltage source $u_c$ (assuming no electrons have tunneled yet), can be calculated using (8):

$$u_c C_c = -u_x C_j \tag{12}$$

Only the amplitude of the output voltage depends on the capacitance of $C_j$ and $C_c$, so we are free to choose those. To get a signal which is as high as possible the capacitance values should be chosen as small as possible.

# 5 Simulations

The digital EXOR function described in Sect. 4 is tested with the help of the simulator SIMON [1]. The test conditions were: $T = 0K$, the amplitude of the input voltage $u_{in}$ is $u_{in,max} = 1mV$, $C_{in,1} = C_{in,2} = C_{in} = 80aF$ and $u_x = 0.25mV$. For simplicity we chose for the junction capacitance $C_j$ and the extra bias capacitance $C_c$ the same capacitive value as the input capacitors, $C_j = C_c = C_{in}$.

Because the electron will not tunnel directly when the critical voltage is reached but rather when the voltage across the SET junction is a bit higher, we compensate for those small errors. This compensation can be done by choosing $u_c$ slightly lower than $u_x$, so $u_c = -0.24mV$.

The simulation results as function of time are shown in Fig. 5 and confirm the expected output of an EXOR function.

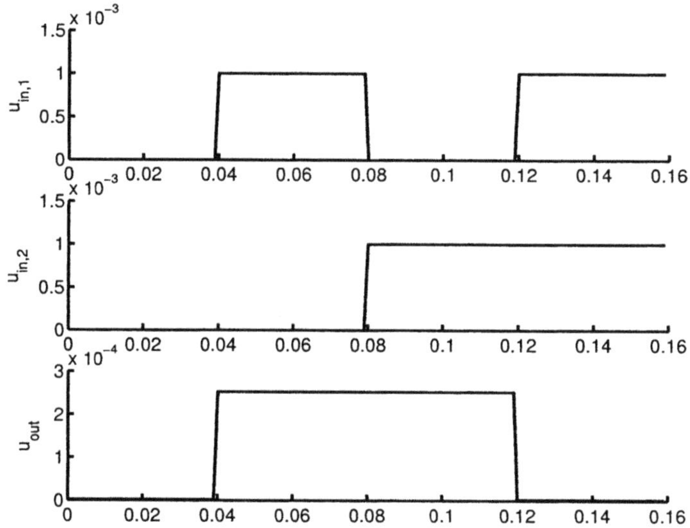


**Fig. 5.** The simulated response of a two input electron-box biased as an EXOR function as function of time.

The rise and fall time of Fig. 5 depend at this moment only on the simulator used, and are therefore not discussed here. The simulator used is based on an existing theory of tunneling that assumes a zero tunnel time. As long as the tunnel time is neglected the transition from high to low and visa versa, are in principle very short.

## 6 Conclusions

In this paper we have shown the possibility to use a basic SET structure, the electron-box, to implement modulo functions. We showed the binary digital EXOR operation as an example of a modulo function. Simulations proved its behaviour.

More complex and larger circuits can be implemented as well. The periodic behaviour of a SET sub-circuit makes the implementation of a modulo function more natural than in standard CMOS.

## Acknowledgement

We gratefully acknowledge the financial support of the European Mel-ari/NID "ANSWERS" project, and the Delft Inter-faculty Research Center "Novel Computational Structures based on Quantum Devices".

## References

1. C. Wasshuber, H. Kosina, and S. Selberherr. SIMON - A Simulator for Single Electron Tunnel Devices and Circuits. *IEEE Transactions on Computer Aided Design of Integrated Circuits and Systems*, 16(9):937–944, September 1997.
2. A. van Roermund and J. Hoekstra. Design Philosophy for Nanoelectronic Systems, from SETs to Neural Nets. *International Journal of Circuit Theory and Applications*, 28(6):563–584, 2000.
3. K.K. Likharev. Single-Electron Devices and Their Applications. In *Proceedings of the IEEE*, volume 87, pages 606–632. IEEE, April 1999.
4. M.J. Goossens, J.H. Ritskes, C.J.M. Verhoeven, and A.H.M. van Roermund. Neural Networks with Periodic Single-Electron Tunneling Transistors. In *ECCTD'97*, pages 936–941, Budapest, September 1997.
5. K.K. Likharev and A.N. Korotkov. Ultradense Hybrid SET/FET Dynamic RAM: Feasibility of Background Charge Independent Room Temperature Single Electron Digital Circuits. In *Proceedings of ISDRS*, pages 355–359, Charlottesville, Virginia, October 1995.
6. R.H. Klunder, J. Hoekstra, and A.H.M. van Roermund. Variable Capacitively Coupled Single Electron Tunneling Transistor. In *Third Workshop on Innovative Circuits and Systems for Nano Electronics*, pages B4/1 – B4/9, Munich, October 1998.
7. R.H. Klunder and J. Hoekstra. Circuit Architecture Solution for Background Charge Fluctuations in Single Electron Tunneling Transistors. In *ECCTD'01*, volume III, pages 213–216, Espoo, Finland, August 2001. European Conference on Circuit Theory and Design, ISBN: 951-22-5571-5.
8. M.A. Soderstrand, W.K. Jenkins, G.A. Jullien, and F.J. Taylor. *Residue Number System Arithmetic: Modern Applications in Digital Signal Processing.* IEEE press, ISBN: 0-87942-205-X, New York, 1986.
9. R. Klunder, K. van Hartingsveldt, and J. Hoekstra. Modelling of Independent Node Charges in Metallic Single Electron Tunneling Circuits. In *Proceedings of the Ninth Workshop on Nonlinear Dynamics of Electronic Systems*, pages 217–220, Delft, The Netherlands, 21-23 June 2001. NDES, ISBN 90-407-2194-7.
10. R.H. Klunder and J. Hoekstra. The Derivation and Calculation of the Critical Voltage in Single Electron Tunnel Circuits. In *ProRISC/IEEE Workshop*, pages 446–451. STW, ISBN 90-73461-29-4, November 2001.
11. R.H. Klunder and J. Hoekstra. Programmable Logic Using a SET Electron Box. In *ICECS*, volume 1, pages 185–188, Malta, September 2001. IEEE International Conference on Electronics, Circuits and Systems, ISBN: 0-7803-7058-9.
12. M.Y. Jeong, Y.H Jeong, S.W. Hwang, and D.M. Kim. Performance of Single Electron Transistor Logic Composed of Multi-Gate Single Electron Transistors. *Japanese Journal of Applied Physics*, 36(11):6706–6710, November 1997.

# Field-Circuit Coupling by Means of the Woodbury Formula

Stefan Kurz and Volker Rischmüller

Robert Bosch GmbH, P.O. Box 10 60 50, 70049 Stuttgart, Germany

**Abstract.** For the coupling of the magnetic field and the electric circuit equations, there are different approaches. In any case, the flux linkage has to be taken into account by augmenting the finite element system by additional equations. Recently, it has been proposed to eliminate the circuit part by taking the Schur complement, which results in symmetric and positive definite matrices. The rank of the circuit's contribution to the Schur complement equals the number of linear independent coupling variables. Field-circuit coupling therefore introduces a low rank correction into the equations of the field problem. The consequences of this key observation are discussed in the paper. If a direct solution of the finite element system is considered, the circuit coupling can be treated elegantly by using the Woodbury formula. The Woodbury formula gives an explicit expression for the inverse of a matrix with low rank correction in terms of the inverse of the original matrix. In the framework of a preconditioned conjugate gradient solver it turns out that it is sufficient to include the circuit equations into the matrix-by-vector product, while the finite element preconditioner can be retained. These considerations will be illustrated by numerical results that have been obtained from a simple model problem.

## 1 Introduction

A typical feature of mechatronic systems is the close interaction between various physical domains. Even if we restrict ourselves to electromagnetic phenomena, a coupling between magnetic field and electric circuit simulation is unavoidable in most cases to describe the system's behaviour correctly.

Field-circuit coupling can be grouped into three main categories. One approach consists of parameter extraction, where the electromagnetic field device is described by an equivalent circuit and the system simulation is carried out on the network level. Field simulations can be employed to obtain both the topology and the parameters of an equivalent circuit. A good account on this topic can be found in [8].

Another common approach is the so-called direct coupling, where the field and the circuit equations are collected in one overall matrix and solved together. This can be done either under control of the field or the circuit simulator. Typically, a finite element (FE) matrix is augmented by the circuit equations [11], or the FE equations are represented in the circuit simulation as a multiport device [12]. In contrast, indirect coupling keeps both simulations separated. They communicate with each other by coupling matrices. The overall problem has to be solved by iteration in this case [1].

In this paper direct coupling inside the field simulator shall be pursued further. Typically, a large sparse FE matrix has to be augmented by a small dense circuit matrix and some coupling matrices. The main goal is to work towards a symmetric positive definite (s.p.d.) overall description, while keeping the sparse FE matrix untouched.

The paper is organized as follows. In Sect. 2, the subject is exposed by means of an eddy current problem driven by a voltage source, which will serve as a model problem. In Sect. 3, it is shown how the Woodbury formula can be used in connection with a direct solver to deal with the field-circuit coupling. Next, the field-circuit coupling is studied in the framework of a preconditioned conjugate gradient (PCG) method. Some conclusions are drawn in Sect. 4.

## 2 Setting of the Problem

The fundamental equation of the eddy current problem is

$$\operatorname{curl} \nu \operatorname{curl} \mathbf{A} + \sigma \frac{\partial \mathbf{A}}{\partial t} = \mathbf{j}_{\mathrm{S}}, \tag{1}$$

where $\nu$ is the reluctivity, $\mathbf{A}$ the magnetic vector potential, $\sigma$ the conductivity and $\mathbf{j}_{\mathrm{S}}$ an impressed current density. In general, $\nu$ depends on the flux density $\mathbf{B} = \operatorname{curl} \mathbf{A}$ due to iron saturation. We restrict ourselves to magnetization curves with $\mathrm{d}\nu/\mathrm{d}B \geq 0$.

The simplest case of field-circuit coupling appears if a single coil driven by a voltage source is taken into account. It is assumed that the coil consists of stranded conductors, i.e. the wires of the coil are so thin that skin effect can be disregarded. However, this is sufficient to show all the principles to be discussed in this paper. A simple 3D model problem which falls into this category has been proposed in [7] and is shown in Fig. 1. A circular coil is located between two square aluminium plates. A step voltage is applied to the coil and the evolution of the current in the coil is considered.

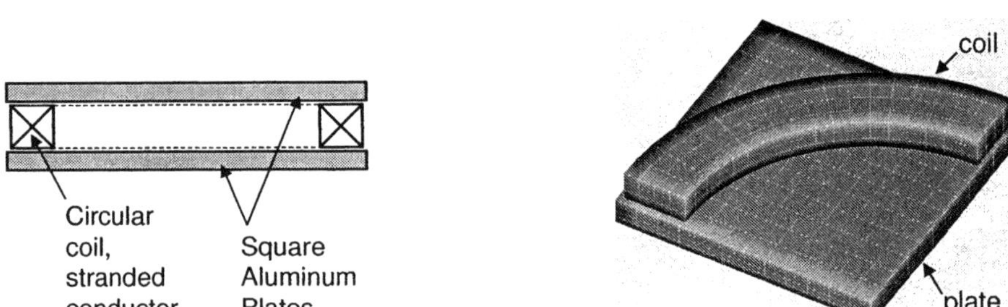


**Fig. 1.** The model problem: Circular coil between two aluminium plates. On the right, the mesh of one eighth of the problem is displayed, which was used for the numerical analysis.

The equation of the electric circuit is $R \cdot i + \psi' = u_{12}$, where $R$ is the coil resistance, $i$ the exciting current, $\psi'$ the time derivative of the coil's flux linkage and $u_{12}$ its terminal voltage. The actual distribution of the wires in the coil, the air between them and their insulation shall not be taken into account here. Usually it is sufficient to assume that the current is distributed homogeneously over the cross section of the coil. The geometry of the winding is described by introducing the winding density $\boldsymbol{\tau} = \mathbf{j}_S / i$, $\operatorname{div} \boldsymbol{\tau} = 0$. Consider one filamentary conductor with closed contour $C$ and cross section $\Delta a$, which gives a contribution

$$\Delta\psi = \oint_C \mathbf{A} \cdot \mathrm{d}\mathbf{l} \tag{2}$$

to the coil's flux linkage. The vectorial line element goes into the direction of the current density and can therefore be expressed as $\mathrm{d}\mathbf{l} = (\mathbf{j}_S / |\mathbf{j}_S|)\mathrm{d}l = \boldsymbol{\tau}\mathrm{d}l \Delta a$. The total flux linkage can be obtained by summing up the contributions from the individual series-connected filamentary conductors, $\psi = \sum \Delta\psi$. In the limit of infinitely thin filaments $\Delta a \to \mathrm{d}a$, $\Delta\psi \to \mathrm{d}\psi$, and the sum can be expressed by an integral over the cross section $a_{\mathrm{Coil}}$ of the coil according to

$$\psi = \int_{a_{\mathrm{Coil}}} \mathrm{d}\psi = \int_{a_{\mathrm{Coil}}} \oint_C \mathbf{A} \cdot \boldsymbol{\tau} \, \mathrm{d}l \, \mathrm{d}a = \int_{\Omega_{\mathrm{Coil}}} \mathbf{A} \cdot \boldsymbol{\tau} \, \mathrm{d}\Omega. \tag{3}$$

The solution of the coupled field-circuit-problem is thus given by solving simultaneously

$$\left. \begin{aligned} &\operatorname{curl} \nu \operatorname{curl} \mathbf{A} + \sigma \frac{\partial \mathbf{A}}{\partial t} - i \cdot \boldsymbol{\tau} = 0, \\ &\int_{\Omega_{\mathrm{Coil}}} \frac{\partial \mathbf{A}}{\partial t} \cdot \boldsymbol{\tau} \, \mathrm{d}\Omega + R \cdot i = u_{12}. \end{aligned} \right\} \tag{4}$$

After application of the FE method, time discretization by the implicit Euler method, and linearization by the Newton-Raphson method, the discrete counterpart of (4) can be written in the following matrix form [10]

$$\begin{pmatrix} [L] & -\{U\} \\ -\{U\}^T & -\Delta t R \end{pmatrix} \begin{pmatrix} \{\delta A\} \\ \delta i \end{pmatrix} = \begin{pmatrix} \{F_\delta\} \\ e_\delta \end{pmatrix}. \tag{5}$$

Matrix $[L]$ is the sparse symmetric Jacobian matrix resulting from the FE discretization, $\{U\}$ is the field-circuit coupling vector and $\{\delta A\}$ and $\delta i$ are the nonlinear increments of the vector potential and the coil current, respectively. If there are non-conducting subdomains, the kernel of the double curl operator might give rise to a singular matrix $[L]$, depending on the numerical discretization scheme. In this case, the line of reasoning of the paper can be applied to the quotient space. For the sake of simplicity, we assume $\sigma > 0$ throughout, even in the coil, resulting in $[L]$ being s.p.d. The complete system (5) is sparse and symmetric, but in any case not positive definite, due to the negative diagonal entry $-\Delta t R$.

In [10] it was proposed to remedy this problem by eliminating $\delta i$ from the first equation of (5). The same algorithm has been adopted in [2]. This corresponds to the solution of the system

$$\begin{pmatrix} [\widetilde{L}] & 0 \\ -\{U\}^T & -\Delta t R \end{pmatrix} \begin{pmatrix} \{\delta A\} \\ \delta i \end{pmatrix} = \begin{pmatrix} \{\widetilde{F}_\delta\} \\ e_\delta \end{pmatrix}, \tag{6}$$

$$\left.\begin{aligned} [\widetilde{L}] &= [L] + \tfrac{1}{\Delta t R}\{U\}\{U\}^T, \\ \{\widetilde{F}_\delta\} &= \{F_\delta\} - \tfrac{1}{\Delta t R}\{U\}e_\delta. \end{aligned}\right\} \tag{7}$$

The problem boils down to the solution of the first equation of (6) with the s.p.d. Schur complement matrix $[\widetilde{L}]$, and subsequent determination of $\delta i$ by means of the second one. Be aware that the coil yields a contribution with all its degrees of freedom (DOF) coupled to each other. This means that $[\widetilde{L}]$ loses sparsity compared to $[L]$. This effect depends on the ratio between the number of coil DOF and the total number of DOF. If the considered problem happens to be a coil only, $[\widetilde{L}]$ will be even a dense matrix. Note also that in the asymptotic case $h \to 0$ ($h$ being the mesh parameter) the memory requirement for $[\widetilde{L}]$ will be totally dominated by the coil's contribution.

This obvious disadvantage can be easily circumvented by observing that the coil introduces a rank-1 update to the original matrix $[L]$. In the general case of $p$ independent coils we would obtain a similar rank-$p$ update. This is the key observation for the construction of efficient solution schemes.

## 3 Solution of the Schur Complement System

In this section we consider the slightly more general problem

$$\begin{pmatrix} [L] & -[U] \\ -[U]^T & -[R] \end{pmatrix} \begin{pmatrix} \{\delta A\} \\ \{\delta i\} \end{pmatrix} = \begin{pmatrix} \{F_\delta\} \\ \{e_\delta\} \end{pmatrix}, \tag{8}$$

with $[L] \in \mathbb{R}^{n\times n}$, $[U] \in \mathbb{R}^{n\times p}$ and $[R] \in \mathbb{R}^{p\times p}$. The time step $\Delta t$ has been absorbed in the matrix $[R]$. The number of DOF is denoted by $n$ and the number of coupling variables by $p \ll n$. Now the Schur complement system reads

$$\begin{pmatrix} [\widetilde{L}] & 0 \\ -[U]^T & -[R] \end{pmatrix} \begin{pmatrix} \{\delta A\} \\ \{\delta i\} \end{pmatrix} = \begin{pmatrix} \{\widetilde{F}_\delta\} \\ \{e_\delta\} \end{pmatrix}, \tag{9}$$

$$\left.\begin{aligned} [\widetilde{L}] &= [L] + [U][R]^{-1}[U]^T, \\ \{\widetilde{F}_\delta\} &= \{F_\delta\} - [U][R]^{-1}\{e_\delta\}. \end{aligned}\right\} \tag{10}$$

The solution of the system

$$[\widetilde{L}]\{\delta A\} = \{\widetilde{F}_\delta\} \tag{11}$$

in the context of time dependent, non-linear eddy current problems will be considered, this is the first equation of (9). Once this system has been solved, the remaining unknowns $\{\delta i\}$ can be gained from the second equation of (9).

In the general case of field-circuit coupling the situation becomes more involved. Besides stranded conductors there are also solid conductors which are subject to eddy currents [11]. Moreover, a general circuit contains voltage and current sources, resistors, capacitors and inductors joined together in a complicated way. However, circuit analysis can be tailored such that the matrix structure (8) is conserved. Once again, a s.p.d. Schur complement system (11) will be received by elimination of the circuit part. For details of the general case the reader is referred to [3,4].

### 3.1 Direct Solution by Cholesky Decomposition

Under certain circumstances a direct solution of (11) by Cholesky decomposition can be useful. For example, in the code described in [6] a coupling of the FE method with the boundary element (BE) method based on domain decomposition (DD) is employed. A solution of the FE problem with Dirichlet boundary condition is followed by a solution of the BE problem with Neumann boundary condition. This iteration is accelerated by a superimposed Krylov subspace method. During the course of the iteration, the FE problem must be solved over and over again. Since the ressource requirements for the BE-FE coupling are usually totally dominated by the BE, it pays off to factorize the FE matrix in (11) to facilitate the repeated solution.

The basic idea is to solve the problem first without taking into account the reaction of the field on the circuit part. The "wrong" solution will be corrected afterwards by including the effect of the rank-$p$ update which can be done at low computational costs. This approach is known as *Sherman-Morrison formula* in the case $p=1$ and as *Woodbury formula* in the case $p>1$ [9]. The Woodbury formula allows to express $[\widetilde{L}]^{-1}$ in terms of $[L]^{-1}$ as follows

$$[\widetilde{L}]^{-1} = [L]^{-1} - [\widetilde{U}][W]^{-1}[\widetilde{U}]^T, \tag{12}$$

$$[\widetilde{U}] = [L]^{-1}[U] \in \mathbb{R}^{n\times p}, \qquad [W] = [R] + [U]^T[L]^{-1}[U] \in \mathbb{R}^{p\times p}. \tag{13}$$

If $[L]$ and $[R]$ are s.p.d., then $[\widetilde{L}]$ and $[W]$ are also s.p.d. In the sequel we shall restrict ourselves to that case. Note that the matrix $[U]^T[L]^{-1}[U]$ can be regarded as the multiport device representation of the FE equations [12].

The field-circuit coupled problem can be solved by the following algorithm:

- Set up $[U]$. This matrix can be computed once and for all, because it only depends on the geometry of the conductors and the numerical formulation.
- After the FE matrix $[L]$ has been set up and factorized, $[\widetilde{U}]$ is determined according to (13). This corresponds to applying backsubstitution $p$ times to the columns of $[U]$. Afterwards, $[W]$ can be computed from $[R]$, $[U]$ and $[\widetilde{U}]$. Note that $[\widetilde{U}]$ and $[W]$ remain unchanged as long as the FE Jacobi matrix $[L]$ can be retained, at least during each non-linear iteration step.
- The FE problem is solved as if there was no reaction of the field on the circuit part, i.e. formally $[U]^T$ is omitted in (8). Equation (10) reveals $[\widetilde{L}] = [L]$ in this case, while $\{\widetilde{F_\delta}\}$ remained unaltered. The solution of (11) with $[L]$ instead of $[\widetilde{L}]$ is corrected afterwards by using the Woodbury formula (12).

We applied the algorithm to the model problem shown in Fig. 1. The coil contributed with about 30% to the total number of 4315 unknowns. A sequence of

70 time steps yielded altogether 1470 DD iterations. A naïve implementation with the Schur complement $[\widetilde{L}]$ being actually constructed and factorized was compared to the proposed algorithm. The FE matrix $[L]$ together with its Cholesky factor required 5,8 MB of memory, the Schur complement $[\widetilde{L}]$ 10,2 MB, together with its Cholesky factor. This extra memory could be saved by the application of the Woodbury formula. The CPU time for the solution dropped from 309 s to 230 s. This is mainly due to the fact that the repeated solution of the FE problem based on the Cholesky factor of $[L]$ is faster, even if the Woodbury correction introduces some overhead. The computed solution is in very good agreement with measured values [7].

## 3.2 Iterative Solution by the Conjugate Gradient Method

In many practical applications, the system (11) will be solved by iteration. As its matrix is s.p.d., the PCG method is the best choice. During the linear iterative solution process, we have to deal with matrix-by-vector products. From (10) it can be seen that such products consist of a contribution of $[L]$, which is the same as if no circuit was present, and a correction by $[U][R]^{-1}[U]^T$. The Schur complement needs not be explicitly formed [3]. The price to be paid for this are two $n \times p$ matrix-by-vector multiplications and the solution of a $p \times p$ system.

In [3] the question was raised how the s.p.d. Schur complement system (11) could be effectively preconditioned. The simplest solution would just neglect the circuit part and apply a good preconditioner for the FE matrix $[L]$. This can also be understood from the theoretical point of view. Suppose a good preconditioner for the FE matrix $[L]$ is available, say $[K]$, which should be easily invertible and spectrally equivalent to $[L]$. Then the spectrum of $[K]^{-1}[L]$ is neatly clustered, yielding a rapid convergence of the PCG iteration. This feature will be inherited by the spectrum of $[K]^{-1}[\widetilde{L}]$ with the exception of at most $2p$ outlying eigenvalues resulting from the circuit part. It is well known that the convergence of PCG is not much affected by a few eigenvalues off the bulk of the spectrum [5, Ch. 9]. More precisely, after $2p$ steps the PCG iteration will recover the same speed as if it had been applied to $[L]$ with preconditioner $[K]$.

Alternatively, one could be tempted to include the circuit part into the preconditioner and use

$$[\widetilde{K}] = [K] + [U]\,[R]^{-1}[U]^T \tag{14}$$

instead of $[K]$. Practically, the preconditioner $[K]$ would be applied as if no circuit was present and afterwards a correction by means of the Woodbury formula took place. However, it is not clear if $[\widetilde{K}][\widetilde{L}]$ has better spectral properties than $[K][\widetilde{L}]$.

To shed light on that point we conducted some numerical experiments. We considered the model problem shown in Fig. 1 and the first step of the DD iteration. The iteration starts with a zero vector potential. This gives rise to a FE problem with homogeneous Dirichlet boundary condition. We used an incomplete Cholesky (IC) factorization as preconditioner and varied the drop tolerance to examine the influence of the preconditioner's quality. Three different problems have been studied: FE problem without circuit coupling (matrix $[K]^{-1}[L]$), FE problem with circuit coupling (matrix $[K]^{-1}[\widetilde{L}]$) and FE problem with circuit coupling and modified preconditioner (matrix $[\widetilde{K}]^{-1}[\widetilde{L}]$) according to (14). The results are summarized in Table 1.

**Table 1.** Performance of ICCG applied to the FE part of the model problem shown in Fig. 1 with homogeneous Dirichlet boundary condition. The termination criterion for the relative residual norm was set to $10^{-10}$. The table gives the number of required iterations for various levels of the drop tolerance of the IC factorization.

| Drop tolerance | $\leq 10^{-5}$ | $10^{-4}$ | $10^{-3}$ | $10^{-2}$ | $10^{-1}$ | $\geq 10^{0}$ |
|---|---|---|---|---|---|---|
| W/o circuit coupling | 1 | 3 | 4 | 7 | 17 | 20 |
| With circuit coupling | 2 | 3 | 4 | 7 | 16 | 20 |
| With circuit coupling, modified preconditioner | 1 | 3 | 4 | 6 | 17 | 18 |

It can be seen that the rank-1 circuit coupling has hardly any influence on the convergence, according to the above theoretical considerations. A drop tolerance $\leq 10^{-5}$ yields a complete Cholesky factorization and the CG scheme degenerates to a direct solver in the case without circuit coupling. This behaviour is recovered in the case with circuit coupling, provided that the circuit part is included in the preconditioner. In all other cases it can be seen that the modification (14) has hardly any influence. In fact, $[\widetilde{K}]$ is a low rank perturbation of $[K]$ so that this finding is in accordance with theory. We also repeated the experiment for other field-circuit coupled problems and found always similar behaviour. This suggests that the observations from Table 1 are of quite general nature and it is not worth while using the modified preconditioner $[\widetilde{K}]$ instead of $[K]$.

## 4 Conclusions

In this paper, a direct field-circuit coupling technique has been addressed. The circuit part has been eliminated from the overall matrix by taking the Schur complement. The solver has to deal with s.p.d. matrices only. This has already been pointed out in [10]. The key point is that the Schur complement matrix $[\widetilde{L}]$ differs from the FE matrix $[L]$ by a low rank correction. In case of a direct solver the situation can be handled elegantly by using the Woodbury formula. As far as iterative solution by the PCG method is concerned, only the matrix-by-vector product needs an additional correction. The overhead of this correction is $O(n)$. The original FE preconditioner can be retained without deterioration of the convergence rate. Numerical evidence for these findings has been given by means of a simple model problem.

## References

1. Bedrosian, G.: A new method for coupling finite element field solutions with external circuits and kinematics. IEEE Trans. Magn. **29** (1993) 1664–1668.

2. Fetzer, J., Kurz, S., Lehner, G., Rucker, W.M.: Transient BEM-FEM coupled analysis of 3-d electromechanical systems: A watch stepping motor driven by a thin wire coil. IEEE Trans. Magn. **34** (1998) 3154–3157.
3. De Gersem, H., et.al.: Solution strategies for transient, field-circuit coupled systems. IEEE Trans. Magn. **36** (2000) 1531–1534.
4. De Gersem, H., et.al.: A topological method used for field-circuit coupling. IEEE Trans. Magn. **34** (1998) 3190–3193.
5. Hackbusch, W. *Iterative Solution of Large Sparse Systems of Equations*, volume 95 of *Applied Mathematical Sciences*. Springer-Verlag, Berlin, 1993.
6. Kurz, S., Fetzer, J., Lehner, G., Rucker, W.M.: Numerical analysis of 3D eddy current problems with moving bodies using BEM-FEM coupling. *Surveys on Mathematics for Industry*, 9:131–150, 1999.
7. Leonard, P.J., Rodger, D.: Voltage forced coils for 3D finite-element electromagnetic models. IEEE Trans. Magn. **24** (1988) 2579–2581.
8. Munteanu, I., Ioan, D.: A survey on parameter extraction techniques for coupling electromagnetic devices to electric circuits. LNCSE **18** (2001) 337–357, Springer-Verlag, Berlin.
9. Press, W.H., Vetterling, W.T., Teukolsky, S.A., Flannery, B.P.: *Numerical Recipes*. Cambridge University Press, Cambridge, 2. edition, 1992.
10. Spasov, V., Noguchi, S., Yamashita, H.: An edge element field-circuit coupling approach resulting in symmetric positive definite matrices. In *Record of the 13th COMPUMAG Conference on the Computation of Electromagnetic Fields*, volume 4, pages 220–221, Evian, France, July 2001.
11. Tsukerman, I.A., Konrad, A., Meunier, G., Sabonnadière, J.C.: Coupled field-circuit problems: Trends and accomplishments. IEEE Trans. Magn. **29** (1993) 1701–1704.
12. Väänänen, J.: Circuit theoretical approach to couple two-dimensional finite element models with external circuit equations. IEEE Trans. Magn. **32** (1996) 400–410.

# Simulation of Complex Electrode-Spacer Configurations Used in HV Systems for Accurate Electric Field Calculation

A. Lahiri and S. Chakravorti

Electrical Engineering Department
Jadavpur University
Kolkata 700 032
India

**Abstract.** In practice the shape of the electrodes and the spacers in high voltage equipments are so designed that the electric stress on the electrodes, mainly on the live electrode, and on the spacers are well within the limits. This necessitates that the practical contours of the insulators have complex geometries. The complexities are often increased by the constraints imposed by the mechanical considerations in the Gas Insulated Systems (GIS). Suitable techniques are required for accurate yet efficient simulation of such complex geometries. This paper highlights a modified algorithm for computation of electric field distribution by indirect boundary element method (indirect BEM) around a complex electrode-spacer configuration used in practise in high voltage system arrangements.

## 1 Introduction

The advent of modern digital computers has encouraged researchers in the area of electrostatic field calculation to concentrate on developing numerical techniques for the purpose. Efficient asymmetric field calculation in high voltage equipments by numerical techniques based on integral equation methods such as Charge Simulation Method (CSM) [1-2] or, Boundary Element Method (BEM) [3-4] have been carried out successfully over the last years. But these numerical methods involve accurate simulation of the electrodes and the spacer geometries. Hence more emphasis should be given for simulating the complex geometries of the arrangements that are used in practise.

For several years, BEM has been applied over planer triangular and rectangular elements [5-6]. But the practical problem with these elements lies in the fact that approximation of curved surfaces with these planer elements often results in inaccuracy in surface simulation leading to distortion of field on the surfaces.

Use of curvilinear triangles in parametric form with quadratic approximation of surfaces were reported in some literatures [7-8]. But for these types of elements, the computation time required is large as has been reported in [4]. The method applied in [4] requires a large number of elements leading to a greater time requirement for efficient handling of realistic field problems.

In the recent past, approximation of the curved surfaces bi-cubic spline function has been reported [9]. But meshing the boundary surface in [9], requires rigorous computational strategies. Hence a more precise and convenient approach is required to solve complicated electric field problems.

# 2 Indirect Boundary Element Method

The basis of indirect BEM lies in discritizing the electrodes and the insulator surfaces into small large number of surface elements, termed as boundary elements, either in form of curvilinear rectangles or triangles. The equivalent charge densities are considered to reside at the vertices of each of these boundary elements having a non-linear distribution inside the elements but linearly distributed along the edges of the elements. These charge densities are solved from the boundary conditions:

1. Given potential value on the conductor surfaces with known potential are maintained.
2. Condition on the interface between two dielectrics are satisfied i.e., the normal components of electric flux density on both sides of the dielectric interface is continuous.

and using the following equations

$$\sum_{j=1}^{N} P_{ij}\sigma_j \; - \; V_i \; = \; 0 \; i = N_c + 1, ...., N_c + N_j \tag{1}$$

$$\sum_{j=1 j\neq i}^{N} F_{ij}\sigma_j \; + \; (F_{ij} - 1)\sigma_i \; = \; 0 \; i = N_c + N_f + 1, ...., N \tag{2}$$

where,
$N$ = Total number of nodes, $N_c$ = Total number of nodes on the conductor surface with known potential, $N_f$ = Total number of nodes on the conductors of free potential $P_{ij}$ = Potential at the $i^{th}$ point due to $j^{th}$ unit charge, $F_{ij}$ = Electric stress at the $i^{th}$ point due to $j^{th}$ unit charge, $\sigma_i$ = Charge density at the $i^{th}$ node and $\sigma_j$ = Charge density at the $j^{th}$ node. The second boundary condition imposed over the dielectric-dielectric interface yields

$$\varepsilon_1 E_{1n}(i) \; = \; \varepsilon_2 E_{2n}(i) \quad i.e., \; D_{1n}(i) \; = \; D_{2n}(i) \tag{3}$$

where, $\varepsilon_k$ is the permittivity of the medium k, $k = 1, 2, 3 ....$
$E_{jn}(i)$ and $D_{jn}(i)$ denote the normal component of the electric intensity and normal component of electric flux density respectively at any point $i$ in medium j, $j = 1, 2, 3 ...$

# 3 Applied Technique

Practical electrodes and insulators are complex combinations of some basic geometries which include sphere, cylinder, cone, toroid and disc. These geometries may also have special features such as inclusion of holes on their surfaces or truncation at the edges. Hence, these elements are coded in computer and embedded in a software named "ASYMBEM" to cast the final electrode-spacer arrangement.

The entire boundary surface of the arrangement is then subdivided into curvilinear triangular boundary elements for generating mesh on the surface. These boundary surface elements are of first order and are transformed into local co-ordinates ($\xi$, $\eta$, $\zeta$) to obtain the master triangle. The co-ordinate of any arbitrary point P inside such an element can now be represented as

$$\alpha = \alpha_1\xi + \alpha_2\eta + \alpha_3\zeta \tag{4}$$

where, $\alpha$ stands for $x$, $y$, $z$, $\alpha_i$, ($i = 1, 2, 3$), are the co-ordinates of the vertices of the triangle and $\zeta = 1 - \xi - \eta$. The equivalent surface charge density has been approximated by a linear function of the charge densities residing at the nodes of the surface elements. i.e.

$$\sigma = \sigma_1\xi + \sigma_2\eta + \sigma_3\zeta \tag{5}$$

Thus the potential at any point Q ($x$, $y$, $z$) due to each boundary surface is calculated as

$$\Phi = \begin{pmatrix} a_{Q1} & a_{Q2} & a_{Q3} \end{pmatrix} \begin{pmatrix} \sigma_1 \\ \sigma_2 \\ \sigma_3 \end{pmatrix} \tag{6}$$

where, $\begin{pmatrix} a_{Q1} & a_{Q2} & a_{Q3} \end{pmatrix}$ is the coefficient matrix for the potential function.

$$\begin{aligned} a_{Q1} &= \frac{1}{4\pi\varepsilon_0} \int_0^1 \int_0^{1-\xi} \frac{\xi}{r_{PQ}} J \, d\xi \, d\eta \\ a_{Q2} &= \frac{1}{4\pi\varepsilon_0} \int_0^1 \int_0^{1-\xi} \frac{\eta}{r_{PQ}} J \, d\xi \, d\eta \\ a_{Q3} &= \frac{1}{4\pi\varepsilon_0} \int_0^1 \int_0^{1-\xi} \frac{1-\xi-\eta}{r_{PQ}} J \, d\xi \, d\eta \\ r_{PQ} &= \sqrt{\sum}(x - (x_1\xi + x_2\eta + x_3\zeta))^2 \end{aligned} \tag{7}$$

The boundary condition at the dielectric-dielectric interface yields

$$\frac{(\varepsilon_1 - \varepsilon_2)}{(\varepsilon_1 + \varepsilon_2)} E_n(Q) + \frac{\sigma(Q)}{2\varepsilon_0} = 0 \tag{8}$$

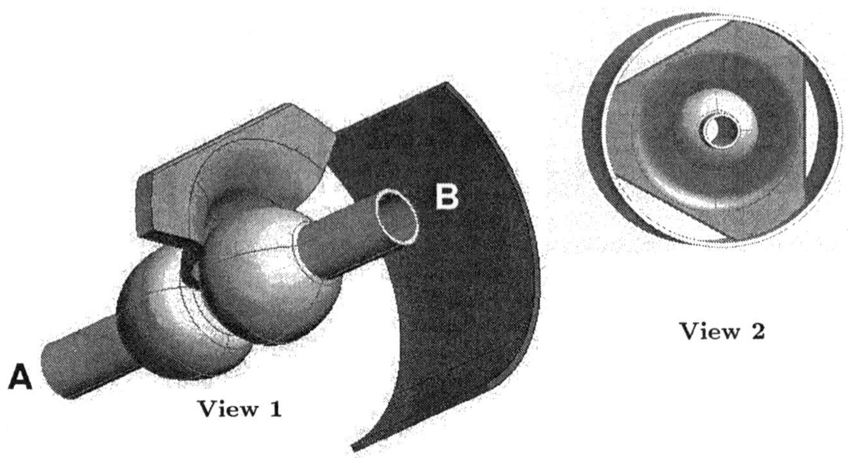


**Fig. 1. Electrode-Spacer Configuration**

The normal component of electric stress at any point Q on the dielectric-dielectric interface due to a charge residing at P is obtained from

$$\mathbf{E}_n(Q) = -\nabla\Phi\, cos(\mathbf{r}_{PQ}, \mathbf{n}_Q) \tag{9}$$

Combining Equations (6), (8) and (9), the coefficient matrix for the entire distribution of charge densities is obtained from which the fictitious charge density residing at each node of the entire arrangement can be evaluated by solving a set of equations represented in a matrix form. With the now known values of the fictitious charge densities, the entire solution of the electrostatic field problem at any arbitrary point on the boundary surfaces can be obtained.

Evaluation of the coefficient matrix has been carried out by integrating equation set (7) numerically using Gaussian formula over triangular domain [10] given by

$$\int\int \chi\, dA = \beta \sum_{i=1}^{N} w_i\, \chi(\xi_i, \eta_i, \zeta_i) \tag{10}$$

where, $\beta$ = area of the triangle, $w_i$ = the weight associated with each of the integrating point and $N$ = number of integrating points. Since a master triangle is a right angled triangle having the length of each of the two sides including the right angle equal to 1 hence for such a triangle $\beta = 0.5$

For solving potential, 7 points Gaussian integration has been carried out. In order to reduce the error due to the derivative of the potential function, 13 points Gaussian integration has been adapted for the calculation of electric field intensity.

The major advantage of using such a numerical technique lies in the fact that the

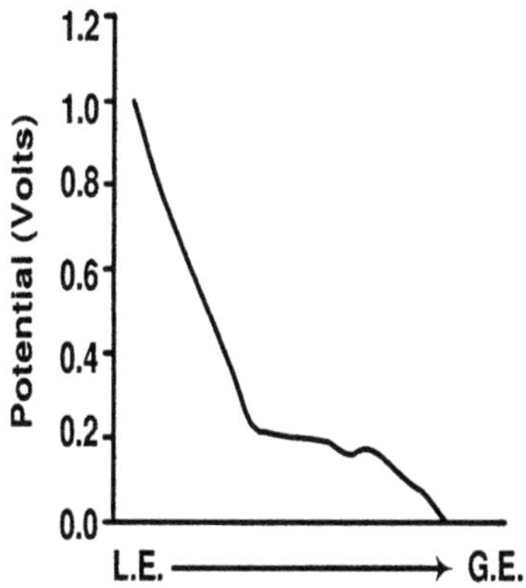


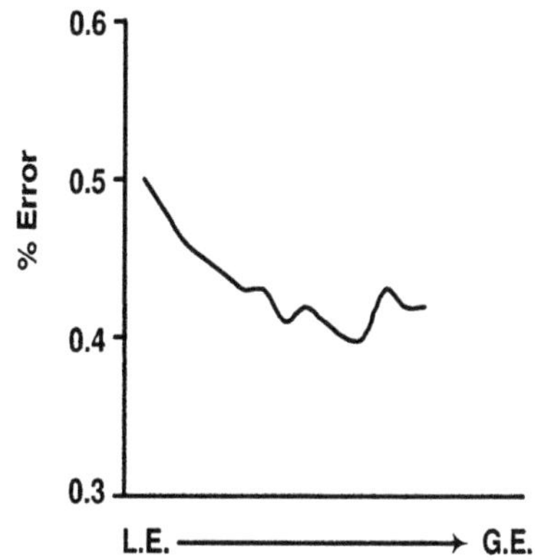


**Fig. 2. Distribution of $\Phi$ along the insulator surface**
**Fig. 3. Error in $D_n$ on insulator surface**

time and the requirement of the computer memory is much less for the evaluation of the entire field solution.

# 4 Application Example

The electrode-spacer arrangement shown in Fig. 1 has been taken as an application example in this paper. This arrangement is typically used in a 420 kV Gas Insulated System (GIS). The inner electrode is the live electrode and is maintained at a potential of 1 volt (normalized voltage has been considered for ready reference) and the outer cylinder is the ground electrode. Both the electrodes are made of extruded aluminium. The insulator is composed of epoxy cast resin whose relative permittivity is taken as 5.3 and the gas mixture used for purpose of insulation is a mixture of $SF_6/N_2$ (80 % $N_2$ and 20 % $SF_6$) whose relative permittivity is 1.005 [11] .

# 5 Results and Discussions

The potential plot along the insulator surface, as shown in Fig.2, shows that the potential along the insulator surface decreases uniformly from 1 V to 0 V as one moves from the live electrode towards the ground electrode which is expected because the live electrode is maintained at a potential of 1 V and the ground electrode at 0 V.

For the dielectric-dielectric interface, the boundary condition for the normal component of the electric flux density has to be maintained as is seen from Equation

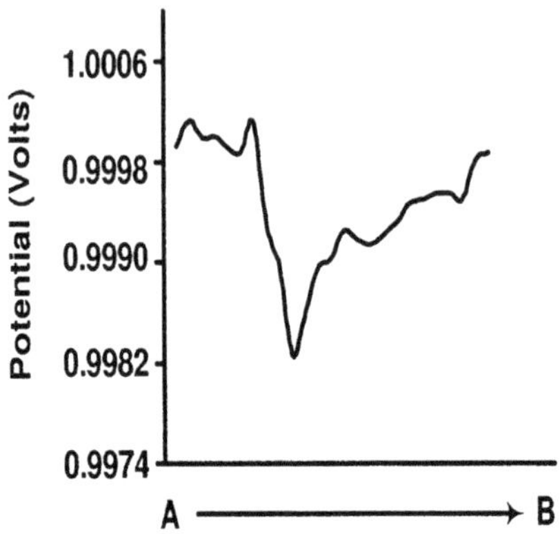


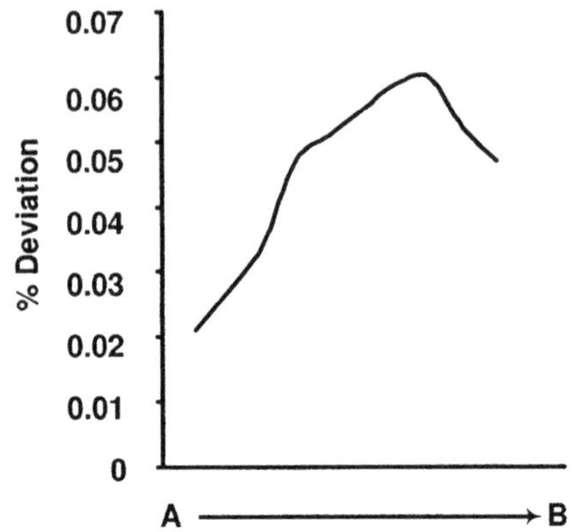


**Fig. 4. Distribution of $\Phi$ along the live electrode surface**
**Fig. 5. Percentage deviation in angle from** $90^0$

3. Fig.3 shows the percentage error on the spacer surface for normal flux densities calculated from both sides of the spacer surface i.e.

$$\%\,Error = \frac{D_{1n} - D_{2n}}{D_{1n}} \times 100 \qquad (11)$$

A total of 330 nodes on the insulator surface restricts this error within 0.5 %, which demonstrates the utility of the simulation tool developed.

Another boundary condition that must also be satisfied is that the given values of the potential are maintained on the conductor surfaces. Fig.4 shows that the potential at any point on the live electrode varies from 1 V to 0.9982 V. This is in good agreement with the boundary condition because the given voltage on the live electrode has been considered as 1 V, as mentioned earlier.

Fig. 5 shows the percentage deviation of the angle which the resultant stress makes with the electrode surface. The percentage deviation has been calculated from $90^0$ due to the fact that electric stress is always normal to the electrode surface. Fig. 5 reflects that this percentage deviation is well within 0.06 percent. This is in fairly good resonance with the expected result.

Fig.6 shows the distribution of resultant stress along the live electrode and the variation of the resultant stress along the spacer surface is presented in Fig. 7. Both of these stress distributions are in good agreement with the shape of the live electrode and the geometry of the spacer respectively. The electrode-spacer arrangement considered in this paper consist of 2400 elementary boundary surfaces

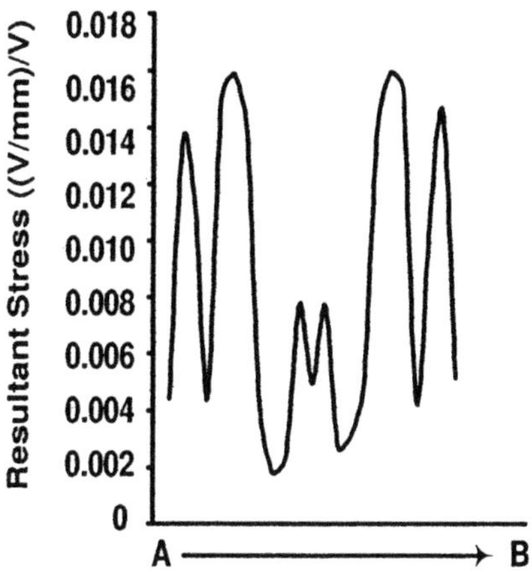


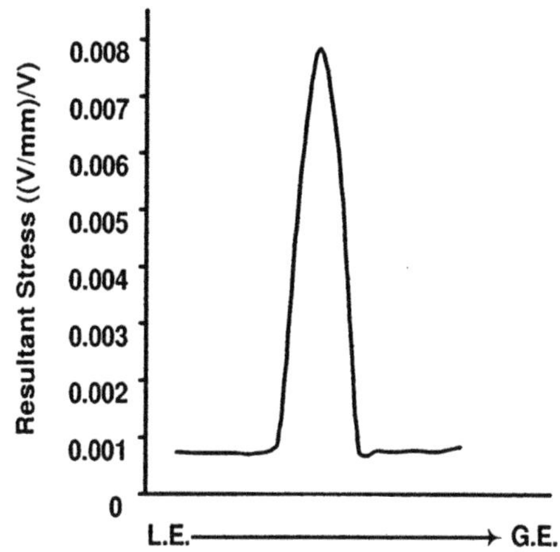


**Fig. 6. Stress Distribution along the live electrode surface**
**Fig. 7. Stress Distribution along the insulator surface**

with 975 nodes on the electrode surface and 330 nodes on the insulator surface. These are the optimized values of the respective parameters for the arrangement under study obtained by varying them over a wide range and observing the effect on the different significant quantities related with the electrostatic field computation as described in this paper.

For the configuration under study, the execution time required for obtaining the results, including the matrix solution of the set of system of equations, is 40 minutes on a Pentium III PC with 550 MHz CPU clock speed and 128 MB RAM.

# 6 Conclusions

The results of a 3-D field analysis presented in this paper reflects the utility of the developed software "ASYMBEM" in computing the complex electrode-spacer configurations used in high voltage system arrangements.

The main advantage of "ASYMBEM" lies in the fact that it is capable of analyzing electric field on and around the surfaces of complex high voltage system arrangements in a faster time and with a lesser memory requirement of the computer. The reason behind this is that "ASYMBEM" considers first order approximation of the boundary elements instead of a second order approximation. This first order approximation to curvilinear triangles approximates triangles of smaller areas to a fairly good degree of accuracy. Hence, "ASYMBEM" is made capable of generating meshes over the surfaces to produce boundary elements of desired areas.

## 7 Acknowledgements

A. Lahiri expresses his thanks to CSIR (India) for granting him Senior Research Fellowship to carry out this work. Dr. S. Chakravorti is thankful to AICTE for providing financial support to this work through the project grant 8019/RDII/R & D/ELE(181)/2000-01. The authors also wish to express their sincere thanks to Dr. Andreas Blaszczyk of ABB Corporate Research Center, Ladenburg, Germany for his continuous encouragement and valuable suggestions.

## References

1. H. Singer, H. Steinbigler and P. Weiss," A Charge Simulation Method for the C alculation of High Voltage Fields", IEEE Transactions on PAS, Vol. PAS-93, 1974, pp 1660-1668.
2. A. Blaszczyk, "Computation of Quasi-static Electric Fields with R egion Oriented Charge Simulation", IEEE Transactions on Magnetics, Vol. 32, 1996, pp 821-831.
3. Z. Andjelic, B. Krstajic, S. Milojkovic, H. Steinbigler, H. Singer and R. Witzmann, "Comparative Analysis of the BEM and CSM in 2D and 3D Electrostatic Field Calculation", 6th International Symposium on HV Engineering, New Orleans, 1989, Paper No. 24.09.
4. F. Gutfleisch, H. Singer, K. Foerger and J. A. Gomollon, "Calculation of High Voltage Fields by means of The Boundary Element Method", IEEE Transactions on PWRD, Vol. 9, 1994, pp 743-749.
5. J. H. McWhirter, J. J. Oravec, "Three-dimensional Electrostatic Field Solutions in a Rod Gap by a Fredholm Integral Equation", 3rd International Symposium on HV Engineering, Milan, 1979, Paper No. 11.14.
6. E. E. Okon, "Potential Integrals Associated with Quadratic Distribution in a Triangular Domain", IJNME, Vol. 21, 1985, pp. 197-207.
7. T. Misaki, H. Tsuboi, K. Itaka, T. Hara, "Computation of Three-Dimensional Electric Field Problems By A Surface Charge Method and its Application to Optimum Insulator Design", IEEE Transactions o Power Apparatus and Systems, Vol. PAS-101, 1982, pp 627-634.
8. Z. Andjelic, "Optimal Shape Investigation ofr a Disconnecting Element in $SF_6$ Gas Insulated Equipment", Proceedings of International AMSE Conference, Modelling and Simulation, Athens, Vol. 2.1, 1984, pp 299-307.
9. C. Vetter, "Adaptive Strategies for Numerical Integration in Electrostatic Field Computation", 10th International Symposium on HV Engineering, India, Vol. 1, 2001, pp 1-4.
10. G. R. Cowper, "Gaussian quadrature Formulas for Triangules", IJNME, Vol. 7, 1973, pp. 405-408.
11. L. G. Christophorou and R. J. Burnt, "$SF_6/N_2$ Mixtures, Basic and HV Insulation Properties", IEEE Transactions on Dielectrics and Electrical Insulation, Vol. 2, No. 5, 1995, pp 925-1002.

# A New Analysis Method For Subway Grounding Systems

A.O.A. Souza and L. Lebensztajn

EPUSP - Escola Politécnica da Universidade de São Paulo
LMAG - Laboratório de Eletromagnetismo Aplicado
Av. Prof. Luciano Gualberto, Travessa 3, no 158, 05508-900, São Paulo SP Brazil

**Abstract.** This work deals with the calculation of touch voltages and leakage current density distribution in direct-current subway grounding systems.

A new approach for direct-current subway grounding systems calculations was developed. In this approach, all system components are modelled as multi-input, multi-output blocks, which are interconnected using appropriate electrical equation.

Results obtained using the proposed approach show good agreement with results reported in previous works.

*Keywords* - **Grounding Systems, Subway, EMC**

## 1 Introduction

On subways, the feed system operates, on general, in direct current, using one of the two rails as a negative terminal.

The positive terminal can be a catenary or a lateral rail, called third rail (3T). In this work we will always call the positive terminal third rail, but the method can also be applied on subways that possess a catenary.

Three different metallic parts compose the subway grounding system as Fig. 1 shows:

**TV** - the railway grounding

**TT** - the tunnel grounding

**TE** - the external grounding

It is interconnected system, and Fig. 1 shows that a dangerous voltage between the platform and the train could be applied to a passenger. In order to decrease the voltage between the train and the platform, TV and TT could be interconnected. However, this solution implies a leakage density current increase and, as consequence, the system will suffer from corrosion. Another solution to this problem is isolate TV but this implies a touch voltage increase.

The solution for these two antagonistic solutions is the installation of the rail on a material with high electric resistance, which grounds TV to the structural hardware of the tunnel. TT is electrically connected, through concrete (high electric resistivity 600 $\Omega.m$) to the metallic rings that make the locking of the tunnel, called external grounding TE.

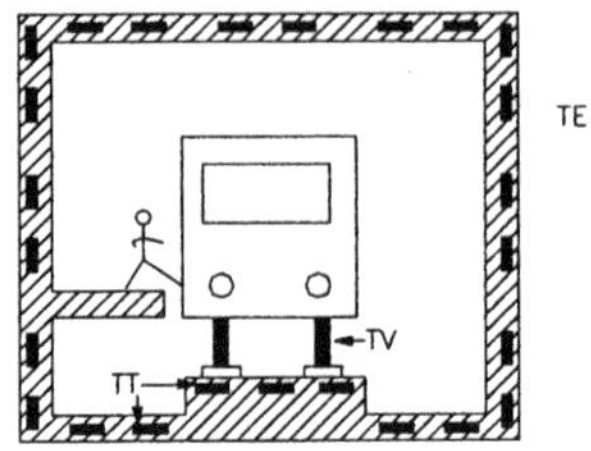


**Fig. 1.** The subway grounding system

## 2 Methodology

Figure 2 shows an elementary length of the railway system (length $\Delta x$).

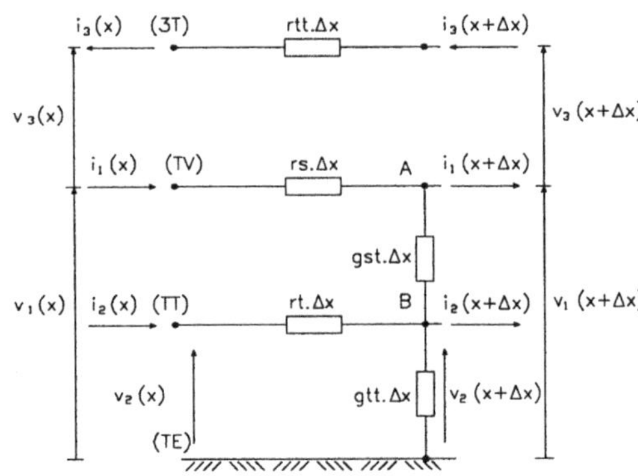


**Fig. 2.** An elementary length of the subway grounding system

Kirchhoff's Laws could be applied to this circuit in order to obtain:

$$\frac{d}{dx}\begin{bmatrix} v_1 \\ v_2 \\ v_3 \\ i_1 \\ i_2 \\ i_3 \end{bmatrix} = \begin{bmatrix} 0 & 0 & 0 & -r_S & 0 & 0 \\ 0 & 0 & 0 & 0 & -r_T & 0 \\ 0 & 0 & 0 & r_S & 0 & r_{TT} \\ -g_{ST} & g_{ST} & 0 & 0 & 0 & 0 \\ g_{ST} & -(g_{ST}+g_{TT}) & 0 & 0 & 0 & 0 \\ 0 & 0 & 0 & 0 & 0 & 0 \end{bmatrix} . \begin{bmatrix} v_1 \\ v_2 \\ v_3 \\ i_1 \\ i_2 \\ i_3 \end{bmatrix} = A. \begin{bmatrix} v_1 \\ v_2 \\ v_3 \\ i_1 \\ i_2 \\ i_3 \end{bmatrix}$$

This set of equations could be solved using an equivalent circuit as proposed by (Pereira, [1997]). This approach is based upon on the model of transmission lines proposed by (Stevenson, [1982]).

In this work, we use a more straightforward approach: the state variable method. The solution of the set of equations, when all initial currents and voltages are given, could be written as:

$$\begin{bmatrix} V_{1S} \\ V_{2S} \\ V_{3S} \\ I_{1S} \\ I_{2S} \\ I_{3S} \end{bmatrix} = e^{A.L} \begin{bmatrix} V_{1E} \\ V_{2E} \\ V_{3E} \\ I_{1E} \\ I_{2E} \\ I_{3E} \end{bmatrix}$$

where $L$ is the length of the railway.

This equation shows us that each length of a subway system could be analyzed as a multi-input, multi-output block.

The electrical substation could be analyzed as a voltage source as Fig. 3 shows. The train is, on general, modeled as a current source as Fig. 4 shows.

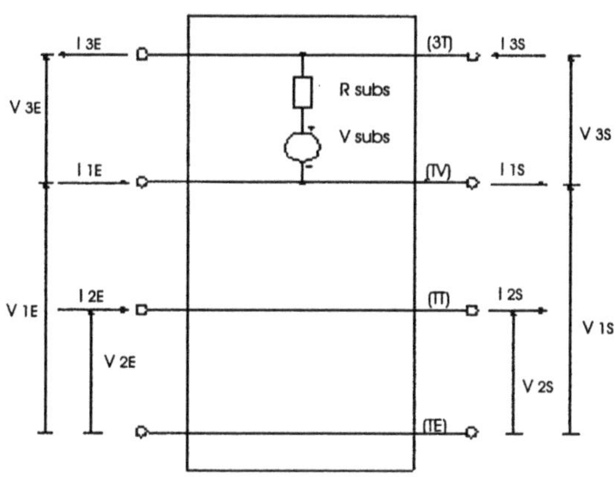


**Fig. 3.** The electrical substation and the grounding system

This allows us to write two sets of equations, the first for substation and the second to the train:

$$\begin{bmatrix} V_{1S} \\ V_{2S} \\ V_{3S} \\ I_{1S} \\ I_{2S} \\ I_{3S} \end{bmatrix} = \begin{bmatrix} 1 & 0 & 0 & 0 & 0 & 0 \\ 0 & 1 & 0 & 0 & 0 & 0 \\ 0 & 0 & 1 & 0 & 0 & 0 \\ 0 & 0 & G_{subs} & 1 & 0 & 0 \\ 0 & 0 & 0 & 0 & 1 & 0 \\ 0 & 0 & G_{subs} & 0 & 0 & 1 \end{bmatrix} \cdot \begin{bmatrix} V_{1E} \\ V_{2E} \\ V_{3E} \\ I_{1E} \\ I_{2E} \\ I_{3E} \end{bmatrix} + \begin{bmatrix} 0 \\ 0 \\ 0 \\ -G_{subs}.V_{subs} \\ 0 \\ -G_{subs}.V_{subs} \end{bmatrix}$$

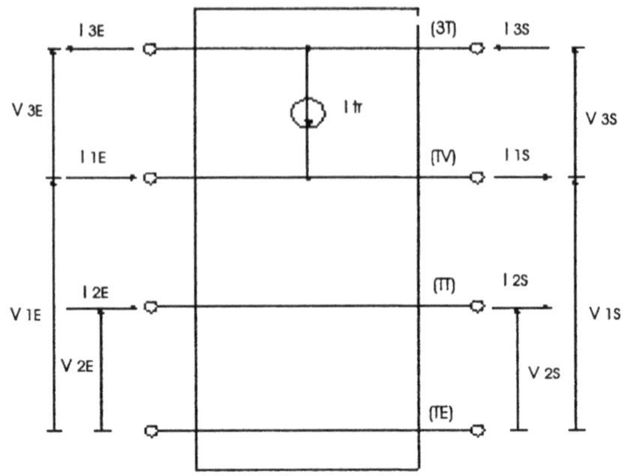


**Fig. 4.** The train of the grounding system

$$\begin{bmatrix} V_{1S} \\ V_{2S} \\ V_{3S} \\ I_{1S} \\ I_{2S} \\ I_{3S} \end{bmatrix} = \begin{bmatrix} V_{1E} \\ V_{2E} \\ V_{3E} \\ I_{1E} \\ I_{2E} \\ I_{3E} \end{bmatrix} + \begin{bmatrix} 0 \\ 0 \\ 0 \\ I_{tr} \\ 0 \\ I_{tr} \end{bmatrix}$$

So any part on subway grounding system could be analyzed as a set of equations for a multi-input, multi-output block.

$$[\gamma_s] = [k].[\gamma_E] + [C] \tag{1}$$

where $[\gamma_s]$ is the output state variable and $[\gamma_E]$ is the input state variable.

In order to calculate the currents and the voltages, it is imposed that the first and the last length have always about $500m$. Thus, the currents on left and right end of the whole system could be set to zero.

To solve the problem all the blocks are reduced to a single block on a recursive process. This single block has 6 equations and 6 unknowns (the voltages at the boundaries), because the currents at the boundaries are known. Equation (1) and the association of blocks allow us the calculation of the currents and voltages on every point of the grounding system.

## 3 Results

The model was used to analyze Paulista subway line at São Paulo, Brazil. Figure 5 show the substations and the position of the trains. In order to validate it, three cases where analyzed. The first case is a steady state problem. A short-circuit in the line is also analyzed. A bad contact between TT and TV was also analyzed.

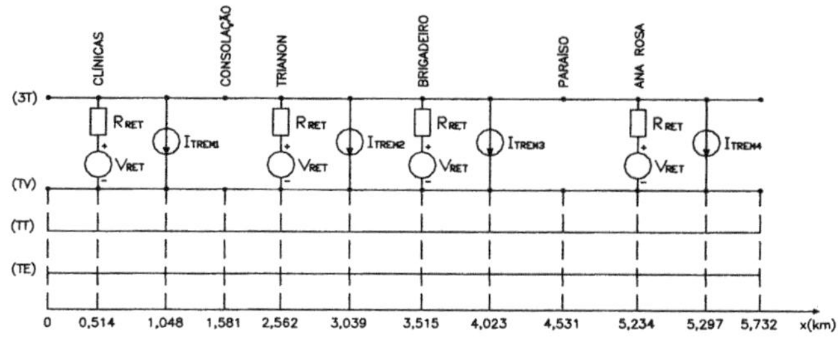


**Fig. 5.** The Paulista Subway Line (São Paulo - Brazil)

The voltage between the third rail and the rail (3T - TV) and the current in the third rail on steady state were calculated as a function of a distance, as Fig. 6 shows. In this condition, the voltage between TV and the external grounding TE as a function of a distance is shown on Fig. 7. The maximum measured voltage between TV and TE at this condition is $10V$ and the minimum measured voltage is $-7V$. Figure 7 shows a agreement between the adopted model and the measurements. It must be remarked that there are uncertainties on the model parameters and on the measurements, so only a qualitative comparison could be done. All the obtained values follows the VDE standard. A short-circuit on the middle of the line was

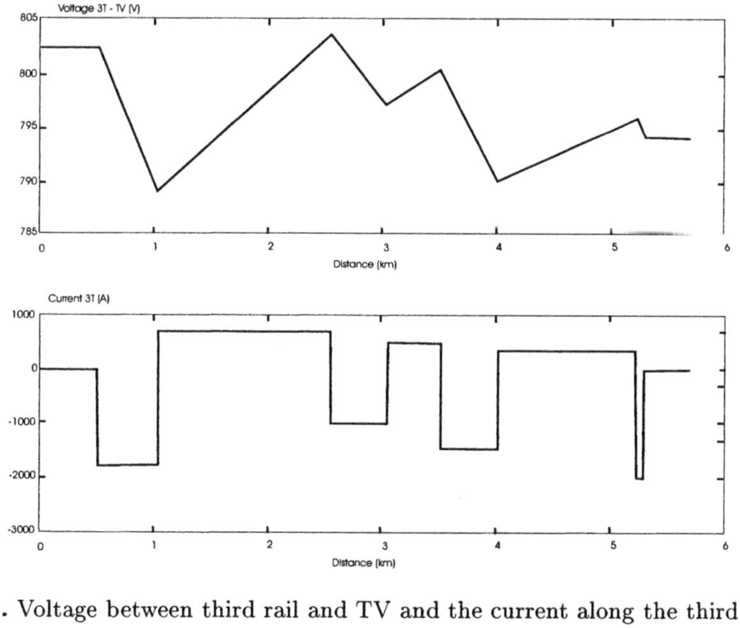


**Fig. 6.** Voltage between third rail and TV and the current along the third rail

also simulated. Figure 8 shows the voltage between (TV-TE). Figure 9 shows the

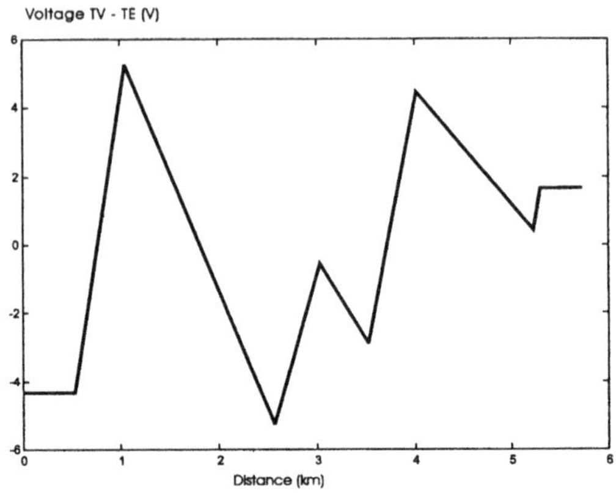


**Fig. 7.** Voltage between TV and TE

voltage between the third rail and (3T-TV) and the current on the third rail at this condition.

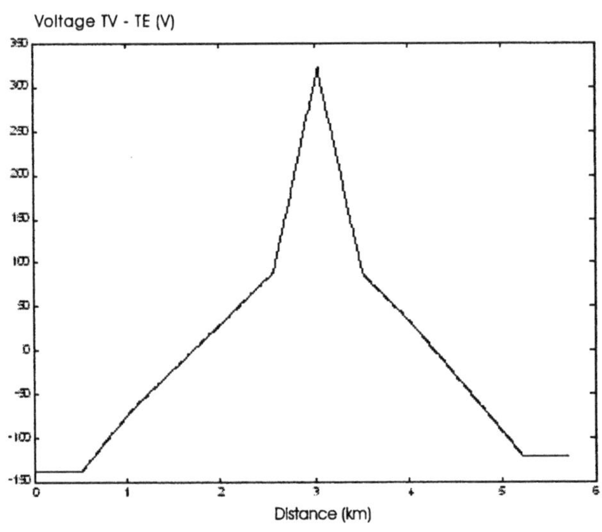


**Fig. 8.** Voltage TV-TE during a short-circuit

A poor grounding condition, e.g., a bad contact between TT and TV was also analyzed. Figure 10 shows current density distributions along the subway line. The bad contact between TT and TV is located between Clinicas station and Consolação station and shows that a corrosion process will begin, because the leakage current density has a very high local value.

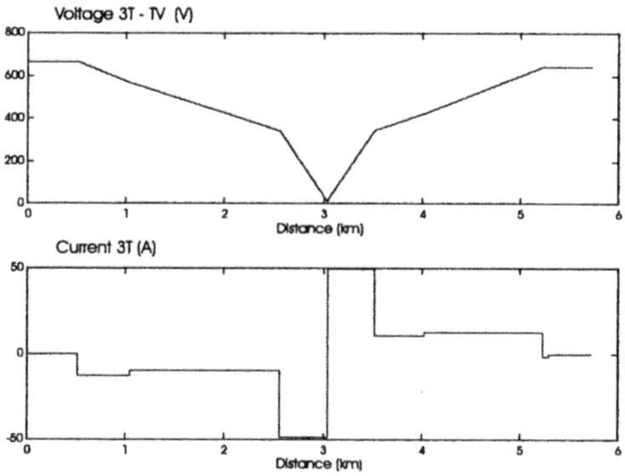


**Fig. 9.** Voltage between third rail and TV and the current along the third rail during a short circuit

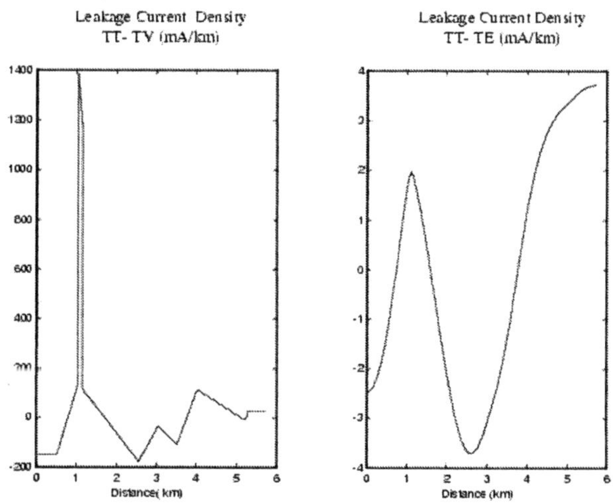


**Fig. 10.** Leakage Current Density along the railway: (TT-TV) and (TT-TE)

# 4 Conclusion

A new analysis method for subway grounding systems was proposed. It is based on state variables and multi-input, multi-output blocks. The proposed method is a powerful approach to subway grounding systems design because it is able to analize steady-state conditions, short-circuit conditions and even a poor grounding condition.

## References


[1997] Pereira da Silva, J.A.: Uma nova metodologia para avaliação de sistemas de aterramento metro-ferroviario, São Paulo, EPUSP, PhD Thesis (in portuguese), (1997)

[2001] De Souza, A.O.A.: Métodos de análise de sistemas de aterramento de linhas metroviárias, São Paulo, EPUSP, MSc Thesis (in portuguese), (2001)

[1982] Stevenson W.D.: Elements of Power Systems Analysis, McGraw Hill, fourth edition, New York, (1982)

[1995] VDE 0115/3.65 Bestimmungen für elektrische Bahnen

# Ab-Initio Calculation of Substrate Currents Using Ghost Field Gauging

Peter Meuris, Wim Schoenmaker, Wim Magnus, and Bert Maleszka

IMEC, Kapeldreef 75, B-3001 Heverlee, BELGIUM

**Abstract.** Recently a new approach was presented to determine the high-frequency electromagnetic behavior of on-chip passives and interconnects. The method solves the electric scalar and magnetic vector potentials in a prescribed gauge. The latter one is included by introducing an additional independent scalar field, whose field equation needs to be solved. This additional field is a mathematical aid that allows for the construction of a gauge-conditioned, regular matrix representation of the curl-curl operator acting on edge elements. This paper reports on the convergence properties of the new method and shows the first results of this new calculation scheme for VLSI-based structures at high frequencies. The high-frequent behavior of the substrate current, the skin effect and current crowding is evaluated.

## 1 Introduction

One of the important simulation challenges in VLSI design is the adequate characterization of high-frequency interconnects and on-chip passives. Important effects are substrate currents, current crowding at the edges of the interconnects due to skin effect and the proximity effect. The characteristic electrical length at the frequencies under consideration (GHz range) is rather large (cm scale). However, the mesh scale required for an accurate field calculation is determined by the very fine geometrical details (sub-micron scale) of the metal lines.
Although the physics of these problems is understood for a long time, detailed and fast calculation schemes are still lacking. We recently introduced an approach to simulate high-frequency effects of on-chip interconnects, dedicated to the specific geometry of the problem. The detailed description of the method is presented elsewhere [1-5].
The frequency domain is addressed. The meshing is Cartesian, suitable to show the validity of the model and to simulate on-chip interconnects, because in a first approximation, interconnects can be regarded as parallel to the axes of a Cartesian frame. However, this is not an essential restriction and the technique can be extended to unstructured meshes.
This paper is organized as follows: In the second section we discuss the need for a gauge condition. In the next section the essential properties of the solution method are given, emphasizing the novel aspects of the approach. Then the influence of the gauge fixing on the eigenvalue spectrum is discussed In the next section several VLSI-based benchmark structures are calculated. Finally we reach our conclusions.

## 2 Do We Need A Gauge?

To describe the electrodynamical environment, different approaches can be pursued. The electric and magnetic field variables can be used as independent variables. Since these variables are gauge invariant, no gauge condition is required. This is the case in most finite-difference schemes. However, to comply with the needs of IC designers who work in the (quasi-) static regime, a formulation that uses the potentials $V$ and $\mathbf{A}$ as independent variables simular to the FIT approach [6–8] is preferred. The electric potential $V$ is associated to the nodes (vertices), while the magnetic potential ($\mathbf{A}$) is put on the links (connections, edges) between the nodes of the mesh. The electrodynamic description results into a Poisson equation for the electric scalar potential and a curl-curl equation for magnetic vector potential respectively. These potentials are not uniquely defined which results in a singular matrix representation. In order to arrive at a unique solution for the potentials we need to introduce a gauge condition. The inclusion of a gauge condition, such as the Lorentz gauge or Coulomb gauge, is occasionally referred to as 'gauging'. The curl-curl equation can be regularized by eliminating the unknown vector potentials assigned to the edges of a spanning tree. However, this kind of gauging leads to a slow convergence of the Krylov-subspace iterative solvers [9].

But do we really need to carry out the extra work of fixing the gauge? Let us start with a matrix representation of a singular linear system:

$$M\,\mathbf{x} = \mathbf{b} \qquad\qquad det(M) = 0 \tag{1}$$

It has been shown that if $\mathbf{b}$ is in the range of $M$, then the standard conjugate gradient like methods are successful [9–11]. However, if $\mathbf{b}$ contains a component outside the range of $M$, then the problem is ill-posed and no convergence is reached. The iterative solution methods without gauging is effective if the right-hand side of the curl-curl equation can be constructed in such a way that its divergence vanishes, i.e. if there is no component outside the range of the curl-curl operator. This can be understood by realizing that the Krylov space $K_n$=span$\{\mathbf{x}, M\mathbf{x}, M^2\mathbf{x}, M^3\mathbf{x}, \ldots, M^{(n-1)}\mathbf{x}\}$ and that the search for the solution fully takes place in the range of $M$.

However such a construction is not always possible. Whereas in magnetostatic calculations with metallic conductors one can easily realize that at the start of solving the equation $\nabla \times \nabla \times \mathbf{A} = \mu_0 \mathbf{J}$, the condition $\nabla \cdot \mathbf{J} = 0$ is satisfied, this is much less easy if a non-linear dependence of the current $\mathbf{J}$ on the vector potential exists. This is the case for non-linear media such as semiconductors as well as for time-dependent fields. Furthermore, numerical errors are inevitable and especially for very large systems or systems that need many iterations, a small component of $\mathbf{b}$ outside the range of $M$ may be amplified and leading to lack of convergence. Therefore, a gauging may be preferred. Also for particular problems in time domain [12] or multigrid we must look for an adequate gauge construction. This might be the tree-cotree gauging [13], the grad-div gauging [7], or the ghost-field gauging [1].

## 3 Ghost-Field Solver

The method of [1-5] introduces an additional scalar field that needs to be obtained as part of the solution method. The solution for this additional field does not carry energy. Therefore, we have named it a 'ghost field', being a mathematical aid that allows for the construction of a gauge-fixed, regular matrix representation of the curl-curl operator acting on edge elements. With the use of the ghost-field gauging technique, the Maxwell problem also results into a Poisson problem for the scalar potential

$$-\nabla \cdot (\varepsilon \nabla V) = \rho, \tag{2}$$

and a curl-curl equation for the magnetic vector potential that is solved *together* with a gauge equation for the ghost field $\chi$:

$$\nabla \times \nabla \times \mathbf{A} - \gamma \nabla \chi = \mu_0 \mathbf{J} - \mu_0 \varepsilon \frac{\partial}{\partial t} \left( \nabla V + \frac{\partial \mathbf{A}}{\partial t} + \frac{\partial \nabla \chi}{\partial t} \right), \tag{3}$$

$$\nabla \cdot \mathbf{A} + \nabla^2 \chi = 0. \tag{4}$$

An extra parameter $\gamma$ (with dimension $m^{-2}$) is introduced in order to account for the dimensions of the system. So instead of the curl-curl operator combined with the gauge condition

$$\begin{pmatrix} \nabla \times \nabla \times \\ \nabla \cdot \end{pmatrix}, \tag{5}$$

that lead to matrices $M$ that are sparse, well-posed, yet *not* square, the operator

$$\begin{pmatrix} \nabla \times \nabla \times & -\gamma \nabla \\ \nabla \cdot & \nabla^2 \end{pmatrix}, \tag{6}$$

is considered. This operator leads to matrices $M$ that are also sparse, regular *and* square. Moreover, the resulting matrices are semi-definite and therefore well suited for iterative solvers.

## 4 Eigenvalue spectrum

The major drawback of the existing gauging methods is their influence on the convergence rate of iterative Krylov-based methods. Tree-cotree gauging for instance does not affect the maximum eigenvalue, but the lowest nonzero eigenvalue is reduced resulting in a reduction of the condition number of the matrix and therefore (much) slower convergence [9].

Predicting the convergence behavior of linear iterative systems is a very difficult task to carry out. Only for CGS there are exact proofs that the condition number is a good indication for convergence. When the condition number is *high* (*low*), CGS will be converging *slowly* (*fast*).

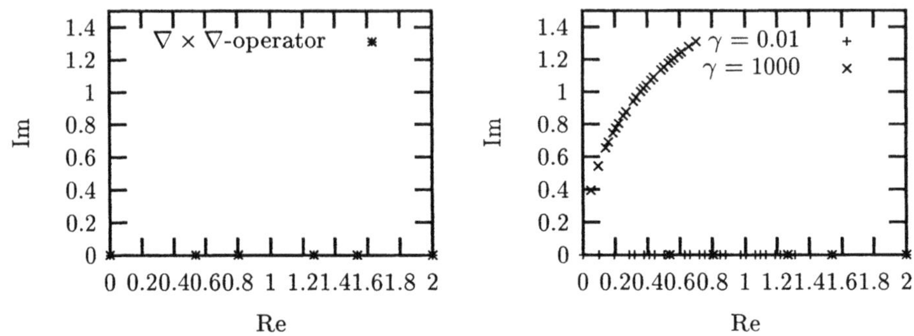


**Fig. 1.** Spectrum of the singular $\nabla \times \nabla$-operator and the regularized version for different values of $\gamma$ before pre-conditioning.

In order to evaluate the ghost-field gauging we look at the eigenvalue spectrum of the matrix generated by discretizing the system (3)-(4), and the influence of the tuning factor $\gamma$. A regular mesh with 7 nodes in each direction is examined. This accounts for 450 degrees of freedom for the singular curl-curl matrix, and 575 degrees of freedom for (3)-(4). The results are shown in Figure 1 before preconditioning and in Figure 2 after preconditioning.

- *Before preconditioning:* Figure 1 shows that eigenvalues of the curl-curl operator and the system (3)-(4) before preconditioning. The curl-curl matrix is symmetric and therefore the corresponding eigenvalues are all real. Furthermore because of the singular character of the matrix we find multiple (125) zero eigenvalues.
  When $\gamma$ is increased, the zero eigenvalues transform in small eigenvalues, while the nonzero eigenvalues remain unchanged. The effect on the condition number is dramatic however. While for $\gamma = 0$ the condition number is not affected by the zero eigenvalues (only the lowest nonzero eigenvalue is important), this is no longer the case for finite values of $\gamma$. In the latter case, the lowest nonzero eigenvalue is very small, and therefore the condition number large. For very large values of $\gamma$, we find 250 eigenvalues at the upper end of the spectrum. The rest of the eigenvalues are still the same 325 of the curl-curl operator. So the smallest eigenvalue for large $\gamma$ is the same is the smallest nonzero eigenvalue for $\gamma = 0$.
- *After preconditioning:* Figure 2 shows the eigenvalues after ICLU preconditioning for the curl-curl operator and the regularized one. While the eigenvalues for the singular operator are close to zero, the regularized version accounts for extra imaginary parts in the eigenvalue spectrum.

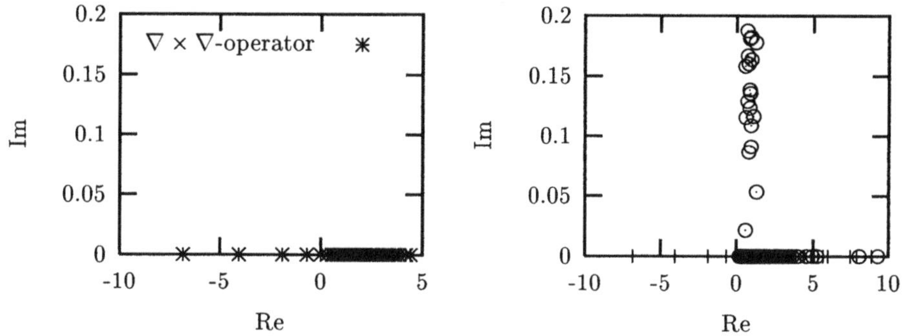


**Fig. 2.** Spectrum of the singular ∇ × ∇-operator and the regularized version for different values of $\gamma$ after preconditioning (circles for $\gamma$=1000, plusses for $\gamma$=0.01)

# 5 Simulation results

## 5.1 Skin effect

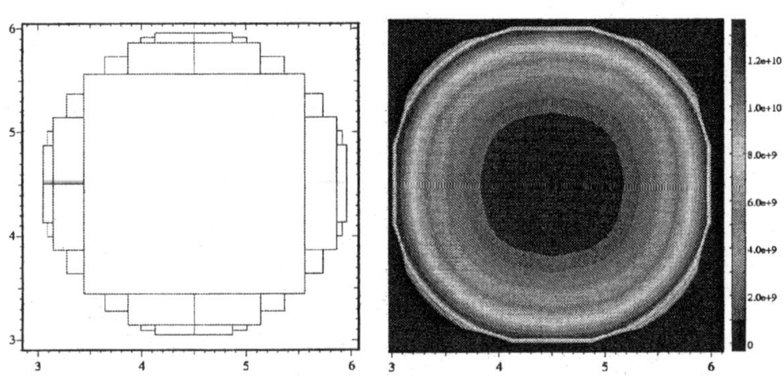


**Fig. 3.** Cartesian approximation of a circle, and the current density at 100GHz.

The quantitative description of skin effect in a cylindrical wire can be found in many text books. Although the discovery of the effect took place more than 125 year ago, nowadays still papers appear on resistance and inductance calculations [14,15]. The internal impedance of a cylindrical wire with radius $a$ and skin depth

$\delta$ is given by [16],

$$Z_{int} = \frac{1+j}{2\pi a \delta \sigma} \frac{I_0\left[(1+j)a/\delta\right]}{I_1\left[(1+j)a/\delta\right]}. \tag{7}$$

with the use of the Bessel functions $I_n$. This is an excellent benchmark problem for high-frequency solvers. We start with a brick representation for the circular form (Figure 3). The current density at 100 GHz is also shown, and for an aluminum cylinder, the skin depth becomes 0.28 $\mu m$ as can be verified in Figure 3.

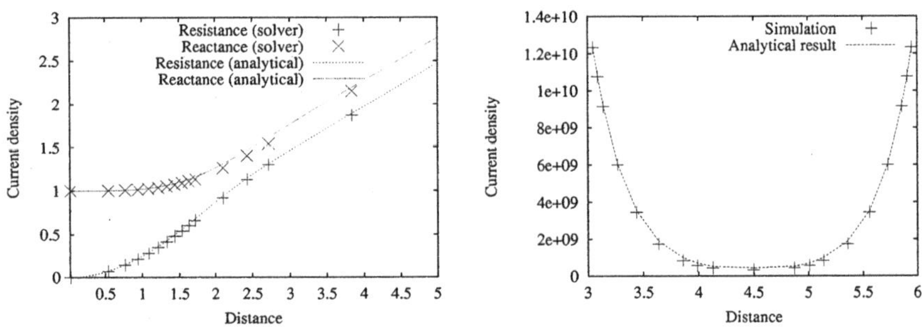


**Fig. 4.** The internal impedance of a cylindrical wire and the current density in the wire (analytical and simulated).

The impedances calculated with this solver compared with the analytical solution, can be seen in Figure 4. For the line resistance (upper curves), the relative error is less than 0.08, while for the reactance (lower curves), the results are much better, and the two curves match closely.

## 5.2 Ring Structure

To illustrate the method, we show the results of an aluminum ring embedded in silicon oxide, on top of a moderately conducting substrate (Fig 4) at 500 MHz. The ring dimensions are 90 x 50 $\mu$m, the substrate is modeled as a low conductive metal ((=0.01 ($\rho$m) and we used a 20 x 20 x 20 mesh. An AC (electric) voltage is put on one of the output ports with an amplitude of 0.01 V, forcing a current in the ring.

- The skin effect of the currents in the lines that connect the output ports and the ring is shown in Fig 6a. The currents is crowding at the inner edges of the conductor.
- Fig 6b show the current density in the substrate, and the circular Eddy currents in the substrate. In a vector plot of the current density we observe that the current is flowing in the opposite direction of the current in the aluminum ring. A region of higher substrate current occurs under the input ports due to a higher magnetic field and hence a higher induced current.

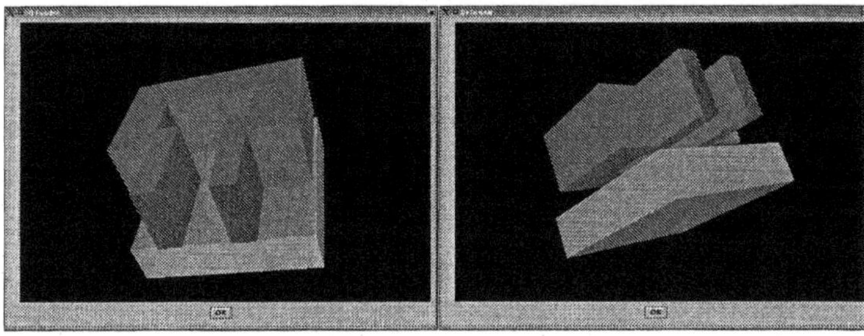

**Fig. 5.** Aluminum ring above a low conductive substrate

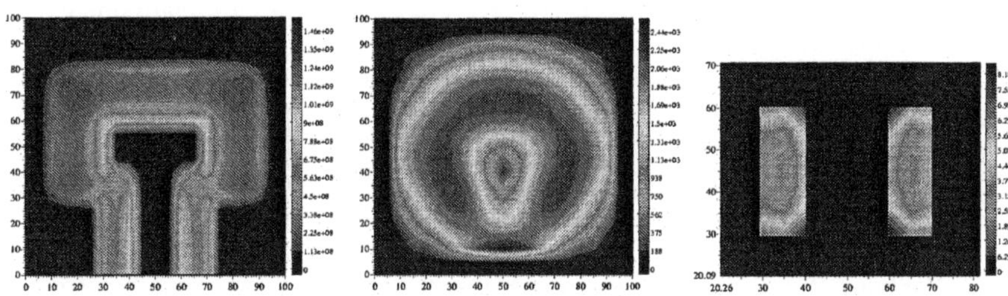

(a) (b) (c)

**Fig. 6.** Top view of the current densities in the ring (a), of the substrate (b) and side view of current densities of the ports (c) at 500 MHz.

- The side view on the current density of the ports (Fig 6c) shows the proximity effect. Because the currents are flowing in the opposite directions for the two ports, these currents will attract each other and the highest current density can be found in the inner corners. Fig. 6 also shows that the current density is the highest in the inner part of the ring, due to the proximity effect.

# 6 Conclusions

In this work we showed that a new approach of ghost field gauging can make the curl-curl operator square and regular. The complex values of the eigenvalues depend on the choice of the additional parameter $\gamma$, and depending on this parameter, different convergence patterns can be expected.

High-frequency effects are recovered with the use of the newly developed approach dedicated to spiral inductors and passive structures [1-5].

## 7 Acknowledgments

Henk Van der Vorst and Domenico Lahaye are acknowledged for sharing their knowledge on linear solvers. This work is partly financially supported by the Flemish Institute of Science and Technology (IWT) and the European Commission.

## References

1. P. Meuris, W. Schoenmaker and W. Magnus, Stategy for electromagnetic interconnect modeling, *IEEE Trans. on CAD*, 20, 6, 753–762, June 2001
2. W. Schoenmaker, P. Meuris, Electromagnetic interconnects and passives Modeling: Software Implementation Issues, *IEEE Trans. on CAD* , 21, 5, 534–543, May 2002,
3. W. Schoenmaker, W. Magnus, P. Meuris, Ghost fields in Classical Gauge Theories, *Phys. Rev. Lett.*, **88**, 18, May 2002
4. P. Meuris, W. Schoenmaker, W. Magnus, Inductance Calculations based on Lattice-Gauge Electromagnetic Modeling, submitted to IEEE Trans. on advanced packaging, 2002
5. W Schoenmaker, Wim Magnus, Peter Meuris and Bert Maleszka, Renormalization Group Meshes and the Discretization of TCAD, *IEEE Trans. on CAD*, 1425–1433, december 2002
6. T. Weiland, Time domain electromagnetic field computation with finite difference method, *Int. J. Num. Mode.: ENDF*, **9**, 259–319, 1996
7. M. Clemens and T. Weiland, Regularization of Eddy-Current Formulations Using Discrete Grad-Div Operators, *IEEE Trans. on Magnetics*, **38**, 2, March 2002
8. M. Clemens and T. Weiland, Magnetic Field Simulations Using Conformal FIT Formulations, *IEEE Trans. Magn.*, **38**, 2, March 2002
9. H. Igarashi, On the Property of the curl-curl matrix in finite element analysis with edge elements, *IEEE Trans. Magn.*, **37**, 5, September 2002
10. H. Igarashi and T. Honma, On Convergence of ICCG Applied to Finite-Element Equation for Quasi-Static Fields, *IEEE Trans. Magn.*, **38**, 2, March 2002
11. A. Kameari and K. Koganezawa, Convergence of ICCG Method in FEM Using Edge Elements without Gauge Condition, *IEEE Trans. Magn.*, **33**, 2, March 1997
12. Alain Bossavit, 'Stiff' Problems in Eddy-Current Theory and the Regularization of Maxwell's equations, *IEEE Trans. on Magn.*, **37**, 5, September 2002
13. J.B. Manges, Z.J. Cendes, A generalized tree-cotree gauge for magnetic field computation, *IEEE Trans. Magn.*, 31, 1342–1345, 1995
14. M. J. Tsuk and J. A. Kong, A hybrid method for the calculation of the resistance and inductance of transmission lines with arbitrary cross sections, IEEE Trans. Microwave Theory Tech., **39**, pp. 1338-1347, 1991.
15. F. Medina, R. Marques, Comments on "Internal Impedance of Conductors of Rectangular Cross Section", *IEEE Trans. Microwave Theory Tech.*, **49**, 1511–1512, 2001
16. M. J. Tsuk, The internal impedance of conductors with arbitrary cross-sections, 2000

# Numerical Simulation of Resonant Tunneling Diodes with a Quantum-Drift-Diffusion Model

Stefano Micheletti[1], Riccardo Sacco[1], and Paolo Simioni

MOX - Modeling and Scientific Computing, Dipartimento di Matematica "F. Brioschi", Politecnico di Milano, via Bonardi 9, I-20133 Milano, Italy

**Abstract.** We deal with a Quantum-Drift-Diffusion (QDD) model for the description of transport in semiconductors which generalizes the standard Drift-Diffusion model (DD) through extra terms that take into account some quantum dispersive corrections. We also study numerically the influence on the I-V curve of the electron effective mass, the barrier height and width, and of the ambient temperature. The performance of several linearization algorithms, i.e. a two Gummel-type iterations and the fully-coupled Newton method are also compared.

## 1 The QDD Model

The unipolar QDD model for a semiconductor occupying the open bounded region $\Omega \subset \mathbb{R}^d$, $d = 1, 2, 3$, comprises the following set of equations in the space-time cylinder $\Omega \times (0, t_f]$:

$$\begin{cases} \dfrac{\partial n}{\partial t} + \nabla \cdot (\mu n \nabla F) = 0 \ , \\ F = V \underbrace{+V_{\mathrm{bar}}}_{\text{(I)}} - V_{\mathrm{th}} \ln \dfrac{n}{n_{\mathrm{i}}} \underbrace{+ \dfrac{\hbar^2}{6mq} \dfrac{1}{\sqrt{n}} \nabla^2 \sqrt{n}}_{\text{(II)}} \ , \\ \nabla^2 V = \dfrac{q}{\epsilon} (n - N_{\mathrm{d}}) \ , \end{cases} \tag{1}$$

where $n$ is the electron concentration, $V$ the electric potential, $F$ the quasi-Fermi potential, $N_{\mathrm{d}}$ the doping profile, $V_{\mathrm{bar}}$ the barrier potential, $V_{\mathrm{th}}$ the thermal voltage, $n_{\mathrm{i}}$ the electron intrinsic concentration, $\mu$ the mobility, while the quantities $\hbar, q, \epsilon, m$ are the reduced Planck constant, the (positive) electron charge, the semiconductor permettivity and the electron effective mass, respectively. The above system, supplemented with suitable initial and boundary conditions, is to be solved for the unknowns $n, V, F$. The electron current density is given by the constitutive law $\underline{J} = -q\mu n \nabla F$, so that $(1)_1$ represents the classical continuity equation but with a nonclassical constitutive law for the current density. Moreover, like in the DD case, the temperature is supposed to be constant, having neglected all the energetic exchange phenomena. In $(1)_1$-$(1)_3$ we have singled out the two terms that characterize the QDD model with respect to the DD model, i.e.

(I) is the potential due to the presence of the heterostructure corresponding to barriers and wells between the different materials, while
(II) is the dispersive term, modeling a typical quantum effect [1].

The bipolar QDD model has been studied in [7] where the conditions under which existence and uniqueness of the solution at thermodynamic equilibrium hold are discussed, basically, via a variational argument consisting in minimizing the total energy of the system. In nonequilibrium conditions, the uniqueness of the solution is proved to hold only for moderate external voltages ([7]). This fact has a real counterpart in RTD's which exhibit a hysteresis cycle for particular values of the external voltage, as proved experimentally, as well as numerically ([3]).

# 2 Iterative Maps and Numerical Algorithms

In this section we address both the linearization techniques and the numerical issues necessary to obtain an approximate solution to the QDD equations.

## 2.1 Iterative Maps

To solve the (nonlinear) system $(1)_1$-$(1)_3$ it is necessary to employ some linearization techniques. In particular, we have compared three different functional iterative algorithms:

1. The fully-coupled Newton method;
2. A two-step fixed-point map for $F$ which can be summed up as: Given $F^{(k)}$, $V^{(k)}$, $n^{(k)}$, first
   - solve $(1)_2$-$(1)_3$ for $n^{(k+1)}, V^{(k+1)}$ with coupled Newton's method, then
   - solve the (linear) equation $(1)_1$ for $F^{(k+1)}$;
3. A three-step generalized Gummel's map such that: Given $F^{(k)}, V^{(k)}, n^{(k)}$
   - solve the (nonlinear) equation $(1)_2$ for $n^{(k+1)}$, then
   - solve the nonlinear version of $(1)_3$ for $V^{(k+1)}$

$$\nabla^2 V^{(k+1)} = \frac{q}{\epsilon}\left(n^{(k+1)}\exp\left(\frac{V^{(k+1)}-V^{(k)}}{V_{\mathrm{th}}}\right) - N_{\mathrm{d}}\right) \quad , \text{ finally}$$

   - solve the (linear) equation $(1)_1$ for $F^{(k+1)}$.

The three algorithms above are ordered in decreasing degree of coupling: algorithm 1. is a very well-known general purpose method, algorithm 2. is theoretically and numerically studied in [7], whereas algorithm 3. is our novel contribution, first devised in [8]. This last algorithm generalizes the classical Gummel map to the QDD model, whereby three successive subproblems are to be solved at each step. In particular, it consists of two nonlinear steps for $n^{(k+1)}$ and $V^{(k+1)}$, respectively, and a linear step for $F^{(k+1)}$. We have applied the three algorithms to the solution of a RTD and some numerical results are shown in Sect. 3. Hereafter, we shall study the stationary QDD model, i.e. we shall assume $\frac{\partial \cdot}{\partial t}$ for all the variables. Let us dwell on algorithm 3. which can be considered as a generalized Gummel map, see e.g. [4].

From $(1)_2$ and letting $\alpha = \frac{\hbar^2}{6mq}$, we have $n = n_\mathrm{i} \exp\left(\frac{\alpha \frac{1}{\sqrt{n}} \nabla^2 \sqrt{n} + V + V_\mathrm{bar} - F}{V_\mathrm{th}}\right)$, and substituting in $(1)_3$, we obtain

$$\nabla^2 V = \frac{q}{\epsilon}\left(n_\mathrm{i} \exp\left(\frac{\alpha \frac{1}{\sqrt{n}} \nabla^2 \sqrt{n} + V_\mathrm{bar}}{V_\mathrm{th}}\right) \exp\frac{V-F}{V_\mathrm{th}} - N_\mathrm{d}\right) , \tag{2}$$

which is the nonlinear version of the Poisson equation. Given $V^{(k)}, F^{(k)}$, $k \geq 0$, the first step of the algorithm consists in solving the following boundary value problem for $n = n^{(k+1)}$

$$\begin{cases} \alpha \frac{1}{\sqrt{n}} \nabla^2 \sqrt{n} - V_\mathrm{th} \ln\frac{n}{n_\mathrm{i}} + V_\mathrm{bar} + V^{(k)} - F^{(k)} = 0 \text{ in } \Omega , \\ n = n_\mathrm{eq} \qquad \text{on } \Gamma_\mathrm{D} , \\ \nabla n \cdot \underline{\nu} = 0 \qquad \text{on } \Gamma_\mathrm{N} , \end{cases} \tag{3}$$

where $n_\mathrm{eq}$ is the equilibrium value of the concentration, while $\Gamma_\mathrm{D}, \Gamma_\mathrm{N}$ are two subsets of the boundary $\partial\Omega$ such that $\partial\Omega = \overline{\Gamma}_\mathrm{D} \cup \overline{\Gamma}_\mathrm{N}$, with $\Gamma_\mathrm{D} \neq \emptyset$, and $\underline{\nu}$ is the unit outward normal vector to $\partial\Omega$. Then from (3) we formally obtain $V_\mathrm{th} \ln \frac{n^{(k+1)}}{n_\mathrm{i}} - V^{(k)} + F^{(k)} = \alpha \frac{1}{\sqrt{n^{(k+1)}}} \nabla^2 \sqrt{n^{(k+1)}} + V_\mathrm{bar}$ , which used in (2) allows us to write the second step of the algorithm, to solve for $V = V^{(k+1)}$, as

$$\begin{cases} \nabla^2 V = \frac{q}{\epsilon}\left(n^{(k+1)} \exp\left(\frac{V - V^{(k)}}{V_\mathrm{th}}\right) - N_\mathrm{d}\right) \text{ in } \Omega , \\ V = V_\mathrm{eq} + V_\mathrm{ext} \qquad \text{on } \Gamma_\mathrm{D} , \\ \nabla V \cdot \underline{\nu} = 0 \qquad \text{on } \Gamma_\mathrm{N} , \end{cases} \tag{4}$$

where $V_\mathrm{eq}, V_\mathrm{ext}$ are the equilibrium value of the potential and the external applied voltage, respectively. We point out that this equation can be regarded as the QDD counterpart of the DD nonlinear Poisson equation for $V$, i.e.

$$\nabla^2 V = \frac{q}{\epsilon}\left(n_\mathrm{i} \exp\left(\frac{V-F}{V_\mathrm{th}}\right) - N_\mathrm{d}\right) ,$$

where $n_\mathrm{i}$ is replaced by $n^{(k+1)} \exp\left(\frac{F^{(k)} - V^{(k)}}{V_\mathrm{th}}\right)$. Finally, the last step of the algorithm requires solving the linear problem for $F = F^{(k+1)}$

$$\begin{cases} \nabla \cdot (\mu n^{(k+1)} \nabla F) = 0 \text{ in } \Omega , \\ F = F_\mathrm{eq} + V_\mathrm{ext} \qquad \text{on } \Gamma_\mathrm{D} , \\ \nabla F \cdot \underline{\nu} = 0 \qquad \text{on } \Gamma_\mathrm{N} , \end{cases} \tag{5}$$

where $F_\mathrm{eq}$ is the equilibrium value of the quasi-Fermi level. The sequence of problems (3)-(5) defines our version of the Gummel map applied to the iterative solution of the QDD model. We remark that the nonlinear steps (3)-(4) are both linearized with Newton's method. Moreover, computationally, it is advantageous to solve problem (3) for $w = \sqrt{n}$ instead of $n$, which yields the following problem for $w = w^{(k+1)}$

$$\begin{cases} \alpha \nabla^2 w - 2V_{\mathrm{th}} w \ln \dfrac{w}{w_{\mathrm{i}}} + w(V_{\mathrm{bar}} + V^{(k)} - F^{(k)}) = 0 \text{ in } \Omega \ , \\ w = w_{\mathrm{eq}} \qquad \text{on } \Gamma_{\mathrm{D}} \ , \\ \nabla w \cdot \underline{\nu} = 0 \qquad \text{on } \Gamma_{\mathrm{N}} \ , \end{cases} \tag{6}$$

where $w_{\mathrm{i}} = \sqrt{n_{\mathrm{i}}}$ and $w_{\mathrm{eq}} = \sqrt{n_{\mathrm{eq}}}$.

### 2.2 Numerical Algorithms

As far as the discretization is concerned, all of the three subproblems (6), (4)-(5), after linearization, are solved using piecewise linear finite elements. Special care is necessary for (5) which requires using piecewise harmonic averages for the terms $\mu n^{(k+1)}$. Important issues to deal with are the scaling of the unknowns and the computational cost required to solve all of the algebraic systems arising after the discretization. With reference to a one-dimensional finite element mesh whose characteristic size is $h$, the following scaling has been used throughout:

$$\overline{n} = Nh^4 \ , \qquad \overline{V} = 10^{-2} \frac{\alpha}{h^2} \sqrt{\frac{\overline{n}}{N}} \ , \qquad \overline{F} = \overline{V} \ ,$$

where $N$ is the value of the doping at the contacts. We have carried out an extensive numerical validation which proves that this scaling guarantees better conditioning of the algebraic linear systems and more equilibrated coefficient matrices. Finally, to compare the three algorithms in terms of computational cost, we have implemented all the numerical codes in Matlab and we have simulated a one-dimensional RTD under different conditions. The results, extensively discussed in [8], show that algorithm 1. is the cheapest one when the sparse structure of the matrices is taken into account, otherwise algorithm 3. performs better. Of course, the algorithms might perform other ways when going to multi-dimensional problems.

## 3 Numerical Results

We show in this section several numerical results referring to a one-dimensional RTD, i.e. a heterostructure based on two AlGaAs barriers and a quantum GaAs well. The model correctly reproduces the Negative Differential Resistance (NDR) of the I-V characteristic of the device. We aim at studying physical phenomena like the dependence of the I-V characteristics on the electron effective mass, ambient temperature, barrier height and width, as well as to test the numerical algorithms, as already discussed in the previous section. We have considered a RTD whose geometry is shown in Fig. 1(left): The device length is 75 nm, the doping profile is $5 \times 10^{21}\ \mathrm{m}^{-3}$ in the channel (well and barriers) and $10^{24}\ \mathrm{m}^{-3}$ elsewhere, while all the other parameters vary according to the following simulations. The mobility is assumed to be temperature dependent (cfr. [6]).

**Dependence on the Effective Mass**
The following figures show some numerical results for a RTD at 77 K. The barrier profile is shown in Fig. 1(right): the height and width are 0.3 V and 5 nm, respectively, while the quantum well is 5 nm wide. Throughout, $J_{\max}, J_{\min}$ denote the current density at the peak and at the valley, respectively, and the Peak to Valley Ratio (PVR) is defined as PVR= $J_{\max}/J_{\min}$. The next four figures in Fig. 2

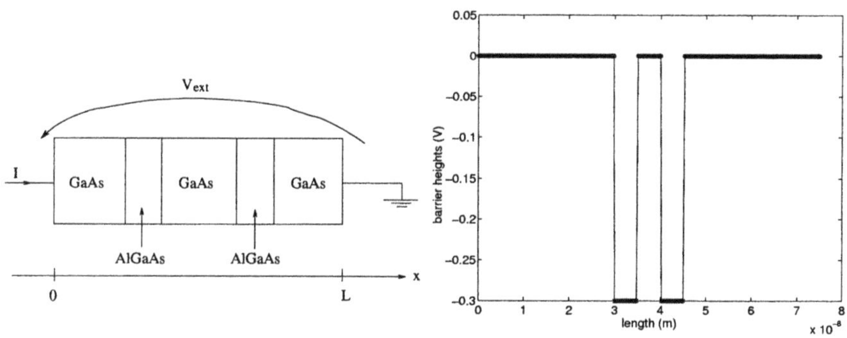


**Fig. 1.** Geometry (left) and barrier profile (right)

display the I-V characteristic for the following values of the effective mass $m$ (from left to right and top-down): $[0.067, 0.126, 0.1675, 0.201]m_0$, where $m_0$ is the free-electron mass. Table 1 (left) summarizes the main results. Notice that the higher the effective mass, the stronger the NDR phenomenon, while the PVR increases progressively. Notice that for the lowest value of the effective mass the NDR does not occur. The last pair of figures in Fig. 3 show the electron concentration at the peak and at the valley of the current for the fixed value of the effective mass $m = 0.126m_0$. In particular, notice the large values of the concentration inside the quantum well, especially at the valley which confirm the resonant phenomenon, in accordance with the basic theory of the device ([5]).

**Dependence on the Temperature**
The second series of simulations is carried out with barrier height and width of 0.35 V and 5 nm, respectively, while the quantum well width is 5 nm and the effective mass is $m = 0.126m_0$. Several values of the ambient temperature are considered, i.e. $[77, 100, 150, 200, 300]$ K. Table 1 (right) collects the results (in I.S. units). Observe that the NDR is a typical low temperature phenomenon, weakening as the temperature rises. This behavior can be explained theoretically by considering the different dependence of the current at the valley and at the peak on the temperature. In the first case, the resonant level is inside the energy gap, thus the higher the temperature, the larger the current, because more electrons occupy the energy levels near the conduction band. In the second case, the resonant level coincides with the conduction band, thus the current is less sensitive to the temperature because all the levels near the conduction band have large tunneling probabilities,

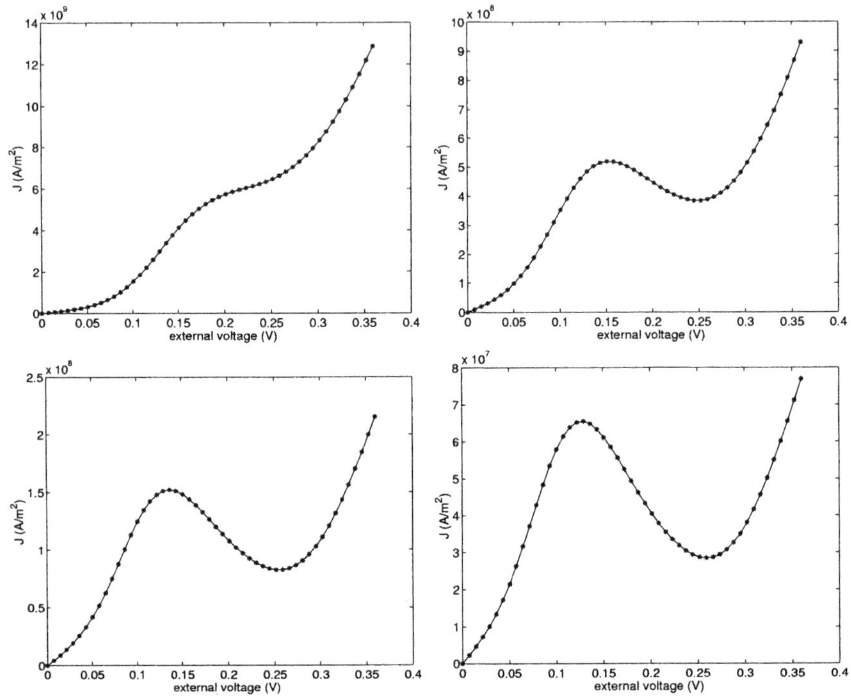


**Fig. 2.** I-V characteristic: $m = 0.067m_0$ (top left), $m = 0.126m_0$ (top right), $m = 0.1675m_0$ (bottom left) and $m = 0.201m_0$ (bottom right)

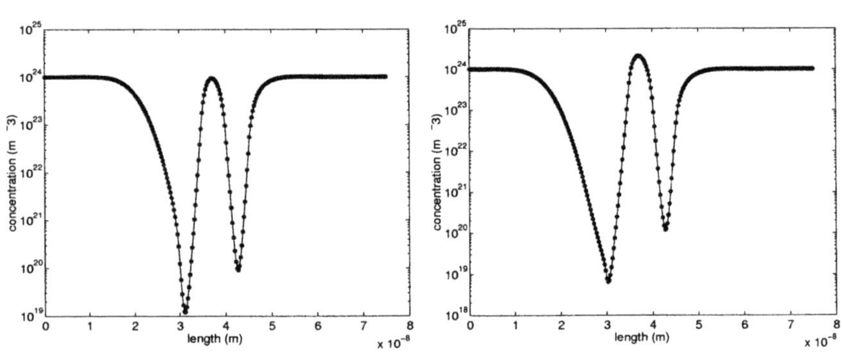


**Fig. 3.** Electron concentration at the peak (left) and at the valley (right) for $m = 0.126m_0$

almost independently of their occupation. As a consequence, the PVR decreases as the temperature rises, until the NDR disappears eventually.

**Table 1.** I-V characteristics as a function of $m$ (left) and $T$ (right)

| $m \times m_0$ | $J_{\max} \times 10^8$ | $J_{\min} \times 10^8$ | PVR |
|---|---|---|---|
| .067 | 6.22 | 6.22 | // |
| .126 | 5.19 | 3.84 | 1.35 |
| .1675 | 1.52 | .827 | 1.84 |
| .201 | .654 | .284 | 2.30 |

| $T$ | $\mu$ | $J_{\max} \times 10^8$ | $J_{\min} \times 10^8$ | PVR |
|---|---|---|---|---|
| 77 | 2.5 | 2.41 | 1.23 | 1.96 |
| 100 | 2.61 | 3.97 | 2.6 | 1.53 |
| 150 | 2.41 | 9.07 | 8.45 | 1.07 |
| 200 | 2.32 | 20.3 | 20.3 | // |

**Dependence on the Barrier Height**
For this block of simulations we have taken a barrier profile with barrier width of 5 nm and a quantum well width of 5 nm, while the effective mass is $m = 0.126m_0$ and $T = 77$ K. Several values of the barrier height are considered, i.e. $[0.1, 0.2, 0.25, 0.32, 0.4]$ V and the results are summarized in Table 2 (left) (in I.S. units). In this case, the higher the barrier height, the lower the tunneling probability, thus we expect a decrease of the current as the height increases. The PVR should increase with the height because of the contribution to the total current of those electrons drifting thermoionically, which become more important at the valley ([2]).

**Dependence on the Barrier Width**
For the last series of simulations we have considered a barrier height of 0.325 V, a quantum well width of 5 nm, a fixed device length of 75 nm, an effective mass $m = 0.126m_0$, and $T = 77$ K. The following values of the barrier width have been studied: $[5, 8, 9, 10]$ nm, and Table 2 (right) collects the simulation results (in I.S. units). As the barrier width gets larger the current decreases due to the reduced width of the peak value of the transmission coefficient. Moreover, the PVR, starting with a barrier width equal to the width of the quantum well, should first increase and then decrease ([9]).

## 4 Conclusions

We have studied and applied the QDD model to the simulation of a RTD. It turns out that the QDD model is able to reproduce typical quantum effects of the device, such as the NDR and correctly reproduces the physical quantities such as electron concentration, electric potential and quasi-Fermi level, even over a wide range of variations of the parameters. We have also carried out a parametric study of the I-V characteristic of the RTD as a function of the electron effective mass, the width and height of the barriers, and of the temperature. Moreover, we have compared three numerical algorithms and we have proposed a suitable scaling of the equations in order to have numerically stable problems.

**Table 2.** I-V characteristics at different barrier heights (left) and widths (right)

| Height | $J_{\max} \times 10^9$ | $J_{\min} \times 10^9$ | PVR | Width $\times 10^{-9}$ | $J_{\max} \times 10^7$ | $J_{\min} \times 10^7$ | PVR |
|---|---|---|---|---|---|---|---|
| 0.2 | 3.23 | 3.23 | // | 5 | 35.0 | 21.7 | 1.62 |
| 0.25 | 1.24 | 1.21 | 1.03 | 8 | .536 | .299 | 1.79 |
| 0.32 | .378 | .243 | 1.56 | 9 | .144 | .103 | 1.39 |
| 0.4 | .12 | .041 | 2.95 | 10 | .041 | .037 | 1.09 |

## References

1. Ancona, M.G., Iafrate, G.J.: Quantum correction to the equation of state of an electron gas in a semiconductor. Phys. Rev. B **39** (1989) 9536–9540
2. Capasso, F.: Physics of quantum electron devices. Springer Verlag, New York, 1989
3. Chen, Z., Cockburn, B., Gardner, C.L.: Quantum hydrodynamic simulation of hysteresis in the resonant tunneling diode. J. Comp. Phys. **117** (1995) 274–280
4. Jerome, J.W.: Analysis of charge transport. Springer-Verlag, New York, 1996
5. Kluksdahl, N.C., Kriman, A.M., Ferry, D.K., Ringhofer, C.A.: Self-consistent study of the resonant tunneling diode. Phys. Rev. B **39** (1989) 7720–7735
6. Lengeler, B.: Semiconductor devices suitable for use in cryogenic environments. Low temperature electronics (R.K. Kirschman ed.) IEEE, New York (1986) 62–70
7. Pinnau, R., Unterreiter, A.: The stationary current-voltage characteristics of the quantum drift-diffusion model. SIAM J. Numer. Anal. **37** (1999) 211–245
8. Simioni, P.: Modelli macroscopici con correzione quantistica per il trasporto di carica nei semiconduttori. Degree Thesis, Politecnico di Milano, 2001
9. Vinter, B., Chevoir, F.: Scattering processes, coherent and incoherent transport in resonant tunneling structures. Resonant Tunneling in Semiconductors (L. Chang ed.), Plenum Press, New York (1991) 201–211

# Global DC-Analysis with the Aid of Standard Network Analysis Programs

Tobias Nähring* and Albrecht Reibiger**

Technische Universität Dresden, Germany

**Abstract.** We present applications of homotopy methods, which make it possible to compute multiple dc-operating points of transistor circuits with standard network analysis programs. It is possible to capture all dc-operating points at least of smaller transistor networks with the help of one- and two-parametric homotopies. Uniqueness criteria of network theory can help to find a parameterization of the homotopy path. As an example for appropriate uniqueness criteria a well known theorem of Nielsen and Willson is applied. Bounds for the parameter space can be found by the no-gain property of transistor circuits.

## 1 Introduction

Most of the standard network analysis programs are provided with some Newton Raphson like algorithm for dc-analysis computing just one dc-operating point (e. g. Newton Raphson algorithm with source stepping). But often it is important to know whether a circuit can exhibit other dc-operating points than the computed one. The reader is reminded only of the unwanted latch-up effect of the operational amplifier $\mu$A709 (see [2]).

In [8] new classes of homotopy methods for the computation of multiple dc-operating points have been introduced which can be realized by means of standard network analysis programs. Here we give a short review of the driving-point-characteristic method (dpc-method), discuss its feasibility, and show some applications in greater detail.

## 2 Some network theoretic preliminaries

In this paper a transistor circuit is modeled as a resistive network $\mathcal{N}$. The topology of $\mathcal{N}$ is described by an oriented graph $\mathcal{G}$ with some branch set $\mathcal{Z}$ (e. g. one branch per resistor, two branches per transistor). Normalized voltage- and current assignments to the branches of $\mathcal{G}$ are the elements $(v, i)$ of the set[1] $\mathcal{S} := \mathbb{R}^{\mathcal{Z}} \times \mathbb{R}^{\mathcal{Z}}$. The $v$-$i$-relation $\mathcal{V} \subseteq \mathcal{S}$ models the behavior of the devices contained in the circuit. It is most often described by equations. E. g., to model bipolar transistors we use the Ebers-Moll equations

$$i_c = I_{\mathrm{CS}}\,(\exp(v_c/V_{\mathrm{T}}) - 1) - \alpha_{\mathrm{f}} I_{\mathrm{ES}}\,(\exp(v_e/V_{\mathrm{T}}) - 1)\,,$$

* e-mail naehring@iee.et.tu-dresden.de

** e-mail reibiger@iee.et.tu-dresden.de

[1] An element $x \in \mathbb{R}^{\mathcal{Z}}$ assigns to each branch $b \in \mathcal{Z}$ a value $x_b \in \mathbb{R}$.

$$i_e = I_{\mathrm{ES}}\,(\,\exp{(v_e/V_{\mathrm{T}})} - 1\,) - \alpha_{\mathrm{r}} I_{\mathrm{CS}}\,(\,\exp{(v_c/V_{\mathrm{T}})} - 1\,)$$

where $e$ is the branch from base to emitter and $c$ is the branch from base to collector, and where the constants $V_{\mathrm{T}}, I_{\mathrm{CS}}, I_{\mathrm{ES}}, \alpha_{\mathrm{f}}, \alpha_{\mathrm{r}}$ have the usual meaning.

For a network $\mathcal{N}$ with graph $\mathcal{G}$ and $v$-$i$-relation $\mathcal{V}$ we write condensed $\mathcal{N} = (\mathcal{G}, \mathcal{V})$. The electrical junctions between the parts of the circuit are modeled by the graph $\mathcal{G}$ and the corresponding mesh and cutset equations. The set of voltage-current-assignments $(v, i) \in \mathcal{S}$ obeying Kirchhoff's laws is named Kirchhoff's set and denoted by $\mathcal{H}$. Finally, we have $\mathcal{L} := \mathcal{V} \cap \mathcal{H}$, the solution set of $\mathcal{N}$ which stands for the set of the dc-operating points of the entire circuit. With these notations the following two statements are direct consequences of set theory.

**Intersection theorem:** Let $\mathcal{N} = (\mathcal{G}, \mathcal{V})$, $\overline{\mathcal{N}} = (\mathcal{G}, \overline{\mathcal{V}})$, $\widetilde{\mathcal{N}} = (\mathcal{G}, \widetilde{\mathcal{V}})$ be networks with the same graph $\mathcal{G}$. If $\mathcal{V} = \overline{\mathcal{V}} \cap \widetilde{\mathcal{V}}$ then $\mathcal{L} = \overline{\mathcal{V}} \cap \widetilde{\mathcal{L}}$.

**Covering theorem:** Let $\mathcal{N} = (\mathcal{G}, \mathcal{V})$ be a network and $(\mathcal{N}^x)_{x\in X}$ a family of networks $\mathcal{N}^x = (\mathcal{G}, \mathcal{V}^x)$ with the same graph as $\mathcal{N}$. If $\mathcal{V} = \bigcup_{x\in X} \mathcal{V}^x$ then $\mathcal{L} = \bigcup_{x\in X} \mathcal{L}^x$.

Throughout this paper we assume that networks are connected and do not have cutsets of independent current sources or loops of independent voltage sources. The transistor models have the no-gain property (see [6]). So the absolut value of any branch voltage (current) of a solution of a network $\mathcal{N}$ does not exceed the sum of the absolut values of the voltages (currents) of the independent sources in $\mathcal{N}$.

Let $\mathcal{N}$ be a network consisting of resistors, independent voltage and current sources and transistors. We say that $\mathcal{N}$ is *reducible to the feedback structure* if 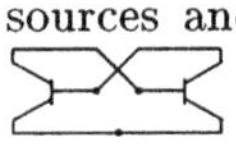 (the feedback structure) is a possible outcome of the following three-step algorithm:

1. choose two transistors (npn and/or pnp), replace them by the generalized transistor symbol 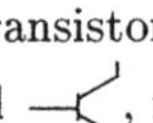, replace all others by the resistor network 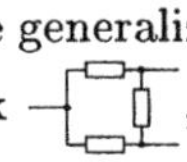,
2. choose some resistors and remove them; remove all current-sources,
3. contract[2] all voltage-source branches and all remaining resistor branches.

If $\mathcal{N}$ is not reducible to the feedback structure then $\mathcal{N}$ has at most one dc-operating point. This uniqueness statement is a consequence of the well known fundamental theorem of NIELSEN and WILLSON for bipolar transistor networks (cf. [4]). We will utilise it in section 3.

In certain examples, where additional controlled sources were involved, the method of HASLER an NEIRYNCK (see [7]) has been successfully employed as a replacement for the theorem of NIELSEN and WILLSON.

# 3 The dpc-method for the global dc-analysis of transistor circuits

In this section we introductorily describe an application of the dpc-method to the simple flip-flop network in figure 1.

[2] 'Contracting a branch' means identifying its incident nodes and removing it.

For the analysis of this network an open circuit $a$ is inserted which does not influence the solutions of the network. The so modified network $\mathcal{N} = (\mathcal{G}, \mathcal{V})$ is depicted in figure 2. As we will see below it is possible to determine all dc-operating points of $\mathcal{N}$ with the help of this additional branch.

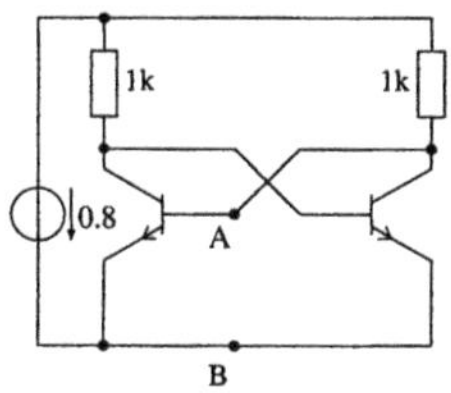


Figure 1

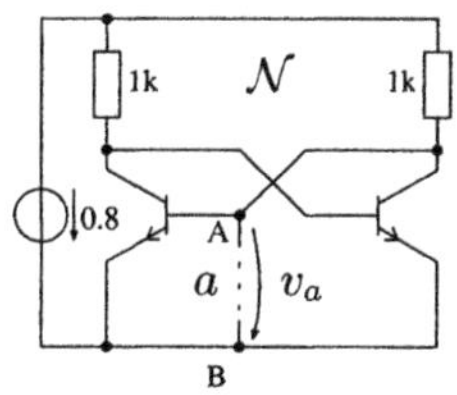


Figure 2

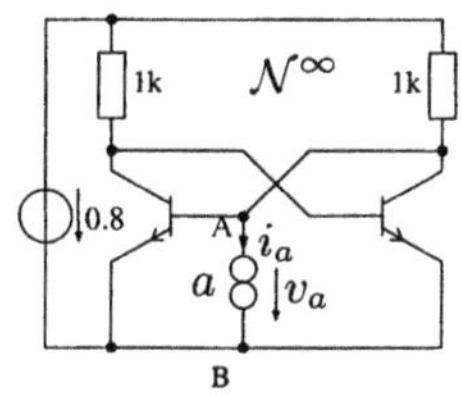


Figure 3

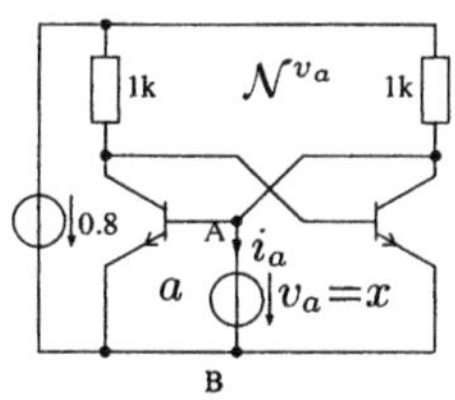


Figure 4

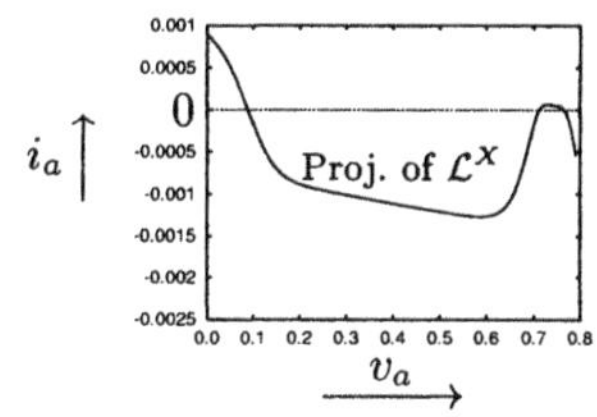


Figure 5

In this example the dpc-method is applied to the interconnection of the flip-flop network with the open circuit $a$ at the terminals $A$ and $B$.

The $v$-$i$-relation $\mathcal{V}$ of $\mathcal{N}$ is represented as the intersection $\mathcal{V} = \mathcal{V}^{\infty} \cap \mathcal{V}^0$ of the $v$-$i$-relations of two modified networks $\mathcal{N}^{\infty}$ and $\mathcal{N}^0$. The network $\mathcal{N}^{\infty}$ is constructed from $\mathcal{N}$ by replacing the open circuit by a norator[3] (cf. fig. 3), i.e. the equation $i_a = 0$ is removed from the system of behavioral equations for this network. In the system of defining equations for the solution set $\mathcal{L}^{\infty}$ of $\mathcal{N}^{\infty}$, the number of variables exceeds the number of equations by one. Thus, we expect that $\mathcal{L}^{\infty} \subset \mathcal{S}$ is an one-dimensional manifold. We will use a partial curve $\mathcal{L}^X \subset \mathcal{L}^{\infty}$ as homotopy path. The projection of $\mathcal{L}^X$ onto the $(v_a, i_a)$-plane is shown in figure 5.

The other network $\mathcal{N}^0$ results from $\mathcal{N}$ by exchanging all branches except the open circuit for norators. Thus, the only behavioral equation left for this network is the equation $i_a = 0$ and $\mathcal{V}^0$ is a hypersurface of the space $\mathcal{S}$.

With $\mathcal{V} = \mathcal{V}^{\infty} \cap \mathcal{V}^0$ the intersection theorem implies $\mathcal{L} = \mathcal{L}^{\infty} \cap \mathcal{V}^0$, i.e., the intersection points of $\mathcal{L}^{\infty}$ with the hypersurface $\mathcal{V}^0$ are the dc-operating points of $\mathcal{N}$. To find some of them we search the homotopy path $\mathcal{L}^X$.

The homotopy path can be computed via a dc-sweep of `spice`: For any $x \in \mathbb{R}$ let $\mathcal{N}^x = (G, \mathcal{V}^x)$ be the network that results from replacing the norator in $\mathcal{N}^{\infty}$ by a voltage source with source voltage $x$ (see fig. 4). For $x \in \mathbb{R}$ the $v$-$i$-relation $\mathcal{V}^x$ of $\mathcal{N}^x$ differs only from $\mathcal{V}^{\infty}$ by the additional condition $v_a = x$ of the inserted voltage source and can therefore be written as $\mathcal{V}^x = \{(v,i) \in \mathcal{V}^{\infty} | v_a = x\}$. Thus, $\mathcal{V}^{\infty} = \cup_{x \in \mathbb{R}} \mathcal{V}^x$ and by the covering theorem we have $\mathcal{L}^{\infty} = \cup_{x \in \mathbb{R}} \mathcal{L}^x$. If for each

[3] For norators all pairs of voltages and currents are admissible (see e. g. [7]).

$x \in \mathbb{R}$ the solution of the network $\mathcal{N}^x$ is unique (i.e., the solution set $\mathcal{L}^x$ has just one element), we obtain a map $x \in \mathbb{R} \mapsto (u,i) \in \mathcal{L}^x$ which parameterizes $\mathcal{L}^\infty$. Actually, this is the motivation for the insertion of a voltage source at the nodes $A$ and $B$ (review fig. 1 to 4). The inserted voltage source *breaks all feedback structures* of $\mathcal{N}$, i.e. the network $\mathcal{N}^x$ is not reducible to the feedback structure and the theorem of Nielsen and Willson ensures that the network $\mathcal{N}^x$ has a unique solution.

In general, it is only possible to compute a numerical approximation of a partial curve $\mathcal{L}^X := \cup_{x \in X} \mathcal{L}^x$ of $\mathcal{L}^\infty$ which is defined on a bounded parameter interval $X \subset \mathbb{R}$. Still we have $\mathcal{L}^X \cap \mathcal{V}^0 \subset \mathcal{L}^\infty \cap \mathcal{V}^0 = \mathcal{L}$ and therefore the intersection points of the homotopy path $\mathcal{L}^X$ with the hypersurface $\mathcal{V}^0$ are dc-operating points of $\mathcal{N}$. Since we want to find all dc-operating points of $\mathcal{N}$ we have to estimate bounds for the parameter interval $X$ such that $\mathcal{L}^X$ contains all solutions of $\mathcal{N}$. Often, the no-gain property of transistor circuits can be applied for this task. Because of this property the branch voltages $v_a$ of all solutions $(v,i)$ of the network $\mathcal{N}$ cannot exceed the boundaries of the interval $[0, 0.8]$ which is determined by the voltage of the power supply[4]. Thus, with $X := [0, 0.8]$ the solutions of $\mathcal{N}$ are all contained in $\mathcal{L}^X$.

Often, more than one independent source is needed, to break all feedback structures of a transistor network. Let $\mathcal{N} = (G, \mathcal{V})$ be such a network and let $\overline{\mathcal{Z}}_V$ and $\overline{\mathcal{Z}}_I$ be minimal sets of branches in $\mathcal{N}$ that break all feedback structures of $\mathcal{N}$ if they are replaced by independent voltage and current sources, resp. Analogous to the introductory example we construct a network $\mathcal{N}^\infty$ by replacing all the branches in $\overline{\mathcal{Z}} := \overline{\mathcal{Z}}_V \cup \overline{\mathcal{Z}}_I$ for norators and a network $\mathcal{N}^0$ by replacing all branches in the complement $\mathcal{Z} \setminus \overline{\mathcal{Z}}$ for norators. Again, we have $\mathcal{V} = \mathcal{V}^\infty \cap \mathcal{V}^0$ and therefore $\mathcal{L} = \mathcal{L}^\infty \cap \mathcal{V}^0$. As in the introductory example we parameterize $\mathcal{L}^\infty$ and search it for the intersection points with $\mathcal{V}^0$. The above source replacements deliver new networks $\mathcal{N}^x = (G, \mathcal{V}^x)$ with $\mathcal{V}^x := \{(v,i) \in \mathcal{V}^\infty | \forall b \in \overline{\mathcal{Z}}_V : v_b = x_b, \forall b \in \overline{\mathcal{Z}}_I : i_b = x_b\}$ for each prescribed source value assignment $x \in \mathbb{R}^{\overline{\mathcal{Z}}}$. By Nielsen and Willson the networks $\mathcal{N}^x$ have unique solutions and the map $x \in \mathbb{R}^{\overline{\mathcal{Z}}} \mapsto (u,i) \in \mathcal{L}^x$ parameterizes $\mathcal{L}^\infty$. In most cases it is possible to find a reasonable bounded parameter set $X \subset \mathbb{R}^{\overline{\mathcal{Z}}}$ such that $\mathcal{L}^X := \cup_{x \in X} \mathcal{L}^x$ covers all the solutions of $\mathcal{N}$. In the case of two replaced branches the set $\mathcal{L}^X$ can be computed by means of a parameterized dc-sweep of `pspice` and the intersection points with $\mathcal{V}^0$ can be found via the graphic program `gnuplot`. An example follows in section 4.

Since the computational effort grows exponentially with the number of replaced branches it is quite desirable to stick to the case of one replaced branch even if uniqueness cannot be ensured for that case. But then one has to be aware that $\mathcal{L}^\infty$ can consist of several components which may even have turning points. Problems with turning points can be avoided if one parameterizes the curves by arc length via a path-following algorithm. Because of lack of space we can only give a simplified sketch here (for a more detailed presentation see [3,5,9]). The path following algorithm is realized with a transient analysis of `spice` applied to an auxiliary dynamical network $\widetilde{\mathcal{N}} = (\mathcal{G}, \widetilde{\mathcal{V}})$. We use here $\mathrm{C}^1$-time functions $u, i \in \mathrm{C}^1(T, \mathbb{R}^{\mathcal{Z}})$ of voltage- and current assignments defined on some time interval $T = [0, t_{\mathrm{end}}]$ and introduce $\widetilde{\mathcal{S}} := \mathrm{C}^1(T, \mathbb{R}^{\mathcal{Z}}) \times \mathrm{C}^1(T, \mathbb{R}^{\mathcal{Z}})$, Kirchhoff's set $\widetilde{\mathcal{H}} := \{(u,i) \in \widetilde{\mathcal{S}} | \forall t \in T :$

[4] In the case that the power supply voltage is considerably higher than the threshold voltage of a basis-emitter diode one can take the Ebers-Moll equations into consideration to estimate narrow bounds for the homotopy parameter.

$(u(t), i(t)) \in \mathcal{H}\}$, the behavioral relation $\widetilde{\mathcal{V}} := \{(u,i) \in \widetilde{\mathcal{S}} \,|\, \forall t \in T : (u(t), i(t)) \in \mathcal{V}^\infty, \|\mathrm{D}_t\, u(t)\|^2 + \|\mathrm{D}_t\, i(t)\|^2 = 1\}$ with the time derivative $\mathrm{D}_t$ and an appropriate norm $\|\bullet\|$ on $\mathbb{R}^{\mathcal{Z}}$, and finally the solution set $\widetilde{\mathcal{L}} = \widetilde{\mathcal{V}} \cap \widetilde{\mathcal{H}}$. Obviously, we obtain the identity

$$\widetilde{\mathcal{L}} = \{(u,i) \in \widetilde{\mathcal{S}} | \forall t \in T : (u(t), i(t)) \in \mathcal{L}^\infty, \|\mathrm{D}_t\, u(t)\|^2 + \|\mathrm{D}_t\, i(t)\|^2 = 1\}.$$

The additional behavioral equation $\|\mathrm{D}_t\, u(t)\|^2 + \|\mathrm{D}_t\, i(t)\|^2 = 1$ forces, that the curves $t \in T \mapsto (u(t), i(t))$ in $\mathcal{L}^\infty$ are traced with constant speed and therefore are parameterized by some equivalent of arc length. This equation can be described by a `spice` netlist with entries for capacitors and controlled sources. The set $\mathcal{L}^T := \{(u(t), i(t)) \in \mathcal{S} | (u,i) \in \widetilde{\mathcal{L}}, t \in T\}$ which consists of traces of the curves $t \in T \mapsto (u(t), i(t))$ is a subset of $\mathcal{L}^\infty$ and can be used as a replacement for $\mathcal{L}^X$.

# 4 Examples

• Analogous to the introductory example, the dpc-method is applied to the network in figure 6 and the associated homotopy path is computed by a dc-sweep of `spice`. Therefore, a family of networks $\mathcal{N}^x$ is constructed by exchanging the voltage source for a current source with source current $x \in X$ (replacing the behavioral equation $v_a = 5$ with $i_a = x$). The networks $\mathcal{N}^x$ have unique solutions since they are not reducible to the feedback structure. The bounds of the parameter interval $X = [0, 0.011]$ can easily be estimated by the cutset equation for the branches $R_1$, $R_2$, $R_3$, $a$ and by the fact that no branch voltage exceeds the value 5. From the `spice`-computed dc-sweep in figure 7 one sees that the behavioral equation $v_a = 5$ of $\mathcal{N}$ is satisfied for five parameter values $x$ (which equal $i_a$). Therefore $\mathcal{N}$ exhibits five dc-operating points. This example suggests that sometimes it may be useful to use our method instead of the usual voltage source stepping for global dc-analysis.

• Next we sketch the transfer-characteristic method from [8]. Consider again the

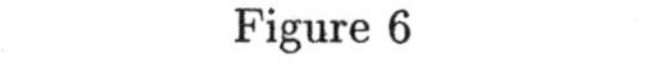


Figure 6

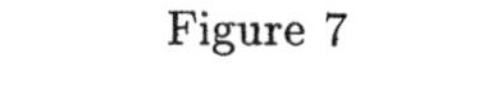


Figure 7

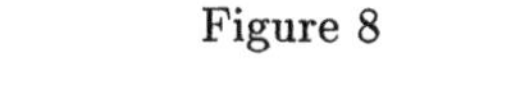


Figure 8

network $\mathcal{N}$ in figure 6 but now with the transistor $\mathcal{N}_Q$ substituted by the terminal equivalent subnetwork $\mathcal{N}_\mathrm{K}$ (such a kind of substitution is due to Kronenberg [1]). For a global dc-analysis a family $(\mathcal{N}^x)_{x \in X}$ of networks is constructed. In each $\mathcal{N}^x$ the controlled source is replaced by an independent voltage source with source voltage $x$ (see the corresponding subnetwork $\mathcal{N}_\mathrm{K}^x$ in figure 6). Uniqueness of the solution of $\mathcal{N}^x$ can be proven by the methods given in [7]. The parameter interval

$X = [0, 0.84]$ is found with the help of the estimation for the current of the power supply (see above) and the Ebers-Moll equations. From the result of a `spice` dc-sweep (see figure 8 for the relevant section of it) one reads off the parameter values $x$ (equal to $v_c$) which fulfill the behavioral equation $v_b = v_c$ of the controlled source. These parameter values correspond to the solutions of $\mathcal{N}$ (cf. [8] for details).

• At least two voltage sources have to be inserted into the network $\mathcal{N}$ of figure 9 to break all feedback structures. With the aid of a parametric dc-sweep of `pspice` it is possible to compute the solutions of the resulting networks $\mathcal{N}^x_{a,b}$ (see figure 10) in dependency of the source voltages $x_a, x_b \in [0, 5]$. The bounds of the two-

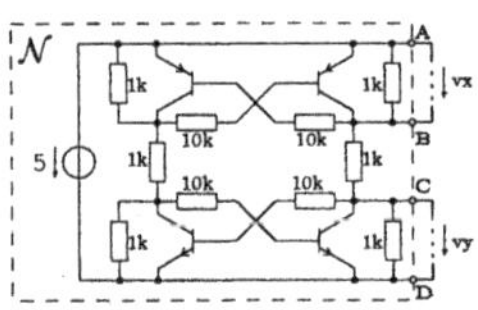

Figure 9

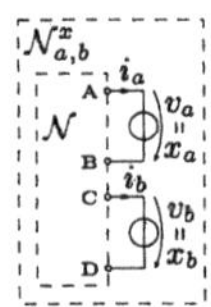

Figure 10

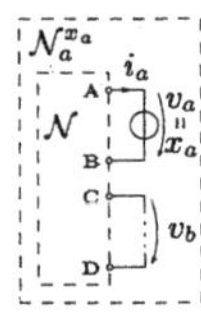

Figure 11

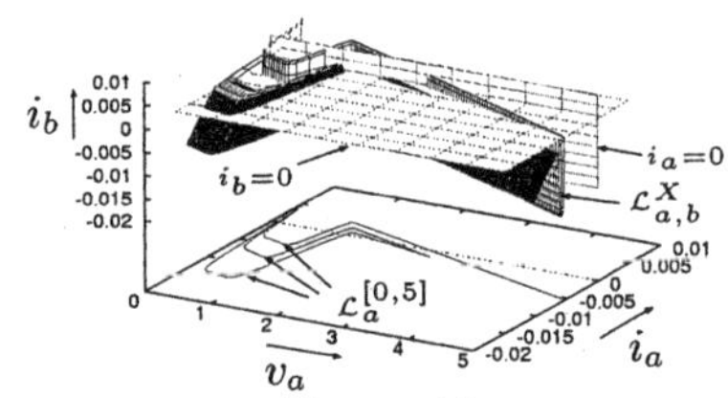


Figure 12

dimensional parameter set $X := [0, 5]^{\{a,b\}}$ are induced by the source voltage of the power supply of $\mathcal{N}$. The set $\mathcal{L}^\infty_{a,b}$ of all solutions of the networks $\mathcal{N}^x_{a,b}$ ($x \in \mathbb{R}^{\{a,b\}}$) is a two-dimensional manifold which can be parametrized by the source voltages $x_a$ and $x_b$. The projection of the relevant part $\mathcal{L}^X_{a,b} \subset \mathcal{L}^\infty_{a,b}$ onto the $v_a$-, $i_a$- and $i_b$-components is depicted in figure 12. The points of $\mathcal{L}^X_{a,b}$ satisfying $i_a = 0$ and $i_b = 0$ are the solutions of $\mathcal{N}$. They are determined in two steps. At first the set $\mathcal{L}^{[0,5]}_a \subset \mathcal{L}^X_{a,b}$ of points with $i_b = 0$ is computed via a contour plot of the well-known freely available program `gnuplot`. This set is the union of the solution sets $\mathcal{L}^{x_a}_a$ ($x_a \in [0, 5]$) of networks $\mathcal{N}^{x_a}_a$ with only one inserted voltage source at branch $a$ (see figure 11). The projection of $\mathcal{L}^{[0,5]}_a$ onto the $(v_a, i_a)$-plane is also shown in figure 12. From this projection the parameters $v_a$ with $i_a = 0$, i.e. the $v_a$-values corresponding to solutions of $\mathcal{N}$, can be read off. The corresponding $v_b$ values are similarly determined. In this way one obtains the parameter values $v_a$ and $v_b$ for the nine dc-operating points of the network $\mathcal{N}$. Figure 12 shows that a dc-analysis of the networks $\mathcal{N}^{x_a}_a$ (see figure 11) with $x_a$ as the sweep parameter will not succeed since $\mathcal{L}^\infty_a$ cannot be parameterized by $x_a$.

• The last example shows the application of the dpc-method to a somewhat larger transistor network. The circuit in figure 13 represents a simple operational amplifier which models the latch-up effect of the $\mu$A 709 (see [2]). This amplifier is applied in network $\mathcal{N}$ (see figure 14) as a voltage follower with input connected to ground by a 10k resistor. For the computation of dc-operating points of this transistor network the load $R$ has been replaced by a voltage source and a dc-analysis was carried out with the source voltage as the dc-sweep parameter. In the sense of the proof above this is interpreted as exchanging the load for a norator and then covering the $v$-$i$-relation of the resulting network with the $v$-$i$-relations of a family of networks where the norator is replaced with voltage sources. The $i_o$-component of the corresponding solution is shown as a bold curve in figure 15. The light straight line in figure 15 is the $v$-$i$-characteristic of a resistor with value 50. If the load resistance is larger than

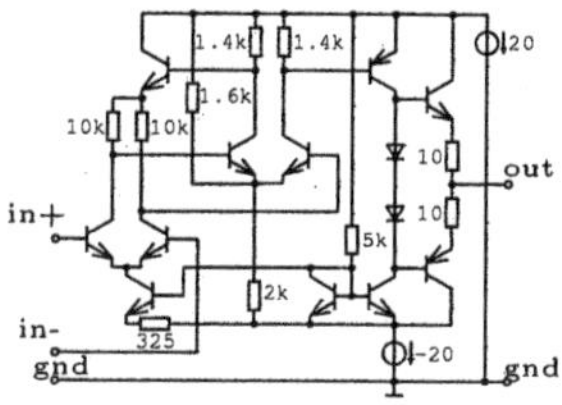


Figure 13

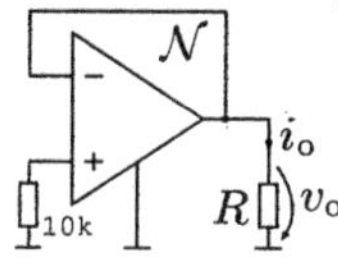


Figure 14

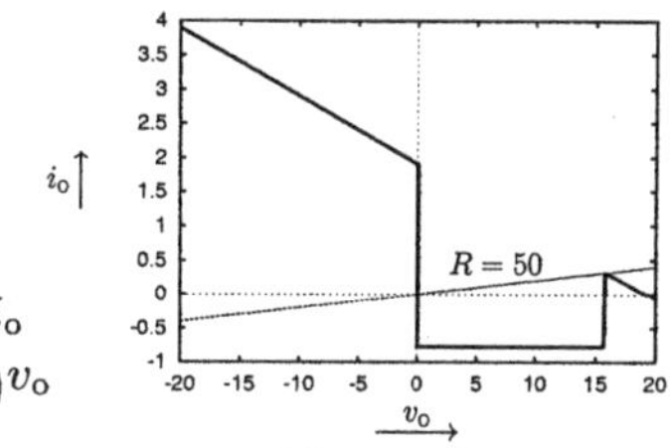


Figure 15

50 then the circuit $\mathcal{N}$ has at least three dc-solutions, i.e. it exhibits the latch-up effect.

Note that we did not break all feedback structures in network $\mathcal{N}$ by inserting the voltage source. Therefore the parameterization of the solution set $\mathcal{L}^\infty$ by $v_o$ is not necessarily unique and it is possible that there exist some other solution components of $\mathcal{N}^\infty$ which have been skipped by the dc-analysis (cf. previous example). In particular, vertical line segments in a `spice`-computed dc-plot as they occur in figure 15 can indicate that the solution manifold consists of more than one component. It may be that such a line segment does not belong to the solution manifold $\mathcal{L}^\infty$ but connects two separate components of $\mathcal{L}^\infty$. In order to exclude this case, the characteristic in figure 15 has been verified with the help of the curve tracing algorithm by J. HAASE which can also be implemented as a `spice` netlist. This last example shows that the dpc-method can be useful for detecting multiple dc-solutions even if well defined parameterization is not ensured.

## 5 Conclusions

With the help of homotopies, global search for all dc-operating points of a network becomes global search of a low-dimensional parameter space. Interesting projects for larger networks are **(1st)** an effective topological algorithm for the highly complex task to place a minimal number of independent sources needed to break all feedback structures, **(2nd)** an algorithm for the global dc-analysis basing on the proposed homotopy method that can compete with existing global algorithms.

We thank the anonymous reviewers very much for their recommendations.

## References

1. Kronenberg, L., Trajković, Lj., and Mathis, W.: Analysis of feedback structures and their effect on multiple dc-operating points. *Proc. ECCTD'99*, Stresa, Italy, Aug. 1999, pp. 683–686.
2. Widlar, R. J.: Design Techniques for Monolithic Operational Amplifiers. IEEE Journal of Solid-State Circuits, vol. SC-4, No. 4, August 1969
3. Haase, J.: Computation of transfer characteristics of multivalued resistive non-linear networks. Proc. SSCT82, Part: Short Communications. (1982) 286–272
4. Nielsen, R. O., Willson, A. N.: A Fundamental Result Concerning the Topology of Transistor Circuits with Multiple Equilibria. Proceedings of the IEEE on Circuits and Systems. 68 (1980) 196–208

5. Ushida, A., Chua, L. O.: Tracing solution curves of non-linear equations with sharp turning points. International Journal of Circuit Theory and Applications. 12 (1984) 1–22
6. Willson, A. N.: The no-gain property for networks containing three-terminal elements. IEEE Trans. Circuits and Systems. CAS-22 (1975) 678–687
7. Hasler, M., Neirynck, J.: Nonlinear Circuits. Artech House, inc. Norwood, 1986.
8. Reibiger, A., Mathis, W., Nähring, T., Trajković, Lj., Kronenberg, L.: Mathematical Foundations of the TC-Method for Computing Multiple DC-Operating Points. XI. ISTET'01 preprints CD-ROM, Linz–Austria, 2001.
9. Nähring, T., Reibiger, A.: Beiträge zur Arbeitspunktberechnung resistiver Netzwerke. Kleinheubacher Berichte, 45 (2001) 262–265

# Fast Calculation of Space Charge in Beam Line Tracking by Multigrid Techniques

Gisela Pöplau[1]*, Ursula van Rienen[1], Marieke de Loos[2], and Bas van der Geer[3]

[1] Rostock University, D-18051 Rostock, Germany
[2] Eindhoven University of Technology, NL-5600 MB Eindhoven, The Netherlands
[3] Pulsar Physics, NL-3762 XA Soest, The Netherlands

*Dedicated to Professor Manfred Tasche on the occasion of his 60th birthday*

**Abstract.** Numerical prediction of charged particle dynamics in accelerators is essential for the design and understanding of these machines. The calculation of space charge forces influencing the behaviour of a particle bunch is still a bottleneck of existing tracking codes.

We report on our development of a new 3D space-charge routine in the General Particle Tracer (GPT) code. It scales linearly with the number of particles in terms of CPU time, allowing over a million particles to be tracked on a normal PC. The model is based on a non-equidistant multigrid Poisson solver that is used to solve the electrostatic fields in the rest frame of the bunch.

A reliable multigrid scheme for the tracking of particles should be very fast, stable and show good convergence for a great variety of meshes. Numerical results demonstrate the effect of the choice of the multigrid components. Further, the values of physical quantities show good agreement compared to the values calculated by a well-tested 2D routine in the GPT code.

## 1 Introduction

Nowadays, particle accelerators play an important role for scientific research as well as for medical and industrial applications. Demanding applications such as colliders and free electron lasers (FELs) require very high quality electron bunches, where any anomaly severely degrades the final performance. Hence, design and operation of accelerators require efficient numerical simulations.

A powerful tool widely used for the study of the behaviour of charged beams is the General Particle Tracer (GPT) [3]. It calculates the trajectories of a large number of sample particles through the combined external and self-induced fields generated by the charged particles (the so-called space-charge forces). Depending on charge density and energy, a direct point-to-point model can not be used to calculate space-charge forces because of granularity problems and the inherent $\mathcal{O}(N^2)$ scaling between the number of sample particles and CPU time [2].

* supported by a research grant from DESY, Hamburg

A method to stabilize the calculations and to rigorously save CPU time is to restrict calculations to 2D, while assuming symmetry properties of the particle bunch. In this paper we introduce a 3D model for the fast calculation of space-charge. The algorithm is a further development of a method given in [9]. The main idea is that space-charge fields are computed in the rest frame of a particle bunch by a non-equidistant multigrid scheme. Hence, the numerical effort scales linearly with the number of particles in terms of CPU time. The new model is well suited for a variety of applications, especially for the handling of non-linear fields (see Figure 1).

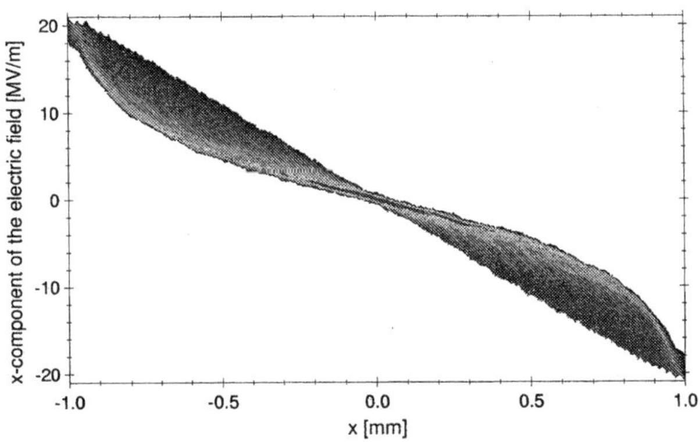


**Fig. 1.** Density plot of the projection of the non-linear transverse electric field of a hard-edged bunch with a radius of 1 mm, a length of 0.1 mm and a charge of 1 nC, as calculated by the new 3D space-charge routine in GPT. The calculation is based on one million particles on a 129 × 129 × 129 mesh.

The new 3D multigrid based space-charge routine enables the tracking of a million particles on a normal PC. Important questions for the efficiency of the algorithm are the construction of the grid adapted to the particle distribution and the choice of a reliable multigrid scheme.

## 2 The 3D Space-Charge Model

The particle tracking is performed by solving the relativistic equations of motion for a set of macro particles (sample particles) representing the distribution of the particles of a bunch. In the GPT code a $5^{\text{th}}$ order embedded Runge–Kutta scheme with adaptive step size control is implemented for the numerical integration of these equations [2]. In each time step of the numerical integration the space-charge fields have to be taken into account. The space-charge calculation with the 3D model is performed as follows:

1. Transformation of the particles from the laboratory frame to the rest frame by Lorentz transformation. The rest frame is here defined as frame where the average particle momentum is zero.

2. Determination of a non-equidistant 3D Cartesian grid in correspondence to the charge density of the bunch (see subsection 2.1).
3. Approximation of the charge at the grid points.
4. Calculation of the electrostatic potential at the grid points via Poisson's equation applying a multigrid algorithm. The finite difference scheme is used for the discretization of Poisson's equation (see subsection 2.2).
5. Derivation of the electric field in the rest frame and trilinear interpolation of the field values to the particle positions.
6. Transformation of the field to the laboratory frame by Lorentz transformation.

The efficiency and accuracy of the space-charge calculation mainly depends on the determination of the 3D mesh and the applied multigrid scheme to solve Poisson's equation. We describe both in the next two subsections.

## 2.1 Mesh Generation

The electromagnetic potential is calculated on a 3D Cartesian mesh where an approximation of the charge in the rest frame is stored at the grid points. The 3D mesh is generated in a box surrounding the bunch. To reduce the number of mesh lines needed, and thus to reduce CPU time, the density of the mesh lines is increased if the charge-density increases.

The actual positioning of the mesh lines is an iterative process. The mesh lines are distributed so that they are spaced according to the distribution of the beam charge density. The parameter $fn$ is introduced to maintain a maximum difference in spacing between neighboring mesh lines, to avoid the creation of a non-optimal mesh line distribution for the Poisson solver. If, e. g. $fn = 0.25$, then the difference in spacing between neighboring mesh lines cannot vary by more than 25%. To span the mesh over the bounding box additional mesh lines are added left and right of the bunch. The spacing of these additional mesh lines is increased, restricted to $fn$, to add as few lines as possible.

The addition of mesh lines and the effect of $fn$ is shown in Figure 2. When $fn = 0$, the spacing between all neighboring mesh lines is allowed to vary by 0%, creating an equidistant mesh. Such a mesh is most stable for the multigrid Poisson solver, but it will create many empty mesh boxes. On the other extreme, setting $fn = 0.5$ results in a dense sampling of the electron bunch and sparse sampling of the surrounding area.

## 2.2 The Multigrid Poisson Solver

After creating the mesh and approximating the charge of the particles on the mesh points, the space-charge forces can be calculated by means of Poisson's equation given by

$$-\Delta\varphi = \frac{\varrho}{\varepsilon_0} \qquad \text{in } \Omega \subset \mathbb{R}^3 .$$

Here, $\varphi$ denotes the potential, $\varrho$ the charge density and $\varepsilon_0$ the dielectric constant. The domain $\Omega$ is a box surrounding the particle bunch. On the boundary Dirichlet

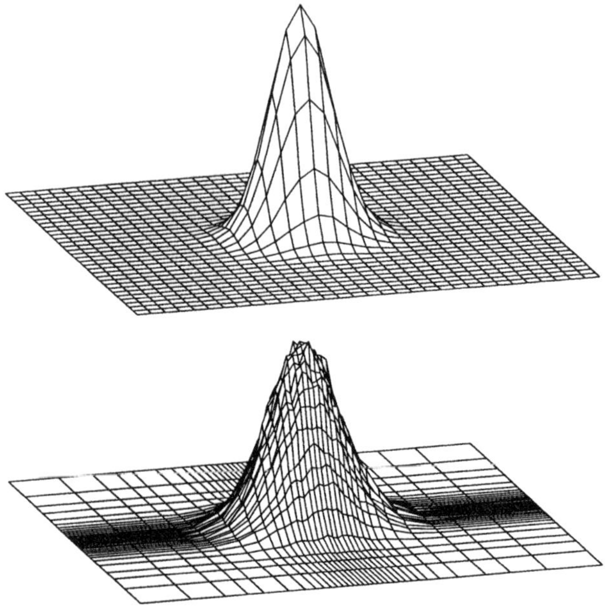

**Fig. 2.** Mesh line positions ($(x, y)$-plane) for a Gaussian charge density with $fn = 0$ (top) and $fn = 0.5$ (bottom). The vertical axis shows the total charge in each mesh box, where the height of the top has been normalized in both plots.

or open boundary conditions are assumed. First, Poisson's equation is discretized by finite differences using the non-equidistant mesh described in the previous section. The solution of the resulting system of equations (with up to 1 million degrees of freedom) requires a fast and robust solver.

State-of-the-art is the application of a multigrid method as Poisson solver. In model cases the numerical effort scales with the number of mesh points. The multigrid algorithm operates on a certain number of grids starting with the mesh given by the discretization of Poisson's equation. Then a sequence of coarser grids is generated by cutting mesh lines. On an equidistant mesh every second mesh line is removed. Now iteratively, a raw approximation of the solution of the systems of equations is obtained by the application of a few steps of a relaxation scheme (e. g. Gauss–Seidel) which is called pre-smoothing. This approximation is then improved by a correction vector obtained on the coarser grids (the so-called coarse grid correction) where restriction and interpolation work as grid transfer operators. After applying interpolation another few steps of relaxation are necessary (post-smoothing). For more details see [4,1].

As shown in [7,8] the coarsening strategy is crucial for the convergence of the multigrid algorithm on non-equidistant grids. The generation of coarse grids with every second grid line removed as suggested in [1] would lead to meshes with large aspect ratio. Hence the convergence of a multigrid scheme on such grids would considerably slow down. Here, the removal of mesh lines follows the rule: Two neighboring steps $h_1$ and $h_2$ remain also in the next coarser grid as long as either

$h_1 \geq sh_{min}$ or $h_2 \geq sh_{min}$, where $h_{min}$ denotes the overall minimal step size of the corresponding fine level. The factor $s$ is chosen as $s = 1.6$ or $s = 1.7$ with the objective to obtain a decreasing aspect ratio of the mesh spacing.

On the other hand the choice of the multigrid parameters such as number of pre- and post-smoothing steps, the application of full or half restriction considerably influence the performance of the multigrid scheme. If the convergence of multigrid algorithm turns out to be not sufficient, multigrid can be applied as a preconditioner for the conjugate gradient algorithm. This method leads to a better convergence at least in cases where a plain multigrid scheme converges too slowly [6,5].

# 3 Numerical Examples

## 3.1 Numerical Tests with the Multigrid Algorithm

The multigrid method has been tested with a sphere filled with charged particles with Gaussian distribution. The space charge effect for only one time step has been computed.

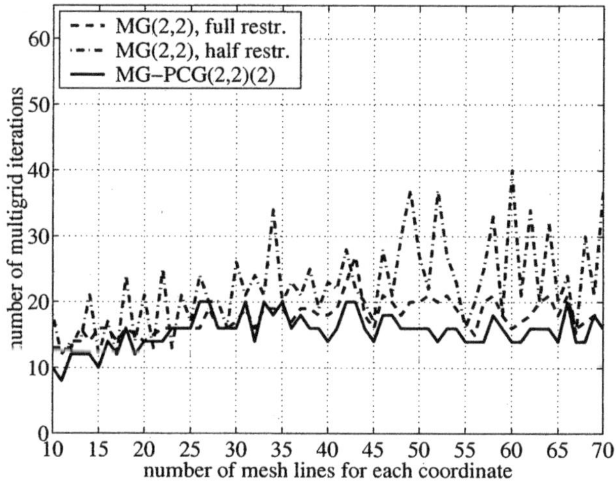


**Fig. 3.** Number of multigrid iterations for several multigrid schemes (MG: multigrid, MG–PCG: multigrid preconditioned conjugate gradients). The calculations have been performed with a spherical bunch of 10,000 particles, where the particles have a Gaussian distribution. The multigrid iteration has been performed to a relative residual less than $10^{-13}$ in the maximum norm.

In order to find out the best performance of the multigrid algorithm for the particle tracking several components of the scheme have been tested. The multigrid algorithm is performed as V-cycle. The red-black Gauss–Seidel relaxation has been taken as smoothing operator with 2 pre- and 2 post-smoothing steps (MG(2,2)). First, full restriction has been tested versus half restriction. Figs. 3 and 4 show that the multigrid scheme performed with half restriction is very sensitive to certain grids, i. e. converges slowly. These are meshes where the sequence of coarse grids has no descending aspect ratio.

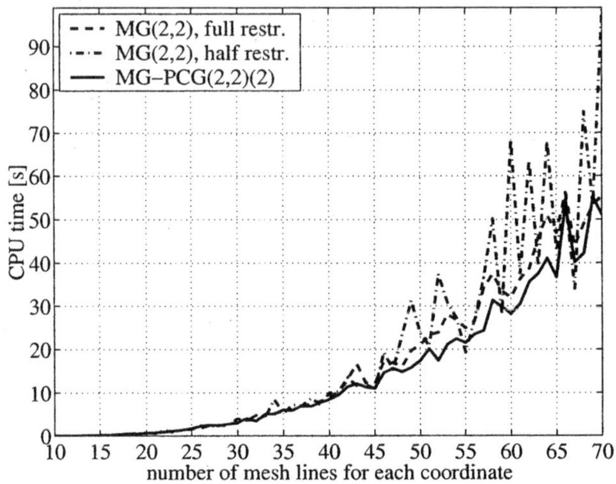


**Fig. 4.** CPU time for several multigrid schemes (MG: multigrid, MG–PCG: multigrid preconditioned CG) on a 800 MHz Pentium PC. The calculations have been performed with a spherical bunch of 10,000 particles, where the particles have a Gaussian distribution.

The application of multigrid as preconditioner for the conjugate gradient method requires that the multigrid scheme is a symmetric and positive operator [5,6]. Thus the components have been chosen as follows: two pre-smoothing steps with red-black Gauss–Seidel relaxation, two post-smoothing steps with black-red Gauss–Seidel relaxation and full restriction. Two V-cycles have been performed per CG-iteration step (MG–PCG(2,2)(2)).

### 3.2 Tracking Example

During a drift, an electron bunch expands both longitudinally and radially due to the space-charge forces. Figure 5 shows a simulation of a hard-edged bunch, starting with an energy corresponding to a Lorentz factor of $\gamma = 5$. The bunch has a total charge of 1 nC, a radius of 1 mm and a length of 0.1 mm ('pancake' bunch). After 100 ps, the hard edges have become smooth, and the density at the head of the bunch is larger than at the tail due to relativistic effects.

The expansion calculated with the new 3D space-charge routine has been compared to the expansion simulated with the well-tested cylindrically symmetric 2D space-charge routine of GPT [3]. The result for the bunch length is shown in Figure 6. It demonstrates the perfect agreement between the 2D and the 3D routine, even with 1000 particles.

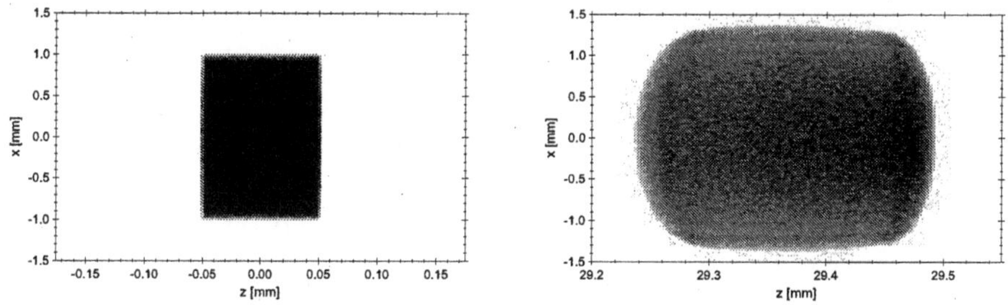


**Fig. 5.** Initial (left) and final (right) projections of the charge density of an expanding 'pancake' bunch in the $(x, z)$-plane. One million particles are used on a $65 \times 65 \times 65$ mesh.

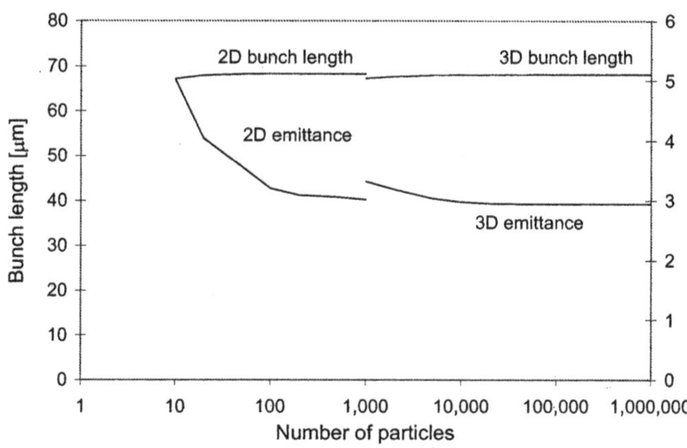


**Fig. 6.** Bunch length (expressed as standard deviation) and final emittance after tracking a 'pancake' bunch with total charge of 1 nC during 100 ps. Tracking a million particles with the 3D model takes only 30 minutes CPU time on a 1.6 GHz Pentium PC.

Although the emittance, a measure of transverse focusability of the bunch, converges a little bit slower, Figure 6 shows already with 1000 particles good agreement with the 2D routine. The value with 10,000 particles is identical to the 2D case, where about 1000 'rings' are sufficient. It should be noted that this example is a quite extreme case where the emittance explodes from 0 to about 3 $\mu$m in just 30 mm.

## 4 Conclusions

A new 3D space-charge routine implemented in the GPT code has been described in this paper. The new method allowing 3D simulations with a large number of particles on a common PC is based on a multigrid Poisson solver for the calculation of the electrostatic potential in the rest frame. Numerical results of the 3D routine show perfect agreement with the standard 2D space-charge model of the GPT code.

## References

1. Briggs, W.L., van Emden, H., McCormick, S.: A Multigrid Tutorial. 2nd edition, SIAM, Philadelphia, 2000.
2. Van der Geer, S.B., de Loos, M.J.: The General Particle Tracer Code. Design, implementation and application. PhD thesis, Einhoven, 2001.
3. General Particle Tracer (GPT), release 2.52, Pulsar Physics, De Bongerd 23, Soest, The Netherlands, `www.pulsar.nl/gpt`.
4. Hackbusch, W.: Multi-Grid Methods and Applications, Springer, Berlin, 1985.
5. M. Jung, U. Langer, Applications of multilevel methods to practical problems. Surv. Math. Ind. **1** (1991), 217–257.
6. M. Jung, U. Langer, A. Meyer, W. Queck, M. Schneider, Multigrid preconditioners and their applications. In the Proceedings of the 3rd GDR Multigrid Seminar held at Biesenthal, 1989, 11–52.
7. Pöplau, G., van Rienen, U., Staats, J., Weiland, T.: Fast Algorithms for the Tracking of Electron Beams, Proceedings of EPAC 2000, Vienna, 2000, 1387–1389.
8. Pöplau, G., van Rienen, U.: Multigrid Solvers for Poisson's Equation in Computational Electromagnetics. Proceedings of the 3rd Conference on Scientific Computating in Electrical Engineering (SCEE–2000), (U. van Rienen, D. Hecht, M. Günther, eds.), LNSCE **18**, Springer, Berlin, 2001, 169–176.
9. Staats, J., Weiland, T., Kostial, S., Richter, A.: Tracking of Electron Beams with Numerically Determined Space Charge Forces. Proceedings of the 1999 Particle Accelerator Conference PAC'99, New York, 1999, 2740–2742.

# Numerical Techniques for Solving Multirate Partial Differential Algebraic Equations

R. Pulch

Institut für Wissenschaftliches Rechnen und Mathematische Modellbildung, Universität Karlsruhe (TH), 76128 Karlsruhe, Germany

**Abstract.** In electric circuits, signals often include widely separated frequencies. Thus numerical simulation demands a large amount of computational work, since the fastest rate restricts the integration step size. A multidimensional signal model yields an alternative approach, where each time scale is given its own variable. Consequently, underlying differential algebraic equations (DAEs) change into a PDAE model, the multirate partial differential algebraic equations (MPDAEs). A time domain method to determine multiperiodic MPDAE solutions is presented. According discretisations rest upon the specific information transport in the MPDAE system along characteristic curves. In contrast, general time domain methods produce unphysical couplings. Hence enormous savings in computational time and memory arise in the linear algebra part. This technique is applied to driven oscillators including two periodic time scales as well as to oscillators, where one periodic rate is forced and the other is autonomous.

## 1 Introduction

In radio frequency applications, effects like filtering, mixing and frequency modulation produce signals with largely differing time scales. Hence transient analysis of the corresponding circuit's equations becomes tedious, because the fastest rate limits the integration step size, whereas the slowest rate determines the total time interval. Numerical techniques, which compute directly the steady state response of a circuit, e.g. shooting methods or harmonic balance, are often time-consuming in view of this multirate behaviour, too.

Alternatively, the introduction of independent variables for each time scale enables an efficient representation of multitone signals and thus creates a useful post-processing tool. Since a network approach yields differential algebraic equations (DAEs) to model a circuit [5], this strategy results in a PDAE description, the multirate partial differential algebraic equations (MPDAEs). Brachtendorf et al. [1] proposed this PDAE model and successfully applied it in frequency domain by a multidimensional generalisation of harmonic balance. In this connection, a multiperiodic MPDAE solution identifies the quasi-periodic response of a circuit. Therefore sophisticated numerical PDAE methods are required to solve the system efficiently.

The arising time rates may be driven or autonomous. The presence of only autonomous scales makes the computation of multiperiodic MPDAE solutions difficult, cf. [9]. However, the MPDAE model enables the efficient calculation of envelope-modulated signals also in the purely autonomous case [2]. A mixture of driven and autonomous rates permits frequency modulation. Therefore the PDAE concept has been modified for this case by Narayan and Roychowdhury [12], which leads to the warped MPDAE model with stretched time scales.

The MPDAE system implies an information transport along characteristic curves, see [8]. This performance allows the construction of efficient time domain methods, which is realised for driven oscillators in [10].

In this paper, we outline the design of numerical techniques in time domain using the structure of the characteristic curves. Sect. 2 gives a brief description of the MPDAE model. Then the characteristic system of the MPDAE is introduced, which yields an appropriate semi-discretisation scheme. In Sect. 4, we apply this approach to solve the MPDAE of a driven oscillator, namely a transistor modulator circuit. Finally, the method of characteristics is generalised to determine MPDAE solutions with a driven as well as an autonomous time scale, where a forced van der Pol oscillator demonstrates the interactions of the given time rates.

## 2 Multivariate Functions and MPDAE Model

Oscillatory signals with widely separated time scales arise in radio frequency applications. Now the initiation of an own variable for each time scale yields an alternative multidimensional model. A simple example is the following transition

$$b(t) = \sin(\frac{2\pi}{T_1}t)\sin(\frac{2\pi}{T_2}t) \longrightarrow \hat{b}(t_1,t_2) = \sin(\frac{2\pi}{T_1}t_1)\sin(\frac{2\pi}{T_2}t_2) \quad (T_1 \gg T_2). \quad (1)$$

The new representation $\hat{b}$ is called the *multivariate function* (*MVF*) of the signal $b$. Since the MVF is biperiodic here, we have just to consider the rectangle $[0,T_1[\times[0,T_2[$ in time domain. Note that only one sinusoid occurs per coordinate direction in this rectangle for any periods $T_1, T_2$. Hence the MVF owns the advantage of being easy to sample. Still we can reconstruct the original signal completely from its MVF, because $b(t) = \hat{b}(t,t)$ holds.

In circuit simulation, Kirchhoff's laws yield implicit systems of differential equations, cf. [6], which we write in the form

$$\tfrac{d}{dt}\mathbf{q}(\mathbf{x}(t)) = \mathbf{f}(\mathbf{b}(t),\mathbf{x}(t)), \quad (2)$$

where $\mathbf{x}(t) \in \mathbb{R}^k$ denotes unknown time-dependent node voltages or branch currents and $\mathbf{q} \in \mathbb{R}^k$ represents a charge term. In general, the Jacobian matrix of $\mathbf{q}$ will be singular, which makes (2) a *differential algebraic equation* (*DAE*). Such systems raise theoretical and numerical problems like the index concept or the demand of consistent initial values [4]. The right-hand side $\mathbf{f} \in \mathbb{R}^k$ depends on input signals $\mathbf{b}(t) \in \mathbb{R}^p$. Assuming the said multirate behaviour for $\mathbf{b}$, we introduce corresponding MVFs $\hat{\mathbf{b}} \in \mathbb{R}^p$ . In the following, we restrict to the most frequent case of two widely separated time scales. Extensions to more rates are straightforward. The introduction of multidimensional functions transforms the DAE model (2) of the circuit into the above mentioned PDAE formulation, the *multirate partial differential algebraic equation (MPDAE)*

$$\frac{\partial \mathbf{q}(\widehat{\mathbf{x}}(t_1,t_2))}{\partial t_1} + \frac{\partial \mathbf{q}(\widehat{\mathbf{x}}(t_1,t_2))}{\partial t_2} = \mathbf{f}(\widehat{\mathbf{b}}(t_1,t_2),\widehat{\mathbf{x}}(t_1,t_2)) \tag{3}$$

with the MVF $\widehat{\mathbf{x}}$ of $\mathbf{x}$. It is easy to show that a solution $\widehat{\mathbf{x}}$ of (3) produces a solution of (2) via $\mathbf{x}(t) = \widehat{\mathbf{x}}(t,t)$. Thereby, a biperiodic MPDAE solution yields a two-tone quasi-periodic DAE solution, i.e. a function of the form

$$\mathbf{x}(t) = \sum_{j_1,j_2=-\infty}^{+\infty} \mathbf{X}_{j_1,j_2} \exp[i2\pi(\frac{j_1}{T_1} + \frac{j_2}{T_2})t] \qquad \text{with} \quad \mathbf{X}_{j_1,j_2} \in \mathbb{C}^k . \tag{4}$$

Sufficient and necessary existence theorems for multiperiodic MPDAE solutions in connection with quasi-periodic DAE solutions can be found in [11]. Therefore the MPDAE model provides an alternative approach for the calculation of quasi-periodic signals in radio frequency applications. Although a PDAE has to be solved now, the simple behaviour of MVFs causes savings in computational time, if efficient numerical techniques are used.

## 3 Characteristic System of MPDAE

A detailed analysis of the system (3) exhibits a hyperbolic structure, cf. [8]. Accordingly, information transport proceeds along characteristic curves, which are just the straight lines in diagonal direction, i.e. $(\tau + c, \tau)$ with a parameter $\tau \in \mathbb{R}$ for any fixed $c \in \mathbb{R}$. Consequently, we define a characteristic system of (3) by

$$\tfrac{d}{d\tau}\mathbf{q}(\overline{\mathbf{x}}_c(\tau)) = \mathbf{f}(\widehat{\mathbf{b}}(\tau + c, \tau), \overline{\mathbf{x}}_c(\tau)), \tag{5}$$

which represents a family of DAEs for the restrictions $\overline{\mathbf{x}}_c(\tau) := \widehat{\mathbf{x}}(\tau + c, \tau)$. Now a function $\widehat{\mathbf{x}}$ is a solution of (3), if and only if the restrictions $\overline{\mathbf{x}}_c(\tau)$ satisfy the characteristic system (5) for all $c \in \mathbb{R}$, cf. [10].

Our aim is to compute numerical approximations of a biperiodic MPDAE solution with periods $T_1$ and $T_2$, without loss of generality $T_1 \geq T_2$. To obtain an appropriate semi-discretisation of the MPDAE (3), we choose a finite number of DAEs from the characteristic system (5). Owing to the periodicity condition in the first coordinate direction, we consider an initial manifold on the line $[0, T_1[\times\{0\}$ and select the characteristic curves with $c = (j-1)h_1$; $h_1 = T_1/n_1$ for $j = 1, \ldots, n_1$. Fig. 1 illustrates this strategy. Hence a system of $n_1$ independent DAEs

$$\tfrac{d}{d\tau}\mathbf{q}(\widetilde{\mathbf{x}}_j(\tau)) = \mathbf{f}(\widehat{\mathbf{b}}(\tau + (j-1)h_1, \tau), \widetilde{\mathbf{x}}_j(\tau)) \qquad j = 1, \ldots, n_1 \tag{6}$$

arises in this semi-discretisation. Furthermore, the periodicity condition in the second coordinate direction yields boundary conditions for the system (6). For the

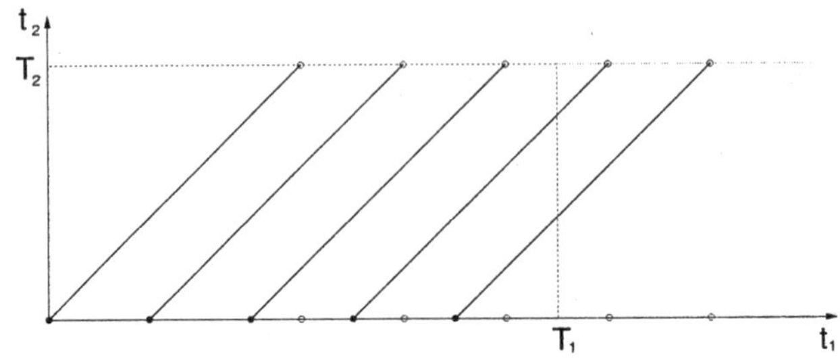


**Fig. 1.** Information transport for MPDAE system.

initial values $\widetilde{\mathbf{x}}_j(0)$, we have the requirement $\widetilde{\mathbf{x}}_j(0) = \widehat{\mathbf{x}}((j-1)h_1, T_2)$. The corresponding point $((j-1)h_1, T_2)$ does not necessarily lie on the selected characteristic curves. Nevertheless, we are able to interpolate the value in this point by the quantities $\widetilde{\mathbf{x}}_j(T_2)$ on the line $t_2 = T_2$. Thus the boundary condition results in

$$(\widetilde{\mathbf{x}}_1(0), \ldots, \widetilde{\mathbf{x}}_n(0))^T = B(\widetilde{\mathbf{x}}_1(T_2), \ldots, \widetilde{\mathbf{x}}_n(T_2))^T \tag{7}$$

with a constant matrix $B \in \mathbb{R}^{n_1 k \times n_1 k}$, whose entries depend on the used interpolation formula. Hence the constants $T_1, T_2$ and $h_1$ also influence $B$. Moreover, polynomial interpolation of a low degree creates a sparse matrix $B$.

By solving the boundary value problem (6),(7) for DAEs, we obtain an approximation to a biperiodic solution of the MPDAE (3) in the underlying parallelogram, see Fig. 1. Using the periodicities, the solution can be interpolated everywhere.

## 4 Application to Driven Time Scales

In this section, we assume an oscillator, which is driven by input signals with two separate time scales. If a corresponding two-tone quasi-periodic solution $\mathbf{x}(t)$ of (2) exists, then we know its rates $T_1, T_2$ by the frequencies of the input signals in $\mathbf{b}(t)$. Hence we choose appropriate MVFs $\widehat{\mathbf{b}}(t_1, t_2)$ and apply the MPDAE model (3) with biperiodic boundary conditions for the given periods $T_1$ and $T_2$.

Consequently, a numerical solution can be computed via the boundary value problem of DAEs (6),(7). As test example, we consider a transistor modulator circuit. Fig. 2 displays its circuit diagram. Using nodal analysis and the Ebers-Moll model for the transistors, a system of four differential equations

$$\begin{aligned}
-C\tfrac{du_1}{dt} &= \tfrac{u_1}{R_L} + i_2 + I_s\left[\exp(\lambda(u_1 - U_{op1})) - 1\right] - \alpha I_s\left[\exp(\lambda u_4) - 1\right] \\
-L\tfrac{di_2}{dt} &= -u_1 \\
0 &= -I_s\left[\exp(\lambda(u_4 + U_{in2})) - 1\right] + \alpha I_s\left[\exp(\lambda(U_{in2} - U_{op1})) - 1\right] \\
&\quad -I_s\left[\exp(\lambda u_4) - 1\right] + \alpha I_s\left[\exp(\lambda(u_1 - U_{op1})) - 1\right] \\
&\quad -I_s\left[\exp(\lambda(u_4 + U_{in1} - U_{op2})) - 1\right] + \alpha I_s\left[\exp(\lambda u_3) - 1\right] \\
0 &= -I_s\left[\exp(\lambda u_3) - 1\right] + \alpha I_s\left[\exp(\lambda(u_4 + U_{in1} - U_{op2})) - 1\right] \\
&\quad + \tfrac{1}{R_b}\left[U_{op1} - U_{op2} + U_{in1} - u_3\right]
\end{aligned} \tag{8}$$

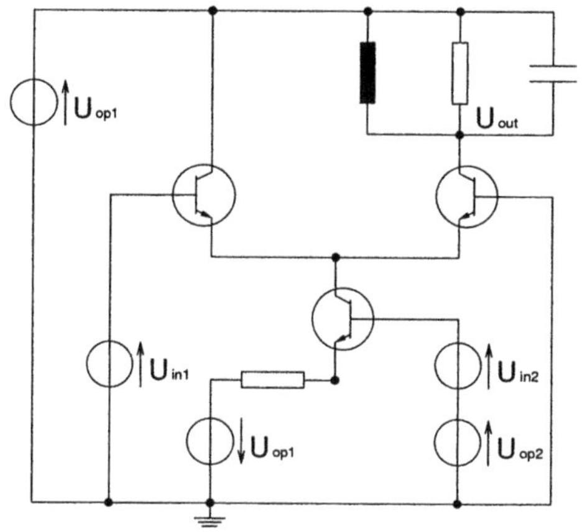


**Fig. 2.** Circuit diagram of transistor modulator.

emerges, which is a DAE of index 1. More details of the derivation and the technical parameters can be found in [3]. We choose input signals $U_{in1}(t) = 4\text{V}\cos(2\pi t/T_1)$ and $U_{in2}(t) = 0.1\text{V}\cos(2\pi t/T_2)$ with the slightly differing periods $T_1 = 0.1\text{msec}$ and $T_2 = 0.01\text{msec}$ in order to obtain a detailed visualisation of the whole signals. A finite difference method is suitable to solve the boundary value problem (6),(7), since it can be easily applied to DAEs. We use a discretisation with centred differences and linear interpolation in the boundary conditions. Fig. 3 and Fig. 4 show the resulting MPDAE solutions for the output voltage $u_1$ as well as for the node voltage $u_4$ and the corresponding DAE solutions, which are interpolated from the MVFs. For comparison, we solve an initial value problem of the DAE system (8) using the RADAU5 integrator. Indeed, an excellent agreement of both solutions can be observed in Fig. 3 and Fig. 4, respectively. Simulations for widely separated periods $T_1 \gg T_2$ produce the same shape for the MPDAE solutions.

A finite difference method, which is directly applied to the MPDAE (3), generates unphysical couplings and thus increases the computational effort. On the other hand, a solution of the problem (6),(7) leads to enormous savings with respect to computational time and memory in the linear algebra part, since the technique regards the information transport in the system (3). This gain of efficiency is investigated in [10], where the ring modulator circuit serves as test example, i.e. a DAE of index 2 consisting of fifteen equations.

## 5 Application to Autonomous Time Scales

Now we consider an autonomous oscillator with a periodic limit cycle. If an independent input signal with another period is added, then a solution with two separate

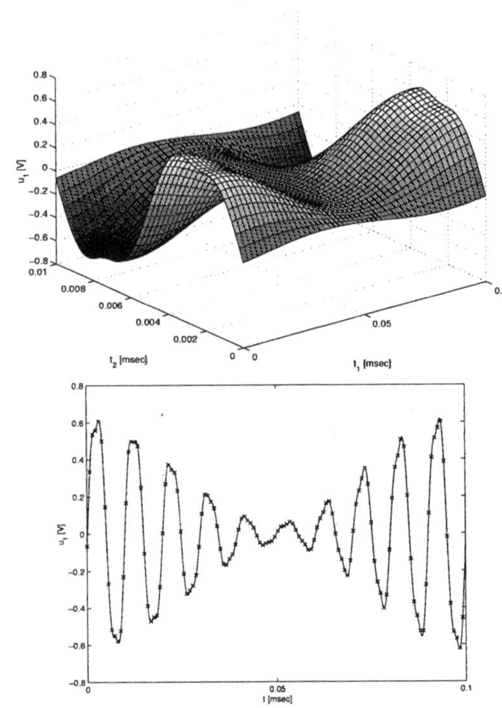


**Fig. 3.** MPDAE solution for output voltage $u_1$ (left) and corresponding DAE solution (right, $\times$) together with integrated DAE solution (right,—) from RADAU5.

time scales may arise. As benchmark, we observe the equation of a forced van der Pol oscillator

$$\frac{d^2x}{dt^2} + \mu(x^2 - 1)\frac{dx}{dt} + k^2 x = A\sin(\frac{2\pi}{T_1}t), \tag{9}$$

which represents an ordinary differential equation (ODE). For the autonomous oscillator ($A = 0$), a limit cycle with a priori unknown period $T_0$ exists. One can apply time or frequency domain techniques to calculate this solution and its period, which is described in [7], for example. The forcing term on the right-hand side ($A \neq 0$) introduces another time rate $T_1$, which leads to a two-tone solution of (9) with a second time scale $T_2$. Let us assume $k = 2\pi$ and $T_1 \gg T_2$, then the behaviour of the fast rate $T_2$ of the new solution depends on the parameter $\mu > 0$. On the one hand, for small values, e.g. $\mu < 0.1$, it holds $T_2 \approx T_0 \approx 1$. On the other hand, if $\mu$ is large, frequency modulation occurs and the rate $T_2$ is not constant any more. Accordingly, we call $T_2^{-1}$ the local frequency.

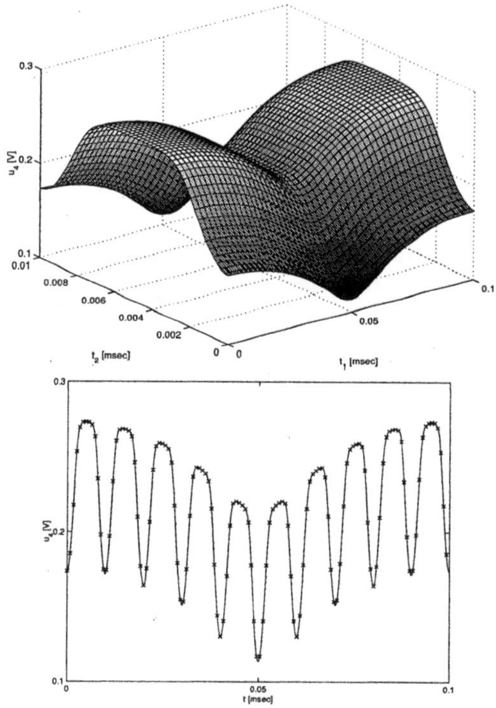


**Fig. 4.** MPDAE solution for voltage $u_4$ (left) and corresponding DAE solution (right, $\times$) together with integrated DAE solution (right,—) from RADAU5.

The MPDAE model has been adapted to the case of frequency modulation in [12]. For a DAE (2) with a single-tone input signal $\mathbf{b}$, the *warped multirate partial differential algebraic equation (WaMPDAE)* is defined by

$$\frac{\partial \mathbf{q}(\widehat{\mathbf{x}}(t_1, t_2))}{\partial t_1} + \nu(t_1) \frac{\partial \mathbf{q}(\widehat{\mathbf{x}}(t_1, t_2))}{\partial t_2} = \mathbf{f}(\mathbf{b}(t_1), \widehat{\mathbf{x}}(t_1, t_2)), \tag{10}$$

where $\nu$ is the local frequency. Therefore the function

$$\mathbf{x}(t) = \widehat{\mathbf{x}}(t, \Phi(t)) \quad \text{with} \quad \Phi(t) = \int_0^t \nu(t_1) dt_1 \tag{11}$$

satisfies the DAE (2). A $(T_1, 1)$-periodic solution $\widehat{\mathbf{x}}$ gives us a two-tone solution $\mathbf{x}(t)$ with local frequency $\nu(t)$. The $T_1$-periodic function $\nu(t_1)$ is unknown a priori and thus (10) is underdetermined. Hence we add a suitable phase condition at the boundary like $\frac{\partial \widehat{\mathbf{x}}_1}{\partial t_2}(t_1, 0) = 0$ for the first component and all $t_1$. The necessity of a phase condition is closely related to the fact that for a solution $\hat{\mathbf{x}}(t_1, t_2)$ of (10) also $\hat{\mathbf{x}}(t_1, t_2 + \gamma)$ satisfies the WaMPDAE for all $\gamma \in \mathbb{R}$, since the system is autonomous in $t_2$. Therefore the solution of (10) yields more information than a special two-tone solution of (2), because we can reconstruct a continuum of DAE solutions by just one WaMPDAE solution.

In contrast to the original MPDAE (3), the characteristic curves are no longer straight lines. They are given by the relation $t_2 = \Phi(t_1) + c$ with $c \in \mathbb{R}$ and thus $\nu(t_1) > 0$ implies monotonic functions. If $\nu$ in (10) is constant, then the characteristic curves are again straight lines with gradient $\nu$ and the WaMPDAE (10) can be transformed into the original MPDAE (3).

We apply the WaMPDAE model to the equivalent first order system of the forced van der Pol oscillator (9) with the parameters $\mu = 10$, $k = 2\pi$, $A = 30$ and $T_1 = 1000$ (dimensions are omitted). Since (9) is an ODE, we call (10) a partial differential equation (PDE) now. The resulting WaMPDE can be directly solved by a finite difference method with centred differences on a uniform grid, for example. Although this discretisation does not regard the information transport and thus increases the computational effort, it allows an easy implementation and yields a robust scheme. Thereby, we calculate a biperiodic solution together with a local frequency function for the above mentioned phase condition. Fig. 5 illustrates the computed solution. Obviously, the local frequency is not constant here and exhibits the mean value $< T_2 > \approx 1.25$ with variations about 10%.

Nevertheless, we are able to tailor the strategy from Sect. 3 to the case of constant frequency $\nu(t_1) = \nu_2$, which arises for small parameters $\mu$ in the forced van der Pol oscillator (9). As mentioned above, a linear transformation changes the WaMPDAE (10) into the original MPDAE (3) here. Hence we encounter again the boundary value problem (6),(7) of DAEs, which has proved suitable for purely forced oscillators. In contrast to this case, the period $T_2 = \nu_2^{-1}$ is unknown in view of the autonomous second time scale. However, the numerical solution of (6),(7) together with the time rate $T_2$ is similar to the computation of a periodic limit cycle of an autonomous ODE. Following [7], we add $T_2$ to the list of unknowns in a numerical method. Furthermore, a scalar phase condition has to be fixed as an extra equation in order to extract a unique function from the continuum of MP-DAE solutions. Solving this extended boundary value problem yields a numerical approximation for the biperiodic MPDAE solution and its second period.

Accordingly, we apply this approach to the forced van der Pol oscillator with $\mu = 0.1$. Since Eq. (9) represents an ODE, the resulting system (6) is also an ODE system and the arising boundary value problem can easily be solved by means of a shooting method. The periodic limit cycle of the completely autonomous oscillator, i.e. $A = 0$ in (9), serves as initial values in the corresponding Newton iteration. We use the trapezoidal rule in the forward integration of the ODE system. The Newton iteration yields $T_2 = 1.0014$ for the second time scale. Fig. 6 shows the result for the biperiodic MPDE solution $\hat{x}$. In comparison to the case $\mu = 10$ in Fig. 5, we observe a greater extend of amplitude modulation and less steeper gradients. The first few cycles of the corresponding ODE solution are also illustrated in Fig. 6, where interpolated values from the MVF agree with the integrated solution using trapezoidal rule. During further cycles, a phase shift between the exact ODE solution and the values from the MPDE solution emerges, because the fast rate $T_2$ includes a numerical error here. Nevertheless, the MPDE model captures completely the qualitative behaviour of the quasi-periodic signal.

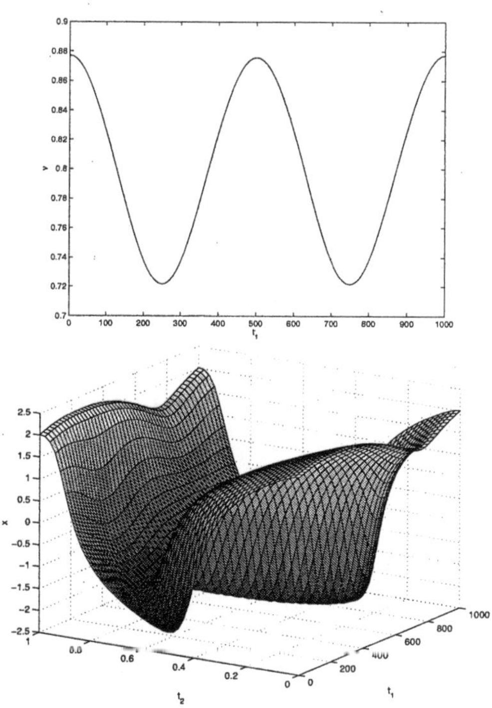


**Fig. 5.** Local frequency $\nu$ (left) and biperiodic solution $\widehat{x}$ (right) of WaMPDE for the forced van der Pol oscillator with $\mu = 10$.

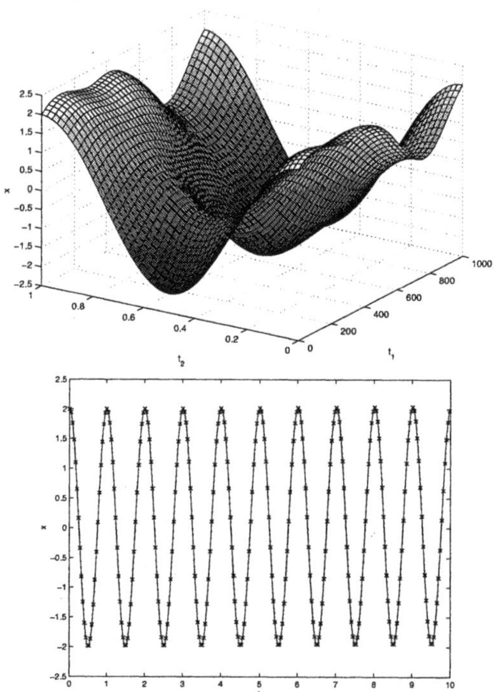


**Fig. 6.** Biperiodic solution $\widehat{x}$ (left) of MPDE for the forced van der Pol oscillator with $\mu = 0.1$ and corresponding ODE solution (right, $\times$) together with integrated ODE solution (right, —) using trapezoidal rule.

# 6 Conclusions

The multivariate signal model demands sophisticated techniques to solve the respective partial differential algebraic equations efficiently. An approach has been developed in time domain, which employs the specific structure of corresponding characteristic curves. Accordingly, significant savings in computational time and memory arise in comparison to common discretisation schemes. Test results confirm that this method is suitable for simulating driven oscillators. Furthermore, the method of characteristics can be applied to driven autonomous oscillators with quasi-periodic signals. Again numerical calculations evidence a good behaviour. These promising results suggest to tailor the technique to frequency modulated signals.

# Acknowledgements

This research has been supported by the Federal BMBF within the programme 'Neue mathematische Verfahren in Industrie und Dienstleistungen' (Project No. 03REM3KA). The author would like to thank Prof. Dr. P. Rentrop and Priv.-Doz. Dr. M. Günther for helpful comments and remarks.

# References

1. Brachtendorf, H. G.; Welsch, G.; Laur, R.; Bunse-Gerstner, A.: Numerical steady state analysis of electronic circuits driven by multi-tone signals. Electrical Engineering **79** (1996), pp. 103-112.
2. Brachtendorf, H. G.; Laur, R.: Multi-rate PDE methods for high Q oscillators. In: Mastorakis, N. (ed.): Problems in Modern Applied Mathematics. CSCC 2000, MCP 2000, MCME 2000 Multi-conference, Athens, July 2000. World Scientific 2000, pp. 391-398.
3. Chua, L. O.; Ushida, A.: Algorithms for computing almost periodic steady-state response of nonlinear systems to multiple input frequencies. IEEE Trans. CAS **28** (1981) 10, pp. 953-971.
4. Günther, M.; Rentrop, P.: The differential-algebraic index concept in electric circuit simulation. Z. angew. Math. Mech., Berlin **76** (1996) Suppl. 1, pp. 91-94.
5. Günther, M.; Feldmann, U.: CAD based electric circuit modeling in industry I: Mathematical structure and index of network equations. Surv. Math. Ind. **8** (1999), pp. 97-129.
6. Kampowsky, W.; Rentrop, P.; Schmitt, W.: Classification and numerical simulation of electric circuits. Surv. Math. Ind. **2** (1992), pp. 23-65.
7. Kundert, K. S.; Sangiovanni-Vincentelli, A.; Sugawara, T.: Techniques for finding the periodic steady-state response of circuits. In: Ozawa, T. (ed.): Analog methods for computer-aided circuit analysis and diagnosis, New York (1988), pp. 169-203.
8. Pulch, R.; Günther, M.: A method of characteristics for solving multirate partial differential equations in radio frequency application. To appear in: Appl. Num. Math.
9. Pulch, R.: PDE techniques for finding quasi-periodic solutions of oscillators. Preprint 01/09, IWRMM, University Karlsruhe (2001).

10. Pulch, R.: A finite difference method for solving multirate partial differential algebraic equations. Preprint 02/04, IWRMM, University Karlsruhe (2002).
11. Roychowdhury, J.: Analysing circuits with widely-separated time scales using numerical PDE methods. IEEE Trans. CAS I **48** (2001) 5, pp. 578-594.
12. Narayan, O.; Roychowdhury, R.: Multi-time simulation of voltage-controlled oscillators. Proceedings Design Automation Conference, 1999, pp. 629-634.

# The Application of Preconditioned Jacobi-Davidson Methods in Pole-zero Analysis

J. Rommes[1], C.W. Bomhof[2], H.A. van der Vorst[1], and E.J.W. ter Maten[3]

[1] Utrecht University, The Netherlands, jrommes@cs.uu.nl
[2] Plaxis BV, Delft, The Netherlands,
[3] Philips Research Laboratories, Eindhoven University of Technology, The Netherlands

**Abstract.** The application of Jacobi-Davidson style methods in electric circuit simulation will be discussed in comparison with other iterative methods (Arnoldi) and direct methods ($QR$, $QZ$). Numerical results show that the use of a preconditioner to solve the correction equation may improve the Jacobi-Davidson process, but may also cause computational and stability problems when solving the correction equation. Furthermore, some techniques to improve the stability and accuracy of the process will be given.

## Introduction

Pole-zero analysis is used in electrical engineering to analyse the stability of electric circuits [10,11,14]. For example, if a circuit is designed to be an oscillator, pole-zero analysis is one of the ways to verify that the circuit indeed oscillates. Another application is the verification of reduced order models over a wide frequency range [9]. Because the complexity of the circuits designed nowadays grows, together with the frequency range of interest, there is need for faster algorithms, not neglecting the accuracy. In this paper, Sect. 6 introduces the pole-zero problem. Section 6 describes the Jacobi-Davidson style methods as alternatives of conventional methods for the pole-zero problem. In Sect. 6, some typical numerical problems and techniques are discussed. In Sect. 6 the methods will be compared by numerical results, concluding with some future research topics.

## Pole-zero Analysis in Circuit Simulation

The Kirchhoff Current Law and the Kirchhoff Voltage Law describe the topology of an electric circuit. Together with the Branch Constitutive Relations, which reflect the electrical properties of the branches, the two Kirchhoff Laws result in a system of differential algebraic equations [10]:

$$\frac{\mathrm{d}}{\mathrm{d}t}\mathbf{q}(t,\mathbf{x}) + \mathbf{j}(t,\mathbf{x}) = 0, \tag{1}$$

where $\mathbf{x} \in \mathbb{R}^n$ contains the circuit state and $\mathbf{q}, \mathbf{j} : \mathbb{R} \times \mathbb{R}^n \to \mathbb{R}$ are functions representing the reactive and resistive behaviour, respectively. The way (1) is solved depends on the kind of analysis (DC-analysis, AC-analysis, transient analysis, pole-zero analysis). In every analysis, the capacitance matrix $C \in \mathbb{R}^{n \times n}$ and the conductance matrix $G \in \mathbb{R}^{n \times n}$ appear:

$$C(t, \mathbf{x}) = \frac{\partial \mathbf{q}(t, \mathbf{x})}{\partial \mathbf{x}}, \qquad G(t, \mathbf{x}) = \frac{\partial \mathbf{j}(t, \mathbf{x})}{\partial \mathbf{x}}.$$

Both matrices are real, non-symmetric and sparse.

Starting from a linearization round the DC-operating point, the time-domain formulation is as follows:

$$\begin{cases} C \frac{\mathrm{d}\mathbf{x}(t)}{\mathrm{d}t} + G\mathbf{x}(t) = \mathbf{e}(t) \\ \mathbf{x}(0) = 0, \end{cases} \tag{2}$$

where $\mathbf{e}(t)$ models the excitation. Because not all properties can be computed in the time domain, the problem is transformed to the frequency domain by applying a Laplace transform:

$$(sC + G)\mathcal{X}(s) = \mathcal{E}(s), \tag{3}$$

where $\mathcal{X}, \mathcal{E}$ are the Laplace-transforms of the variables $\mathbf{x}, \mathbf{e}$ and $s$ is the variable in the frequency domain. The response of the circuit to a variation of the excitation is given by the transfer function

$$\mathcal{H}(s) = (sC + G)^{-1}. \tag{4}$$

The elementary response of circuit variable $\mathcal{X}_o$ to excitation $\mathcal{E}_i$ is given by

$$\mathcal{H}_{oi}(s) = \mathbf{e}_o^T (sC + G)^{-1} \mathbf{e}_i. \tag{5}$$

The poles are the values $p_k \in \mathbb{C}$ that satisfy $\Delta t(p_k C + G) = 0$, hence $(G + p_k C)\mathbf{x} = 0$ for some $\mathbf{x} \neq 0$, which leads to a generalised eigenproblem ($\lambda = -p_k$):

$$G\mathbf{x} = \lambda C\mathbf{x}, \qquad \mathbf{x} \neq 0. \tag{6}$$

Because the problem of computing the zeroes is similar to the problem of computing the poles, the rest of this paper will consider the problem of computing the poles.

Especially for large circuits ($n > 10^4$), robust, iterative methods for the generalised eigenvalue problem (6) with sufficient accuracy and acceptable computational costs are needed. Furthermore, all right half-plane poles and no false right half-plane poles are desired. The dominant poles and zeroes must be accurate enough to produce correct Bode-plots for the frequency range of interest.

Two kinds of pole-zero methods are known in literature [10]: combined and separate pole-zero computation. This paper focuses on separate pole-zero computation.

## Jacobi-Davidson Style Methods

Because of its accuracy and robustness, the full-space $QR$-method (and the $QZ$-method to a less degree) is a popular choice as solver for the eigenproblem (6). However, the total costs of $O(n^3)$ and the necessity of an $LU$-decomposition, which destroys the sparsity of $G$ and causes numerical inaccuracies and maybe instabilities, become unacceptable for larger problems. Iterative methods like the implicitly restarted Arnoldi method also need the $LU$-decomposition and are designed to compute only a few ($m \ll n$) eigenvalues [11].

The Jacobi-Davidson method [13], on the other hand, is designed to converge fast to a few selected eigenvalues. Based on the Jacobi-Davidson method, the JDQR-method [8], which computes a partial Schur form, and the JDQZ-method [8], which computes a partial generalised Schur form, are developed. Without going into much detail, the basic idea behind the Jacobi-Davidson methods is as follows. For the problem $A\mathbf{x} = \lambda\mathbf{x}$, given the eigenpair approximation $(\vartheta_k, \mathbf{u}_k)$:

- Search a correction $\mathbf{v} \in \mathbf{u}_k^\perp$ for $\mathbf{u}_k$ such that

$$A(\mathbf{u}_k + \mathbf{v}) = \lambda(\mathbf{u}_k + \mathbf{v}).$$

- Solve $\mathbf{v}$ from the *correction equation*, with $\mathbf{r}_k = A\mathbf{u}_k - \vartheta_k\mathbf{u}_k$:

$$(I - \mathbf{u}_k\mathbf{u}_k^*)(A - \vartheta_k I)(I - \mathbf{u}_k\mathbf{u}_k^*)\mathbf{v} = -\mathbf{r}_k.$$

- Orthogonally expand the current basis $V$ with $\mathbf{v}$.

The Ritz-vector $\mathbf{u}_k = V\mathbf{s}$ is obtained by applying a full-space method, for instance the $QR$-method, to the projected matrix $V^*AV$, resulting in the eigenpair $(\vartheta_k, \mathbf{s})$. The Jacobi-Davidson method satisfies a Ritz-Galerkin condition [13].

The correction equation needs more attention. For the JDQR-method, it is

$$(I - QQ^*)(A - \vartheta_k I)(I - QQ^*)\mathbf{v} = -\mathbf{r}_k, \tag{7}$$

where $Q \in \mathbb{R}^{n\times k}$. If the correction equation is solved exactly, the convergence of the Jacobi-Davidson method is quadratic [13]. Besides solving the correction equation exactly, one can use linear iterative methods, like GMRES, with or without preconditioning. Because exact solvers are often not feasible in practice, the focus is on iterative methods with preconditioning. Using a preconditioner, however, is not as easy as it seems. Consider a preconditioner $K \approx A - \vartheta_k I$. There are three major issues. Firstly, the preconditioner is projected afterwards ($\widetilde{K} = (I - QQ^*)K(I - QQ^*)$). Secondly, $A - \vartheta_k I$ becomes more and more ill conditioned as the approximations $\vartheta_k$ become near the eigenvalue $\lambda$. Thirdly, this $\vartheta_k$ changes every iteration, and so does $A - \vartheta_k I$.

## Numerical Problems and Techniques

In [5], a technique is described to reduce the problem size of the ordinary eigenproblem. The idea is to remove the columns (and corresponding rows) from $G^{-1}C$ which are equal to zero, as well as the rows (and corresponding columns) equal to zero. This is justified because rows and columns equal to zero have corresponding eigenvalues of value zero, and removing these rows and columns does not influence the other eigenvalues. Because the product $G^{-1}C$ is not available explicitly, for computational reasons, the $k$ rows and columns to keep are administrated in a matrix $S = [e_{i_1}, \ldots, e_{i_k}]$, where $e_j$ is the $j$-th unit vector of length $n$. The reduced matrix is then defined by $S^T G^{-1} CS$.

Reduction of the problem in this way has a number of advantages. Firstly, Jordan blocks may be removed, thereby improving the stability and accuracy of the computations. Secondly, the computational costs will be reduced. Table 1 shows the degree of reduction for some example problems.

**Table 1.** The size, reduced size and degree of reduction for some example problems. The data is extracted from [5] and [10].

| Problem | Size | Size (reduced) | Reduction |
|---|---|---|---|
| pz_09 | 504 | 365 | 28% |
| pz_28 | 177 | 74 | 58% |
| pz_36_osc | 120 | 86 | 28% |
| pz_agc_crt | 114 | 96 | 16% |
| jr_1 | 815 | 681 | 16% |

Note, however, that the spectral properties of the reduced problem do not differ from the spectral properties of the original problem. As a consequence, the speed of convergence of the eigenmethod used will not be improved significantly. This technique is not applicable directly to the generalised eigenproblem.

Considering the computation of the preconditioner (for the operator in the correction equation), one has to cope with two problems, i.e. the near singularity of the operator to precondition and the continuous change of the operator. These two problems appear in both Jacobi-Davidson $QR$ and Jacobi-Davidson $QZ$. For computational reasons, a new preconditioner should be computed at most once for each new eigenvalue target. This is possible in practice, because the search space and corresponding Ritz-values in general contain good approximations for new eigenvalues. Some additional advantage can be gained by only recomputing the preconditioner for changes in $\vartheta$ larger than a certain threshold. Singularity problems can be dealt with by replacing zero diagonal elements by a small threshold value, a common technique for incomplete $LU$-decompositions. The costs and fill-in can be controlled by using a drop-tolerance for non-diagonal elements, resulting in ILUT [12] decompositions.

## Numerical Results and Conclusions

The data for the test problems was generated by the in-house analog electric circuit simulator Pstar of Philips Research [14]. Both full-space and iterative methods have been used to solve the ordinary and generalised eigenproblem. A small selection of the results presented in [10] has been made to identify the problems which are typical for the different approaches. Implementations of the Jacobi-Davidson methods are based on the algorithms in [4] and are available on the Internet: refer to [1] for JDQR and to [2] for JDQZ. Experiments have been done in Matlab 5.3 [3].

The transformation of the generalised eigenproblem to the ordinary eigenproblem may introduce inaccuracies, as has been mentioned before. Bode-plot (**a**) in Fig. 1 shows an example of this. Before computing the eigenvalues, the problem has been reduced from size $30 \times 30$ to $12 \times 12$ with the method described in [5]. The solution computed by $QR$ differs significantly on two points from the exact solution, which is computed by using (5) for several frequencies $s$. The two notches are caused by non-cancelling poles and zeroes, which do cancel in the original problem. It is conceivable that this is caused by the inversion of $G$. The iterative methods Arnoldi and JDQR suffer even more from inaccuracies. Bode-plot (**b**) in Fig. 1 shows the computed solutions for the generalised eigenproblem. In this case, the $QZ$-method nearly resembles the exact solution, while the iterative schemes still suffer from inaccuracies. The fact that $QZ$ performs better than $QR$, while both methods in theory compute the same eigenvalues, strengthens the argument that the inversion of $G$ introduces critical inaccuracies. A general remark can be made

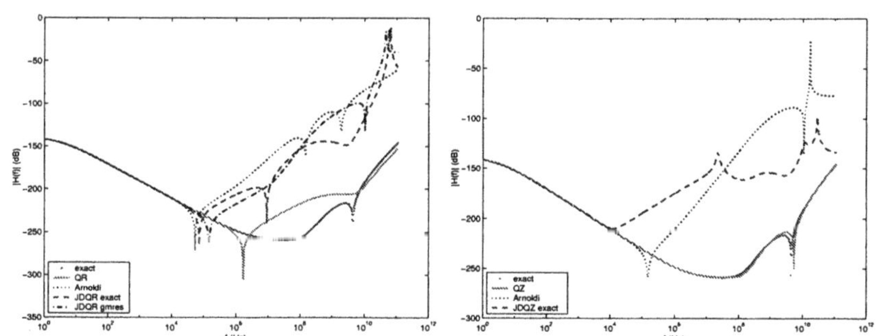


**Fig. 1.** (**a**) Bode-plot computed from the ordinary eigenproblem; (**b**) Bode-plot computed from the generalised eigenproblem.

about the interpretation of Bode-plots. It is not clear how accurate the original data of the circuit is. As a consequence, one may argue that the resulting Bode-plots are only representative up to a certain frequency, depending on the accuracy of the original data. For applications in the RF-area, this issue and the accuracy of eigenvalues near zero play even a more important role.

Using preconditioners when solving the correction equation of the JDQR method does indeed improve the speed of convergence, as can be seen in Fig. 2, where graph (**a**) shows the convergence history when using GMRES as solver, and graph (**b**) when using GMRES with an ILUT preconditioner ($t = 10^{-8}$). The quality of the

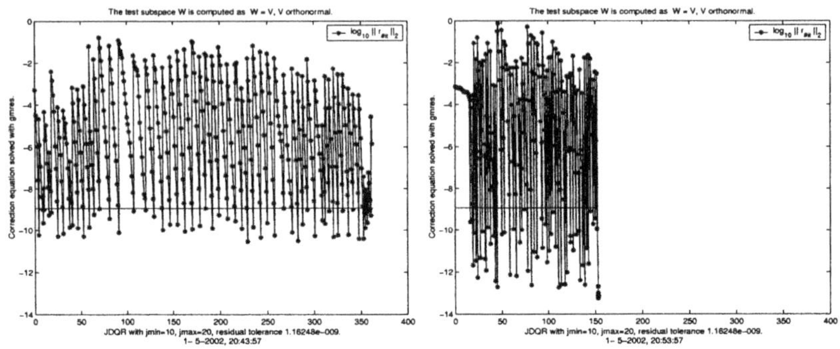


**Fig. 2.** (**a**) Convergence history for JDQR with GMRES; (**b**) Convergence history for JDQR with ILUT-preconditioned GMRES ($t = 10^{-8}$). A convergence history plots the residual against the Jacobi-Davidson iteration number; each drop below the tolerance means an accepted eigenvalue.

improvement strongly depends on the accuracy of the preconditioner. When using an ILUT preconditioner, a drop-tolerance of maximal $t = 10^{-6}$ is acceptable. This shows also one of the difficulties: the preconditioner has to be rather accurate, and in the case of ILU based preconditioners this means in general high costs. Apart from that, the ILU based preconditioners experience problems for singular matrices, and the matrix $A - \vartheta_k I$ becomes more and more singular. This last problem has appeared to be more severe for the JDQZ method. The fact that the preconditioner is projected afterwards has not a significant influence on the quality.

The reduction technique for the ordinary eigenproblem does lead to lower computational costs for both the direct and iterative methods. However, the gain depends on the degree of reduction, and is more pronounced for the iterative methods, because of the dominating matrix-vector products. The computational gained varied from 15% for the $QR$-method to 30% for the iterative methods, with top gains of 50%. Also, the condition number of the problem is improved, and if Jordan blocks are removed, the computed eigenvalues are more accurate. The number of iterations needed to computed the eigenvalues is not decreased, as expected. For more information and result of this reduction technique, refer to [5,11].

The observations launch ideas for future work. One can think of efficient updates for preconditioners [6], model reduction techniques and combinations of Jacobi-Davidson with other iterative methods like Arnoldi or combined pole-zero methods, such as Padé via Lanczos [7].

## References

1. http://www.math.uu.nl/people/sleijpen/JD_software/JDQR.html.
2. http://www.math.uu.nl/people/sleijpen/JD_software/JDQZ.html.
3. http://www.mathworks.com.

4. Bai, Z., Demmel, J., Dongarra, J., Ruhe, A., and van der Vorst, H., Eds. *Templates for the Solution of Algebraic Eigenvalue Problems: a Practical Guide.* SIAM, 2000.
5. Bomhof, C. Jacobi-Davidson methods for eigenvalue problems in pole zero analysis. Nat.Lab. Unclassified Report 012/97, Philips Electronics NV, 1997.
6. Bomhof, C. *Iterative and parallel methods for linear systems, with applications in circuit simulation.* PhD thesis, Utrecht University, 2001.
7. Feldmann, P., and Freund, R. W. Efficient linear circuit analysis by Padé approximation via the Lanczos process. *IEEE Trans. CAD 14* (1995), 639–649.
8. Fokkema, D. R., Sleijpen, G. L., and van der Vorst, H. A. Jacobi-Davidson style QR and QZ algorithms for the reduction of matrix pencils. *SIAM J. Sc. Comp. 20*, 1 (1998), 94–125.
9. Heres, P. J., and Schilders, W. H. Reduced order modelling of RLC-networks using an SVD-Laguerre based method. In *SCEE 2002 Conference Proceedings* (2002).
10. Rommes, J. Jacobi-Davidson methods and preconditioning with applications in pole-zero analysis. Master's thesis, Utrecht University, 2002.
11. Rommes, J., van der Vorst, H. A., and ter Maten, E. J. W. Jacobi-Davidson Methods and Preconditioning with Applications in Pole-zero Analysis. In *Progress in Industrial Mathematics at ECMI 2002* (2002).
12. Saad, Y. *Iterative methods for sparse linear systems.* PWS Publishing Company, 1996.
13. Sleijpen, G. L., and van der Vorst, H. A. A Jacobi-Davidson Iteration Method for Linear Eigenvalue Problems. *SIAM Review 42*, 2 (2000), 267–293.
14. ter Maten, E. J. W. Numerical methods for frequency domain analysis of electronic circuits. *Surv. Maths. Ind.*, 8 (1998), 171–185.

# The Electro–Quasistatic Model in Different Applications

Ute Schreiber*, Jürgen Flehr**, Victor Motrescu***, and Ursula van Rienen

Institute of General Electrical Engineering, Rostock University, Germany

**Abstract.** An electromagnetic field can be considered as slowly varying if the wavelength is large compared to the problem region. In the electro-quasistatic case it then may be assumed that the time-derivative of the magnetic flux is negligible, whereas the displacement currents have to be taken into account. Under these assumptions Maxwell's equations for time harmonic fields reduce to a complex Poisson's equation and discretization yields a complex symmetric system of equations. Krylov-subspace methods with an algebraic multigrid (AMG) preconditioner are used for fast solution. The electro-quasistatic model is applicable in many different constellations. This paper deals with applications from three different fields: high-voltage engineering, neural sensor-actor systems and the influence of slowly varying fields on human tissue.

## Introduction

An electromagnetic field can be considered as slowly varying if the wavelength is large compared to the problem region which means

$$|kR| \ll 1 \qquad \text{with the wavenumber} \qquad k = \omega\sqrt{\mu\epsilon(1 - i\frac{\sigma}{\omega\epsilon})}$$

and some characteristic dimension $R$ of the studied system [1]. For our time-harmonic applications we obtain the estimates $|kR| \approx 2 \cdot 10^{-6}$ for the insulator problem and $|kR| \approx 0.0037$ for 50 Hz fields in the human body. The spatial wavelength is given by $|1/k|$.

For a predominantly electric field, $\partial\mathbf{B}/\partial t \approx 0$, $\partial\mathbf{D}/\partial t \neq 0$ may be assumed in case of $|kR| \ll 1$. Regarding e.g. time harmonic electro-quasistatic fields in anisotropic media these assumptions yield the following set of simplified Maxwell's equations

$$\text{curl}\underline{\mathbf{E}} \approx 0, \quad (1) \qquad \text{div}\underline{\mathbf{D}} = \rho, \quad (3)$$

$$\text{curl}\underline{\mathbf{H}} = i\omega\underline{\mathbf{D}} + \underline{\underline{\sigma}}\,\underline{\mathbf{E}} + \underline{\mathbf{J}}_I, \quad (2) \qquad \text{div}\underline{\mathbf{B}} = 0. \quad (4)$$

Therein, $\mathbf{H}$ and $\mathbf{E}$ denote the magnetic and electric field strength, $\mathbf{D} = \underline{\underline{\epsilon}}\mathbf{E}$ the electric field density, $\mathbf{B} = \underline{\underline{\mu}}\mathbf{H}$ the magnetic induction, $\mathbf{J} = \mathbf{J}_I + \underline{\underline{\sigma}}\mathbf{E} + \delta\text{grad}\rho$ the

* supported by DFG under the contract RI 814/10-2

** supported by DFG under the contract GRK-45

*** supported by DFG under the contract RI 814/12-1 and by CST GmbH current density composed of the impressed current density $\mathbf{J}_I$, the conduction current density $\underline{\underline{\sigma}}\mathbf{E}$ and the convection current density with the charge carrier density $\rho$ and the diffusion constant $\delta$. Here, we have used the common representation $\mathbf{E}(\mathbf{r},t) = \mathrm{Re}(\underline{\mathbf{E}}(\mathbf{r})\mathrm{e}^{i\omega t})$ with the complex amplitude $\underline{\mathbf{E}}(\mathbf{r}) = \mathbf{E}(\mathbf{r})\mathrm{e}^{i\Phi}$. The non-linear, time-dependent rank two tensors $\underline{\underline{\sigma}}$, $\underline{\underline{\epsilon}}$ and $\underline{\underline{\mu}}$ are assumed to be piecewise constant functions with $\underline{\underline{\sigma}} \geq 0$, $\underline{\underline{\epsilon}} > 0$ and $\underline{\underline{\mu}} > 0$ ($\underline{\underline{\nu}} = \underline{\underline{\mu}}^{-1}$) if not stated differently.

We assume that appropriate boundary and interface conditions are defined. According to (1) the electric phasor $\underline{\mathbf{E}}$ is curl-free and thus may be described as the gradient of a scalar potential. Note that this is a *complex* potential: $\underline{\mathbf{E}} = -\mathrm{grad}\underline{\varphi}$. Under these conditions and from (1) - (4) we get the complex divergence equation for the time–harmonic EQS potential

$$\mathrm{div}\left[(i\omega\underline{\underline{\epsilon}} + \underline{\underline{\sigma}})\ \mathrm{grad}\ \underline{\varphi}\right] = \ \mathrm{div}\left(\underline{\mathbf{J}}_I\right). \tag{5}$$

With the Finite Integration Technique (FIT) [15] the continuous equation (5) is transformed into the discretisized one

$$\widetilde{\mathbf{S}}(i\omega\mathbf{M}_\epsilon + \mathbf{M}_\sigma)\widetilde{\mathbf{S}}^T\underline{\Phi}_E = \widetilde{\mathbf{S}}\underline{\mathbf{j}}_0 \tag{6}$$

with the divergence operator $\widetilde{\mathbf{S}}$, the material matrices $\mathbf{M}_\epsilon, \mathbf{M}_\sigma$ and the gradient operator $\mathbf{G} = -\widetilde{\mathbf{S}}^T$ [13]. The system matrix (6) is a complex symmetric almost singular matrix with seven bands. The large condition number mainly results from large differences in the material parameters.

The electro-quasistatic model has been implemented in the software package MAFIA [10], which is based on FIT, on a Cartesian grid. MAFIA is used for geometric modeling[1], creation of the complex symmetric system of equations and post-processing of the examples shown in this paper.

# Application Fields for the EQS Model

## Electric Field on Contaminated High-Voltage Insulators

High-voltage insulators are stressed by the applied electric field as well as by other environmental factors. As a result, the surface of the insulating material gets aged and the dielectric material looses its hydrophobic and insulating characteristics. Contamination of the object with water droplets accelerates the aging process. Experimental investigations have shown that with the increase of applied voltage, droplets vibrate first, they are then extended to the direction of the applied electric field and finally flash-over bridging water droplets occurs. To improve the understanding of the aging phenomenon it seems advisable to observe single droplets on an insulating surface. The shape of the droplets supplies more information about the status of the insulating material [8]. In addition to the experiments [9] the simulation of the electric field strength near the water droplets is necessary. It

[1] The human body is discretized with CST MicrowaveStudio™ [10].

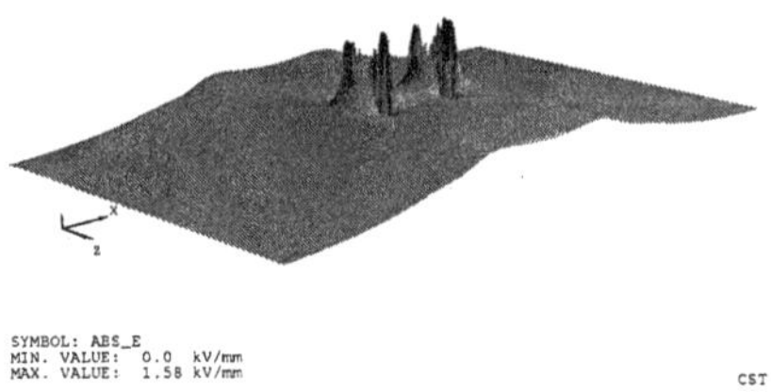


**Fig. 1.** Magnitude of electric field strength on the epoxy resin sample.

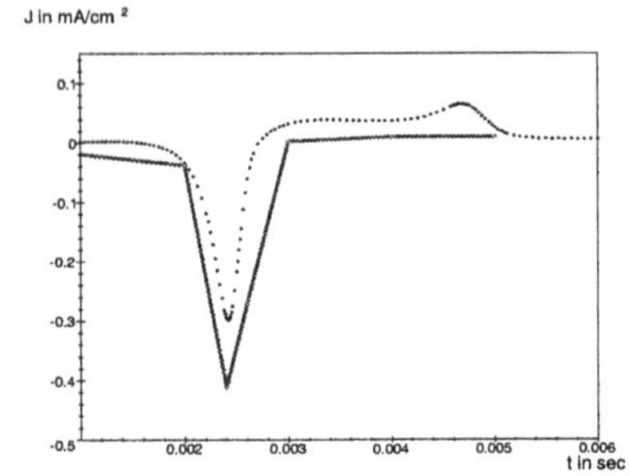


**Fig. 2.** Current $(j_{\mathrm{Na}}(t_i) + j_{\mathrm{K}}(t_i), i = 1, ..., n)$ calculated in MAFIA for $n = 6$ and by (8) for large $n$ (single dots).

allows to calculate the electric forces on the droplet surfaces and thus to find a correlation between the shapes and the droplet movement [14]. For experimental investigations of droplet movements it is necessary to eliminate other parameters which influence the distribution of field strength on the insulating surface.

This is why simplified test specimen (blocks of epoxy resin) are used for experiments [9] and simulations. The considered high–voltage devices are driven with 50 Hz a.c. voltage. The epoxy resin has a relative permittivity of $\varepsilon_{\mathrm{r}} = 4$ and a conductivity of $\sigma = 10^{-12}$ S/m. The water drops have a relative permittivity of $\varepsilon_{\mathrm{r}} = 81$ and a conductivity of $\sigma = 10^{-6}$ S/m. The permittivity of the air surrounding the structure is $\varepsilon_{\mathrm{r}} = 1.000576$. A voltage of 15 kV is used. Figure 1 shows the absolute value of the electric field strength on an insulating surface with two water droplets. The position of the latter is clearly visible by the augmented field strength at the "triple points" with air, water and insulating material; the maxima are in the direction of applied voltage. From the electric field the force density can be computed [14]. The final goal is to simulate the droplet movements in a coupled calculation.

## Electric Field of Living Neurons and the Neuron-Electrode Interface

Another application of the EQS model is the simulation of the electric field of living neurons and the neuron-electrode interface. The possibility to cultivate a living neural network on a microchip opens new chances in the field of neurophysics. With a living neural network on a chip it is possible to capture the signals of neurons, e.g. the action potential of a single neuron.

An action potential is an electrical signal of excitable cells. Such signals can propagate along the axon of the nerve. Hence the action potential is essential for signal propagation. The present study is part of a project where neurons from mice are exposed to a certain liquid and the response of the neuron is recorded. Thus the neurons are used as sensors and facilitate the analysis of the liquid provided that an interpretation of the signals is possible. The range of applications of this kind of cell based sensors is wide, including pharmaceutical screening and toxin detection.

To obtain an increased output signal we want to simulate the electric field of an active neuron firing an action potential and the neuron-electrode interface. The field will be calculated by means of Maxwell's equations using FIT. The results will be compared with practical measurements.

As proposed in the literature [1] electro-quasistatics is a good approximation for the simulation because the wavelength is much larger than the considered geometry. We are talking about a wavelength in the dimension of cm and a geometry of mm in the worst case [5]. So we may assume the EQS model to be a good approximation. Later, a posteriori calculations of the magnetic field will be used for validation.

We base our simulations on transient EQS equations which will be solved for different times $t_i$. Our strategy is to solve this equation first under the assumption that we have a steady state system with $\frac{\partial \mathbf{D}}{\partial t} + \mathbf{J} = 0$. So we can calculate $\mathbf{E}_0 := \mathbf{E}(t_0)$ with the electrostatic equations. After having calculated $\mathbf{E}_0$ and therefore $\mathbf{D}_0 := \mathbf{D}(t_0)$ we calculate the fields $\mathbf{E}_i$ for $i \geq 1$ step by step. For this task we can show that we only need $\mathbf{D}_0$ and Ampère's law. Actually, it is enough if we find a solution for the equation

$$\operatorname{div}(\varepsilon \operatorname{grad}\varphi_{new}) = \operatorname{div}(\mathbf{D}_{old} + \mathbf{J}\Delta t) \tag{7}$$

with $\frac{\partial \mathbf{D}}{\partial t} \approx \frac{\mathbf{D}_{new} - \mathbf{D}_{old}}{\Delta t}$, $\operatorname{curl}\mathbf{E} = \mathbf{0} \Rightarrow \mathbf{E} = -\operatorname{grad}\varphi$. At this point the reader may have recognized that we missed to describe how we get $\mathbf{J}$ for the right hand side of equation (7) yet.

During an action potential of a neuron the conductivity of the membran is changing drastically. Thanks to Hodgkin and Huxley we can model this with the following set of equations, the Hodgkin-Huxley-Equations [16]:

$$\frac{\partial V}{\partial t} = \frac{-1}{C_m}(\overline{g}_{\mathrm{Na}} m^3 h(V - E_{\mathrm{Na}}) + \overline{g}_{\mathrm{K}} n^4 (V - E_{\mathrm{K}}) + \overline{g}_{\mathrm{l}}(V - E_{\mathrm{l}})) \tag{8}$$

$$\frac{\partial n}{\partial t} = \alpha_n(1-n) - \beta_n n, \ \frac{\partial m}{\partial t} = \alpha_m(1-m) - \beta_m m, \ \frac{\partial h}{\partial t} = \alpha_h(1-h) - \beta_h h$$

with the membrane capacity $C_m$, the potential difference $V$ over the membrane, the time $t$, the membrane conductivity for sodium ions $\overline{g}_{\mathrm{Na}} m^3 h$, the membrane conductivity for potassium ions $\overline{g}_{\mathrm{K}} n^4$, the membrane conductivity for all other currents $\overline{g}_{\mathrm{l}}$ (leakage current), the resting potential for sodium $E_{\mathrm{Na}}$, the resting potential for potassium $E_{\mathrm{K}}$, the resting potential for the rest $E_{\mathrm{l}}$ and given constants $\overline{g}_{\mathrm{Na}}, \overline{g}_{\mathrm{K}}, \overline{g}_{\mathrm{l}}$ as well as given functions $\alpha_i, \beta_i$ ($\alpha_i = \alpha_i(V), \beta_i = \beta_i(V)$). We use this set of equations to calculate the conductivity for different states of the system - more precisely, the conductivity of sodium, potassium and the leakage current for the membrane of a neuron. Also, these equations yield the potential difference across the membrane.

Having already coupled the Hodgkin-Huxley-equations with Maxwell's equations, which is an important first result, we are able now to use this for the transient EQS simulation: Figure 2 shows the currents of sodium $j_{\mathrm{Na}}$ and potassium $j_{\mathrm{K}}$ calculated by MAFIA and by (8) as a reference: for only six points in time the agreement is already very good. Based on these currents a 3D field calculation is carried out according to (7). Later, we will of course use finer time steps than in these proof of principle simulations.

Until now the electric field is calculated as described above including the calculation of a potential at some electrode in a simplified model of a neuron-electrode interface: At moment we ignore the effects of electrolysis and consider only capacitive effects. The next steps will then be to implement *all* necessary equations. Furthermore, a comparison with other software packages able to simulate electrolysis is foreseen.

## Slowly Varying Electromagnetic Fields in the Human Body

Human body tissues exposed to low frequency electromagnetic fields present two kinds of conductivity. Some tissue classes are isotropic (i.e. fat tissue) whereas other tissue classes (i.e. skeletal and cardiac muscles) are anisotropic having a conductivity that is higher along the fibers than across, which can be described by a diagonal tensor only in a local coordinate system [17]

$$\underline{\underline{\sigma}}_L = \begin{pmatrix} \sigma_l & 0 & 0 \\ 0 & \sigma_t & 0 \\ 0 & 0 & \sigma_t \end{pmatrix} \tag{9}$$

where $\sigma_l$ is the longitudinal conductivity and $\sigma_t$ is the transverse conductivity. To express this tensor in a global coordinate system, it has to be rotated by means of a rotation matrix R

$$\underline{\underline{\sigma}}_G = R\underline{\underline{\sigma}}_L R^{-1} \tag{10}$$

where $R$ is the product of two other rotation matrices: $R = R_{xy}R_{xz}$

$$\mathrm{R}_{xy} = \begin{pmatrix} \cos\phi & \sin\phi & 0 \\ -\sin\phi & \cos\phi & 0 \\ 0 & 0 & 1 \end{pmatrix}, \quad \mathrm{R}_{xz} = \begin{pmatrix} \sin\vartheta & 0 & \cos\vartheta \\ 0 & 1 & 0 \\ \cos\vartheta & 0 & -\sin\vartheta \end{pmatrix}. \tag{11}$$

The two angles $\phi$ and $\vartheta$ are the rotation angles on z- and y- axes, respectively. Building the product of $R_{xy}$ and $R_{xz}$

$$\mathrm{R} = \begin{pmatrix} \sin\vartheta\cos\phi & -\sin\phi & \cos\vartheta\cos\phi \\ \sin\vartheta\sin\phi & \cos\phi & \cos\vartheta\sin\phi \\ \cos\vartheta & 0 & -\sin\vartheta \end{pmatrix}, \tag{12}$$

substituting (12) into (10) we obtain the conductivity tensor $\underline{\underline{\sigma}}_G$ of muscle tissues expressed in a global coordinate system. This is a full and symmetric tensor.

In order to simulate slowly varying electromagnetic fields in the human body taking advantage of the properties of FIT, the latter has to be extended to fully anisotropic materials. The classical method can handle materials that are isotropic or diagonal anisotropic in a global coordinate system. It was then extended in [18] to gyrotropic materials which satisfy that - although their material tensor has some off-diagonal terms - it is possible to align one of the coordinate axes to one of the main directions in the material. We will follow the basic ideas but in the case of the human body no main direction is ascertainable in the body as muscle fibers have miscellaneous orientations.

The off-diagonal terms from the global conductivity tensor of muscle tissues will connect components of the electric current density and electric voltage which are allocated in different places on the FIT grid. For this reason, in each location of an electric current density component we define three electric voltage components by interpolating each of them among four values situated in the first vicinity on the grid, oriented on the same axis. To keep the symmetry of the material matrices, the diagonal terms from the conductivity tensor of muscle tissues are averaged from primal grid cells to dual facets as in the classical FIT but the off-diagonal elements which couple components of electric current density and electric voltage on two axes, are averaged along a dual edge oriented on the third axis. This approach implies some first order approximation considering that primal edges have about the same length as the corresponding dual edges. The error introduced by this approximation vanishes in case of an equidistant grid which is true for our human body model because it consists of cubic voxels. By introducing the anisotropy in the FIT equation for EQS regime and using a realistic human body model of 1 mm resolution as well the software package PETSc [21], [19], [20] for parallel computation, we hope to obtain very fast and accurate solutions to the problem of simulating low frequency electromagnetic fields in the human body. A major part of the implementation is completed, actually.

## Solving the Linear Systems

In this paper the Krylov-subspace methods BiCGCR (Bi-Orthogonal Conjugate Gradient Conjugate Residual) [3,4], QMR (Quasi Minimal Residual) and so-called CSYM [2] are compared, each combined with Algebraic Multigrid (AMG) or Jacobi as preconditioner. Explicit descriptions of the algorithms, further references or a detailed investigation for complex symmetric systems can be found in [3,4,11]. Both preconditioners are implemented in the software package PEBBLES [11,12], others showed to be too expensive [4].

Results are shown for an insulator sample, an epoxy resin block of 100 mm x 100 mm x 20 mm and two water droplets on top. The test object has horizontally embedded electrodes with a center distance of 35 mm and a radius of 7.5 mm. The droplets' diameter is 6 mm (hemispheres), their center distance is 10 mm according to the accompanying experiments [9].

The characteristic convergence behavior for the methods is shown in Fig. 3. QMR and BiCGCR with the AMG–preconditioner PEBBLES perform very similar with respect to the number of iterations. The assumption that the combination of CSYM with PEBBLES instead of Jacobi improves convergence could not be verified. This fact is also reflected in Table 1 where CPU-times are specified.

The AMG-preconditioner obviously accelerates the iteration process in spite of the relatively large setup times compared to classical iterative solvers. PEBBLES' coarsening factor $\alpha$, which influences the number of necessary levels and the dimensions of the reduced problems, also intensively influences the performance: For

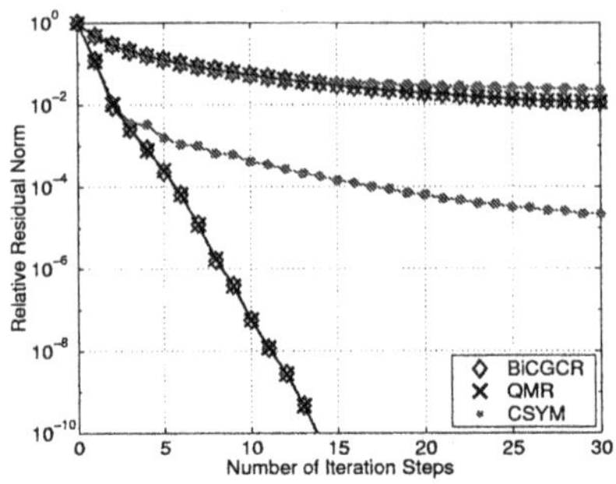


**Fig. 3.** BiCGCR, QMR and CSYM with Jacobi- (upper 3) and AMG-preconditioner (lower 3 curves).

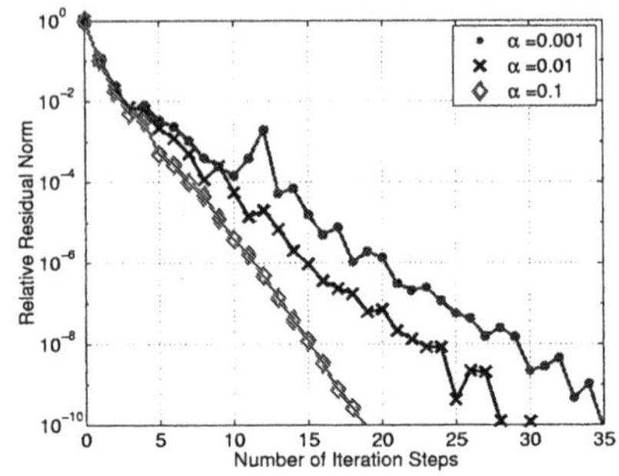


**Fig. 4.** AMG–QMR with different coarsening factors $\alpha$.

| | $\epsilon$ | setup (sec) | solver (sec) | total time (sec) | # iter. |
|---|---|---|---|---|---|
| AMG–BiCGCR | $10^{-10}$ | 187.77 | 703.44 | 891.21 | 15 |
| AMG–QMR | $10^{-10}$ | 186.83 | 493.21 | 680.04 | 15 |
| AMG–CSYM | $10^{-10}$ | 187.14 | >2,000 | >2,000 | >65 |
| Jacobi–BiCGCR | $10^{-2}$ | 10.45 | 856.15 | 866.60 | 41 |
| Jacobi–QMR | $10^{-2}$ | 10.43 | 810.17 | 820.60 | 40 |
| Jacobi–CSYM | $10^{-2}$ | 10.61 | 1,390.01 | 1,400.62 | 90 |

**Table 1.** Number of mesh points $N_h = 450,241$, 6 levels, $\alpha = 0.01$, mesh size ratio $\rho_h = 1$. The calculations were done on a sparc SUN, Ultra–1 with 296 MHz.

smaller $\alpha$ the total CPU-time grows in spite of lower setup times *and* the convergence curve gets slightly oscillatory (see Fig. 4).

**Conclusions and Acknowledgment** The electro–quasistatic model is suitable for many applications in the low frequency regime. Together with powerful algebraic solution methods and the discretization with FIT various new application areas of detailed electromagnetic field simulation are opened.

The authors wish to thank Markus Clemens and Frank Sachse for many helpful discussions, especially on the human body problem.

## References


1. H.A. Haus, J.R. Melcher, *Electromagnetic Fields and Energy*, Prentice-Hall. Inc., 1989.
2. A. Bunse-Gerstner, R. Stöver, *On a Conjugate Gradient-Type Method for Solving Complex Symmetric Linear Systems*, Lin.Alg.Appl., Vol. 287 (1999): 105-123.
3. M. Clemens, R. Schuhmann, U. van Rienen, T. Weiland, *Modern Krylov Subspace Methods in Electromagnetic Field Computation Using the Finite Integration Theorie*, ACES Journal, Vol. 11 (1996): 70 - 84.

4. M. Clemens, T. Weiland, U. van Rienen, *Comparison of Krylov-Type Methods for Complex Linear Systems Applied to High–Voltage Problems*, IEEE-T.Mag., Vol. 34 (1998): 3335 - 3338.
5. J. Dudel, R. Menze, R.F. Schmidt, *Neurowissenschaft*, Springer, 2001.
6. R.W. Freund, N.M. Nachtigal, *An Implementation of the QMR Method Based on Coupled Two–Term Recurrences*, SIAM J.Sci.Comput., Vol. 15 (1994): 297 - 312.
7. J. Keener, J. Sneyd, *Mathematical Physiology*, Springer, $2^{nd}$ printing, 2001.
8. S. Keim, D. König, *Study of the Behavior of Droplets on Polymeric Surfaces under the Influence of an Applied Electrical Field*, Proc. IEEE-CEIDP (1999): 707 - 710.
9. M. Kneuer, Diploma Thesis, TU Darmstadt, 2000.
10. MAFIA 4, CST GmbH, Bad Nauheimer Str. 19, D-64289 Darmstadt, Germany.
11. S. Reitzinger, U. Schreiber, U. van Rienen, *Algebraic Multigrid Methods for Complex Symmetric Matrices and Applications*, Journal for Computational and Applied Mathematics, to appear.
12. S. Reitzinger, *PEBBLES - User's Guide*, 1999, www.sfb013.uni-linz.ac.at.
13. U. van Rienen, *Numerical Methods in Computational Electrodynamics – Linear Systems in Practical Applications*, Springer-LNCSE, Vol. 12, 2000.
14. U. Schreiber, U. van Rienen, S. Keim, *Simulation of Electric Field Strength and Force Density on Contaminated H-V Insulators*, Springer-LNCSE, Vol. 18 (2001): 79 - 86.
15. T. Weiland, *A discretization method for the solution of Maxwell's equation for six–component fields*, Electron. Commun. AEÜ, Vol. 31 (1977): 116 - 120.
16. E.K. Yeargers, R.W. Shonkwiler, J.V. Herod, *An Introduction to the Mathematics of Biology*, Birkhäuser, 1996.
17. F.B. Sachse et al., *Comparison of Solutions to the forward Problem in Electrophysiology with homogeneous, heterogeneous and anisotropic Impedance Model*, Biomedizinische Technik, Vol. 42 (1997): 277 - 280.
18. H. Krüger, *Zur numerischen Berechnung transienter elektromagnetischer Felder in gyrotopen Materialien*, Der Andere Verlag, 2000.
19. S. Balay et al., *PETSc home page* (2001), http://www.mcs.anl.gov/petsc
20. S. Balay, W.D. Gropp, L.C. McInnes, B.F. Smith, *PETSc Users Manual*, Argonne National Laboratory, ANL-95/11 - Revision 2.1.3 (2002).
21. S. Balay, W.D. Gropp, L.C. McInnes, B.F. Smith, *Efficient Management of Parallelism in Object Oriented Numerical Software Libraries*, in E. Arge, A.M. Bruaset, H.P. Langtangen (Eds.): *Modern Software Tools in Scientific Computing*, Birkhauser (1997): 163 - 202.

# Substrate Resistance Modeling by Combination of BEM and FEM Methodologies

E. Schrik and N.P. van der Meijs

Delft University of Technology / DIMES
Mekelweg 4
2628 CD, Delft
The Netherlands

**Abstract.** In present-day IC's, substrate noise can have a significant impact on performance. Thus, modeling the noise-propagation characteristics of the substrate is becoming ever more important. Two ways of obtaining such a model are the Finite Element Method (FEM) and the Boundary Element Method (BEM). The FEM makes a full 3D discretization of the entire substrate and is very accurate and flexible, but, in general, it is also slow. The BEM only discretizes contact areas on the substrate-boundary, and is usually faster, but less flexible, because it assumes the substrate to consist of uniform layers. Sometimes, layout-dependent doping patterns near the top of the substrate may also play a significant role in noise-propagation. The FEM would easily be able to model such patterns, but it can often be too slow. The BEM, on the other hand, might not always be accurate enough. This paper describes a combination between BEM and FEM, which results in a method that is faster than FEM but more accurate than BEM. Through a number of experiments, the method is validated and successfully verified against 2 commercially available tools.

## Introduction

In present-day micro-electronic designs, substrate crosstalk can significantly influence the functionality of the design. In mixed-signal designs, for example, the noise originating from the switching activity in the digital part can propagate through the substrate and have a serious negative impact on the behaviour of the analog part. Similarly, in digital designs, substrate noise can have influence on the clock generator (typically a Phase-Locked Loop, PLL) causing fluctuations in the clock frequency (clock jitter). Thus, modeling the noise-propagation characteristics of the substrate is becoming ever more important [1–3].

Two principal ways of obtaining such a model are the Finite Element Method (FEM, as extensively described in e.g. [4]) and the Boundary Element Method (BEM [5]). The FEM makes a full 3D discretization of the entire substrate, and therefore it is very accurate and flexible, but, in general, it is also slow. The BEM, on the other hand, assumes the substrate to consist of uniform layers and only discretizes contact areas on the substrate-boundary. Therefore it is less flexible, but it can be significantly faster.

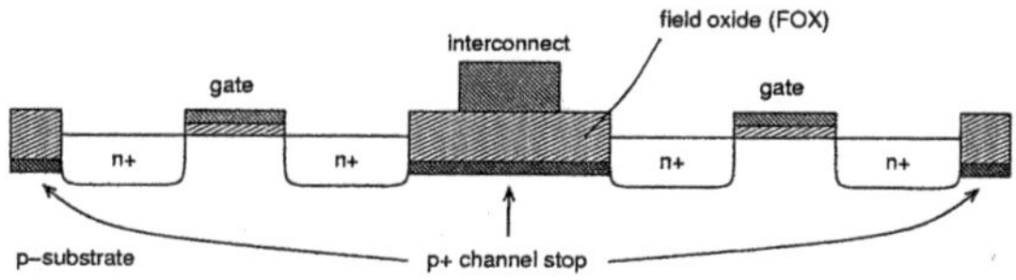


**Fig. 1.** Highly doped channel-stop layer underneath the Field Oxide.

Unfortunately, the assumption of a uniform, layered substrate often does not hold due to the presence of specific, layout-dependent doping patterns near the top of the substrate (e.g. channel-stoppers, trenches, buried layers or sinkers). An example of this is presented in Figure 1, which shows a highly-doped channel-stop layer immediately underneath the Field Oxide. The high doping-level of this layer will cause its resistivity to be low and, as such, it can play an important role in the noise-propagation through the substrate. Even though the FEM would easily be able to incorporate such patterns into the model, it can often be too slow. The BEM, on the other hand, might not always be accurate enough, because the patterns do not form a uniform layer.

In order to circumvent this modeling-dilemma, we have developed a combined BEM/FEM method [6] that is faster than FEM and more accurate than BEM. Even though the concept of a combined BEM/FEM method is not new [5,7], we do apply it in a new context. In this paper, we will briefly summarize the method, its proof and its implementation into the SPACE layout-to-circuit extractor [8], after which we will describe some successful applications and comparison to 2 commercially available tools.

The structure of this paper is as follows: Section 6 will first give a brief background of the BEM and FEM methods and their combination. Section 6 will briefly summarize the proof after which Section 6 will mention some details on the implementation of the prototype. Finally, Section 6 will present some applications, and Section 6 will state our conclusions.

## Background

In our modeling problem we are interested in finding a resistance network that represents the substrate. Since resistance relates potential differences to currents, our modeling approach is to solve a potential problem in a passive 3D domain. In mathematical terms, this is equivalent to solving the Laplace equation

$$\nabla(\sigma\nabla\Phi) = 0 \qquad (1)$$

with $\Phi$ the potential, and $\sigma$ the conductivity of the domain. When imposing continuous boundary conditions on the domain, the resulting potential field will be such that the energy contained within the domain is minimal. In mathematical terms, this is equivalent to minimizing the following energy functional:

$$\mathcal{E} = \int_{\Omega} \sigma||\nabla\Phi(p)||^2 dp \qquad (2)$$

with $\Omega$ representing the entire 3D domain of interest.

**Boundary Element Method** The Boundary Element Method [5] originated from the observation that the Laplace equation can be written as a boundary integral equation by applying Green's second identity. When imposing Dirichlet boundary conditions on the contact areas of the substrate and Neumann boundary conditions on non-contact areas of the substrate, a very simple expression will remain

$$\Phi(i) = \int_{S_1} k(j)G(i,j)dj \quad \text{with} \quad i,j \in S_1 \tag{3}$$

where $S$ is the entire boundary, $S_1 \subset S$ is the entire contact area, $k(j)$ is the continuous current distribution on $S_1$, and $G(i,j)$ is the so-called Green's function.

The Green's function relates the potential in point $i$ to a unit current in point $j$. It is a fundamental solution to the Laplace equation and, as such, it automatically ensures minimization of the energy functional (2). For a uniform homogenous medium, it looks as follows

$$G(i,j) = \frac{1}{4\pi\sigma r} \tag{4}$$

where $r = \sqrt{(x_i - x_j)^2 + (y_i - y_j)^2}$ (because contact areas are usually situated on top of the substrate, the $z$-coordinates have been omitted from this expression). The Green's function basically "encodes" the characteristics of the medium. It exists for media consisting of multiple uniform layers, but its complexity will drastically increase with each added layer, and it will eventually become infeasible and even impossible to compute.

The unknown current distribution $k(j)$ can be approximated by discretization of the entire contact area into smaller panels (see Figure 2; left) and assuming a constant current distribution on each panel. By applying the Method of Moments [9], a set of linear equations can be found from which we can solve a piecewise constant approximation of the current distribution [3]. From this solution, we can then easily find the resistance network we were looking for.

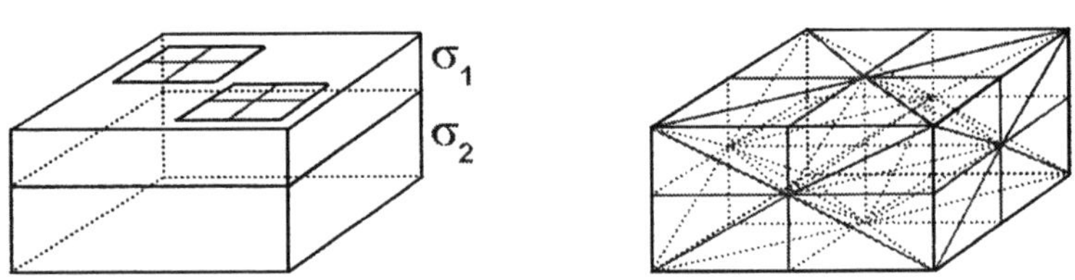


**Fig. 2.** Visual representation of the BEM (left) and the FEM (right). For the BEM, only contact areas on the top boundary of the domain are discretized; for the FEM, the entire domain is discretized

The BEM owes its speed to the fact that the integration only takes place over the contact area $S_1$ of the boundary and to the fact that the Green's function encodes the characteristics of the entire medium. The bottle-neck, however, is solving the system of linear equations, which is typically full and potentially large; dedicated, highly optimized solvers and/or sparsification techniques are required.

**Finite Element Method** The FEM makes a full 3D tetrahedral discretization of the entire domain (see Figure 2; right), and assumes linearity within each tetrahedron. As such, it allows for a very simple mathematical expression to describe the electrical interactions between the nodes at the corners of each tetrahedron. Through an incidence strategy, a typically very large, but sparse system of equations is found for the interactions between the nodes in the system. The field solution is then found by imposing boundary conditions and minimizing a discretized version of the energy functional (2), which acts as an alternative formulation of the Laplace equation.

However, for finding a resistance network, the minimization of the energy does not have to be carried out explicitly, because the linear mathematical relationships between the FEM nodes can be represented by resistors. As such, the FEM discretization is equivalent to a large, sparse resistance network and minimization of energy is ensured (implicitly) by the circuit simulator during simulation.

The 3D discretization will automatically incorporate any inhomogeneities of the domain into the model, which lends the FEM its accuracy and flexibility. However, the size of the model, despite its sparsity, also renders the FEM rather slow.

**Combined BEM/FEM** As presented in [6], a combined BEM/FEM method may be efficient when both speed and accuracy are required. For this approach, we effectively define the specific doping patterns (see the introduction) as FEM domains, and the rest of the underlying substrate as a BEM domain. As such, the BEM and the FEM "communicate" with each other through an interface.

The combination between the methods is immediately allowed when we choose the BEM interface mesh as the Voronoi polygon of the FEM interface nodes [6]. In other words, we construct a duality between the BEM and FEM interface meshes, such that each FEM node will be associated with one BEM panel (see Figure 3; left) and an immediate connection of the resulting networks is allowed (see Figure 3; right).

## Indication of Proof

We will use Figure 4 as a step-by-step guide through the process.

**Figure 4a**: When imposing continuous boundary conditions on a domain that is subject to the Laplace equation, a field $\Phi_m$ with minimum energy will manifest itself in the domain.

**Figure 4b**: Discretization of the boundary conditions results in a new minimum-energy field $\Phi'$ that is close to $\Phi_m$. In terms of energy: $E(\Phi') = E(\Phi_m) \pm \varepsilon$

**Figure 4c**: We now divide the domain with an interface. If the interface preserves energy, the solution will be valid, and there will be a continuous potential match along the interface. The energy will be such that $E(\Phi'_1) + E(\Phi'_2) = E(\Phi')$.

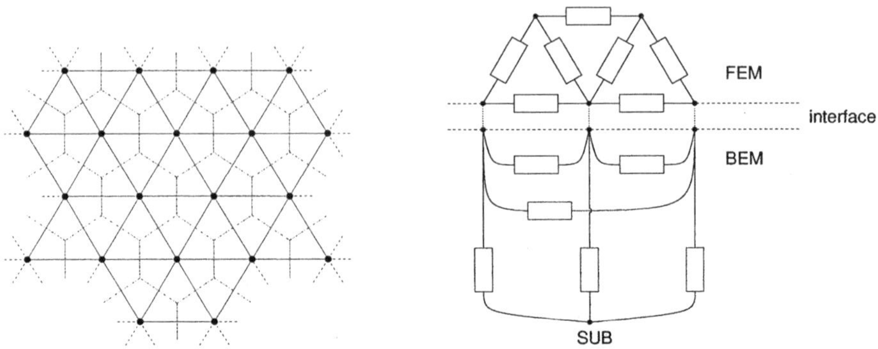


**Fig. 3.** Left: Top view of the BEM/FEM interface, the BEM mesh is the Voronoi polygon of the FEM nodes; Right: Side view of the BEM/FEM interface, Voronoi meshing allows for direct connection of the BEM and the FEM model

**Figure 4d**: Discretization of the interface introduces a discrete potential match along the interface, which causes some "strain" in both fields. Therefore, the total energy will be slightly larger: $E(\Phi_1'') + E(\Phi_2'') = E(\Phi') + \varepsilon$.

**Figure 4e**: If we were to introduce BEM and FEM, their difference in basis functions (piecewise constant and piecewise linear, respectively) would cause a discontinuity in the field. Therefore, we introduce an h-thin layer along the interface across which we define a linear interpolation between both fields.

**Figure 4f**: We now introduce $\Phi_{FEM}$ and $\Phi_{BEM}$. Variable $h$, as defined above, controls both the thickness of the h-thin layer and the granularity of the dual interface meshes (this keeps the gradient of the linear interpolation in proportion as $h \to 0$). Utilizing the convergence properties of BEM and FEM ($O(h^2)$ and $O(h)$, respectively), we can prove that the energy contained within the linear interpolation goes to zero as $O(h)$ (see [10] for a more extensive discussion).

The final observation necessary for proving convergence is that the BEM and FEM fields both minimize energy (by definition) and that the energy contained in the linear interpolation between them contributes a negligible part to the total

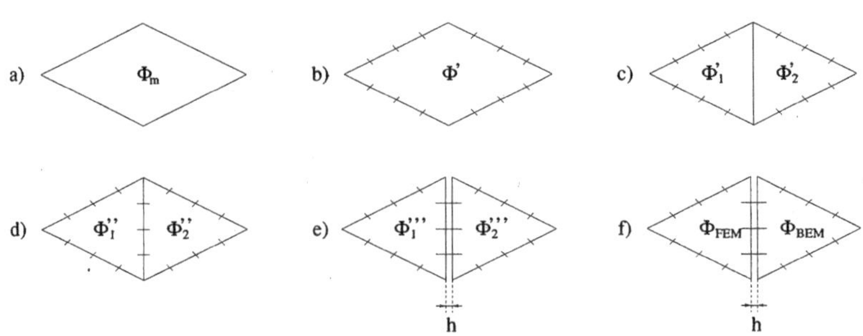


**Fig. 4.** Different stages in the convergence proof of the combined BEM/FEM method

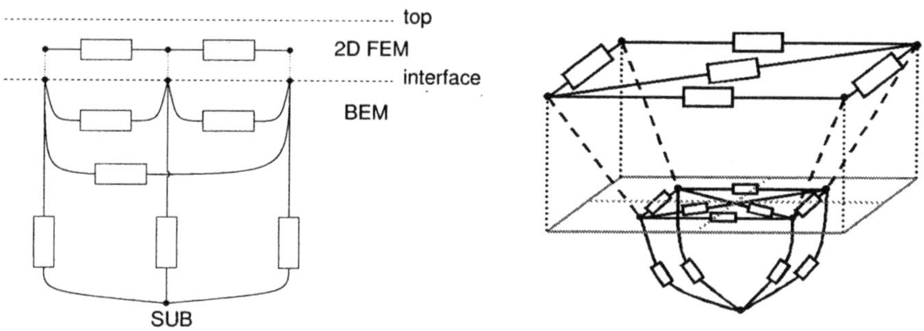


**Fig. 5.** The combined BEM/2D FEM modeling approach. Left: side view of theoretical version based on dual meshing; Right: 3D view of implemented version based on "nodal proximity"

energy. As such, for fine enough interface mesh, the total energy in Figure 4f can always become smaller than the energy in 4d. Additionally, the energy in 4f will never become smaller than the energy in 4b, because 4b is the exact, minimum energy solution for this set of boundary conditions. Situation 4f is now enclosed between 4d and 4b, which proves the convergence.

## Implementation

To test the behaviour of our method in practice, we have implemented a first prototype of our method into the SPACE layout-to-circuit extractor. The prototype utilizes a 2D FEM instead of a 3D FEM (as schematically represented in Figure 5; left). The 2D FEM is a valid modeling methodology, as long as the FEM domain is thin and has a significantly lower resistivity than the BEM domain.

Unfortunately, the concept of dual meshing is not very feasible to implement, because calculating the BEM for hexagonal panels is more difficult. Therefore, in our prototype, the combination between BEM and FEM will not be done based on dual meshing, but on "nodal proximity" (as drawn in Figure 5; right).

## Experiments

Layout (a) from Figure 6 represents a strip of channel-stop that "connects" terminals A and B on top of a substrate consisting of a $250\mu m$-thick epitaxial layer with conductivity 10 S/m and a backplane metallization. The left part Figure 7 compares the results of our method to those found with Momentum (RF simulator, part of the ADS software by Agilent Technologies Inc.) as a function of channel-stop resistivity and for different mesh-settings. Upon increasing the resistivity of the channel stop layer, we see an increase of the error, but mesh refinement in Momentum's mesh reduces the error again.

The layout from Figure 6(b) represents two separate terminals A and B that are $30\mu m$ apart, with an island of channel-stop in between. The whole structure is situated on top of a similar substrate as in the previous experiment. We have varied

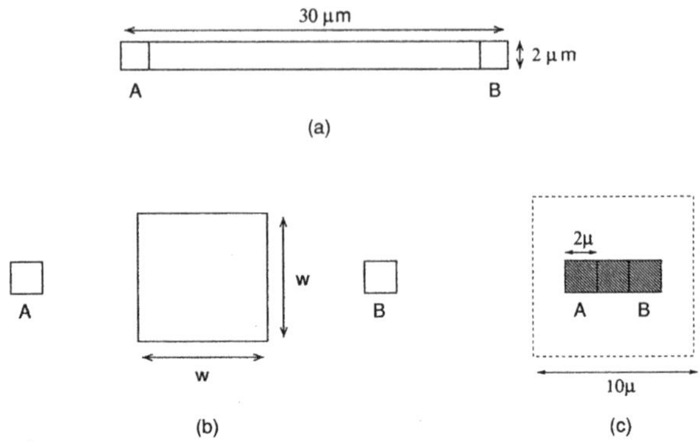


**Fig. 6.** (a) A simple strip of channel-stop with terminals $A$ and $B$ at the ends. (b) Two terminals with an "island" of channel-stop in between. (c) Another simple strip structure with drastically reduced size; the dashed box represents the FEM domain boundary.

**Table 1.** Resistance values and computation times. $R_{AB}$ = resistance between terminals $A$ and $B$, $R_{AS}$ = resistance between terminal $A$ and the backplane.

| | $R_{AB}$ $(k\Omega)$ | $R_{AS}$ $(k\Omega)$ | time (s) |
|---|---|---|---|
| SPACE | 5.22 | 21.1 | 58 |
| Davinci | 4.08 | 33.0 | 382 |

the size of the island with the parameter $w$ (which also keeps the island square) and calculated the resistance network, with both SPACE and Momentum. The results can be found in the right part of Figure 7. We can see clearly that SPACE and Momentum both show similar behaviour. As it appears, there is some offset between SPACE and Momentum, but this is possibly caused by a minor difference in the handling of the terminals ($A$ and $B$).

As a final experiment, we compared our results to those of the 3D FEM device simulator Davinci (integrated into Taurus-Workbench, by Synopsys). We used the layout shown in Figure 6(c), which is again a strip structure, but scaled down to a size that can be handled by Davinci. The experiment setup is similar to the first experiment, except that the epitaxial layer is now only $10\mu m$ thick. The results can be found in Table 1. The table clearly shows that the results of the 3D FEM method and the 2D FEM/BEM method are reasonably close, and that the new method can be considerably faster.

## Conclusions

Specific doping patterns in the top layers of the substrate can play an important role in substrate crosstalk effects. Thus, it is important to be able to incorporate these patterns into our substrate resistance model. Unfortunately, the two main modeling methodologies for the substrate, BEM and FEM, both have their disadvantages in this respect. The BEM is not accurate enough and the FEM is (usually) too slow.

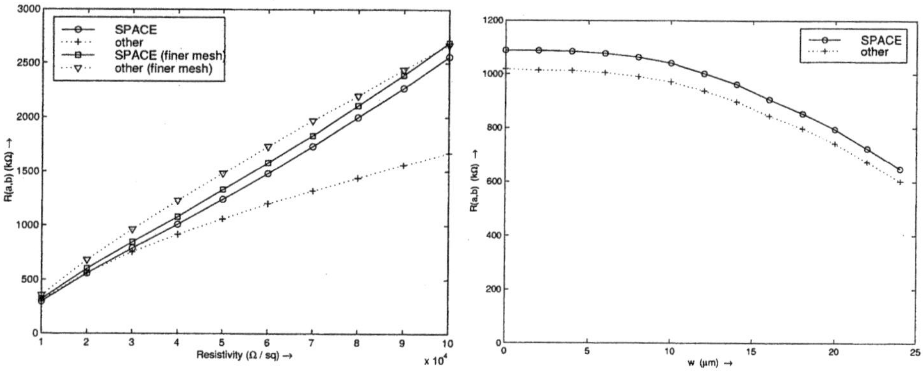


**Fig. 7.** Left: The direct resistance between terminals $A$ and $B$ from Figure 6(a) for increasing values of the resistivity of the channel-stop strip and for different mesh settings. Right: The direct resistance between terminals $A$ and $B$ from Figure 6(b) for increasing size of the channel-stop island.

Therefore, this paper presents a combined BEM/FEM methodology, which is more accurate than BEM, and significantly faster than FEM. The speed and the accuracy of the method perform well with respect to two commercially available tools. Our current research concentrates on model reduction and solution schemes for fast global solution of the combined BEM/FEM equations.

## References

1. N.K. Verghese, T. Schmerbeck, and D.J. Allstot. *Simulation Techniques and Solutions for Mixed-Signal Coupling in ICs.* Kluwer, Boston, 1995.
2. X. Aragones, J.L. Gonzalez, and A. Rubio. *Analysis and Solutions for Switching Noise Coupling in Mixed-Signal ICs.* Kluwer, Boston, 1999.
3. T. Smedes, N. P. van der Meijs, and A. J. van Genderen. Extraction of Circuit Models for Substrate Cross-Talk. In *Proceedings ICCAD*, pages 199–206, San Jose, California, November 1995.
4. G. Strang and G.J. Fix. *An Analysis of the Finite Element Method.* Prentice-Hall, Englewood Cliffs, 1973.
5. C.A. Brebbia. *The Boundary Element Method for Engineers.* Pentech Press, Plymouth, 1978.
6. E. Schrik and N.P. van der Meijs. Combined BEM/FEM Substrate Resistance Modeling. In *Proceedings 39th DAC*, pages 771–776, New Orleans, Louisiana, June 2002.
7. P. M. Dewilde and E. B. Nowacka. Circuit Models for the Hybrid Element Method. In *Proceedings ISCAS*, pages IV 616–619, Atlanta, Georgia, May 1996.
8. F. Beeftink, A.J. van Genderen, N.P. van der Meijs, and J . Poltz. Deep-Submicron ULSI Parasitics Extraction Using Space. In *Design Automation and Test in Europe, Designer Track*, pages 81–86, February 1998.

9. R.F. Harrington. *Field Computation by Moment Methods.* Macmillan, New York, 1968.
10. E. Schrik, P.M. Dewilde, and N.P. van der Meijs. Theoretical and Practical Validation of Combined BEM/FEM Substrate Resistance Modeling. In *Proceedings ICCAD*, pages 10–15, San Jose, California, November 2002.

# Induction Heating of 3D Non-ferromagnetic Metal Bodies By Moving Inductors

Pavel Šolín[1], Ivo Doležel[2], and Bohuš Ulrych[3]

[1] CAAM, Rice University, Houston, Texas
[2] Czech Technical University, Prague, Czech Republic
[3] University of West Bohemia, Pilsen, Czech Republic

**Abstract.** The numerical simulation of the induction heating of three-dimensional metal bodies by moving inductors can be problematic from several points of view when using standard FE or BE schemes. In this paper we describe the main difficulties and propose an alternative modelling approach based on the reformulation of the Maxwell's equations into a system of second-kind Fredholm integral equations. These equations for the eddy currents are coupled with the heat transfer equation with non-linear temperature-dependent material parameters. Mathematical analysis of the existence and uniqueness of solution to the continuous as well as discrete problem is provided and convergence of the numerical scheme is shown. An illustrative numerical example is presented.

## Introduction

Nowadays, the induction heating becomes more and more popular because it belongs to the most efficient, safe and ecological technologies for the heating of metals. It is based on the generation of eddy currents and consequent Joule losses in the metal body. These internal heat sources are capable of producing a convenient time-dependent temperature distribution, minimizing the temperature gradients within the metal body and dramatically reducing the danger of surface damages by oxidation and other chemical changes.

Efficiency and other parameters of the process depend on a number of various factors. Important is, for example, the arrangement of the inductor (that may produce transversal, longitudinal or generally oriented electromagnetic field within the heated metal body), its position with respect to the body, presence or absence of a magnetic circuit, frequency of the field current etc. A good and economical design of such a device, therefore, must be based on a sufficiently accurate modelling of the process.

As the induction heating is driven by the eddy currents which can be derived from the electromagnetic field, one could ask why the standard Maxwell's equations, discretized by standard methods, are not enough to resolve the problem. Let us briefly list the main obstacles making the use of standard schemes difficult.

* Financial support of the Grant Agency of the Czech Republic (projects No. 102/00/0933 and GP102/01/D114) is highly acknowledged.

a) Moving inductors make the use of FEM prohibitive due to the necessity of repeated reconstruction of the finite element mesh. Moving grid techniques do not help here due to the very complex and global nature of the inductor motions.

b) Remeshing can be difficult or infeasible even for problems with stationary inductors. The reason is the geometrical incommensurability of the heated body (bodies), inductors (formed by curved conductors, coils and/or their combinations) and the surrounding air subdomain.

c) Maxwell's equations resolve the complete electromagnetic field in the *whole domain*, particularly also in the surrounding air subdomain where no induction heating occurs. This considerably increases the computational costs.

d) Formulation of boundary conditions for Maxwell's equations can be problematic due to the geometrical incommensurability of the subdomains (e.g. for 3D inductors represented as 1D wires).

**Alternative modelling approach:**

The aim of this presentation is to show how the mentioned obstacles can be overcome for non-ferromagnetic metals by reformulating the Maxwell's equations into a system of complex second-kind Fredholm integral equations describing the behaviour of the eddy currents in the heated body. The eddy currents are used for the evaluation of the Joule losses that enter the heat transfer equation through a source term. Herewith we eliminate the necessity of remeshing the air subdomain, the necessity of resolving the electomagnetic field in it, and effectively allow for computations involving moving inductors. We deal with time-harmonic electromagnetic field and consider general temperature-dependent material properties $\gamma$, $\lambda$ and $\rho c$ in this study.

The example presented in this study comes from a difficult industrial simulation dealing with the heating of a brass workpiece of a nontrivial shape by a tubular water-cooled rotating inductor. This example is not supposed to be easily solvable by means of standard finite element or boundary element methods.

Mathematical background of the method is established by proving the existence and the uniqueness of solution for the continuous as well as discrete problem. Convergence results for the numerical scheme are also provided.

## Description of the Technical Problem

A bounded metal body $\Omega_1$ with a Lipschitz-continuous boundary is heated by an inductor formed by a system of conductors and/or coils $\Omega_2$ (see Fig. 1). For simplicity, let the conductors and coils carry identical harmonic current $I_{\text{ext}}$ of angular frequency $\omega$. The inductor contains no ferromagnetic parts. Due to the absence of any non-linearities within the investigated domain all quantities of the electromagnetic field may be expressed in terms of their phasors. Let us mention, e.g., copper, brass and stainless steel as a few examples of non-ferromagnetic metals.

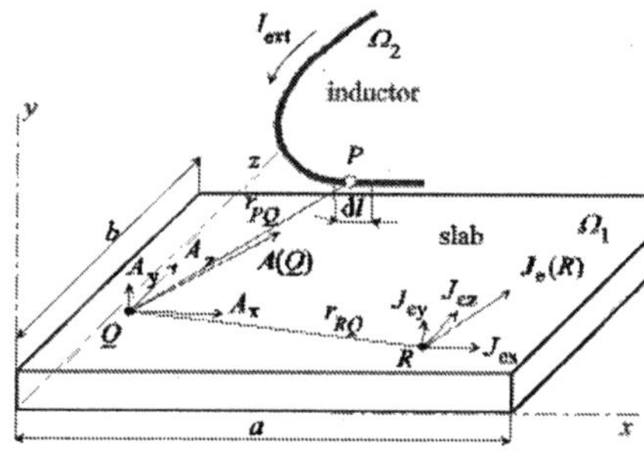


**Fig. 1.** Basic arrangement of the device.

## The coupled electromagnetic-thermal model

Let us consider a point $Q \in \Omega_1$. Using the Coulomb gauge, the phasor $\underline{\boldsymbol{A}}$ of the vector potential at this point is given by the superposition of two components excited by uniform field current $\underline{I}_{\text{ext}}$ and eddy currents $\underline{\boldsymbol{J}}_{\text{eddy}}$ in $\Omega_1$:

$$\underline{\boldsymbol{A}}(Q) = \underline{\boldsymbol{A}}(PQ) + \underline{\boldsymbol{A}}(RQ) = \frac{\mu_0}{4\pi}\left(\underline{I}_{\text{ext}}\int_{\Omega_2}\frac{\mathrm{d}\boldsymbol{l}(P)}{r_{PQ}} + \int_{\Omega_1}\frac{\underline{\boldsymbol{J}}_{\text{eddy}}(R)}{r_{RQ}}\mathrm{d}R\right). \tag{1}$$

Here, $\mu_0$ denotes the permeability of vacuum, $\mathrm{d}\boldsymbol{l}(P) = (dx(P), dy(P))^T$ is a vector denoting the elementary length of conductor of the field coils and $\mathrm{d}R$ means the elementary volume of $\Omega_1$. Remaining symbols follow from Fig. 1.

The second Maxwell equation yields that $\mathrm{rot}\boldsymbol{E} = -\partial\boldsymbol{B}/\partial t = -\partial\mathrm{rot}\boldsymbol{A}/\partial t$. Interchanging the order of the operators, we obtain $\boldsymbol{E} = -\partial\boldsymbol{A}/\partial t - \mathrm{grad}\varphi$ where $\varphi$ denotes the scalar potential. Applying this equation to the body that is not connected to any external source of voltage ($\mathrm{grad}\varphi = 0$) and rewriting it in terms of the corresponding phasor quantities we finally obtain that $\underline{\boldsymbol{E}} = -\mathrm{j}\cdot\omega\underline{\boldsymbol{A}} \Rightarrow \underline{\boldsymbol{J}}_{\text{eddy}} = -\mathrm{j}\cdot\omega\gamma\underline{\boldsymbol{A}}$. Hence,

$$\underline{\boldsymbol{A}}(Q) = \frac{\mathrm{j}}{\omega\gamma}\underline{\boldsymbol{J}}_{\text{eddy}}(Q) \tag{2}$$

where $\gamma = \gamma(T)$ denotes the temperature-dependent electrical conductivity of the metal and $\omega$ the angular frequency of the field current. Substitution of (2) into (1) provides the basic integral equation for $\underline{\boldsymbol{J}}_{\text{eddy}}$

$$\mathrm{j}\,\underline{\boldsymbol{J}}_{\text{eddy}}(Q) - \kappa(Q)\int_{\Omega_1}\frac{\underline{\boldsymbol{J}}_{\text{eddy}}(R)}{r_{RQ}}\mathrm{d}R = \kappa(Q)\,\underline{I}_{\text{ext}}\int_{\Omega_2}\frac{\mathrm{d}\boldsymbol{l}(P)}{r_{PQ}} \tag{3}$$

where $\kappa(Q) = \omega\gamma(T(Q))\mu_0/(4\pi)$. For each bounded and piecewise-continuous temperature distribution $T$ in $\Omega_1$, $\kappa$ is a bounded and piecewise-continuous function greater than a positive constant. The specific average Joule losses $w_{\mathrm{Ja}}$ in the metal body are then given by the formula

$$w_{\mathrm{Ja}} = \frac{\underline{\boldsymbol{J}}_{\mathrm{eddy}} \cdot \underline{\boldsymbol{J}}^{*}_{\mathrm{eddy}}}{\gamma}, \tag{4}$$

where $\underline{\boldsymbol{J}}^{*}_{\mathrm{eddy}}$ is the complex conjugate to $\underline{\boldsymbol{J}}_{\mathrm{eddy}}$. The non-stationary distribution of the temperature in the metal body is generally described by the equation

$$\mathrm{div}(\lambda\, \mathrm{grad}\, T) = \rho c \frac{\partial T}{\partial t} - w_{\mathrm{Ja}}, \tag{5}$$

where $\lambda = \lambda(T)$ denotes the thermal conductivity, $\rho = \rho(T)$ the specific mass of the heated material, $c = c(T)$ its specific heat and $w_{\mathrm{Ja}}$ the specific Joule losses given by (4). The boundary condition along the whole surface of the body reads (radiation is not considered)

$$-\lambda \frac{\partial T}{\partial n} = \alpha (T - T_{\mathrm{ext}}), \tag{6}$$

where $\alpha$ denotes the coefficient of the convective heat transfer, $T_{\mathrm{ext}}$ the temperature of the surrounding medium (moving or quiet air) and $n$ direction of the outward normal.

## Analysis of solvability and uniqueness

The phasor equation (3) may easily be subdivided into three identical equations (for the components in spatial directions $x$, $y$, $z$) of a complex form. For the $x$-component, we obtain

$$\mathrm{j}\, \underline{J}_{\mathrm{eddy,x}}(Q) - \kappa(Q) \int_{\Omega_1} \frac{\underline{J}_{\mathrm{eddy,x}}(R)}{r_{RQ}} \mathrm{d}R = \kappa(Q)\, \underline{I}_{\mathrm{ext}} \int_{\Omega_2} \frac{\mathrm{d}x(P)}{r_{PQ}}. \tag{7}$$

The symbol $\mathrm{d}x(P)$ means $\mathrm{d}P \cdot e_x$ where $e_x$ is the unit vector in the $x$-direction. Using the notation

$$\boldsymbol{L}_{\mathrm{x}}(\mathbf{v}) = \left(-\mathrm{Im}\{\underline{J}_{\mathrm{eddy,x}}\}, \mathrm{Re}\{\underline{J}_{\mathrm{eddy,x}}\}\right)^{T}(\mathbf{v}), \tag{8}$$

$$\boldsymbol{F}_{\mathrm{x}}(\mathbf{v}) = \kappa(\mathbf{v}) \int_{\Omega_2} \frac{\mathrm{d}x(\mathbf{u})}{|\mathbf{u} - \mathbf{v}|} \cdot \left(\mathrm{Re}\{\underline{I}_{\mathrm{ext}}\}, \mathrm{Im}\{\underline{I}_{\mathrm{ext}}\}\right)^{T}, \tag{9}$$

($P, Q \in \Omega_1$ are renamed to $\boldsymbol{u}$, $\mathbf{v}$), we can rewrite (7) into an operator form

$$(\boldsymbol{I}+\boldsymbol{K})\boldsymbol{L}_{\mathrm{x}} = \boldsymbol{F}_{\mathrm{x}} \tag{10}$$

with

$$(\boldsymbol{K}\boldsymbol{L}_{\mathrm{x}})(\mathbf{v}) = \kappa(\mathbf{v}) \int_{\Omega_1} \boldsymbol{k}(\mathbf{v}, \mathbf{w})\boldsymbol{L}_{\mathrm{x}}(\boldsymbol{w})\mathrm{d}\boldsymbol{w}, \tag{11}$$

$$\boldsymbol{k}(\mathbf{v}, \mathbf{w}) = \frac{1}{|\mathbf{v}-\mathbf{w}|} \cdot \boldsymbol{M}, \quad \boldsymbol{M} = \begin{pmatrix} 0 & -1 \\ 1 & 0 \end{pmatrix}. \tag{12}$$

We consider the operator $K$ to be defined as $[L^2(\Omega_1)]^2 \to [L^2(\Omega_1)]^2$. It is easy to see that the operator is antisymmetric. Antisymmetric operators have only purely imaginary eigenvalues. Therefore, as $1, -1$ cannot lie in the spectrum of $\boldsymbol{K}$, we immediately have the solvability, uniqueness and continuous dependence on the right-hand side for (10). An analogous conclusion holds, of course, for the remaining spatial components.

There are no problems with the existence and the uniqueness of solution for the parabolic heat transfer equation (5) in a weak sense as all the temperature-dependent material parameters are Lipschitz-continuous functions.

Analysis of the solvability and uniqueness of the discrete problem is performed in an analogous way.

## Discretization

We discretize the integral equation by piecewise-constant functions on structured hexahedral meshes. The degrees of freedom for the eddy currents correspond to cell centers. The resulting dense systems of linear equations are solved by means of both a problem-optimized Gaussian elimination and ILU preconditioned iterative solvers. The Gaussian quadrature is in some sense more advantageous since it easily can take into account the specific structure of the matrix and one easily implements the simultaneous solution of multiple right-hand sides (corresponding to the three spatial components of the eddy currents).

The equation (5) is semi-discretized in space using the method of lines and integrated in time by means of higher-order multistep methods correcting the position of the inductors and the temperature-dependent material parameters $\lambda$, $\gamma$ and $\rho c$ after each time step. The size of the time step is driven by usual criteria for the time-integration of parabolic equations.

## Convergence of the numerical scheme

Let us consider the continuous problem (10). For simplicity, let us further consider that the domain $\Omega_1$ is covered by the discretization mesh exactly ($\Omega_{1,h} \equiv \Omega_1$).

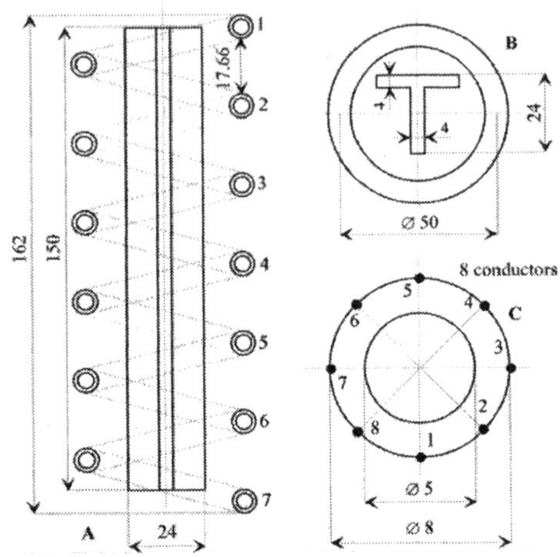


**Fig. 2.** Geometry of the heated workpiece and inductor.

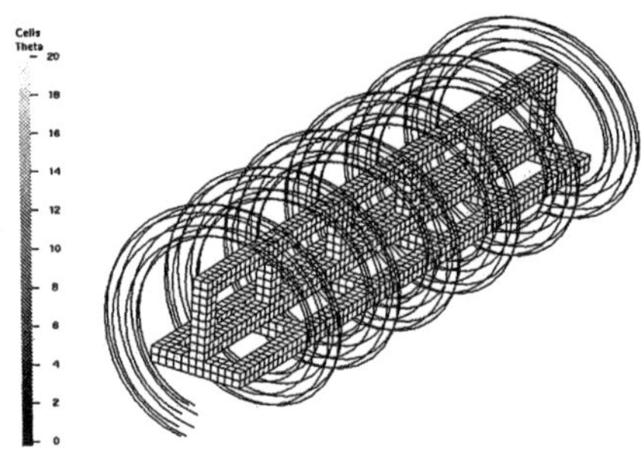

**Fig. 3.** Geometry, mesh and initial temperature distribution.

With a function $\kappa_h$ obtained by elementwise averaging the function $\kappa$ from (7), we can write the discrete problem for the eddy current density $\underline{\boldsymbol{J}}_{\mathrm{eddy}}$ in $\Omega_1$ as

$$(\boldsymbol{I} + \boldsymbol{K}_h)\boldsymbol{L}_{\mathrm{x},h} = \boldsymbol{F}_{\mathrm{x},h}. \tag{13}$$

with

$$(\boldsymbol{K}_h \boldsymbol{L}_{\mathrm{x},h})(\mathbf{v}) = \kappa_h(\mathbf{v}) \int_{\Omega_1} \mathbf{k}(\mathbf{v}, \mathbf{w}) \boldsymbol{L}_{\mathrm{x},h}(\mathbf{w}) \mathrm{d}\mathbf{w}. \tag{14}$$

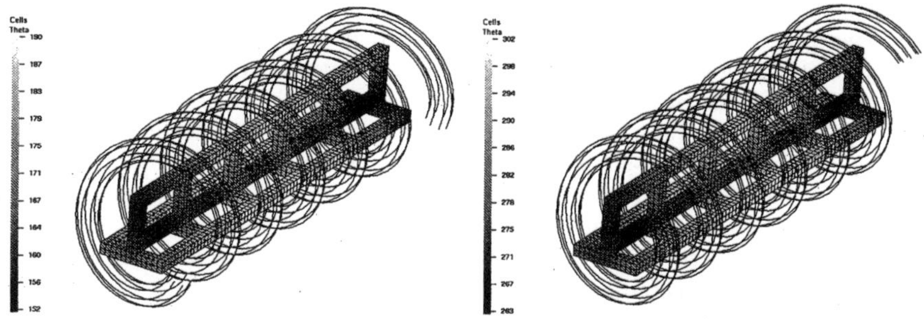


**Fig. 4.** Temperature distribution at $t$ = 60 s (ranges from 152 to 190 °C) and at $t$ = 120 s (ranges from 263 to 302 °C).

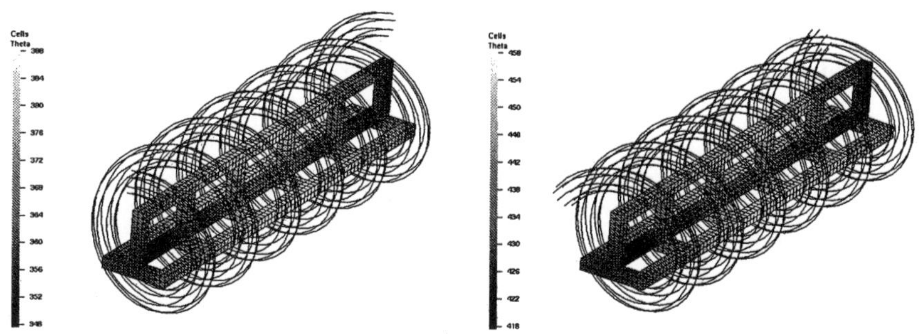


**Fig. 5.** Temperature distribution at $t$ = 180 s (ranges from 348 to 388 °C) and at $t$ = 240 s (ranges from 418 to 458 °C).

Subtracting (13) from (10) we obtain that $\boldsymbol{L}_{\text{err}} = \boldsymbol{L}_{\text{x}} - \boldsymbol{L}_{\text{x},h}$ is governed by

$$\boldsymbol{L}_{\text{err}} = (\boldsymbol{I} + \boldsymbol{K})^{-1} \left[ \boldsymbol{F}_{\text{err}} - (\boldsymbol{K} - \boldsymbol{K}_h) \boldsymbol{L}_{\text{x},h} \right] \tag{15}$$

where obviously $\boldsymbol{F}_{\text{err}} = \boldsymbol{F}_{\text{x}} - \boldsymbol{F}_{\text{x},h} \to 0$ as the grid diameter $h \to 0$, $\boldsymbol{K} - \boldsymbol{K}_h \to 0$ as $h \to 0$ from the definition of $\kappa_h$ and $\boldsymbol{L}_{\text{x},h}$ is bounded from the compactness of $(\boldsymbol{I} + \boldsymbol{K}_h)^{-1}$.

It is reasonable to suppose that the temperature-dependent material parameters $\gamma$, $\lambda$ and $\rho c$ are Lipschitz-continuous functions of the temperature. Then, convergence in the source terms of the heat transfer equation (5) yields also the convergence of the whole discrete coupled model for all finite times.

## Numerical example: Heating of a brass workpiece

A brass workpiece of a nontrivial shape (depicted in Fig. 2) is heated by a coil-shaped inductor formed by a hollow tubular water-cooled conductor. The arrangement of the system is obvious from parts **A** and **B** of Fig. 2 and from the Fig. 3. All cells are hexahedral of a uniform size 2 × 2 × 2 mm. The workpiece stands on the $xy$-plane and the T-shaped face matches the points [0.012, 0.002], [-0.08, 0.002], [-0.08, 0.012], [-0.012, 0.012], [-0.012, -0.012], [-0.08, -0.012], [-0.08, -0.002] and [0.012, -0.002]. The field current in the inductor $I$ = 320 A, its frequency $f$ = 150 kHz. As the inductor is formed by a massive conductor, it was substituted by 8 thinner conductors located at points indicated in part **C** of Fig. 2. Each of the eight conductors carries a current $I_k$ = 40 A, $k$ = 1, 2, ..., 8. The inductor rotates around the heated body with an angular frequency $\omega$ corresponding to one $2\pi$-turn in 72 seconds. The initial temperature $T_{\mathrm{start}}$ of the body and temperature $T_{\mathrm{ext}}$ of the surrounding air are 20 °C. The coefficient $\alpha$ of the convective heat transfer is 25 W/m$^2$. We are interested in the temperature evolution during the first 240 seconds of heating.

In the Figures 4 – 5 we present the evolution of the the temperature within the investigated metal body at times $t = 60, 120, 180$ and 240 seconds. Fig. 6 depicts the temperature along several important surface lines.

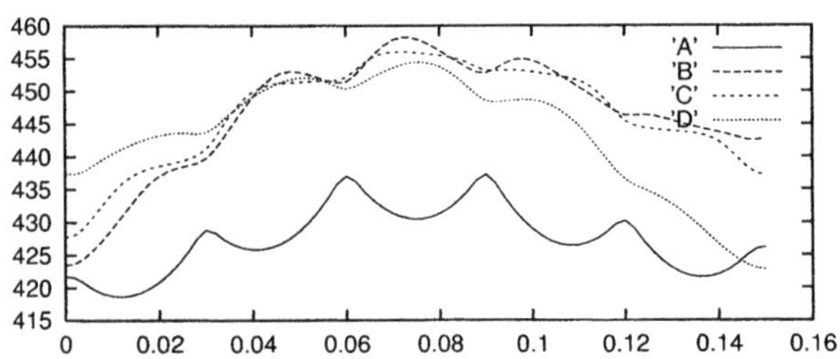


**Fig. 6.** Temperature at the time $t$ = 240 s along the lines A: $(-0.006, 0, z)$; B: $(-0.01, -0.012, z)$; C: $(-0.01, 0.012, z)$ and D: $(0.012, 0, z)$.

**Outlook**:

The presented methodology will be further developed on both the modelling and numerical levels. Currently, the most focussed modelling problem is the extension to ferromagnetic metals. On the numerical side, we develop a higher-order discretization for the integral equations on unstructured tetrahedral grids, and study the possibility to improve the efficiency of the solution of the arising dense systems of linear algebraic equations. Implicit solvers for the heat transfer equation will be implemented.

## References

1. Doležel, I., Šolín, P., Ulrych, B.: On a Fieldless Method for the Computation of Induction-Generated Heat in Three-Dimensional Non-Ferromagnetic Bodies,

Math. Comput. Sim. **61** (2003) 239 - 247
2. Engl, H. W.: Integralgleichungen, Springer, Wien, (1997)
3. Oden, J. T., Demkowicz, L. F.: Applied Functional Analysis, CRC Press Inc., Boca Raton - New York - London - Tokyo (1996)
4. Šolín, P., Doležel, I., Škopek, M., Ulrych, B.: Stationary Temperature Field in a Non-magnetic Thin Plate Heated by Transversal Electromagnetic Field, Acta Technica ASCR **45** (2000) 105–128
5. Šolín, P., Doležel, I., Škopek, M., Ulrych, B.: Induction Heating of Thin Slabs in Nonmagnetic Media, Lecture Notes in Computational Science and Engineering **18** (Springer Verlag, ISBN 3-540-42173-4) (2000) 379-388

# Multigrid for Time-Harmonic Eddy Currents without Gauge

O. Sterz

Interdisziplinäres Zentrum für Wissenschaftliches Rechnen (IWR),
Universität Heidelberg, Im Neuenheimer Feld 368,
69120 Heidelberg, Germany

**Abstract.** The application of multigrid (MG) methods for the solution of electromagnetic problems has attracted attention in recent years (e.g. [8]). These problems are related to bilinear forms $(\operatorname{curl}\cdot, \operatorname{curl}\cdot)_{L^2_\Omega} + \alpha\,(\cdot,\cdot)_{L^2_\Omega}$, $\alpha \in \mathbb{R}$, which require special smoothers as presented in [1,7]. This paper shows by numerical experiments that these ideas also work for the time-harmonic eddy currents, i.e. for *complex* bilinear forms $(\operatorname{curl}\cdot, \operatorname{curl}\cdot)_{L^2_\Omega} + \mathrm{i}\,\alpha\,(\cdot,\cdot)_{L^2_\Omega}$. Furthermore, an approximate projection procedure is presented that allows the application of multigrid to an un-gauged electric formulation even if there are regions with zero conductivity. Numerical results are shown for the TEAM Workshop problem 7.

## Introduction

The time-harmonic eddy-current approximation of Maxwell's equations is usually formulated as a second order PDE (partial differential equation). Two different formulation classes are possible, either the electric or the magnetic one. Here, we choose the electric one with the complex amplitude $\mathbf{E}$ of the electric field as the primal variable.

We consider a bounded domain $\Omega$ that is split into two (open) disjoint subsets $\Omega_C$ and $\Omega_I$ occupied by conductive and perfectly insulating material respectively. $\Omega_C$ is the union of all $N$ conductors $\Omega_{C,i}$, which are not connected to each other (see Fig. 1). For simplicity, we assume that $\Omega_I$ is connected, $\Omega$ is simply connected and none of the $\Omega_{C,i}$ touches the boundary $\partial\Omega$. Note that these restrictions can be relaxed with some technical overhead. Finally, $\Omega_G \subset \Omega_I$ denotes the support of generator current densities $\mathbf{j}_G$ that are assumed to be divergence-free and orthogonal to all harmonic Dirichlet vector fields in $\Omega_I$.

For a weak formulation of the problem, we introduce the (complex) spaces

$$\mathcal{U} = \{\mathbf{E} \in H(\operatorname{curl}, \Omega) : \mathbf{n} \times \mathbf{E}|_{\partial\Omega} = 0\}$$

and

$$\mathcal{V} = \{\lambda \in H^1(\Omega_I) : \lambda|_{\partial\Omega} = 0,\ \lambda|_{\partial\Omega_{C,i}} = \text{ const. } \forall i \in \{1,..,N\}\}$$

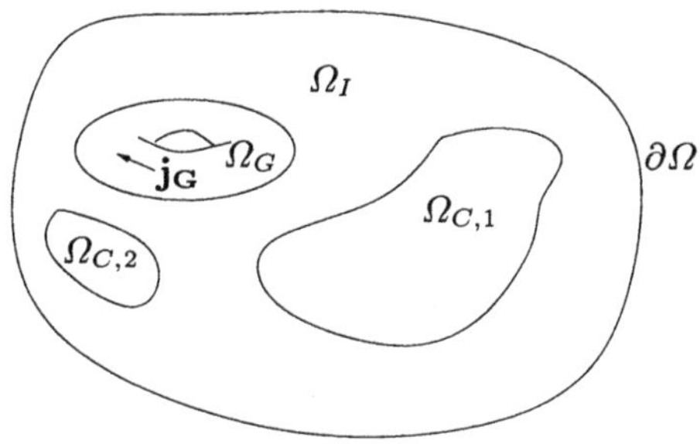


**Fig. 1.** A typical eddy-current setting with $N = 2$ conductors

and define bilinear forms

$$a(\mathbf{E}, \mathbf{E}') = \int_\Omega \frac{1}{\mu} \operatorname{curl} \mathbf{E} \cdot \operatorname{curl} \mathbf{E}' + i\omega \int_{\Omega_C} \sigma \mathbf{E} \cdot \mathbf{E}'$$

$$b(\mathbf{E}, \lambda) = \int_{\Omega_I} \mathbf{E} \cdot \operatorname{grad} \lambda$$

and the source functional

$$f(\mathbf{E}') = -\, i\omega \int_{\Omega_G} \mathbf{j_G} \cdot \mathbf{E}'$$

with the usual symbols $\mu$ for permeability, $\sigma$ for conductivity and $\omega$ for the angular frequency. A variational formulation of the time-harmonic eddy-current problem then reads: Find $\mathbf{E}$ in $\mathcal{U}$ and $\lambda$ in $\mathcal{V}$, such that

$$a(\mathbf{E}, \mathbf{E}') + b(\mathbf{E}', \lambda) = f(\mathbf{E}') \quad \forall \mathbf{E}' \in \mathcal{U}\,, \tag{1a}$$

$$b(\mathbf{E}, \lambda') = 0 \quad \forall \lambda' \in \mathcal{V}\,. \tag{1b}$$

The constraint (1b) enforces weak solenoidality of $\mathbf{E}$ in $\Omega_I$ and, due to the non-local condition in the definition of the space $\mathcal{V}$, it also forces the total charge on every conductor $\Omega_{C,i}$ to zero. This fixes what can be designated as the *electrostatic part of* $\mathbf{E}$.

However, in most instances in eddy-current modeling, we are only interested in the magnetic field and the current densities. Here, we adopt this restricted modeling task. Since the magnetic field and the current densities are independent of wether and how we fix the electrostatic part, equation (1b) plays the role of a gauge, i.e. in general, $\mathbf{E}|_{\Omega_I}$ is not the physical electric field but rather resembles a vector potential for $\mathbf{B}$.

Hence, we drop (1b) without loss of modeling information and switch to the *un-gauged* formulation: Find $\mathbf{E}$ in $\mathcal{U}$, such that

$$a(\mathbf{E}, \mathbf{E}') = f(\mathbf{E}') \quad \forall \mathbf{E}' \in \mathcal{U}\,. \tag{2}$$

This seems attractive since we do not end up with a saddle-point formulation.

# Multigrid Method

For eddy-current problems, the application of multigrid (MG) methods (see e.g. [6]) is appropriate, since these are the fastest methods currently available for solving linear systems of equations derived from a discretization of second-order PDEs. For the implementation of the finite element discretization with the lowest-order edge elements on simplices and an MG scheme, we use the $\mathcal{UG}$ (shorthand for *unstructured grids*) simulation environment, which provides very general tools for the generation and manipulation of unstructured meshes as well as a flexible data layout. $\mathcal{UG}$ also includes a lot of numerical algorithms already implemented (see [2]).

To take the different natures of $\mathbf{E}$ in $\Omega_C$ and $\Omega_I$ into account, we first restrict our view to MG for a conductive domain.

## Multigrid in Conductive Regions

It is well known (see [1,7]) that the standard smoothers do not work for problems involving the curl curl-operator. In the cited papers, special smoothers are constructed to deal with operators of the form (curl curl $+\alpha$ id), $\alpha$ real and positive. Operators of this form arise in every time-step for the time-dependent eddy-current problem if we have a positive conductivity in the whole domain $\Omega$. For a convergent MG scheme, a special overlapping block smoother is suggested in [1], whereas in [7] a "hybrid" smoother with an additional smoothing step in the space of scalar potentials is presented; see the latter reference for algorithmic details.

Although the time-harmonic eddy-current problem, associated with *complex* operators of the form (curl curl $+\mathrm{i}\,\alpha$ id), is not yet covered by the theory, our experiments show that, using the proposed smoothers, MG will even work for the complex case.

The following experiments are done with the hybrid smoother. MG is used together with Krylov acceleration, since this is the relevant case for practical applications:

**Experiment 1 (2D)** We consider a hierarchy of grids on a conductive unit square. The grid on level 0 (base level) consists of 2 triangles. The grids at higher levels are constructed by regular refinement. Thus, the number of elements is multiplied by 4 from level $l$ to $l+1$, see Figure 2. Natural boundary conditions are imposed, the angular frequency and material coefficients are chosen such that the bilinear form can be written as

$$a(\mathbf{E},\mathbf{E}') = \int_{\Omega} \operatorname{curl}\mathbf{E}\cdot\operatorname{curl}\mathbf{E}' + \mathrm{i}\alpha\int_{\Omega_C} \mathbf{E}\cdot\mathbf{E}'$$

with $\alpha = 7.16$ leading to a penetration depth $\delta \approx 0.5$.

A multigrid V(1,1)-cycle with the hybrid smoother as the pre- and post-smoother is used as a pre-conditioner for a BiCGStab [10] solver. Note that, for problems with a complex symmetric structure like the one under consideration, a transpose-free QMR (see e.g. [4]) would be more appropriate, but is not available in the $\mathcal{UG}$ environment yet.

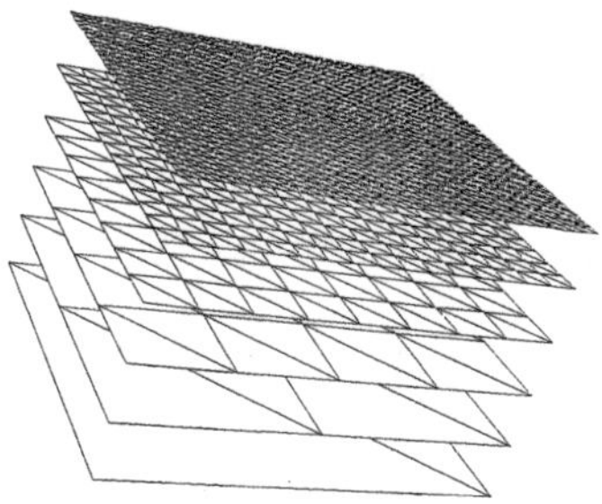

| no. of levels | 2 | 3 | 4 | 5 | 6 | 7 | 8 |
|---|---|---|---|---|---|---|---|
| no. of iterations | 4 | 4 | 5 | 5 | 5 | 5 | 5 |

**Fig. 2.** Grids on levels 0 to 5 of experiment 1 and number of iterations of the BiCGStab solver with MG pre-conditioner

In the table in Fig. 2, the number of iterations of the BiCGStab solver for a defect reduction by a factor $10^{-8}$, measured in the Euclidian norm, is displayed (vanishing RHS, random initial-guess). The experiment shows that the number of iterations is independent of the grid size for the higher levels.

**Experiment 2 (3D)** We adopt the parameters of experiment 1 and consider the unit ball in 3D. As we use a boundary projection of the surface nodes on refined levels, we do not end up with nested finite element spaces. Note that a boundary projection is important to avoid coarse grids that are too fine on the one hand (then the application of MG does not pay off) and geometric discretization errors that are too large on the other hand if we deal with curved boundaries. The grids of the first 4 levels are shown in Figure 3. Again, the number of BiCGStab iterations

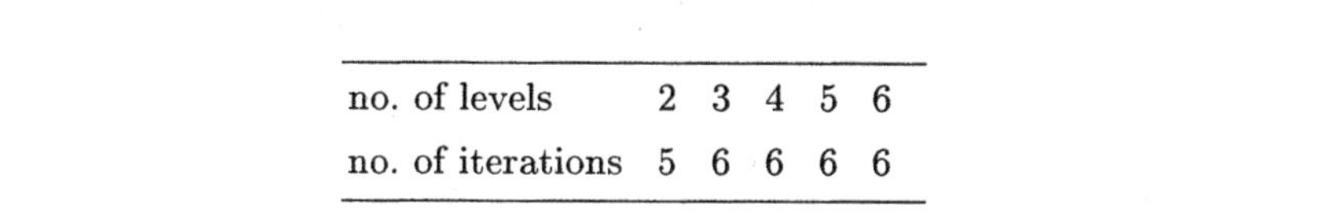

| no. of levels | 2 | 3 | 4 | 5 | 6 |
|---|---|---|---|---|---|
| no. of iterations | 5 | 6 | 6 | 6 | 6 |

**Fig. 3.** Grids on levels 0 to 3 of experiment 2 and number of iterations of the BiCGStab solver with MG pre-conditioner

shows no dependence on the mesh width for fine grids (Figure 3).

## Multigrid and the Presence of Insulators

As we use the un-gauged formulation (2) instead of the full saddle point problem (1), we unfortunately have to deal with a singular stiffness matrix if insulating regions are present. In order to guarantee solvability, we assume discrete consistent right hand sides. These can be constructed by an appropriate a priori computation of the source current densities.

**Projection** As mentioned above, without the gauge (1b), the solution $\mathbf{E}$ in $\Omega_I$ is unique except for an *electrostatic part.* During an MG solution procedure, the $L^2$-norm of this part can reach very large values, possibly leading to cancellation errors. To prevent this, a procedure is employed that fixes the electrostatic part of the solution by projection on the subspace defined by (1b), leaving the defect unchanged. Hence, one has to find a $\phi$ in $\mathcal{V}$, such that

$$\int_{\Omega_I} \operatorname{grad} \phi \cdot \operatorname{grad} \phi' = -b(\mathbf{E}, \phi') \quad \forall \phi' \in \mathcal{V} \tag{3}$$

and update the electric field

$$\mathbf{E} \leftarrow \mathbf{E} + \operatorname{grad} \phi$$

after one or several MG cycles.

Since the procedure basically fixes the weak divergence of $\mathbf{E}$ in $\Omega_I$ at zero, the algorithm can be referred to as a "weak divergence correction scheme". However, it also has to remove admixtures of harmonic Dirichlet vector fields (cohomology vector fields) that are also ruled out by (1b). Therefore, the matrix associated with the discretization of (3) will not be completely sparse, giving rise to the application of some Schur-complement solver. However, for the implementation in the $\mathcal{UG}$ environment, we decide to use a different method that will lead to completely sparse matrices. In this way, it is possible to use all solver and smoothing procedures available in $\mathcal{UG}$ without any modification. We have to pay for this comfortable coding feature by an a priori computation of a (weakly) solenoidal basis of the harmonic Dirichlet vector fields:

The implemented projection procedure consists of a pre-processing step and the projection step itself, see Algorithm 6. In the pre-processing step, a (weak) divergence operator $\widetilde{B}$ and the discrete Laplacian $L$ are assembled. Furthermore, a scalar potential for a Dirichlet basis and the Maxwell potential coefficients $p_{ij}$ (e.g. see [9]) are computed. This requires solving $N$ Laplace problems with Dirichlet boundary conditions. $N$ is the Betti number of dimension 2 of $\Omega_I$, i.e. the number of "holes" in the insulator (occupied by the $N$ conductors). The $i$-th Laplace problem reads: Find $\phi_{D,i}$ in $\{u \in H^1(\Omega_I) : u|_{\partial\Omega_{C,j}} = \delta_{ij}\}$, such that

$$\int_{\Omega_I} \operatorname{grad} \phi_{D,i} \cdot \operatorname{grad} \phi' = 0 \quad \forall \phi' \in H_0^1(\Omega_I) .$$

The Laplace problems are solved using standard MG.

```
ProjectingIteratorPreProcess() {
    assemble B̃                              // discrete weak divergence operator
    assemble L                                          // discrete Laplacian
    compute Dirichlet basis {grad φ_{D,i}}_n
    compute Maxwell's potential coefficients p_ij
}

ProjectingIteratorStep(EdgeVector c) {
    d_node ← B̃ c                                        // weak divergence
    solve (or sweep(s) with initial value c_node = 0 on)
    L c_node = −d_node                             // weak divergence corr.
    compute n conductor fluxes Q_j
    for i ← 1(1)n
        c_node ← c_node + (Σ_{j=1}^n p_ij Q_j) φ_{D,i}   // total fluxes elimination
    c ← G_{Ω̄_I} c_node                                 // discrete gradient
}
```

In the projection step itself, one Poisson problem must be solved. This can be done *approximately* (resulting in an *approximate projection*) by doing one or several sweeps of an iterative solver (typically MG). After this first correction step, all Dirichlet vector fields are removed from $c$ using the a priori computed potentials of a Dirichlet basis. $G_{\overline{\Omega}_I}$ denotes the node-to-edge incidence matrix, restricted to nodes in $\overline{\Omega}_I$ (the discrete gradient operator in the insulating region).

## TEAM 7 benchmark problem

To validate the proposed algorithm and its implementation, we considered the TEAM benchmark problem 7 (see [5]). The problem consists of an aluminum plate ($\sigma = 3.526\,10^7$ S/m) with a hole and an excitation coil above the plate with a time-harmonic driving current of 2742 AT (see Fig. 4). The driving current reaches the maximum at $\omega t = 0$ and is directed anticlockwise in Fig. 4. The computational domain $\Omega$ was artificially restricted to a cube with 1 m edge length with boundary conditions $\mathbf{n} \times \mathbf{E} = 0$ at $\partial\Omega$.

To generate a hierarchy of grids at the base level $l = 0$, uniform refinement was applied; see Fig. 5 for the base level grid. The grid levels $l > 1$ are generated by local refinement with the help of a residual error estimator (see [3]). Additionally, the grid was refined along the line $\overline{AB}$ in Fig 4. By means of this adaptive algorithm, grids up to level 5 were generated.

For the solution of the linear systems of equations, we applied BiCGStab again as the outer iteration. The pre-conditioner was composed of one V(1,1)-MG cycle with the hybrid smoother and one iteration of Algorithm 6 with a single V(1,1)-MG sweep on the Poisson problem.

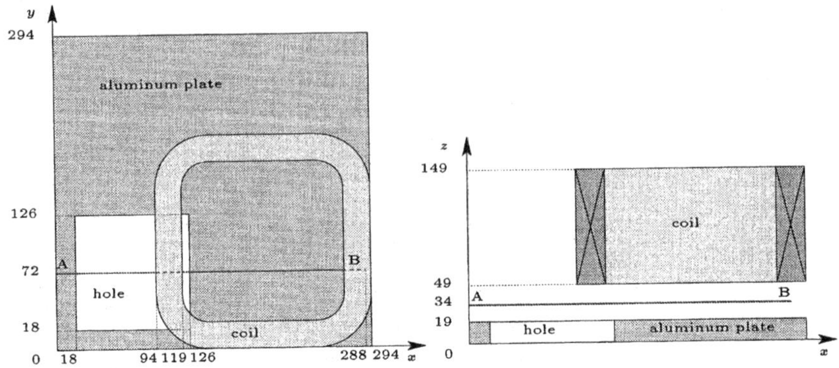


**Fig. 4.** Team 7 benchmark problem

Note that *local* MG (e.g. see [11]) was applied here, i.e. at each level, smoothing had been done only for the refined region and not over the whole grid. This is essential for an adaptive algorithm since otherwise the overall complexity may not be optimal. The table on the right of Fig. 5 shows that the number of itera-

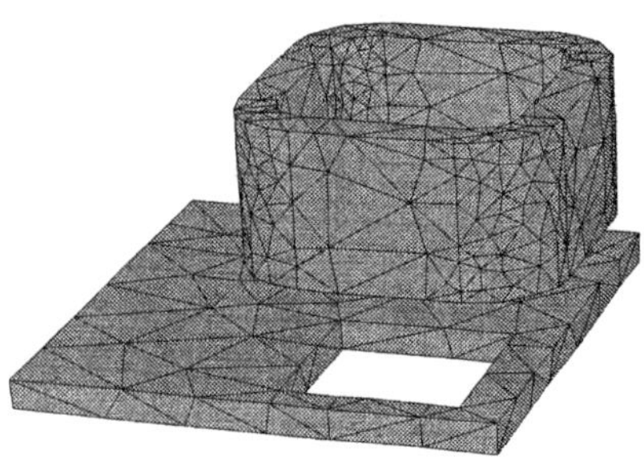

| level | ♯elements | ♯edges | ♯iterations |
|---|---|---|---|
| 0 | $5.6\,10^3$ | $6.7\,10^3$ | — |
| 1 | $4.5\,10^4$ | $5.3\,10^4$ | 15 |
| 2 | $7.2\,10^4$ | $8.8\,10^4$ | 15 |
| 3 | $8.0\,10^4$ | $1.0\,10^5$ | 16 |
| 4 | $1.6\,10^5$ | $2.0\,10^5$ | 15 |
| 5 | $3.8\,10^5$ | $4.7\,10^5$ | 16 |

**Fig. 5.** Left: Base level grid (level $l = 0$), the grid in the air region is not displayed. Right: Number of BiCGStab iterations for a defect reduction by factor $10^{-8}$ using a V(1,1)-multigid pre-conditioner

tions remains constant during the refinement. Thus, MG works very well for the benchmark problem. The solution at level 5 is displayed in Fig. 6 and shows good agreement with measured results.

**Acknowledgments** The author would like to thank R. Hiptmair for helpful conversations.

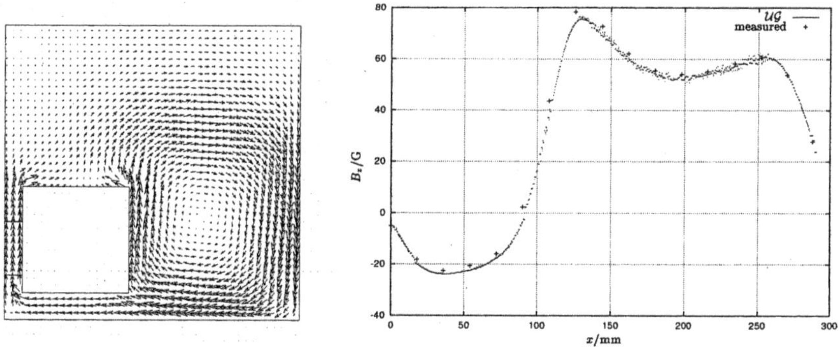


**Fig. 6.** Left: Real part of the current density **J** in the aluminum plate. Right: sign $(\mathrm{Re}\,B_z)|B_z|$ along line $\overline{AB}$ in Fig. 4. Note that the computed **B** is piecewise constant and the grid is very fine at level 5.

# References


1. D. Arnold, R. Falk, and R. Winther. Multigrid in $H(\mathrm{div})$ and $H(\mathbf{curl})$. *Numer. Math.*, 85(2):175–195, 2000.
2. P. Bastian, K. Birken, K. Johannsen, S. Lang, N. Neuss, H. Rentz-Reichert, and C. Wieners. UG—a flexible software toolbox for solving partial differential equations. *Computing and Visualization in Science*, 1:27–40, 1997.
3. R. Beck, R. Hiptmair, R. Hoppe, and B. Wohlmuth. Residual based a-posteriori error estimators for eddy current computation. M$^2$AN, 34(1):159–182, 2000.
4. R. Freund. Transpose-free quasi-minimal residual methods for non-Hermitian linear systems. *IMA Volumes in Mathematics and its Applications*, 60(69):69–93, 1994.
5. K. Fujiwara and T. Nakata. Results for benchmark problem 7. *COMPEL*, 9(3):137–154, 1990.
6. W. Hackbusch. *Multi-grid Methods and Applications.* Springer–Verlag, Berlin, 1985.
7. R. Hiptmair. Multigrid method for Maxwell's equations. *SIAM J. Numer. Anal.*, 36(1):204–225, 1999.
8. M. Kaltenbacher, S. Reitzinger, M. Schinnerl, J. Schöberl, and H. Landes. Multigrid methods for the computation of 3D electromagnetic field problems. *COMPEL*, 20(2):581–594, 2001.
9. E. Martensen. *Potentialtheorie.* Teubner, Stuttgart, 1968.
10. H. van der Vorst. BI-CGSTAB: A fast and smoothly converging variant of BI-CG for the solution of nonsymmetric linear systems. *SIAM J. Sci. Stat. Comput.*, 13(2):631–644, 1992.
11. C. Wieners. Local multigrid methods on hierarchical meshes. In *Proceedings of the 17th GAMM-Seminar*, pages 1–8, Leipzig, 2001.

# Oscillator Modeling Using the Averaging Principle

Piet Vanassche, Georges Gielen, and Willy Sansen

K.U.Leuven, Department of Electrical Engineering - ESAT, Kasteelpark Arenberg 10, B-3001 Leuven, Belgium — e-mail: piet.vanassche@esat.kuleuven.ac.be

**Abstract.** This article introduces a method for modeling and analysis of perturbed oscillator behavior, i.e. the behavior of "ideal" oscillators subjected to weak interactions with the outside world. These interactions can e.g. involve an amplitude-regulating mechanism, as in harmonic oscillators, couplings to other oscillators, as in quadrature-type oscillators, and noise, both white and colored. The method is grounded on perturbation theory and averaging. Perturbation techniques allow us to separate the analysis of the unperturbed, ideal, oscillator from the analysis of the perturbed one. Averaging is used to separate the fast-varying and the slow-varying components of the oscillator's behavior. Applications of this method include oscillator phase noise analysis and the construction of compact behavioral models for harmonic oscillators.

## Introduction

Oscillators are key building blocks in almost all of today's communication systems. Because of their importance, there is a need for methods that generate compact behavioral oscillator models. These models can be used for verification and trade-off analysis at the architectural level and for obtaining insight into the oscillator's behavior and functioning.

This work presents a method for modeling perturbed oscillator behavior, i.e. the changes in the behavior of an "ideal" oscillator when subjected to weak interactions with the outside world. These interactions can e.g. involve an amplitude-regulating mechanism, as illustrated in Fig. 1, couplings to other oscillators and noise, both white and colored.

The modeling method performs a sequence of transformations on the original set of circuit equations and is solidly grounded on perturbation theory [7] and averaging [2,5]. A first transformation applies perturbation techniques to model the impact perturbations have on the behavior of an ideal, unperturbed oscillator. A second, averaging, transformation separates the fast-varying from the slow-varying components of the oscillator's behavior. The averaging transformation presented in this work deals with both deterministic and noisy systems. The transformed oscillator equations make up a valuable starting point for further analysis or for the construction of compact behavioral models. The equations are valid over an infinite time-interval and can be integrated using a time step of the order $T/\epsilon$, where $T$ is the oscillation period and $\epsilon$ a perturbation scaling parameter.

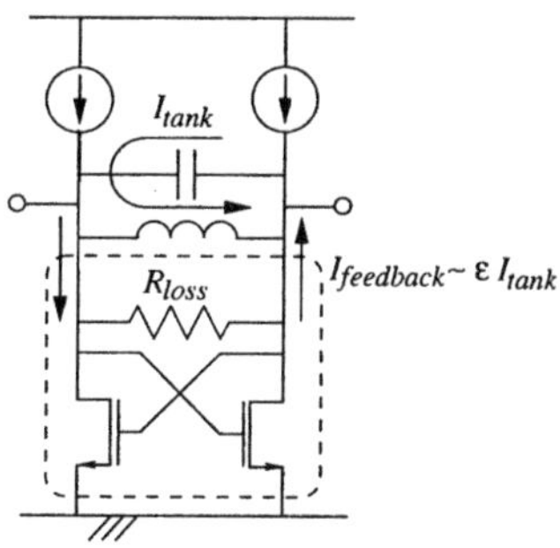


**Fig. 1.** An ideal oscillator, in this case a lossless LC tank, weakly interacts with a cross-coupled transistor pair. This transistor pair serves as an amplitude-regulating mechanism.

The remainder of this text is organized as follows. Section 6 discusses characterizing the unperturbed oscillator and use of perturbation techniques to capture the behavior of oscillators subjected to weak perturbations. In Sect. 6, we show how to handle these equations efficiently using averaging. Section 6 applies these results to phase noise analysis and construction of compact behavioral models for harmonic oscillators. Finally, Sect. 6 provides some conclusions

# Perturbed Oscillator Behavior

This section concerns the structure of the solutions of the —properly normalized— system of equations,

$$\frac{d\mathbf{x}}{dt} = \mathbf{f}\left(\mathbf{x}(t)\right) + \epsilon \mathbf{b}\left(\mathbf{x}(t), t\right) \quad . \tag{1}$$

This system describes the behavior of a weakly perturbed oscillator. Here, $\mathbf{x} \in \mathbb{R}^N$ represents a vector of unknown state variables, $\mathbf{f}(\mathbf{x})$ models the behavior of the unperturbed oscillator and $\epsilon \mathbf{b}(\mathbf{x}, t)$ represents a perturbation term disturbing the oscillator's behavior. This perturbation term can be either deterministic or stochastic. We furthermore assume the perturbation variable $\epsilon$, characterizing the magnitude of the perturbation term, to be small, i.e. $|\epsilon| \ll 1$. The analysis that follows generalizes the results in [6], which discusses oscillator phase behavior, while establishing the link with the theory of motions over stable manifolds [1].

In solving (1), we first characterize the set of steady-state solutions of the unperturbed system, i.e. for $\epsilon = 0$. Assuming that all solutions converge, in the limit $t \to \infty$, to a $P$-dimensional, stable manifold $M$, each steady-state solution $\mathbf{x}(t)$, i.e. each solution of (1) which lies entirely on $M$, can be characterized as

$$\mathbf{x}(t) = \mathbf{x}_s\left(t, \mathbf{p}\right) \quad . \tag{2}$$

Here $\mathbf{x}_s(t, \mathbf{p}) \in \mathbb{R} \times \mathbb{R}^P \to \mathbb{R}^N$ with $P \leq N$ and $\mathbf{p} \in \mathbb{R}^P$ a vector specifying the initial conditions of the solutions of the unperturbed system that are located on $M$.

As illustrated by the applications in section 6, the components of $\mathbf{p}$ correspond, for example, with amplitudes, phases and phase differences.

Due to action of the perturbation term $\epsilon\mathbf{b}(\mathbf{x},t)$, solutions $\mathbf{x}_\epsilon(t)$ of the perturbed system, i.e. for $|\epsilon| > 0$, deviate from the manifold $M$. However, stability of $M$ ensures that the $\mathbf{x}_\epsilon(t)$ remain bounded to its neighborhood. Solutions can therefore be written as

$$\mathbf{x}_\epsilon(t) = \mathbf{x}_s(t, \mathbf{p}(t,\epsilon)) + \epsilon\Delta\mathbf{x}(t,\epsilon) \ . \tag{3}$$

They contain two components. The first one, $\mathbf{x}_s(t,\mathbf{p}(t,\epsilon))$, describes the motion over the manifold $M$. The second component, $\epsilon\Delta\mathbf{x}(t,\epsilon)$, describes the departure from the attracting manifold. Note that the processes $\mathbf{p}(t,\epsilon)$ and $\Delta\mathbf{x}(t,\epsilon)$ depend on both $t$ and $\epsilon$. For notational convenience, the dependence on $\epsilon$ will never be mentioned explicitly unless when relevant.

While the motion over the manifold may grow unbounded, i.e. $\|\mathbf{p}(t) - \mathbf{p}(0)\| \to \infty$ as $t \to \infty$, the manifold's stability guarantees that $\mathbf{p}(t)$ can be chosen as for the departure $\epsilon\Delta\mathbf{x}(t)$ to remain $O(\epsilon)$ and this $\forall t$. Substituting (3) into (1) therefore yields

$$\begin{aligned}\tfrac{\partial \mathbf{x}_s}{\partial \mathbf{p}}(t,\mathbf{p}(t))\tfrac{d\mathbf{p}}{dt} &+\epsilon\tfrac{d}{dt}\Delta\mathbf{x}(t)\\ &= \epsilon\tfrac{\partial \mathbf{f}}{\partial \mathbf{x}}(\mathbf{x}_s(t,\mathbf{p}(t)))\,\Delta\mathbf{x}(t)\\ &\quad+\epsilon\mathbf{b}(\mathbf{x}_s(t,\mathbf{p}(t)),t) + O(\epsilon^2) \ .\end{aligned} \tag{4}$$

Here we used the fact that, for arbitrary but fixed $\mathbf{p}$, $\mathbf{x}_s(t,\mathbf{p})$ solves (1) for $\epsilon = 0$.

The motion $\mathbf{x}_s(t,\mathbf{p}(t))$ over the manifold $M$ typically contains that part of the behavior which is of greatest interest. It is completely determined by the process $\mathbf{p}(t)$. The equations governing this process are obtained by projecting (4) onto the tangent space of $M$ spanned by $\mathbf{U}(t,\mathbf{p}) = \frac{\partial \mathbf{x}_s}{\partial \mathbf{p}}(t,\mathbf{p}) \in \mathbb{R}\times\mathbb{R}^P \to \mathbb{R}^{N\times P}$. Clearly, $\mathbf{U}(t,\mathbf{p})$ satisfies

$$\frac{\partial \mathbf{U}}{\partial t}(t,\mathbf{p}) = \frac{\partial \mathbf{f}}{\partial \mathbf{x}}(\mathbf{x}_s(t,\mathbf{p}))\,\mathbf{U}(t,\mathbf{p}) \ . \tag{5}$$

With $\mathbf{V}(t,\mathbf{p}) \in \mathbb{R}\times\mathbb{R}^P \to \mathbb{R}^{N\times P}$ a solution of

$$\frac{\partial \mathbf{V}^T}{\partial t}(t,\mathbf{p}) = -\mathbf{V}^T(t,\mathbf{p})\frac{\partial \mathbf{f}}{\partial \mathbf{x}}(\mathbf{x}_s(t,\mathbf{p})) \ , \tag{6}$$

it is readily seen that the product $V^T(t,\mathbf{p})U(t,\mathbf{p})$ remains constant with respect to $t$. Here, we are interested in the bi-orthonormality case: $\mathbf{V}^T(t,\mathbf{p})\mathbf{U}(t,\mathbf{p}) = \mathbf{I}_P, \forall t$ with $\mathbf{I}_P \in \mathbb{R}^{P\times P}$ the unity matrix. Note, however, that if $P < N$, (6) and the bi-orthonormality condition do not uniquely characterize $\mathbf{V}^T(t,\mathbf{p})$. An additional constraint, related to the requirement that $\epsilon\Delta\mathbf{x}(t)$ must remain bounded, needs to be imposed. In practice, this almost always requires $\mathbf{V}^T(t,\mathbf{p})$ to be (quasi-)periodic with respect to $t$. Multiplying both sides of (4) with $\mathbf{V}^T(t, \mathbf{p})$ while using (6) and $\mathbf{V}^T\mathbf{U} = \mathbf{I}_P$ then yields

$$\frac{d\mathbf{p}}{dt} = \epsilon \mathbf{V}^T\left(t, \mathbf{p}(t)\right) \mathbf{b}\left(\mathbf{x}_s\left(t, \mathbf{p}(t)\right), t\right) - \epsilon \frac{d}{dt}\left(\mathbf{V}^T\left(t, \mathbf{p}(t)\right) \Delta\mathbf{x}(t)\right) + O(\epsilon^2) \ . \quad (7)$$

In order to explicitly separate the motion over the manifold $M$, characterized by the process $\mathbf{p}(t)$, from the deviation from the manifold, characterized by $\Delta\mathbf{x}(t)$, we impose $\Delta\mathbf{x}(t)$ to satisfy

$$\mathbf{V}^T\left(t, \mathbf{p}(t)\right) \Delta\mathbf{x}(t) = 0 \ . \quad (8)$$

With (8), the second right-hand side term in (7) drops out, leaving

$$\frac{d\mathbf{p}}{dt} = \epsilon \mathbf{V}^T\left(t, \mathbf{p}(t)\right) \mathbf{b}\left(\mathbf{x}_s\left(t, \mathbf{p}(t)\right), t\right) + O(\epsilon^2) \ . \quad (9)$$

This set of ordinary differential equations (ODEs), which only depends upon $\mathbf{p}$ and $t$, completely characterizes the motion $\mathbf{x}\left(t, \mathbf{p}(t)\right)$ over the manifold $M$.

## Averaging

The system of ODEs (9), governing the motion of the perturbed oscillator over the manifold $M$, belongs to the class of ODEs structured as

$$\frac{d\mathbf{p}}{dt} = \epsilon \sum_{k=1}^{K} \mathbf{h}_k(\mathbf{p}(t), t) \ . \quad (10)$$

Comparing with (9), we find $\sum \mathbf{h}_k(\mathbf{p}, t) = \mathbf{V}^T\left(t, \mathbf{p}\right) \mathbf{b}\left(\mathbf{x}_s\left(t, \mathbf{p}\right), t\right)$. We make it clear later on why it is sometimes useful to partition this right-hand side into a number of different components. Equation (10) can be solved efficiently using averaging [2,5]. Averaging exploits the fact that solutions of (10) will only vary substantially on a time scale $t/\epsilon$, i.e. they have a dominant slow-varying component. Use of averaging techniques allows us to extract that part of (10) producing this dominant slow-varying component. The averaged system of ODEs that results, serves as an excellent starting point for noise and stability analysis or for extraction of compact behavioral models.

Technically, averaging involves constructing of a near-identity, time-varying transformation of variables $\mathbf{p} \to \overline{\mathbf{p}}$[1], whereby values of the original vector $\mathbf{p}$ and the newly introduced vector $\overline{\mathbf{p}}$ are related to each other through

$$\mathbf{p} = \overline{\mathbf{p}} + \epsilon \sum_{k=1}^{K} \widetilde{\mathbf{h}}_k(\overline{\mathbf{p}}, t) + O(\epsilon^2) \ . \tag{11}$$

The terms $\widetilde{\mathbf{h}}_k(\overline{\mathbf{p}}, t)$ in this transformation should be chosen such that (a) the transformed ODEs are easier to solve than the original ones, and (b) $\widetilde{\mathbf{h}}_k(\overline{\mathbf{p}}, t)$ remains bounded over the time interval of interest, which, in our case, is an infinite time interval. Both conditions are met for the $\widetilde{\mathbf{h}}_k(\overline{\mathbf{p}}, t)$ satisfying

$$\frac{\partial \widetilde{\mathbf{h}}_k(\overline{\mathbf{p}}, t)}{\partial t} = \mathbf{h}_k(\overline{\mathbf{p}}, t) - \mathbf{M}_k[\mathbf{h}_k](\overline{\mathbf{p}}, t) \ , \tag{12}$$

where the operator $\mathbf{M}_k[\cdot]$ represents some suitable, low-pass, averaging operator.

Substituting (11) into (10) and using (12) shows the transformed *process* $\overline{\mathbf{p}}(t)$ to satisfy

$$\frac{d\overline{\mathbf{p}}}{dt} = \epsilon \sum_{k=1}^{K} \overline{\mathbf{h}}_k(\overline{\mathbf{p}}(t), t) + O(\epsilon^2) \ . \tag{13}$$

with $\overline{\mathbf{h}}_k(\overline{\mathbf{p}}, t) = \mathbf{M}_k[\mathbf{h}_k](\overline{\mathbf{p}}, t)$. With careful choice of the $M_k[\cdot]$, this transformed equation is often much easier to solve than (10). It is called the *averaged equation* corresponding to (10). This averaged equation makes up a valuable starting point for further analysis or for the construction of compact behavioral models.

Selecting the averaging operator $\mathbf{M}_k[\cdot]$ is a degree of freedom which can be used to optimize the properties of the averaged ODEs (13). For each component $\mathbf{h}_k$ in (10), a different averaging operator can be selected. This can, for example, be used to exploit particular properties of the different components $\mathbf{h}_k$. Some convenient averaging operators are

$$\mathbf{M}_k[\mathbf{h}_k] = \frac{1}{T} \int_{t-T_k/2}^{t+T_k/2} \mathbf{h}_k(\overline{\mathbf{p}}, \tau) d\tau \tag{14}$$

$$\mathbf{M}_k[\mathbf{h}_k] = \frac{1}{T_k} \frac{\sin(\pi t/T_k)}{\pi t/T_k} \otimes \mathbf{h}_k(\overline{\mathbf{p}}, t) \ , \tag{15}$$

[1] At this point, it might be useful to stress the distinction between a vector $\mathbf{p} \in \mathbb{R}^P$ and a process $\mathbf{p}(t) \in \mathbb{R} \to \mathbb{R}^P$. As a vector, $\mathbf{p}$ should be treated as having an arbitrary but fixed value. As a process, $\mathbf{p}(t)$ links a vector in $\mathbb{R}^P$ to each instance of time. To avoid confusion, we will always write $\mathbf{p}$ to indicate an arbitrary but fixed value and $\mathbf{p}(t)$ to indicate a time-dependent process.

with $\otimes$ denoting the convolution operator and $T_k$ a suitable time constant. Here, (14) is the classical averaging operator [2] and (15) is the ideal low-pass filter over the frequency range $f = [-1/2T_k, 1/2T_k]$. Both averaging operators (14) and (15) ensure that $\overline{\mathbf{p}}(t)$ contains almost all the low-frequent content of $\mathbf{p}(t)$ while $\epsilon\widetilde{\mathbf{h}}_k(\overline{\mathbf{p}}, t)$ only contains high-frequent components. This way, slow-varying and fast-varying components of the oscillator's behavior are explicitly separated.

In order to illustrate how averaging helps in simplifying noise analysis, let's assume that, for a fixed $\overline{\mathbf{p}}$, $\mathbf{h}_k(\overline{\mathbf{p}}, t)$ is a Gaussian, $T$-periodic *cyclostationary* noise source with its autocorrelation equal to $\Phi_k(\overline{\mathbf{p}}, t_1, t_2) = E\{\mathbf{h}_k(\overline{\mathbf{p}}, t_1)\mathbf{h}_k^*(\overline{\mathbf{p}}, t_2)\}$. Using results from cyclostationary noise theory [8], it is then possible to show that, with $\mathbf{M}_k[\cdot]$ being the ideal low-pass filter defined in (15), $\overline{\mathbf{h}}(\overline{\mathbf{p}}, t) = \mathbf{M}_k[\mathbf{h}_k](\overline{\mathbf{p}}, t)$ is a Gaussian, *stationary* noise process with its autocorrelation determined by

$$\overline{\Phi}_k(\overline{\mathbf{p}}, \tau) = \frac{\sin(\pi\tau/T)}{\pi\tau} \otimes \frac{1}{T}\int_{-T/2}^{T/2} \Phi_k(\overline{\mathbf{p}}, t - \frac{\tau}{2}, t + \frac{\tau}{2})\, dt \tag{16}$$

where $\tau = t_2 - t_1$. This result is readily verified by considering it from a frequency-domain point of view. Typically, this reduction from cyclostationary to stationary noise processes greatly facilitates further analysis. An interesting application of this result involves solving the oscillator phase noise equation.

## Applications and Results

### Phase Noise Analysis

Modeling oscillator phase noise behavior, as discussed in [3,4,6], turns out to be a special case of the method presented in this paper. Here, the noise-free oscillator corresponds with the unperturbed system while the noise sources are associated with the perturbation part. Assuming the oscillator to be mono-stable, the manifold $M$, spoken of in Sect. 6, coincides with the oscillator's orbit in the phase space. The solutions of the unperturbed system that are located on this 1-dimensional manifold, are characterized as

$$\mathbf{x}_s(t, \mathbf{p}) = \mathbf{x}_{osc}(t + \vartheta) \ . \tag{17}$$

Here, $\mathbf{x}_{osc}(t)$ is a $T$-periodic solution of the noise-free oscillator equations while $\mathbf{p} = [\vartheta]$ is a 1-dimensional parameter vector. The parameter $\vartheta$ represents the oscillator's phase, i.e. a measure for the position of oscillator's initial state $\mathbf{x}_0 = \mathbf{x}_s(0, \mathbf{p}) = \mathbf{x}_{osc}(\vartheta)$ on its orbit in the phase space.

Given a noise source $n(t)$ and using the perturbation techniques outlined in Sect. 6, we obtain a phase noise equation that is structured as

$$\frac{d\vartheta}{dt} = \epsilon v\,(t + \vartheta(t))\, n(t) \ . \tag{18}$$

where $v(t)$ is $T$-periodic. This phase noise equation describes the oscillator's stochastic phase behavior, i.e. its random motion along its own orbit.

In what follows, we assume $n(t)$ to be a Gaussian, stationary noise source with autocorrelation $\Phi(\tau)$. We furthermore impose that either $\Phi(\tau) \to 0$ on the scale $T/\epsilon$ or that the spectrum of $\Phi(\tau)$ is mainly contained within the frequency band $[-1/2T, 1/2T]$. The results in Sect. 6 then allow us to prove that the solution of (18) is, up to zero-th order in $\epsilon$, equal to that of the averaged equation

$$\frac{d\overline{\vartheta}}{dt} = \epsilon\overline{n}(t) \ . \tag{19}$$

Here, $\overline{n}(t)$ is a Gaussian, *stationary* noise source with its autocorrelation determined by

$$\begin{aligned}\overline{\Phi}(\tau) &\approx \frac{1}{T}\int_{-\frac{T}{2}}^{\frac{T}{2}} v(t+\overline{\vartheta}+\frac{\tau}{2})v(t+\overline{\vartheta}-\frac{\tau}{2})\Phi(\tau)dt \\ &\approx \frac{1}{T}\int_{-\frac{T}{2}}^{\frac{T}{2}} v(t+\frac{\tau}{2})v(t-\frac{\tau}{2})\Phi(\tau)dt \ .\end{aligned} \tag{20}$$

Note how the dependence on $\overline{\vartheta}$ disappears due to the periodicity of $v(t)$. This averaged noise signal $\overline{n}(t)$ turns out to be the stationary component of the cyclostationary noise signal $v(t)n(t)$.

With cyclostationarity being removed, (19) is much easier to solve than (18). The results obtained in this way correspond with those in [4] but avoid the necessity to construct the often complicated (modified) Fokker-Planck equations.

## Behavioral Modeling of Harmonic Oscillators

Constructing behavioral models for (coupled) harmonic oscillators [9] constitutes another useful application of the method discussed above. These behavioral models can, for example, be used for system-level simulations and trade-off analysis.

For a system of $K$ coupled harmonic oscillators, the $K$ lossless resonance tanks can be considered as corresponding to the unperturbed system. Currents coming from the feedback transistors are associated with the perturbations. This partitioning is illustrated in Fig. 1 for a single harmonic oscillator. The manifold $M$, containing the steady-state solutions of the unperturbed system, therefore is $2K$-dimensional, i.e. two degrees of freedom per resonance tank. A compact behavioral model can now be obtained using the averaged ODE (13) with its right-hand side approximated through, for example, polynomial regression or table lookup and interpolation. The advantages of such a behavioral model is twofold. Firstly, the averaged ODE can be integrated using a time step that is proportional to $T/\epsilon$ compared to a time step $T$ needed for solving the original circuit equations. Secondly, evaluating the approximated right-hand side involves a computational cost that well below the one corresponding with evaluation of a set of complex transistor equations.

The behavioral model extraction algorithm was implemented in Matlab. For the harmonic oscillator in Fig. 1, we constructed a behavioral model based upon table lookup and interpolation of the averaged ODEs (13). The details of the model extraction algorithm are outlined in [9]. Constructing the model takes 3 minutes of CPU time. Fig. 2 compares the startup behavior of the harmonic oscillator as predicted by the behavioral model (envelope signal) with results from SPICE-like integration of the original set of circuit equations. As can be seen, results are in excellent agreement (within 1%). The behavioral model, however, evaluates 2 orders of magnitude faster as compared to the SPICE-like simulations, i.e. 0.4 seconds versus 37.2 seconds for a simulation run over 40 periods of oscillation. Use of the behavioral model hence significantly speeds up simulations which is, for example, useful for lengthy and repetitive simulations as in system-level verification and trade-off analysis.

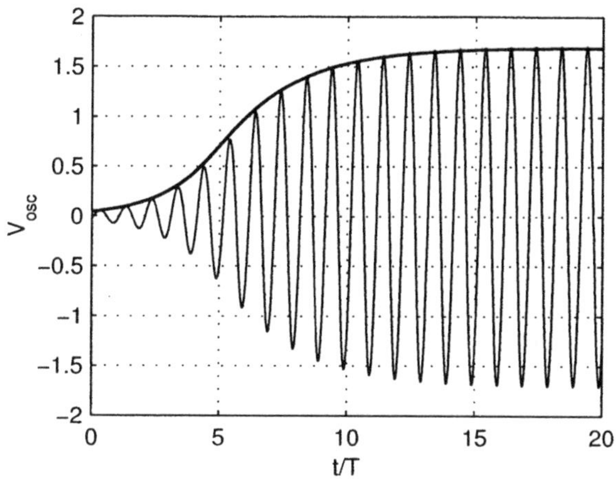


**Fig. 2.** Oscillator startup behavior. The fast-oscillating line is computed by solving the original set of circuit equations using a SPICE-like algorithm. The envelope is computed by solving the behavioral model equations.

## Conclusions

In this article, we have discussed a method for modeling and analysis of perturbed oscillator behavior. This method deals with perturbations that can be both deterministic, e.g. an amplitude-regulating mechanism as in harmonic oscillators, and stochastic, e.g. noise sources causing oscillator phase noise. The method is founded upon perturbation theory and averaging. The latter allows us to explicitly separate the fast-varying and the slow-varying components of the oscillator's behavior. Use of this method was illustrated for oscillator phase noise analysis and the construction of compact behavioral models for harmonic oscillators.

## References

1. R. Abraham, J. Marsden and T. Ratiu, *Manifolds, Tensors, Analysis and Applications*, 2nd. edition, Springer-Verlag, 1988
2. N.N Bogoliubov and Y.A. Mitropolsky, *Asymptotic methods in the theory of nonlinear oscillations*, Hindustan Publishing Corp., India, 1961
3. A. Demir, A. Mehrotra and J. Roychowdhury, "Phase Noise in Oscillators: A Unifying Theory and Numerical Methods for Characterization", In *IEEE Trans. Circ. and Syst.-I*, vol. 47, no. 5, pp. 655-674, May 2000
4. A. Demir, "Phase Noise in Oscillators: DAEs and Colored Noise Sources", In *Proc. IEEE ICCAD*, pp. 170-177, 1998
5. M.I. Freidlin and A.D. Wentzell, *Random perturbations of dynamical systems*, 2nd. edition, Springer-Verlag, 1998
6. F.X. Kaertner, "Analysis of White and $f^{-\alpha}$ Noise in Oscillators", In *Int. J. Circuit Theory Appl.*, vol. 18, pp 485-519, 1990
7. J. Kevorkian and J.D. Cole, *Perturbation Methods in Applied Mathematics*, Springer-Verlag, New York, 1981
8. J. Roychowdhury, D. Long and P. Feldmann, "Cyclostationary noise analysis of large RF circuits with multitone excitation", In *IEEE J. Solid State Circ.*, vol. 33, pp. 324-335, March 1998
9. P. Vanassche, G. Gielen and W. Sansen, "Modeling of (Coupled) Harmonic Oscillators", In *Proc. IEEE DAC*, Session 35.1, 2002

# 2-D FEM Tuned Analytical Approximation for Fringing Permeances

Vencislav Valchev[1], Alex Van den Bossche[2], and Todor Filchev[1]

[1] Department of Electronics, Technical University of Varna, Studenska 1, Bulgaria
[2] Department of Electrical Energy, Systems and Automation, Ghent University, Sint-Pietersnieuwstraat 41, Belgium

**Abstract.** This paper presents an analytical approach for calculating fringing permeances in gapped inductors. For most of the gapped inductors, the permeance of other field paths out of the air gap (the fringing paths) is not negligible. Existing three-dimensional modelling techniques using finite element analysis for magnetic components are accurate, but require prohibitive amount of simulation time. Two-dimensional models are often used, but the accuracy is low as a 2D simulation fails taking into account important 3D effects. We present analytical approximations for fringing permeance calculation for the most usual field patterns, denoted as basic cases. The proposed fringing coefficients can be used to present all symmetrical cases and cases with multiple air gaps. The derived equations are sufficient for a normal engineering accuracy.

## Introduction

Gapped inductors are widely used in power electronics equipment. In most of the designs the additional inductance because of the fringing flux path around the air gaps is important and should be taken in account.
There are many discretisation methods such as Finite Difference Method (FDM), Finite Difference in Time Domain (FDTD), Finite Elements Method (FEM), Boundary Element Method (BEM) which allow an accurate presentation and calculation of the field in a gapped core inductor. The advantages of the numerical methods are the flexibility for modelling irregular field geometry and boundaries and the possibility to handle non-linearly of the material. The air gap effects have been discussed by discretisation methods in several papers [1–3] and 2-D and 3-D solutions have been given. An improved computer-aimed optimisation of inductor design considering air gap effects has been proposed in [4]. Other general methods for calculating gapped inductors are presented in [5–8], all of them having their advantages and disadvantages. Analytical solutions based on the Schwartz-Christoffel transformation [9] have been proposed. Although the mathematical base is promising, the accuracy is low as a field problem is solved where the conductors are placed far away from the air gap, which is usually not the case in practical arrangements. The fringing effects are also important for the design of electrical machines and the correction factors and slotting factors (Carter factor) have been discussed analytically in [10,11]. The analytical methods have their advantages: possibility of generating diagrams, solving the reverse problems and optimising more complex problems.

Neglecting the fringing flux one could use a simple equation for an inductance of a coil:

$$L = \mu_0 N^2 + \frac{A_g}{\sum l_g + l_m/\mu_r} \tag{1}$$

where:

- $\Sigma l_g$: sum of the air gap lengths in the flux path;
- $A_g$: cross section of the air gap, equal to core cross section;
- $l_m$: length of the flux path in a core;
- $\mu_r$: permeability of the core material;
- $N$: number of turns.

For large air gaps, the permeance of other field paths (fringing paths) out of air gap is not negligible. This result in much larger values for$L$ than predicted by (1). In almost all designs of gapped inductors for power electronics the fringing field should be considered and thus the expression (1) gives a poor approximation. A better approximation for a centre gapped UU and centre gapped EE cores is McLyman's formula [12]:

$$L' = L\, X_f\,,\ \ X_f = 1 + \frac{l_g}{\sqrt{A_g}}\ \ln(\frac{2\, w}{l_g}) \tag{2}$$

in which

- $L'$: inductance corrected for fringing; $X_f$: fringing factor;
- $w$: total height of the winding; $l_g$: air gap length.

The expression (2) does not make difference between air gaps with rectangular or round cross- section and is limited to small air gaps.
In this paper, to obtain a better accuracy we proposed the following approach:

- First, investigate and provide basic analytical approximations for the fringing coefficient for several basic cases of different air gap configurations.
- Secondly, these results will be used to solve usual cases.

The derived fringing coefficients are compared and tuned using FEM calculations. The proposed approach is valid for round wires or litz wires. The extension of the proposed approach to 3-D is given in the next work of the authors [13].

## Proposed Analytical Approach: Basic Cases

It is clear that, in a 2D problem, the fringing permeance is proportional to the third dimension. The fringing permeance leads mainly to a correction on the permeance of the air gap and it is of a little influence to the main permeance.
The permeance of an air gap is:

$$\Lambda_g = \mu_o\, \frac{A_g}{l_g} + \mu_o\, C_g\, F \tag{3}$$

where

- $\Lambda_g$ : permeance of the air gap;
- $A_g$: surface of the air gap;
- $C_g$: the third dimension;
- $l_g$: total length of air gap;
- $F$: fringing coefficient (see later).

We will derive the expressions for the fringing coefficients for the basic cases.

## Basic case 1

In the basic case 1, the conductors are surrounded by magnetic material except at the air gap (see Fig. 1). The permeability of the magnetic material is assumed to

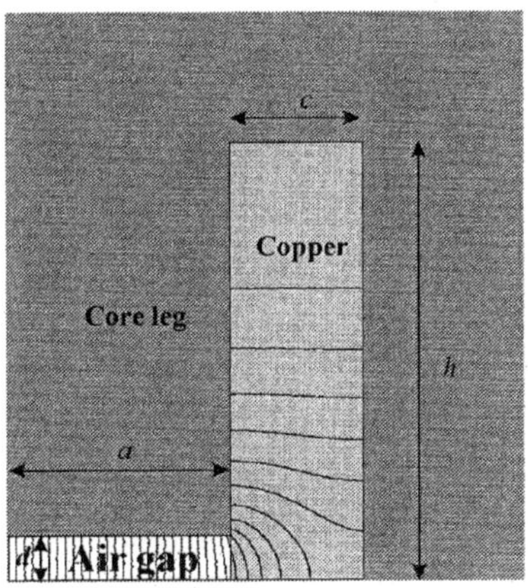


**Fig. 1.** Magnetic field in the basic case 1 (conductors surrounded by a core).

be infinite. We propose the following approximation for the fringing coefficient $F_1$:

$$F_1(d,c,h) = \frac{2}{\pi}\ \ln\left(\frac{\frac{1}{c}+\frac{1}{d}}{\frac{1}{c}+\frac{1}{h}}\right) + \frac{(h-d)^2(h-0.26\,d-0.5\,c)}{3\,c\,h^2} + \frac{c}{3\,h} \qquad (4)$$

in which

- $F_1$: fringing coefficient for the basic case 1;
- $d$: air gap distance to the reference;
- $c$: thickness of winding;
- $h$: height of the winding compared to the reference plane.

The tuning constants (0.26 and 0.5) have been fitted, using the software Finite Element Method Magnetics (FEMM3.1) [14]. The normalised values of $F_1$ ($h$=1), obtained by the proposed approximation (4) and by FEM simulation, are shown in Fig. 2(a).

Fig. 2(b) shows the deviation between the proposed approximation and the finite element simulation. The error is below 3% (the relative error on $F_1$).

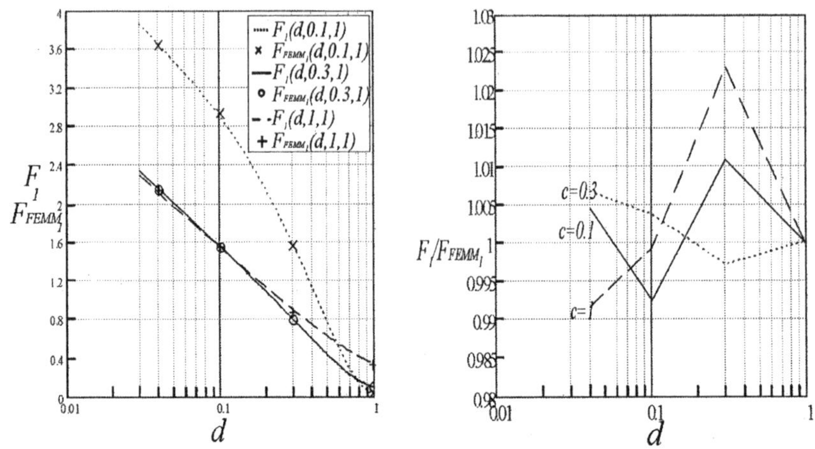


**Fig. 2.** (**a**) Fringing coefficient $F_1$ of (4) as a function of $d$; $h$=1; $c$ is a parameter; $F_{FEMM1}$ is the values found by FEM simulation. (**b**) Ratio of the values of $F_1$ obtained by the proposed approximation and by FEMM; $h$=1; $c$ is a parameter.

## Basic case 2

In the basic case 2, the winding touches the core but all other sides of the winding are surrounded by air, see Fig. 3. In this case the fringing field is also mainly concentrated near the air gap, the field lines are almost parts of circles. We propose the following expression for the fringing coefficient $F_2$:

$$F_2(d,c,h) = \frac{2}{\pi} \ln\left[\frac{0.44(h^2+c^2) - 0.218\,d\,h + 0.67\,c\,d + 0.33\,h\,c + 0.7825\,d^2}{d^2}\right]^{1/2} \quad (5)$$

For small $d$, the expression (5) is symmetrical in respect to $c$ and $h$. The normalised values of $F_1(h{=}1)$, obtained by the proposed approximation (5), and by the FEM simulation, are shown in Fig. 4(a). Fig. 4(b) shows the deviation between the proposed approximation and the finite element simulation results. The matching is about 1% (relative error on $F_2$).
Note that basic case 1 and basic case 2 are almost equal for $c = h$. The reason for this is that no m.m.f. (magneto-motive force) is present and thus, there is not much field far away from the air gap so that the presence of material does not influence much the result.

## Basic case 3

The basic case 3 represents a problem, where no conductors are present (see Fig. 5 and the m.m.f. can be put in the air gap. This is the case of the outside legs

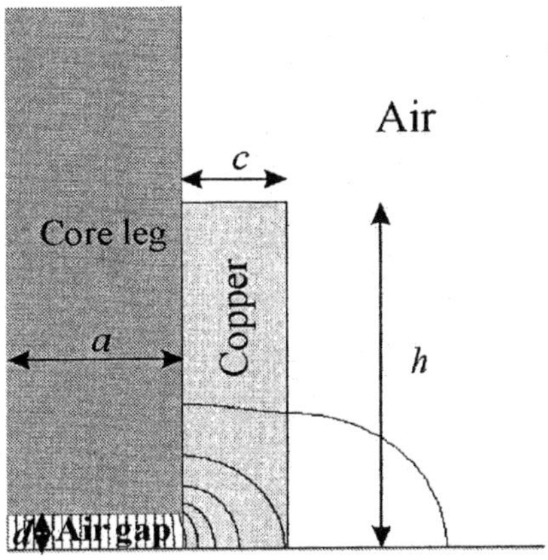


**Fig. 3.** Magnetic field in the basic case 2 (conductors in open area).

(for instance EE, ETD, ER cores). The total height is now larger than the winding height, so we use $g$, instead of $h$. The fringing coefficient is:

$$F_3(d,g) = \frac{1}{\pi}\text{acosh}[3.395(\frac{g}{d})^2 + 0.15\ \frac{g}{d} + 1.1155] \tag{6}$$

For better interpolation in the expression (6) we use the function *acosh* instead of

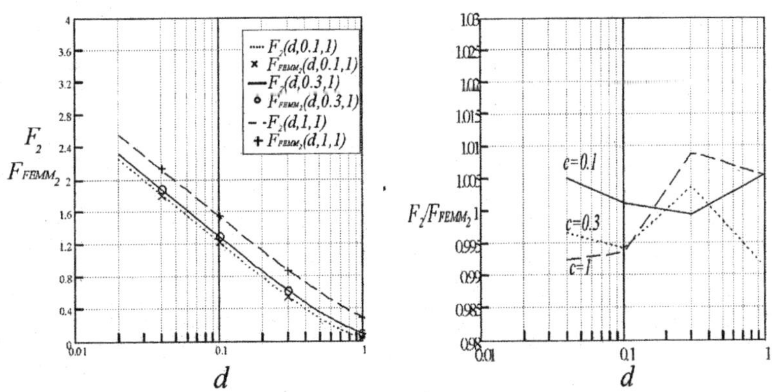


**Fig. 4.** (**a**) Fringing coefficient $F_2$ of (5) as a function of $d$; $h = 1$; $c$ is a parameter. (**b**) Ratio of the values of $F_2$, obtained by proposed approximation and by FEMM as a function of $d$,; $h = 1$; $c$ is a parameter.

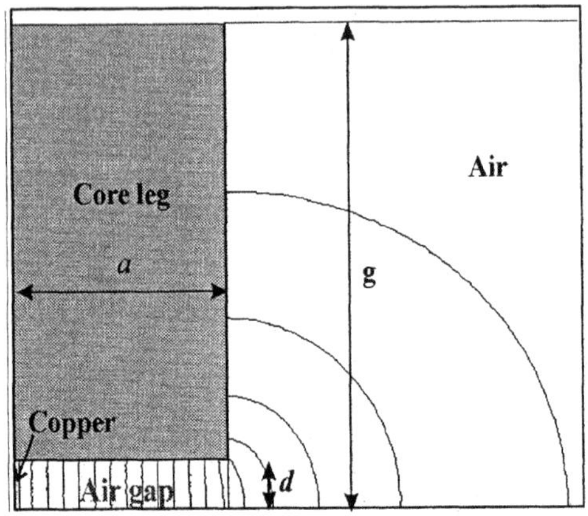


**Fig. 5.** Magnetic field in the basic case 3.

the logarithmic function.
Fig. 8 shows the proposed approximation (6) and the finite element simulation results. The matching is better than 1% (relative error on $F_3$).

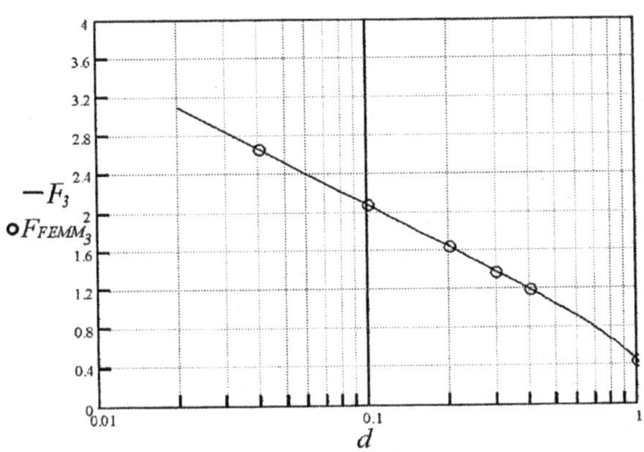


**Fig. 6.** Fringing coefficient $F_3$ of (6) as a function of $d$; $g = 1$.

## Basic case 4

The basic case 4 represents a top to bottom problem, without conductors (Fig. 7). This field pattern occurs when the yoke-to-yoke m.m.f. is not zero. The fringing

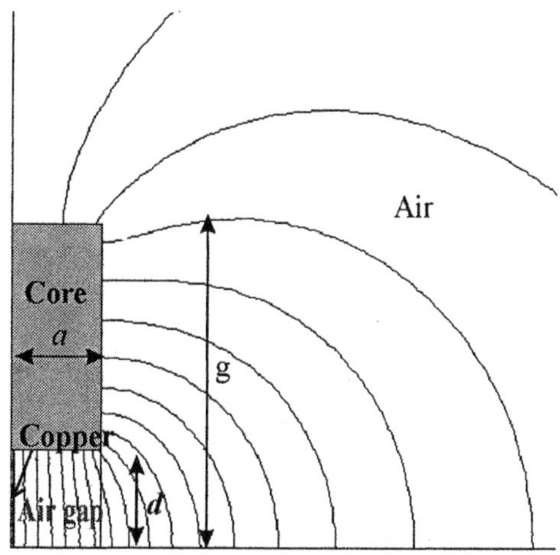


**Fig. 7.** Magnetic field of the combination of the basic cases 3 and 4.

coefficient is:

$$F_4(a,g) = \frac{1}{\pi}\text{acosh}[1.4(\frac{a}{g})^{0.38} + 1] \tag{7}$$

Figure 8 shows the fringing coefficient $F_4$ as a function of $a$; $g$=1. From Fig. 8 it is seen that when $a$ is small the value of $F_4$ decreases and almost vanishes. The cases 3 and 4 are not independent. We did subtract the case 3 to find the case 4. We did see that the result for case 4 was quite independent of $d$ in a range for $d = 0, \ldots, 0.75g$, which confirms that the separation is possible.

# Application of the obtained fringing coefficients

By using the already obtained fringing coefficients describing the basic cases, we can present the fringing field in symmetrical cases.
As an example, let us consider a symmetrical case representing a wound centre leg core with an air gap in the middle, each side has a field pattern identical to the case 1. In that case $d = l_g/2$. The permeance of the air gap should be divided by 2. Thus, the fringing coefficient $F_{1s}$ is:

$$F_{1s}(l_g, c, w) = \frac{F_1(l_g/2, c, w/2)}{2} \tag{8}$$

The proposed approach and approximation can be extended from 2D to 3D. Those calculations can take in account the specific effects as corner effects. That extension of the topic is out of the scope of this paper and the problem is discussed in the next works of the authors.

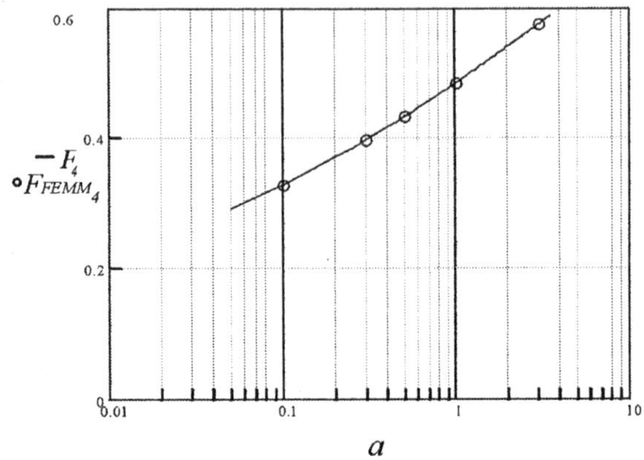


**Fig. 8.** Fringing coefficient $F_4$ as a function of $a$; $g = 1$.

# Conclusion

A quite general approach to calculate 2-D fringing permeances of gapped inductors has been proposed. Analytical expressions for the fringing coefficients of four basic cases are derived and verified by finite element calculation. The proposed fringing coefficients can be used to present all symmetrical cases of gapped inductors as well as multiple air gap cases. A typical accuracy of 1, ..., 2% can be obtained for the fringing permeance by the proposed equations. For small air gaps the obtained accuracy usually exceeds the usual accuracy of the physical dimensions of the cores.

# References

1. Chew, W.M., Evans, P.D.: High-frequency inductor design concepts, in Proc. IEEE PESC'91, (1991) pp. 673–678.
2. Nysveen, A., Hernes, M.: Minimum loss design of a 100kHz inductor with foil windings, EPE93, Brighton, September 13-16, 1993, pp. 106–111.
3. Leonavicisius, V., Duffy, M., Mathuna, C.O., Boeke, U.: Comparison of different techniques to realise PFC boost inductor, APEC 2002, Dallas, USA, 10-14 March 2002, pp. 1183–1189.
4. Walmeier, P.W.: Improved analytical modelling of conduction losses in gapped high-frequency inductors, IEEE Trans. on Ind. Applics, **37-4**, (2001) pp. 1045–1054.
5. Evans, P.D., Saied, B.M.: Calculation of effective inductance of gapped core assemblies, IEE Proceeding, vol. **133-1**, (1986) pp. 41–45.
6. Chandler, P.L., Yan, X., Paterson, D.J.: Novel high-power ferrite inductor design with improved design accuracy and overall performance, IEEE $32^{nd}$ Annual PESC-2001, Vol. **4**, pp. 2090–2094.
7. Rahmi-Kain, A., Keyhani, A., Powell, J.: Minimum loss design of 100kHz inductor with litz wire, IEEE IAS Annual Meeting, New Orleans, October 5-9, 1997, pp. 1414–1420.

8. Hoke, A.F., Sullivan, C.R.: An improved two-dimensional numerical modeling method for E-core transforms, APEC 2002, Dallas, USA, 10-14 March 2002, pp. 151–157.
9. Balakrishnan, A., Joines, W.T., Wilson, Th.G.: Air gap reluctance and inductance calculations for magnetic circuits using a Schwartz-Christoffel transformation, IEEE Trans. on Power Electr., vol. **12-4**, (1997) pp. 654–663.
10. Kofler, H.: Force and torque in single phase pole induction motors with eccentric rotor: an attempt to include slot effects due the eccentricity, ICEM 2000, Espoo, Finland, 20-28 August 2000, pp. 1825–1828.
11. Kulda, J., Adam, J.: The magnetic field in air-gap slotted on side, Elektrotechnische Zeitschrift ETZ A, vol **91-2**, (1970) pp. 109–113
12. Col. McLyman: Transformer and inductor design handbook, Marcel Dekker (1988).
13. Van den Bossche, A., Valchev, V., Filchev, T.: Improved approximation for permeances of gapped inductors, IEEE-IAS 37$^{th}$ Annual Meeting, Pittsburgh, PA, USA, 13-18 October, 2002
14. Meeker, D.: Finite Element Method Magnetics (FEMM), http://femm.berlios.de/.

# Stochastic DAEs in Transient Noise Simulation

Renate Winkler

Humboldt-Universität Berlin, Institut für Mathematik
winkler@mathematik.hu-berlin.de

**Abstract.** In this paper we describe how stochastic differential-algebraic equations (SDAEs) arise as a mathematical model for network equations that are influenced by additional sources of Gaussian white noise. We give the necessary analytical theory for the existence and uniqueness of strong solutions, provided that the systems have noise-free constraints and are uniformly of DAE-index 1. We express these conditions in terms of the network-topology for reasons of use within a circuit simulator. In the second part we analyze discretization methods. Due to the differential-algebraic structure, implicit methods will be necessary. By the examples of the drift-implicit Euler and Milstein schemes we show how drift-implicit schemes for SDEs can be adapted to become directly applicable to stochastic DAEs and prove that the convergence properties of these methods known for SDEs are preserved. For illustration we apply the drift-implicit Euler scheme to an oscillator circuit.

## Problem Formulation

The charge-oriented Modified Nodal Analysis (MNA) represents a standard tool in circuit simulation. The equations are generated automatically by combining the network topology, Kirchhoff's Current Law, and the characteristic equations describing the physical behaviour of the network elements. This results in large systems of DAEs, whose special structure was analyzed in a number of papers, e.g. [5,6,16]. Due to decreasing signal to noise ratios in special applications linear noise analysis around the deterministic solution is no longer satisfactory. The noise influences such systems in an essentially nonlinear way, the solution of the deterministic models differs from the mean of the solution of the stochastic models. Transient noise analysis is necessary. Here we deal with models where Gaussian white noise sources are added to the systems. Thermal noise of resistors and shot noise of *pn*-junctions are modelled as external Gaussian white noise sources in parallel to the original element (see Figures 1 and 2), as will be explained next.
Nyquist's theorem (see e.g. [2–4,18]) states that the current through an arbitrary linear resistor having a resistance $R$, maintained in thermal equilibrium at a temperature $T$, can be described as the sum of the noiseless, deterministic current and a current due to a Gaussian white noise process with spectral density $S_{th} := \frac{2kT}{R}$ where $k$ is Boltzmann's constant. Hence, the additional current is modelled as

$$I_{th} = \sigma_{th} \cdot \xi(t) = \sqrt{\frac{2kT}{R}} \cdot \xi(t) \, ,$$

where $\xi(t)$ is a standard Gaussian white noise process. In [17,18] a thermo-dynamical foundation to apply this model to mildly nonlinear resistors and reciprocal networks is given.

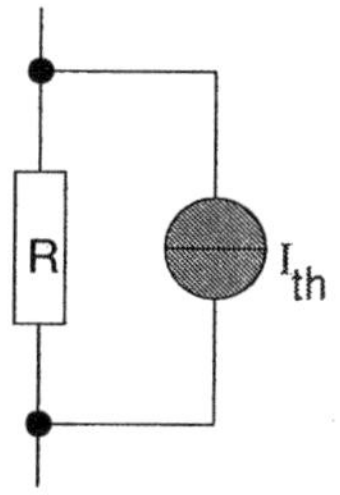


thermal noise of a resistor

Figure 1

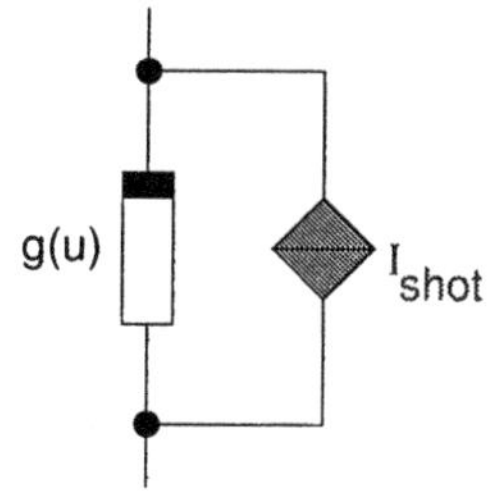


shot noise of a pn-junction

Figure 2

Shot noise of pn-junctions, caused by the discrete nature of current due to the elementary charge, is also modelled by a Gaussian white noise process, where the spectral density is proportional to the current $I$ through the *pn*-junction: $S_{shot} := q|I|$ , where $q$ is the elementary charge. If the current through the *pn*-junction is described by a characteristic $I = g(u)$ , where $u$ is some voltage, the additional current is modelled by

$$I_{shot} = \sigma_{shot}(u) \cdot \xi(t) = \sqrt{q|g(u)|} \cdot \xi(t) \, ,$$

where $\xi(t)$ is a standard Gaussian white noise process. For a discussion of the model assumptions we refer to [2–4,17,18].
We represent the topology of a network by means of the incidence matrix ($A_C$, $A_R$, $A_L$, $A_V$, $A_I$, $A_N$), with indices referring to branches of capacitances, resistances, inductances, possibly controlled voltage and current sources, and $n_N$ additional noise sources, respectively. Then the charge-oriented MNA system has the following structure (see [5,6] for the deterministic case):

$$A_C q' + f_1(e, j_L, j_V, t) + A_N \operatorname{diag}\,(\sigma(A_N^T e, t))\xi(t) = 0 \tag{1}$$

$$\phi' - A_L^T e = 0 \tag{2}$$

$$A_V^T e - v_s(e, j_L, t) = 0 \tag{3}$$

$$q - q_C(A_C^T e, t) = 0 \tag{4}$$

$$\phi - \phi_L(j_L, t) = 0 \, , \tag{5}$$

where $f_1(e, j_L, j_V, t) := A_R g(A_R^T e, t) + A_L j_L + A_V j_V + A_I i_s(e, j_L, j_I, t)$, and $q_C, g, \phi_L, v_s, i_s, \sigma$ are given, noiseless functions. The vector of unknowns describing the system behaviour consists of all node potentials $e$, the branch currents of current-controlled elements (inductances and voltage sources) $j_L, j_V$, and the charges $q$ of capacitances, and the fluxes $\phi$ of inductances. $\xi$ denotes an $n_N$-dimensional vector of independent standard Gaussian white noise processes. In industry-relevant applications one has to deal with a large number of unknowns

and noise sources.
The first block of equations (1) denotes a stochastic integral equation :

$$A_C q(s)\Big|_{t_0}^{t} + \int_{t_0}^{t} f_1(x(s),s)ds + \int_{t_0}^{t} A_N \mathrm{diag}\,(\sigma)(x(s),s)dw(s) = 0\;,$$

where the Itô integral is employed and $w$ denotes an $n_N$-dimensional Wiener process (also known as Brownian motion) given on the probability space $(\Omega, \mathcal{F}, P)$ with a filtration $(\mathcal{F}_t)_{t\geq t_0}$ (see e.g. [7] for the stochastic background). A solution $x = x(t,\omega)$ is now a stochastic process depending on the time $t$ and the chance element $\omega \in \Omega$. The parameter $\omega$ is omitted in the notations above. The solution $x(t) = x(t,\cdot)$ for fixed time $t$ is a vector-valued random variable in $L^2(\Omega)$, a realization $x(\cdot,\omega)$ is called a path.

The equations (1)-(5) form a specially structured **S**tochastic **D**ifferential **A**lgebraic **E**quation (SDAE) of the type

$$Ax(s)\Big|_{t_0}^{t} + \int_{t_0}^{t} f(x(s),s)ds + \int_{t_0}^{t} G(x(s),s)dw(s) = 0\;, \tag{6}$$

which is the subject of this paper. $A$ is a constant singular matrix, $t$ varies over a compact interval $\mathcal{J}$. The short-hand notation

$$Ax'(t) + f(x(t),t) + G(x(t),t)\xi(t) = 0 \tag{7}$$

emphasizes the relations of (6) to its deterministic counterpart but may be misleading for readers who are less familiar with the stochastic background. Though the notation $x'(t)$ is used in (7), a typical realization $x(\cdot,\omega)$ of the solution is nowhere differentiable.
A process $x(\cdot) = (x(t))_{t\in\mathcal{J}}$ is called a strong solution of (6) if it is adapted to the filtration (i.e., it does not depend on future information), and if, with probability 1, its sample paths are continuous, the integrals in (6) exist and (6) is fulfilled.

## Index 1 SDAEs

Due to the singularity of the matrix $A$ the deterministic part of (6)

$$Ax'(t) + f(x(t),t) = 0\;, \tag{8}$$

where the solution $x$ is now a deterministic function of $t$, forms a DAE. Solutions have to fulfil the constraints of the equation. The solution components belonging to $\ker A$ (we call them the algebraic components) do not occur under the differential operator $d/dt$, and the inherent dynamics live only in a lower-dimensional subspace. The DAE (8) is characterized as an index 1 DAE iff the constraints are locally solvable for the algebraic components. Solving an index 1 DAE involves a coupling of an integration task and a nonlinear equation solving task. If a DAE is of higher index, the constraints are not locally solvable for the algebraic components, and there exist solution components that are determined only by a hidden differentiation step, which may cause serious difficulties in the numerical solution of such problems (see e.g. [1,9]).

We assume here that the deterministic part (8) is globally an index 1 DAE in the sense that the constraints are regularly and globally uniquely solvable for the algebraic variables. The globally unique solvability is stronger than the deterministic index 1 condition, which requires only the non-singularity of the corresponding Jacobian and guarantees only local solvability of the constraints for the algebraic variables. The globally unique solvability holds for the MNA-system (1)-(5) if (see [19])

> there are no loops of capacitances and voltage sources and no cut-sets of inductances or current sources, the capacity, conductance, and inductance matrices are symmetric and uniformly positively definite,
> and the controlled sources fulfil certain conditions described in [5].

In [13,14] it is shown that special conditions are needed to ensure solution processes that are not directly effected by white noise. Then the SDAEs are called SDAEs without direct noise, otherwise with direct noise. To avoid a solution process that is directly affected by white noise we have to assume that the noise sources do not appear in the constraints. This means that

$$\operatorname{im} G(x,t) \subseteq \operatorname{im} A \quad \forall (x,t) \in \mathbb{R}^n \times \mathcal{J} \ .$$

This is true for (7) if and only if there are always capacitances in parallel to a noise source. This is quite restrictive in the actual noise modelling (see also the example in Section 4). Nevertheless, one can also handle many situations where this condition is violated. Often noisy constraints are only needed for the determination of algebraic solution components that do not interact with the dynamical ones. Future work should be dedicated to classify such situations.

Under these conditions the constraints of the SDAE can be described by the deterministic equation

$$Rf(x(t),t) = 0 \quad ,$$

where $R$ is a projector along $\operatorname{im} A$, i.e., $R^2 = R$, $\ker R = \operatorname{im} A$. Solving the constraints for the algebraic components

$$Rf(u+v,t) = 0 \ , \ Av = 0 \Longleftrightarrow v = \widehat{v}(u,t) \ ,$$

inserting the result into the differential equations, and scaling the system by a Pseudo-inverse $A^-$ (with $AA^- = I - R, A^-A$ a projector along $\ker A$) leads to a so-called inherent regular SDE in the differential components $u$:

$$u' + A^- f(u + \widehat{v}(u,t),t) + A^- G(u + v(u,t),t)\xi(t) = 0 \qquad (9)$$

It can be shown that (9), together with $x(t) = u(t) + \widehat{v}(u(t), t)$, is theoretically equivalent to (6). Based on this, the following theorem concerning the existence and uniqueness of solutions of (6) is proved in [19]:

**Theorem 1.** *Let the above conditions be fulfilled for (6), and suppose that $f$ and $G$ are globally Lipschitz-continuous with respect to $x$, continuous with respect to $t$, and that $Ax^0$ is $\mathcal{F}_{t_0}$-measurable, independent of the Wiener process $w$, and with finite second moments.*
*Then there exists a strong solution $x(\cdot)$ of the initial value problem*

$$Ax(t) - Ax^0 + \int_{t_0}^{t} f(x(s), s)ds + \int_{t_0}^{t} G(x(s), s)dw(s) = 0 , \tag{10}$$

*which is pathwise unique. Moreover, the solution $x(t)$ is square-integrable.*

## Discretization Schemes for Index 1 SDAEs

Starting with the Euler Maruyama scheme [8] a wide spectrum of numerical methods for SDEs has been developed. However, first decoupling the SDAE numerically and then applying a scheme to the resulting inherent SDE would be an inefficient procedure in general. We aim at numerical methods for SDAEs that should work directly on the given implicit structure, as in the case of deterministic DAEs. Only little previous work has been done in this direction. In [13,14] linear SDAEs are analyzed and the convergence of the drift-implicit (i.e., implicit in the deterministic part, but explicit in the stochastic part) Euler scheme is proved. In [11,12] a scheme with strong order 1 is developed for the specially structured SDAEs arising in transient noise simulation for electronic circuits. Later we will point out its relation to the drift-implicit Milstein scheme.
Our approach applies to nonlinear SDAEs too, and is used here to analyze the drift-implicit Euler scheme, and to derive and analyze a drift-implicit Milstein scheme for SDAEs. Designing the methods such that the iterates $x_\ell$ have to fulfil the constraints of the SDAE at the current time-point $t_\ell$ is the key idea to adapt known methods for SDEs to (6).

### Drift-Implicit Euler Scheme

On the deterministic grid $0 = t_0 < t_1 < \ldots < t_N = T$ the drift-implicit Euler scheme for (6) is given by

$$A\frac{x_\ell - x_{\ell-1}}{h_\ell} + f(x_\ell, t_\ell) + G(x_{\ell-1}, t_{\ell-1})\frac{1}{h_\ell}\Delta w_\ell = 0 , \tag{11}$$

where $h_\ell = t_{\ell+1} - t_\ell$, $\Delta w_\ell = w(t_{\ell+1}) - w(t_\ell)$. Realizations of $\Delta w_\ell$ can be simulated as $N(0, h_\ell I)$-distributed random variables.
The scheme (11) for the SDAE (6) possesses the same convergence properties as

the drift-implicit Euler scheme for SDEs. In general, its order of strong convergence is 1/2, i.e.,

$$\varepsilon_\ell := \|x(t_\ell) - x_\ell\|_{L_2} = (E|x(t_\ell) - x_\ell|^2)^{1/2} \le c \cdot h^{1/2}$$

holds for the mean square norm of the global errors. For additive noise, i.e., $G(x,t) = G(t)$, the order of strong convergence is 1, for small noise, $G(x,t) = \epsilon\widetilde{G}(x,t)$, the error is bounded by $O(h + \epsilon^2 h^{1/2})$, (see [10] for related results).

## Drift-Implicit Milstein Scheme

We intend to design this method in such a way that it realizes the drift-implicit Milstein scheme for the inherent SDE $u' + \widehat{f}(u,t) + \widehat{G}(u,t)\xi(t) = 0$ , i.e.,

$$\frac{u_\ell - u_{\ell-1}}{h_\ell} + \widehat{f}(u_\ell, t_\ell) + \widehat{G}(u_{\ell-1}, t_{\ell-1})\frac{1}{h_\ell}\Delta w_\ell - \sum_{j=1}^{m}(\widehat{g}_j)_u \widehat{G}(u_{\ell-1}, t_{\ell-1})\frac{1}{h_\ell} I_j^\ell = 0\,,$$

where $I_j^\ell = (I_{j,i}^\ell)_{i=1}^m$ , $I_{j,i}^\ell = \int_{t_{\ell-1}}^{t_\ell}\int_{t_\ell}^{s} dw_i(\tau)dw_j(s)$ , $\widehat{G} = (\widehat{g}_1, \ldots, \widehat{g}_m)$ and $\widehat{f}(u,t) := A^- f(u + \widehat{v}(u,t), t)$ , $\widehat{G}(u,t) := A^- G(u + \widehat{v}(u,t), t)$ .
The Milstein scheme possesses strong convergence of order 1. It differs from the Euler scheme by an additional correction term for the stochastic part, which includes double stochastic integrals. For additive noise the additional term vanishes and both schemes coincide.
The Milstein scheme for the inherent SDE is realized by

$$A\frac{x_\ell - x_{\ell-1}}{h_\ell} + f(x_\ell, t_\ell) + G(x_{\ell-1}, t_{\ell-1})\frac{\Delta w_\ell}{h_\ell} - \sum_{j=1}^{m}(g_j)_x x_u A^- G(x_{\ell-1}, t_{\ell-1})\frac{I_j^\ell}{h_\ell} = 0\,,$$

where $G = (g_1, \ldots, g_m)$, which we call the drift-implicit Milstein scheme for (6). The expression $x_u A^-$ in the latter term can be reformulated (cf. [15]) as

$$x_u A^- = (I + \widehat{v}_u)A^- = (A + Rf_x)^{-1}(I - R) = (A + hf_x)^{-1}(I - R) + \mathcal{O}(h)\,.$$

Without changing the order of the scheme the partial derivative $x_u$, which is generally not explicitly known, can be substituted reusing the Jacobian of the system. Penski's approach [11,12] results in a similar approximation to the Milstein scheme in a more specialized setting. The higher order 1 of strong convergence of these schemes has to be paid for by the use of a large number of double stochastic integrals and the use of the derivatives of the diffusion coefficients. In an application with a large number of small noise sources one has to pay much for a mostly theoretical gain in accuracy.

## Numerical Example

We have applied the drift-implicit Euler scheme to simulate a ring-oscillator with simple mosfet-models which was used also in [11]. The equivalent circuit diagram is given in Fig. 3.

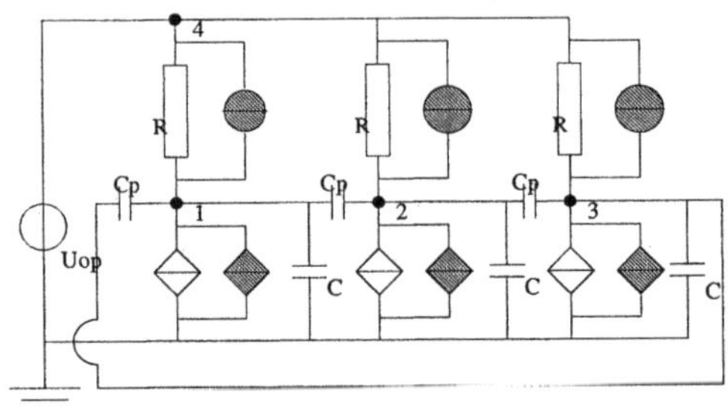


Figure 3: Thermal noise sources in a mosfet ring-oscillator model

The variables in the MNA system are the charges for the six capacities, the four nodal potentials and the current through the voltage source. The system fulfils the assumptions made in Chapter 2 only partly. It is of index 1, but clearly the three thermal resistance noise sources directly affect the current through the voltage source. But, the direct noise occurring in this variable is harmless in the sense that this variable does not influence others. Omitting this critical variable together with the nodal equation for node 4 would lead to a system without direct noise.
In this simple model nearly no differences between the solutions of the noisy and the deterministic problem could be seen. Therefore, we dealt with a system where the diffusion coefficients had been scaled by a factor thousand.

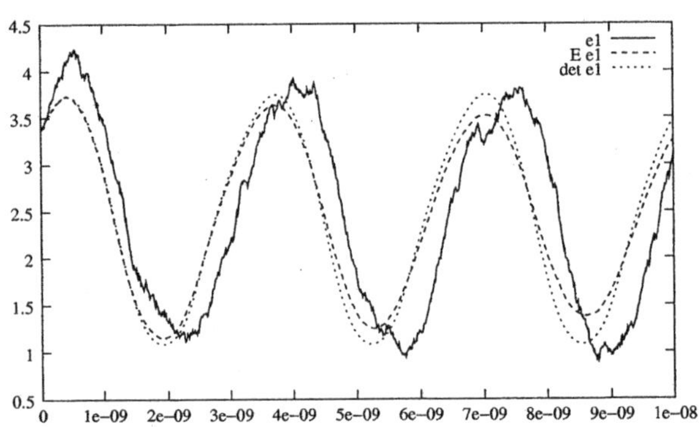


Figure 4: a sample path of the voltage in node 1 (e1) compared to the mean over 100 sample paths (E e1) and the noiseless voltage (det e1)

In Fig. 4 we present numerical results obtained with the drift-implicit Euler scheme and fixed stepsize $h = 10^{-11}$. We plotted the nodal potential at node 1. The mean of the computed 100 sample paths differs considerable from the noiseless solution.

## References


1. U. Ascher and L. Petzold. *Computer methods for ordinary differential equations and differential-algebraic equations.* SIAM, 1998.
2. A. Blum. *Elektronisches Rauschen.* Teubner, 1996.
3. A. Demir. *Analysis and simulation of noise in nonlinear electronic circuits and systems.* PhD thesis, University of California, Berkeley, 1997.
4. A. Demir and A. Sangiovanni-Vincentelli. *Analysis and simulation of noise in nonlinear electronic circuits and systems.* Kluwer Academic Publishers, 1998.
5. D. Estevez Schwarz and C. Tischendorf. Structural analysis for electronic circuits and consequences for MNA. *Int. J. Circ. Theor. Appl.*, 28:131–162, 2000.
6. M. Günther and U. Feldmann. CAD-based electric-circuit modeling in industry I. mathematical structure and index of network equations. *Surv. Math. Ind.*, 8:97–129, 1999.
7. D.J. Higham. An algorithmic introduction to numerical simulation of stochastic differential equations. *SIAM Review*, 43:525-546, 2001.
8. G. Maruyama. Continuous Markov processes and stochastic equations. *Rend. Circolo Math. Palermo*, 4:48–90, 1955.
9. R. März. Numerical methods for differential-algebraic equations. *Acta Numerica*, 141-198, 1992.
10. G.N. Milstein and M.V. Tretyakov. Mean-square numerical methods for stochastic differential equations with small noise. *SIAM J.Sci. Comput.*, 18:1067–1087, 1997.
11. C. Penski. Analysis and numerical integration of stochastic differential-algebraic equations and applications in electronic circuit simulation. Preprint TUM-M9907, Technische Universität, München, 1999, (`www-lit.ma.tum.de/bb`).
12. C. Penski. A new numerical method for SDEs and its application in circuit simulation. *J. Comput. Appl. Math.*, 115:461–470, 2000.
13. O. Schein. *Stochastic differential algebraic equations in circuit simulation.* PhD thesis, Technische Universität Darmstadt, 1999.
14. O. Schein and G. Denk. Numerical solution of stochastic differential-algebraic equations with applications to transient noise simulation of microelectronic circuits. *J. Comput. Appl. Math.*, 100:77–92, 1998.
15. J. Sieber. Local error control for general index-1 and index-2 differential algebraic equations. Preprint 97-21, Humboldt-Universität, Berlin, 1997.
16. C. Tischendorf. Topological index calculation of DAEs in circuit simulation. *Surv. Math. Ind.*, 8(3-4):187–199, 1999.
17. L. Weiß. *Rauschen in nichtlinearen elektronischen Schaltungen und Bauelementen - ein thermodynamischer Zugang.* PhD thesis, Otto-von-Guericke Universität Magdeburg, 1999.
18. L. Weiß and W. Mathis. A thermodynamical approach to noise in nonlinear networks. *Int. J. Circ. Theor. Appl.*, 26:147–165, 1998.
19. R. Winkler. Stochastic differential algebraic equations and applications in circuit simulation. Preprint 01–13, Humboldt-Universität, Berlin, 2001, submitted to J. Comput. Appl. Math., (`www.mathematik.hu-berlin.de/publ/pre/2001/p-list-01.html`).

# Author Index

www.ingramcontent.com/pod-product-compliance
Ingram Content Group UK Ltd.
Pitfield, Milton Keynes, MK11 3LW, UK
UKHW020131290926
15471UKWH00003B/413
*9783642624780*